IMAGE EVALUATION
TEST TARGET (MT-3)

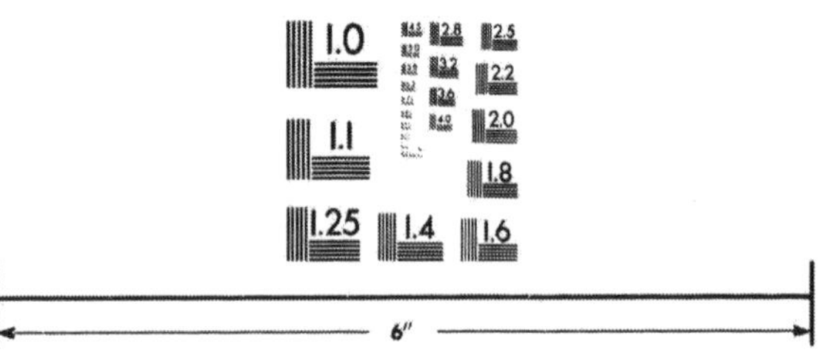

Photographic
Sciences
Corporation

23 WEST MAIN STREET
WEBSTER, N.Y. 14580
(716) 872-4503

Technical and Bibliographic Notes/Notes techniques et bibliographiques

Compiled Drawn and Published from Personal E

BY

H. BELDEN

TORONT

1878.

TORONTO LITH. CO. GORRELL CRAIG & CO. PROPRS.

STRATED

CAL ATLAS

OF

UNTIES

TINGS

AND

EDWARD

ONT.

rom Personal Examinations and Surveys.

BY

DEN & CO.

RONTO.

878.

GENERAL I

MAPS OF TOWNSHIPS AND GENERAL MAPS.

PLANS OF TOWNS AND VILLAGES.

VIEWS.

VIEWS—Continued.

VI

RAL INDEX.

ued.

VIEWS—Continued.

PORTRAITS.

PORTRAITS—Continued.

PATRONS' DIRECTORY.

REFERENCES.

Farm House......
School House......
Church......
Proposed Railroads......
Creeks......
Mills......
Blacksmith Shop......
Cemeteries......
Roads......
Railroads......

HISTORICAL S

OF THE

Counties of Hastings and

Rich in agricultural resources, prolific in the products of the mine, vast in its lumbering operations, and extensive in its manufactures, the well-watered County of Hastings presents to the immigrant one of the finest tracts of territory to be found in the whole Province of Ontario. It is full of natural advantages to the settler. Its fisheries yield, in enormous quantities, bass, pickerel, perch, maskinonge, trout and other fine-flavored denizens of the deep. Its water power is one of the most extensive in all Canada. In geological formation and in mineralogical treasures few counties can boast of like wealth, and indeed, in every way, this commanding county, which is situated on the northern side of the beautiful Bay of Quinte, possesses characteristics which are destined to make it, ere half a century rolls away, the veritable garden county of the Great West. For thirty miles, this magnificent territory sweeps the front of that body of water which successfully rivals in splendor the legendary Bay of Dublin, and which sparkles in the sun like a vast sheet of polished silver. In addition to commanding this frontage, the County of Hastings extends to within fully five miles of the head of the Bay of Quinte. In a geographical way hardly another county in the Province enjoys advantages at all equal to those which Hastings possesses. Its population at the last census amounted to upwards of fifty-five thousand souls, and as the abundant resources of the territory become more largely known, a noticeable increase will assuredly follow.

Its city, towns and villages, are prosperous, and well to do. Its people are thrifty, cautious and hospitable. It is well built up. Its 170 miles of gravel roads are good. Travel is inexpensive. On every side the traveller observes the growth of the country, and can witness the frugality and thrift of the inhabitants. The fields are well cultivated, and the hardy husbandman is a genuine king of the glade. The pioneer has penetrated what were once forest wilds, and instead of a rugged wilderness, we have splendid tracts of land well tilled and yielding rich harvests. The houses are well built, and none can fail to see the success which has attended the labors of this contented and happy people. An excellent system of education has been inaugurated, the County Council, as early as 1860, appointing County Superintendents, and it is gratifying to observe how rapidly the inhabitants of the district take advantage of these opportunities.

Within the limits of Hastings are a prosperous and wealthy city, several thriving townships, and numerous hamlets. Fronting on the Bay are the City of Belleville, the incorporated village of Trenton, and the three townships of Sidney, Thurlow and Tyendinaga. In their rear are Rawdon, Huntingdon and Hungerford, while to the north lie Marmora, Madoc and Elzivir. To the north of these latter are Lakes Tudor and Grimsthorpe. The county has been divided into three Electoral Divisions. These are the East, West and North Ridings. The former consists of the townships of Thurlow, Tyendinaga and Hungerford, the West Riding is composed of the City of Belleville, Sidney, and the village of Trenton, and the North Riding is made up from the townships and municipalities belonging to the northern part of the county. The county is easy of access by land and water, and may be approached by the United States and the Dominion, from the west and the east, by steamer, sailing vessel and railway. Navigation on the Bay is in nowise dangerous nor inconvenient. The vessels running at regular intervals between Belleville, Trenton and Montreal, Belleville and Kingston, and Belleville and Oswego, facilitate commercial operations very much, and afford a cheap and excellent mode of traffic which the inhabitants are quick to take advantage of. The traveller too, will find these means of travel at once comfortable and low in price. From Belleville there are daily stages which run out to the principal villages of Hastings, as well as elsewhere in the back country, and to some of the towns in the County of Prince Edward.

The amateur sportsman will find an abundance of game and fish in almost every part of the county, and at the proper season his rod and gun need not remain idle a single moment. Deer, various birds, squirrels and rabbits people the forests in the interior parts of the county, while the ambitious fisherman will find plenty of opportunities to try his skill with his greenheart rod and favourite line in almost any of the bars in the harbour. In Camp Meeting Cove, Anderson's Cheese factory and Massassaga Point, fine specimens of fat bass abound in quantities, the precise extent of which the veracious historian dare not chronicle, lest his statement be accepted by the non-initiated as a mere fish story unworthy of credence.

The following steamers touch at Belleville and the Bay Ports, "Alexandria," "Utica," "Norfolk," "Armenia," "Pilgrim," and the "Shannon;" and the ferry steamer "Prince Edward," plies from Belleville to Ferry Point, Prince Edward County.

It may not be amiss at this place to speak more definitely of the manufacturing capacities of Hastings. Its immense water power privileges are known the whole country round. Three streams of water wind throughout the county, and numerous tributaries flow into them from many points. The power of these rivers with their tributaries affords a supply sufficiently ample to drive almost every description of machinery. Nearly one hundred miles northward these swelling streams take their rise. The largest is the Trent, and its power is on an enormous scale. It is situated in the west end of the county, is broad and deep, and runs at a whirling pace, as rapid, at times, as the River St. John, in New Brunswick, which flows at a tremendous gait through one of the fairest and most Eden-like portions of that section of the Dominion. The Salmon reminds the tourist of some of the characteristics of the Miramichi. It runs through the eastern portion of the county and empties, at Shannonville, into the Bay of Quinté. It, too, is a stream of much beauty, and its motive power is good. The other river is the Moira—a fine stream of ample and natural advantages, and which flows into the Bay at the city of Belleville. It furnishes many grist and saw mills with a splendid force of water, and factories and machinery in various places are driven by it. On many points on these rivers are factories, mills, &c., in great profusion, and these exhibit in an unmistakable manner the scope and character of the manufacturing interests of the county. Every spring these noble rivers are filled with rafts and drives of timber. On their bosoms lightly ride millions of feet of lumber, thousands of stalwart logs, and vast quantities of square timber, *en route* down the Bay and the swinging St. Lawrence to the great markets beyond the seas. The forests resound with the sound of the woodman's axe, and his cheery voice awakes the echoes in these silent woods. Some of the raw lumber stops at Belleville and undergoes the dressing process at her mills until it becomes as fine as gold, and fills the heart of the lumberman with joy. All along the ... extensive mills are situated, and Trenton, Mill Point, Mill Island, Ameliasburgh and Shannon furnish a very material quota to the whole number in the county. Hastings affords a fine and growing field to the capitalist.

Her privileges are extensive. She can put into the field an ar
workmen. All she wants is an increase of capital to dev
sources which are to be found in every foot of her territory
money are beginning to realize how rich this county is, as ev
industries receive fresh impetus, and as new capital is en
county's products are placed upon the markets.

In mineral wealth Hastings is also rich. Seven miles nor
in Lot 18, in the fifth concession, lies the celebrated Richa
which was discovered in the fall of 1866. The operations here
into history. Twelve years ago the gold fever set in in this pa
only equalled by the tremendous excitement which was cause
came from California that gold had been found in that far-off
county towns were full of miners, speculators, black-legs, a
wells. Bold operators vigorously "salted" various sections o
and for a bribe, "assayers" of questionable honesty, publish
of their bogus assays. The whole populace were up in arms,
ment grew to vast proportions, and the people everywhere fl
mines and the quartz rocks. The cry was for gold, and Ha
went abroad as the new *Eldorado*, and Belleville became a gold
is safe to say that hundreds of thousands of dollars were thro
speculation and in machinery to work the mines.

Shortly after the discovery of this famous Richardson mine,
became a case for the lawyers to settle. A chancery suit in d
lowed, and the mine became a sealed book for a time. This w
of a judge's injunction, and for nearly a year after the mine wa
1867, however, a change took place. The great gold-mini
settled, and a company of ambitious proclivities was instantly
nominal capital was $500,000, based upon one-third of the pro
second company was also instituted on another third of the p
with a nominal capital of $600,000. The proprietors, grossly
even the first principles of mining, began their operations on
scale. A handsome stone structure was erected at an astoundin
newest and most expensive machinery which could be obtained
and the parties calmly awaited the result. But, alas for hu
Alas, for the credulity of mankind! After all this expense
money, it was found that a mistake had been made. The rock
gold was embedded was of a variety that baffled the efforts of all
ical appliances which had been set up by the too confident direc
winter of 1868 the returns were about fifteen dollars to the ton,
found to be a very small return for the enormous outlay whi
made. The mine was closed, the operations ceased, and the w
mating machinery was overhauled. New appliances were put i
began again. But in a short time, to their dismay, the manage
that the new machinery was no better than the old, and the
hardly equalled those under the former *régime*. Then trouble
nest. The directors felt unable to meet the farther demands on
Much of their personal means had been swallowed up in thes
ments," and it was deemed necessary to call in the stockholder
to them for aid. The shareholders were at first indignant, then
and finally savagely angry. They refused to advance more
mill was closed, and it remained in that state for a long time.

At length, after having been laid in for a few years, the mi
up by Mr. McCrae, who worked it for some time without obtai
tory results. He was finally compelled to abandon it, and the
closed. On the eleventh concession Mr. McCrae has taken a
able amount of the precious metal from the old Fiegel mine, w
dormant for some years. This mine was discovered ten year
situated across the line in Marmora, west of Madoc, and the
some excellent specimens of the ore. It, too, had its trials and
two or more chancery suits, and other ups and downs in the
world. Experimental work is still going on in the ninth conce
mora. The Dean & Williams mine and the Gatling property
out some ore which has yielded good value. More capital see
only thing requisite to develop these mines and find the true v
properties. In Elzivir, the township east of Madoc, there are so
fine free gold. At Bridgewater gold was also found, and th
Flint erected a crusher on the spot. In Hungerford more gol
ered, and a further mine exists at Tudor.

In addition to these there are great deposits of iron, whic
into the possession of Pardee & Lloyd and J. B. Mass & Co.,
only the completion of the Belleville & North Hastings Rail
further development. It is firmly believed by leading citizens
and many of the inhabitants of the adjacent towns, that if this
the benefit which will arise to the County of Hastings will
There exist great quantities of white and coloured marbles all
county, and particularly in the Township of Elzivir. The rail
will assist the development of this important branch of indust
it one of the staple exports of the country.

GEOLOGY.

In geological formations Hastings is particularly wealthy.
limestone of the Silurian series extends to Hog Lake, where the
ment of the same meets with the Laurentian and Huronian,
and in the neighbourhood of Stoco, in the Township of Hunger
between the two is strongly marked, and is of deep interest to
. e quote the following full description of the geology o
from Prof. Chapman's excellent report of the same :—

The rock formations present in Hastings County, comprise
ing order ; (1), The Laurentian Series of Canadian geologists ;
the Lower Silurian rocks ; (3), the Drift Formation ; (4), cert
posits of local occurrence.

1. *The Laurentian Formation.*—The rocks of this divisi
the most ancient deposits hitherto recognized on the contin
America. They extend from Labrador along the north sho
Lawrence, to within a short distance of Quebec, from whence
inland, and cross the Ottawa at the Lac des Chats. West
their outcrop sub-divides into two branches, one of which p
the south-east, crossing the St. Lawrence at the Thousand Isl
ing the wild district of the Adirondack Mountains in the state
The other branch sweeps broadly towards the north-west, and
edge runs through the south limits of the Townships of El
and Marmora, in Hastings County, and, continuing its course,
Huron at Matchadash Bay.

·CAL SKETCH

—— OF THE ——

ngs and Prince Edward.

:nsive. She can put into the field an army of skilled
:ants is an increase of capital to develop the re-
be found in every foot of her territory. Men with
o realize how rich this county is, as every year her
sh impetus, and as new capital is employed, the
placed upon the markets.
Hastings is also rich. Seven miles north of Madoc
ı concession, lies the celebrated Richardson Mine,
n the fall of 1866. The operations here have passed
years ago the gold fever set in in this part of Canada.
remendous excitement which was caused when news
hat gold had been found in that far-off region. The
ll of miners, speculators, black-legs, and ne'er-do-
vigorously "salted" various sections of the county,
yers" of questionable honesty, published the result
The whole populace were up in arms. The excite-
portions, and the people everywhere flocked to the
rocks. The cry was for gold, and Hastings' fame
Eldorado, and Belleville became a golden gate. It
ıdreds of thousands of dollars were thrown away in
chinery to work the mines.
scovery of this famous Richardson mine, the property
awyers to settle. A chancery suit in due course fol-
ecame a sealed book for a time. This was the result
, and for nearly a year after the mine was closed. In
nge took place. The great gold-mining suit was
of ambitious proclivities was instantly formed. The
00,000, based upon one-third of the property; and a
lso instituted on another third of the property, also
l of $600,000. The proprietors, grossly ignorant of
s of mining, began their operations on an extensive
me structure was erected at an astounding cost. The
sive machinery which could be obtained was put up,
y awaited the result. But, alas for human hopes!
of mankind! After all this expense of time and
at a mistake had been made. The rock in which the
s of a variety that baffled the efforts of all the mechan-
ad been set up by the too confident directors. In the
rts were about fifteen dollars to the ton, and this was
all return for the enormous outlay which had been
closed, the operations ceased, and the whole amalga-
overhauled. New appliances were put in, and work
short time, to their dismay, the managers discovered
y was no better than the old, and the new returns
under the former *régime*. Then trouble began in ear-
t unable to meet the further demands on their capital.
means had been swallowed up in these "improve-
med necessary to call in the stockholders and appeal
shareholders were at first indignant, then bewildered,
ngry. They refused to advance more money. The
remained in that state for a long time.
ving been laid in for a few years, the mine was taken
o worked it for some time without obtaining satisfac-
finally compelled to abandon it, and the mine is again
th concession Mr. McCrae has taken a not inconsider-
cious metal from the old Fiegel mine, which was also
rs. This mine was discovered ten years ago. It is
in Marmora, west of Madoc, and the veins showed
ns of the ore. It, too, had its trials and difficulties—
suits, and other ups and downs in the metallurgic
work is still going on in the ninth concession of Mar-
Villiams mine and the Gatling property have turned
s yielded good value. More capital seems to be the
develop these mines and find the true value of these
, the township east of Madoc, there are several veins of
idgewater gold was also found, and the Hon. Billa
on the spot. In Hungerford more gold was discov-
ne exists at Tudor.
e there are great deposits of iron, which have come
Pardee & Lloyd and J. B. Mass & Co., which await
f the Belleville & North Hastings Railway for their
It is firmly believed by leading citizens of Belleville,
itants of the adjacent towns, that if this road is built
arise to the County of Hastings will be enormous.
ntities of white and coloured marbles all through the
ly in the Township of Elzivir. The railway facilities
ment of this important branch of industry, and make
ts of the country.

GEOLOGY.

tions Hastings is particularly wealthy. The Trent and
an series extends to Hog Lake, where the great escarp-
ts with the Laurentian and Huronian. Both these
ed of Stoco, in the Township of Hungerford, the line
ngly marked, and is of deep interest to the geologist.
wing full description of the geology of the county
excellent report of the same:—
s present in Hastings County, comprise in an ascend-
aurentian Series of Canadian geologists; (2), some of
ks; (3), the Drift Formation; (4), certain recent de-
es.
Formation.—The rocks of this division constitute
sits hitherto recognized on the continent of North
nd from Labrador along the north shore of the St.
short distance of Quebec, from whence they continue
Ottawa at the Lac des Chats. West of this point,
les into two branches, one of which passes towards
g the St. Lawrence at the Thousand Isles, and form-
the Adirondack Mountains in the state of New York.
ps broadly towards the north-west, and its southern
south limits of the Townships of Elzivir, Madoc,
ings County, and, continuing its course, strikes Lake
Bay.

The Laurentian rocks form also the greater part of the north shore of Lake Superior, and cover an enormous area throughout the northern part of the Province, generally. In popular language they are often, though incorrectly, called *granite*. True granite never occurs in beds or strata, but always in irregular, and generally intrusive masses, or in veins; whereas our Laurentian rocks are always stratified. They are looked upon as altered sedimentary deposits, and belong chiefly to the rocks known as micaceous and hornblendic (or syenitic) gneiss. Micaceous, or common gneiss, is composed of quartz, feldspar, and mica, and has usually a grey or red colour, but is sometimes almost black. Hornblendic or syenitic gneiss consists of quartz, feldspar, and hornblende, and possesses in general a well-marked green colour; or is, otherwise, red and green, or red and black. These rocks, in layers or strata of different colours, alternate with one another, and occasionally by the absence of feldspar, pass into mica slate and hornblende slate. They are frequently traversed by broad bands and veins of white quartz, and in some localities, are interstratified with beds of white, pink, and greyish crystalline limestone or marble. A bed of this substance occurs at the village of Bridgewater, or Troy, in Elzivir Township; and others of fine quality lie in Barrie Township, a little beyond the limits of the county. Marble is likewise found in the Townships of Madoc and Marmora; but it should be mentioned that white quartz is sometimes mistaken for it. Attempts have even been made by persons ignorant of the nature of this latter substance, to burn it into lime. It may not, therefore, be out of place to point out the more salient, distinctive characters of the two, as in the following table:—

MARBLE:	QUARTZ:
Dissolves with effervescence in diluted hydrochloric or nitric acid*. Does not scratch glass, but may be easily scratched by a knife.	Not attacked in any way by acids. Scratches glass easily, and does not yield to the knife.

These Laurentian or gneissoid rocks constitute also the great iron-holding rocks of Canada. This metal occurs in Hastings County in the form of the Black or Magnetic Iron ore, a compound of the oxide and the sesquioxide of iron, containing in percentage values, Iron 72.4, Oxygen 27.6. This valuable mineral forms thick beds, interstratified with the gneiss, in the Townships of Madoc and Marmora; but the ore used at the Marmora smelting works, when these were in operation, came chiefly from the south shore of Crow or Marmora Lake, in the adjoining Township of Belmont. When the ore contains small shining specks or particles (Iron Pyrites) of a brass-yellow colour, it should be made up into heaps and roasted, and afterwards subjected for some time to the action of the atmosphere, before being taken to the furnace. The masses of ore broken out of the rocks and mixed up with the Drift of this locality, are abundant in some places, and of excellent quality, the pyrites having become decomposed, or oxidized by long exposure to atmospheric agencies.

In the north part of Elzivir Township, as well as in adjoining townships beyond the limits of the county, some of the green or hornblendic beds of gneiss contain numerous garnets in well-defined twelve-sided crystals or rhombic dodecahedrons, of a brownish-red colour. These, however, are only of value as mineralogical specimens.

The Laurentian rocks described above occur in highly inclined strata, dipping generally (at least among their more southern outcrop), towards the north-west. The succeeding or over-lying Silurian strata, on the other hand, lie on the upturned edges of the Laurentian rocks, in almost horizontal beds. A good section exhibiting these relations may be seen on the river banks at Marmora village.

Although, as a general rule, where Laurentian rocks prevail the country is not favourably adapted for agricultural occupation, many acres of good and fertile land occur upon this formation in Hastings county. The more rocky portions, also, if useless in other respects, will probably constitute available grazing lands as the country becomes gradually cleared.

2. *The Lower Silurian Formation.*—This formation is sub-divided from the upper part downwards, into the following subordinate groups:—

5. The Hudson River Group.
4. The Utica Slate.
3. The Trenton Group { The Trenton Limestone. The Black River Limestone. The Bird's-eye Limestone. The Chazy Limestone.
2. The Calciferous Sand Rock.
1. The Potsdam Sandstone.

In Hastings County, the three lower members of the formation are alone present; and of these the Potsdam Sandstone and Calciferous Sand-rock are more or less blended together, and are also but slightly developed. Their common representative appears to be a calcareous sandstone of a few feet in thickness, occurring immediately above the Laurentian rocks, or at the extreme base of the Silurian formation. This sandstone is of a light greenish colour above, passing into pale red, or pale red with irregular greenish spots below. It may be seen in horizontal position, or dipping almost imperceptibly towards the south-west, on the river banks at the village of Marmora, and also on the banks of the river Moira, at Tweed Village, in Hungerford township, as well as other places near the outcrop of the Laurentian rocks. It is apparently destitute of fossils. The succeeding Trenton group, properly so-called, is, on the other hand, largely developed, and constitutes the foundation rock of the whole of the South Riding of the County, and also of the southern portion of the North Riding. At its base in the North Riding, a band of fine grey limestone, available as a lithographic stone, is met with. This is succeeded by (in general) a thick-bedded limestone, poor in fossils; and the latter is again followed, in ascending order, by thin-bedded and shaly limestones, containing fossils in great abundance. A list of these fossils, comprising various corals, brachiopods, etc., collected around Belleville, may be seen in a paper by the writer, published in the *Canadian Journal* for January, 1880 [New Series, vol. V.] The Trenton limestone is well displayed along the banks of the Trent, Moira, and Salmon Rivers, and in many places on the shores of the Bay of Quinté. It yields excellent lime, and building-stones of good quality are obtained from some of the thick beds, as at Ox Point, near Belleville, and elsewhere. Some care, however, is required in

* Hydrochloric acid is the muriatic acid or spirit of salt of the stores. For testing limestone rocks it should be diluted with an equal bulk of water, and kept in a small bottle provided with a glass stopper.

their selection, as many of them are apt to crack from minute flaws; but properly selected blocks appear to resist the action of frost remarkably well.

3. *The Drift Formation.* An accumulation of clay, sand and gravel with rounded stones or "boulders," partly of limestone, but chiefly of the more northern gneissoid rocks, is spread over the surface of the greater part of the county. The same deposit extends, indeed, over the larger portion of the Province itself, and reaches far into the United States. Geologically, it is known as the Drift, or Drift and Boulder formation. Its age is much more recent than that of the underlying rocks. Between the deposition of the two an enormous interval of time must have occurred, many intervening formations being absent. It is now universally conceded that after the deposition of our Silurian (or Lower Palæozoic) rocks, this part of Canada was elevated above the sea in which these rocks were deposited, and that it remained dry land for many ages, whilst the succeeding members of the Palæozoic series, with the Secondary and Tertiary rocks (properly so-called), were under process of deposition in the seas, lakes, and estuaries of other localities. Then a movement of depression ensued, and our Province was again covered, or partly covered, by the waters of the ocean. It is also inferred from perfectly trustworthy data, that this period was one of comparative cold. Vast glaciers were formed in northern regions, from whence numerous icebergs laden with earth and stones, drifted southward, and gradually melting or becoming stranded on shoals and islands, deposited their rocky freights over the sea bottom. By the agency of these floating icebergs, also, the limestone ridges were broken down, and the calcareous sediments thus formed were mixed with the more northern deposits. Proofs of this are seen in the polished and striated surfaces of our limestone strata in many localities; in almost all places, indeed, in which a recent removal of the Drift has been effected. The polished rock, when first exposed, is sometimes as smooth as a mirror; and the fine lines which cross it, and which are supposed to have been produced by stones and gravel frozen in to the under side of the icebergs, have almost always a general north and south direction. The same effects of ice action are seen also on most of the exposed gneissoid rocks in the northern part of the country. Finally, the ground must have been again slowly elevated above the sea; and many of our valleys and other surface inequalities were then produced by the action of waves and currents on the yielding materials of the Drift and underlying strata.

4. *Recent Deposits.* These are of very slight extent, and of local occurrence only. They are due to causes now in action, or which have prevailed during comparatively recent periods. So far as regards the County of Hastings, they comprise a few beds of "shell marl," arising from deposits in swamps and partially dried up ponds and lakes. These consist of white and more or less earthy calcareous matter, filled with minute shells of *cyclas*, *planorbis*, and other fresh water genera of mollusca. A deposit of this kind occurs on the high ground above the west bank of the Moira at Belleville; also in the vicinity of Trenton; and at other places. Another recent formation consists of "calcareous tufa," deposited on twigs, moss, stones, etc., in many streams and springs; but frequently, both shell marl and calcareous tufa occur intermixed, and form but one deposit. In addition to the above substances of recent origin, a few subordinate deposits of bog iron ore are said to occur within the limits of the county. Respecting these, however, we have obtained no certain knowledge.

FLORA OF HASTINGS.

A valuable treatise on the Flora of Hastings, written by Mr. John Macoun, has been placed at our disposal. It is of so much interest at this time that we must quote largely from it.

There is not a township in the county which has not its peat beds, though it is in North Hastings where they have their greatest development.

Almost every tamarack swamp is a peat bog; all our beaver meadows, are more or less composed of peat, every cranberry marsh is a real peat bog, while almost every lakelet or pond in the whole county, is surrounded by marsh or bog from two to thirty feet deep; even the marsh below Belleville, on the opposite side of the bay, is an immense peat bed.

The "open tamarack swamps" and cranberry marshes of our county, have the same vegetation, and general aspect of peat bogs throughout North America, and any farmer who has one on his farm, can tell by actual examination how much peat he possesses without going to the expense of sinking a *shaft*.

OUR FOREST TREES.

The productions of our forests, at present, are the most valuable of our natural products. At the head of these stands the white pine (Pinus Strobus), a tree which stands unequalled in North America, for its many uses. It has been and will be for many years, a source of great wealth to the county. Next in value is the sugar maple (Acer Saccharinum), whether we prize it as an article of fuel, for its sugar, or for its use in the arts. Were the wood of this tree properly brought before the manufacturers of Europe, it would be largely exported for veneering purposes.

The red pine (Pinus resinosa), white spruce (Abies alba) and tamarack (Larix Americana), are abundant in many parts of the county, the former, however, being confined almost exclusively to the rear townships of the North Riding. All three are of great value to the ship builders for various purposes.

Another of our forest monarchs—the hemlock (Abies Canadensis) is of great value in the arts, on account of its tanning properties. By a new process this tannin, it is said, can be concentrated and exported in small bulk. If this should be so, it will be a greater source of wealth to our back townships than the manufacture of potash. The extraction of this tannin would be a surer road to fortune than a large investment in our best gold mine.

The white cedar (Thuja Occidentalis) is rapidly assuming an important place amongst our articles of export; for over two years the Moira has been filled at certain seasons with immense quantities which are exported to the neighbouring States.

My limits will not allow of a more extended notice of our remaining forest trees than merely a passing glance at the most prominent. There is not one, however, that is not of use in the arts for some purpose or other.

The balsam and black spruce are abundant at the North, and in some instances, attain a large size. "They are much used when young, as rafters for barns and houses. Both species are highly ornamental, delighting the eye of the most careless, and giving a charm to the most uninviting prospect." The "gum" of the former is much used by the lumbermen, both as an outward application and internal remedy. In fact, it is the "medicine chest" of the backwoodsman.

Red cedar is found in a few localities along the Moira and Bay of Quinte. Its wood is manufactured into bedsteads, which are said to be a terror to bed-bugs. Its wood seems to be indestructible, as whether buried in the ground or exposed to the atmosphere it shows scarcely any symptoms of decay.

We have at least six species of oak, all of which are useful, especially the white and blue, which are cut by the lumbermen for export as squared timber, while the others, cut into lengths and manufactured into staves, are a source of income to those who are unable, through want of means, to engage in the manufacture of squared timber.

Of elms we have three species, and perhaps one or two varieties. Two of these, the rock and swamp or white elm, are exported in large quantities, while the ashes of all three are counted the best for making potash. The bark of the slippery elm is much used by medical men, as it contains a mucilage, which seems to act beneficially in many complaints.

Three species of ash are scattered over the county; the white ash on our uplands, the red or rim ash along our rivers and streams, while the black

ash forms extensive swamps in many local
in the manufacture of carriages, sleighs,
implements.

We have three birches the yellow, th
birch. The two former are much used in c
almost invaluable to the Indian on account
his canoe.

Of our remaining trees, the principal ar
hickory, the bitternut, the butter-nut,
cherry and the red or soft maple, all of whi
ful or ornamental purposes. In all we ha
nearly all of which are used by our own m
every article, and by the farmer for di
our household furniture—in the building o
sels, and in the erection of our houses.

THE ROADS

No finer roads exist in Canada than the
for many years the people have taken espe
condition of their thoroughfares. From t
macadamized roads lead to all sections of
exists on any of them, every road being as t

THE HASTINGS

By the Act 22 Victoria, Cap. 14, 29th
ships near to the Hastings road "were atta
for all purposes whatsoever":

McClure, Bangor, Monteagle, Faraday,
schel, Carlow, Dungannon, Wollaston, Ced

The Hastings road agency also includes
Murchison, and Airley, unsurveyed; six hu
nine thousand one hundred and twenty-
roads, comprise the area of this agency. N
this area are divided into farm lots of one hu
Commissioner of Crown Lands published an
it, the whole proportion which is suitable
about 40 per cent. of the entire area. T
broken and rocky lands, of beaver meadows,

This road begins on the northern bounda
runs north 16 degrees, west for some seventy
line of two ranges of seven townships. Th
with clay. Along this road in a northerly d
of the Madawaska river, the land is of bette
grow here are very abundant. For purposes
authorized free grants of land not to exceed
on application to the local agents, and in c
set forth in the following conditions of settle

"That the settler be eighteen years of ag

"That he take possession of the land all
and put in a state of cultivation at least t
course of four years—build a house, (at leas
the lot until the conditions of settlement a
accomplishment of which only, shall the sett
a title to the property. Families, compris
lands, preferring to reside on a single lot, wil
tion of building and of residence, (except up
provided that the required clearing of the la
non-accomplishment of these conditions wi
the assigned lot of land, which will be sold o

"The road having been opened by the Go
quired to keep it in repair.

"The LOG-HOUSE required by the Gove
a description as can be put up in four days
generally help to build the log-cabin for n
charge, and when this is done, the cost of t
can be covered with bark, and the spaces bet
clay and whitewashed. It then becomes a
stone house.

"The lands thus opened up, and offered f
of Canada West, capable, both as to soil and
and crops of winter wheat, of excellent qualit
crops of every other description of farm pr
longest cultivated districts of that portion
good.

"There are, of course, in such a large ext
to, great varieties in the character and qua
much superior to others; but there is an abu
for farming purposes.

"Water for domestic use is everywhere
throughout, numerous streams and falls of wa
manufacturing purposes.

"The heavy-timbered land is almost als
ashes of three acres—well taken care of, an
duce a barrel of potash, worth from $3 to
quired to manufacture potash is very small,
and easily understood.

"The expense of clearing and enclosing, i
the labour of the settler at the highest rates,
money per acre, which the first wheat crop,
repay. The best timber for fencing is to be

This free grant road joins the new towns
south, and has its southern termination on th

EARLY HISTO

In 1792, Governor Simcoe issued a proclan
into nineteen counties, for purposes of represe
thereby are as follows, in the order named:—
Grenville, Leeds, Frontenac, Ontario, Adding
Hastings, Northumberland, Durham, York, Li
and Kent.

The Proclamation, though very interesti
respective localities to which the various sec
to quote at length. We give below that sect
Hastings:—

"That the eleventh of the said counties b
"of Hastings; which county is to be bound
"most line of the County of Lennox; on the
"meets the boundary on the easternmost lin
"along the said river until it intersects the
"thence by a line running north, sixteen deg
"Ottawa, or Grand River; thence descendin
"meets the north-westernmost boundary of th
"said County of Hastings to comprehend all
"Quinte and River Trent nearest to the said
"part fronting the same."

After describing the metes and bounds of
clamation goes on to give to each its proper di

TORICAL SKETCH OF THE

forms extensive swamps in many localities. All three are of great use he manufacture of carriages, sleighs, and almost all our agricultural lements.

We have three birches—the yellow, the black, and the white or canoe h. The two former are much used in cabinet work, while the latter is st invaluable to the Indian on account of its bark, of which he makes anoe.

If our remaining trees, the principal are the beech, the bass-wood, the ory, the bitter-nut, the butter-nut, the balm of Gilead, the black ry and the red or soft maple, all of which are used more or less for use r ornamental purposes. In all we have forty species of forest trees, ly all of which are used by our own mechanics for the manufacture of y article, and by the farmer for all our conveyances—for nearly all household furniture—in the building of our steamboats and sailing ves and in the erection of our houses.

THE ROADS, &c.

No finer roads exist in Canada than those of Hastings County. Indeed, many years the people have taken especial pride in the character and ition of their thoroughfares. From Belleville admirable gravel and macadamized roads lead to all sections of the county, and not a toll gate is on any of them, every road being as free as the air we breathe.

THE HASTINGS ROAD.

By the Act 22 Victoria, Cap. 14, 30th June, 1858, the following towns near to the Hastings road "were attached to the County of Hastings all purposes whatsoever":—

...deClare, Bangor, Monteagle, Faraday, Mayo, Limerick, Wicklow, Her..., Carlow, Dungannon, Wollaston, Cashel.

The Hastings road agency also includes the townships of Lyell, Sabine, ...chison, and Airley, unsurveyed: one million one hundred and eighty- thousand one hundred and twenty-nine acres, including water and ls, comprise the area of this agency. Nine hundred thousand acres of ...cres are divided into farm lots of one hundred acres each. In 1863 the ...missioner of Crown Lands published an official map, and according to he whole proportion which is suitable for purposes of agriculture is it 40 per cent. of the entire area. The balance consists of rough, ...en and rocky lands, of beaver meadows, ash, swales and cedar swamps. This road begins on the northern boundary line of Madoc township and ...north 16 degrees, west for some seventy-five miles, and forms the centre of two ranges of seven townships. The soil is a sandy loam, mixed ...clay. Along this road in a northerly direction, between the branches ...e Madawaska river, the land is of better quality and the crops which ...here are very abundant. For purposes of settlement the Government ...orized free grants of land not to exceed in each case one hundred acres, ...pplication to the local agents, and in compliance with the provisions ...orth in the following conditions of settlement:—

"That the settler be eighteen years of age.

"That he take possession of the land allotted to him within one month, put in a state of cultivation at least twelve acres of the land in the ...se of four years; build a house, (at least 20 by 18 feet,) and reside on lot until the conditions of settlement are duly performed; after the ...mplishment of which only, shall the settler have the right of obtaining ...le to the property. Families, comprising several settlers entitled to ...s, preferring to reside on a single lot, will be exempted from the obliga... of building and of residence, (except upon the lot on which they live,) ...ided that the required clearing of the land be made on each lot. The ...accomplishment of these conditions will cause the immediate loss of ...ssigned lot of land, which will be sold or given to another.

The road having been opened by the Government, the settlers are re... ...d to keep it in repair.

The LOG-HOUSE required by the Government to be built, is of such ...ription as can be put up in four days by five men. The neighbors ...ally help to build the log-cabin for newly-arrived settlers, without ...ge, and when this is done, the cost of the erection is small: the roof ...e covered with bark, and the spaces between the logs plastered with ...and whitewashed. It then becomes a neat dwelling, and warm as a ...house.

The lands thus opened up, and offered for settlement, are in sections ...nada West, capable, both as to soil and climate, of producing abund... ...rops of winter wheat, of excellent quality and full weight; and also, ...of every other description of farm produce, grown in the best and ...st cultivated districts of that portion of the province, and fully as

There are, of course, in such a large extent of country as that referred ...eat varieties in the character and quality of land—some lots being ...superior to others; but there is an abundance of the very best land ...ming purposes.

Water for domestic use is everywhere abundant; and there are, ...ghout, numerous streams and falls of water, capable of being used for ...acturing purposes.

The heavy-timbered land is almost always the best, and of it the ...of three acres—well taken care of, and covered from wet—will pro... ...barrel of potash, worth from £6 to £7 currency. The capital re... ...to manufacture potash is very small, and the process is very simple ...sily understood.

The expense of clearing and enclosing heavily-timbered lands, valuing ...our of the settler at the highest rates, is about Four Pounds Cur... per acre, which the first wheat crop, if an average one, will nearly

The best timber for fencing is to be had in abundance."

This free grant road joins the new townships in a line nearly north and ...and has its southern termination on the northern boundary of Madoc.

EARLY HISTORY.

...1792, Governor Simcoe issued a proclamation dividing Upper Canada ...neteen counties, for purposes of representation. The counties formed ...are as follows, in the order named:—Glengarry, Stormont, Dundas, ...lle, Leeds, Frontenac, Ontario, Addington, Lennox, Prince Edward, ...gs, Northumberland, Durham, York, Lincoln, Norfolk, Suffolk, Essex, ...nt.

...Proclamation, though very interesting to those belonging to the ...ive localities to which the various sections refer, is too voluminous ...at length. We give below that section referring to the County of ...gs:

"...at the eleventh of the said counties be hereafter called by the name ...astings; which county is to be bounded on the east by the western ...line of the County of Lennox; on the south by Bay Quinte, until it ...the boundary on the easternmost line of the River Trent; thence ...the said river until it intersects the rear of the sixth concession; ...e by a line running north, sixteen degrees west, until it intersects the ...a, or Grand River; thence descending along the said river until it ...the north-westernmost boundary of the County of Addington. The ...ounty of Hastings to comprehend all the islands of the said Bay of ...e and River Trent nearest to the said county in the whole, or greater ...ronting the same."

...r describing the metes and bounds of the various counties, the pro... ...n goes on to give to each its proper distribution as to representation in the Assembly. Glengarry received two members; some received one each, while in other cases two or more were put together to elect a single representative. By this distribution, Hastings, Northumberland, and Lennox (except Adolphustown, which was put with Prince Edward County to form an Electoral Division), were comprised within a single division, with the privilege of returning one member. In all, sixteen members were elected by the nineteen counties to form the first parliament of Upper Canada.

This proclamation was signed by William Jarvis, Private Secretary to the Governor, and countersigned by the Governor affixing his initials "J. G. S."

In the year 1839 Hastings was separated from the Midland district. The county since that date has made gigantic strides. It has increased in every way, and few counties in the Dominion can boast of such rapid and systematic progress. We append here a tabular statement which will show at a glance the splendid position which Hastings enjoys.

COUNTY INFORMATION.

SENATE.—Hon. Billa Flint, Belleville; Hon. Robert Reed, Belleville.

HOUSE OF COMMONS.—John White, East Riding, Roslin; Mackenzie Bowell, North Riding, Belleville; James Brown, West Riding, Belleville.

LEGISLATURE OF ONTARIO.—Nathaniel S. Appleby, East Riding, Shannonville; G. H. Boulter, M.D., North Riding, Stirling; Thos. Wills, West Riding, Belleville.

COUNTY COUNCIL.

MILLPOINT.—Wm. R. Aylsworth, Warden.

SIDNEY.—Clement Armstrong, Reeve; Robert E. Gross, 1st Deputy; R. H. Vandervoort, 2nd Deputy.

THURLOW.—Harford Ashley, Reeve; Owen R. Weese, 1st Deputy; James H. Peck, 2nd Deputy.

TYENDINAGA.—John White, M.P., Reeve; Donald Anderson, 1st Deputy; Thos. Currie, 2nd Deputy.

RAWDON.—James Cook, Reeve; Paul Kingston, Deputy.

HUNTINGDON.—Thomas Emo, Reeve; Matthew Robinson, Deputy.

HUNGERFORD.—Robt. Gordon, Reeve; Patrick Murphy, Deputy.

MADOC.—Peter Vankleek, Reeve; John Caskey, Deputy.

MARMORA AND LAKE.—Thomas P. Pierce, Reeve.

ELZEVIR AND GRIMSTHORPE.—Hon. Billa Flint, Reeve.

TUDOR, WOLLASTON, LIMERICK, AND CASHEL.—Peter M. Grauter, Reeve.

DUNGANNON AND FARADAY.—Dermot Kavanagh, Reeve.

CARLOW AND MAYO.—Wm. D. Parkhurst, Reeve.

BANGOR AND WICKLOW.—Charles McDavitt, Reeve.

MONTEAGLE AND HERSCHEL.—John Fitzgerald, Reeve.

TRENTON.—Chas. Francis, Reeve; Jas. N. Lee, Deputy.

MILL POINT.—Wm. R. Aylsworth, Reeve.

MADOC VILLAGE.—Thomas Cross, Reeve.

STIRLING.—Charles Craige, Reeve.

COUNTY OFFICERS.—Wm. R. Aylsworth, Warden; George Taylor, Sheriff; Hon. Geo. Sherwood, Judge of County and Surrogate Courts, and Chairman Quarter Sessions; Thos. A. Lazier, Junior Judge; Wm. H. Ponton, Registrar; James Ponton, Deputy Registrar; Charles L. Coleman, Clerk of the Peace and County Attorney; Anson G. Northrup, Clerk County Court, Deputy Clerk of the Crown and Registrar Surrogate Court; Prof. James T. Bell, County Clerk; Thos. Wills, M.P.P., County Treasurer; John Taylor, Deputy Sheriff; John Francis, County Surveyor; Timothy Farnham, County Gravel Road Superintendent; E. W. Meyers, Gaoler; John A. M. Gilbert, Turnkey.

PUBLIC SCHOOL INSPECTORS.—John Johnston, South Riding of Hastings; Wm. Mackintosh, North Riding of Hastings.

COUNTY CORONERS.—John P. McDonnell, W. H. Ponton, Dr. Tracy, Dr. Burdett, Dr. Farley, Dr. Hamilton.

INSPECTORS OF POOR.—James Ross, for Belleville.

STREET SURVEYOR.—Richard Taylor.

INSPECTOR OF WEIGHTS AND MEASURES.—John P. McDonnell.

INSPECTORS OF LICENSES.—Lewis A. Appleby, East Hastings; James Ross, West Hastings; Ed. Mounsey, North Hastings.

AGRICULTURAL SOCIETIES.—There are three Agricultural Divisions in the county, viz., North, East and West Ridings. These societies embrace among their active members some of the most prominent names in the county. Officers are elected every January, and consist of president, first and second vice-presidents, directors, treasurer, secretary, and auditors.

BELLEVILLE.

Up to the 31st day of December, 1877, the pushing and enterprising city of Belleville enjoyed the enviable notoriety of being one of the most vigorous towns in all Canada. On that date the lusty town became a city. Besides being a thriving and busy place, it is one of the most beautiful spots in Canada. It is the centre of a perfect hive of industry. Its people are noted far and wide for their integrity, enterprise and spirit. The city is full of natural advantages, and these together with the hospitable character of the inhabitants, make it an eminently desirable place for the tourist and visitor. It is truly the City of the Bay.

In 1787, Louis Kotte surveyed the township of Thurlow, in those days known as the "Kotte town." Lot No. 4, 1st con. of Thurlow, was blocked off for an Indian burial ground, and upon this site, in the year 1816, a village was laid out by Surveyor Wilmot. But though settlement was slow, and for many years uninhabited swamps alone marked the spot in the rude map, there was planted in that little village the germ of a race at once thrifty, strong, and vigorous, who were destined to found a city, and leave behind them children who possessed all their distinguishing characteristics. The townships of Sidney and Thurlow were settled first, and the new village languished for a time until Capt. John Singleton and Lieutenant Ferguson moved into it. These gentlemen, hardy pioneers as they were, settled on lot No. 6, which in turn was respectively owned by John Canniff, sr., and Shubael Foster. In 1789 Singleton died, and in less than three months afterwards Ferguson too past away. In the spring of 1789 fifty Loyalists came to Sidney and Thurlow, and next year John Taylor settled near the mouth of the river on lot No. 5. This is now one of the most important parts of the new city. Capt. John W. Meyers bought the north half of the lot from Taylor, for one hundred dollars, and immediately erected a dam and built a mill. In 1794 Meyers built the first brick house that Canada had ever seen. For 80 years this architectural wonder stood intact on the northern brow of the hill. Asa Wallbridge built one of the earliest houses in Belleville, and in 1798 John Simpson put up the first tavern in Thurlow. It too has stood up to within a few years ago, on its old site, in the south-east corner of Dundas and Front streets, opposite the Hon. Lewis Wallbridge's residence. In 1801 a bridge was built across the narrowest part of the river. Up to 1816 Belleville was known as Meyers' Creek. The river was called Singleton's River, though the Indian tribes named it Saganaska. In 1816 the village was renamed by Governor Gore, who called it Belleville, after his wife Lady Bella. During this same year, Surveyor Wilmot laid out the Indian Reserve, on lot No. 4, into town lots of half an acre each. A post-office was established, and S. McNabb had the honor of being the first postmaster of Belleville.

There is little of moment in the history of Belleville from 1816 to 1836. Its growth while progressive in its way was also very slow. In 1818, it had 150 inhabitants; in 1824, about 400; in 1829, 700; in 1836, no less than 1,700. In this latter year the first flag-stone sidewalk was put down; it extended on Front street, from the south side of Dundas street to the northern side of Pinnacle street. In this same important year of 1836 Belleville became an independent municipality, and the first board of police was organized, and these gentlemen were chosen in the following wards: First Ward —William McCarty and Asa Yeomans; Second Ward—Zenas Dafoe and William Connor; Billa Flint jr., was chosen fifth member, and he was elected by the board as the first president, by a vote of 3 to 2; George Benjamin was made clerk to the board. In 1838 the Court house and gaol were finished at a cost of $23,610. The contractors were Robert Matthews and David Duff, of Kingston.

We will pass over the turbulent times of the year 1837, as the deeds done at that time belong not merely to Belleville, but to the whole territory of Canada. The Mackenzie-Papineau trials are still fresh in the memories of many who are living to-day, and few can forget the sad troubles of that eventful period in our country's history. Belleville took her share in the agony of the moment. Mr. Richard S. Bell, of Belleville, possesses a little volume, containing the minutes of the magistrates of Hastings, during the year 1837, which were kept by the late George Benjamin, and from its pages we are permitted to excerpt these names of the loyal and true men, who distinguished themselves in the dark days of the Canadian rebellion. They are worth placing on record in a work of this kind. They are:—Thomas Parker, Billa Flint, jr., John Turnbull, Anthony Marshall, Henry Baldwin, T. D. Appleby, F. McAnnany, Donald McKenzie, Peter O'Reilly, Elijah Ketcheson, Benjamin Ketcheson, G. N. Ridley, William Ketcheson, Rufus Purdy, William Bowen, George Bleecker, J. H. Bleeker, J. Lambton, Dr. Murcherson, J. McNabb, and E. Fidlar. Right loyal names all of them; they deserve a golden setting.

In 1839 Benjamin Dougall presided at the first assembly of the Quarter Sessions, Edmund Murney was clerk of the Peace, and J. W. D. Moodie, was Sheriff. In 1850 Belleville became an incorporated town, and the following gentlemen were chosen members of the Council for that year:—Samson Ward —C. O. Benson, F. McAnnany, and Jacob Bonter; Ketcheson Ward—B. F. Davy, Dr. E. Holden, and Samuel Stevens; Baldwin Ward—Jonathan McCarty, John Coulter, and Edw. P. Bosely; Coleman Ward—R. F. Coleman, Burleigh Hunt, and John Donoghue; M. Sawyer was the Town clerk; Benjamin F. Davy, chosen by the Council, was the first Mayor of Belleville town. In 1849, the Collector's roll for collection was £749 17s. 6d., which is to-day but the twentieth part of the amount of taxes raised. In 1860 the assessed value of the town was $1,581,320; in 1870, $2,669,786; and in 1877, it amounted to $[illegible]789,131. Gravel roads were begun in 1850, and nine years later, through the instrumentality of George Benjamin, all tolls were taken from the roads. This was a boon which materially assisted the progressive and enterprising capabilities of the people.

In 1850 the town was separated from the county. In 1861 the census returns revealed a population of 6,277, and ten years later the same authority shows a total of 7,305. In 1876 a special census was taken when the increase was found to be wonderfully large. It is no less than 11,120. In 1816 the village occupied 200 acres, to-day the city rests upon an acreage of 1360.

For most of the information contained in the above brief sketch of this successful city, we are indebted to Dr. Canniff, Mr. Richard S. Bell and Prof. J. T. Bell, the efficient county clerk; also to the Rev. Canon Bleasdell, D.C.L., Trenton, and others, for information in connection with the early settlement of the township.

RELIGIOUS.

The first church in Belleville was erected by the Methodists, in 1816. It was a frame building, 36x30, and though service was held within this skeleton, the structure was never completed. The pulpit was composed of rough boards, and the seats were made from the same material, with small blocks of wood for rests. In 1831 a new building was put up on the old site. The first English church west of Adolphustown was built in Belleville in 1820; Rev. Thos. Campbell was the first minister. In 1835 he died and was succeeded by the Rev. Mr. Cochrane. Three years later the Rev. John Grier was appointed pastor, and he remained in charge until his death, which occurred in the year 1873. The first Presbyterian clergyman was Rev. Mr. Ketcham, and through his instrumentality the first Presbyterian church was erected. We append a list of the various churches in Belleville at the present day, together with their localities and the names of their pastors.

EPISCOPALIAN.

Congregation of St. Thomas' Church.—The congregation of St. Thomas' Church meet for Divine worship in the Metropolitan Hall on Sunday at 11 a. m., 7 p. m., and Sunday School at 3 p. m.

Rev. J. W. Burke, B. A., Rector.

St. Paul's Church.—Divine Service at 4.30 o'clock p. m., on Sundays, and celebration of Holy Communion on all Sundays in month (except first Sunday), at 8 o'clock a. m. Rev. J. W. Burke, Clk., Rector.

St. John's Church, West Belleville.—Services Sunday, 11 a. m., and 7 p. m., Sunday School at 3 p. m.; service Wednesday evening, 7 p.m.; Communion, 2nd Sunday in the month.

Rev. R. S. Forneri, B. A., Incumbent.

Christ Church.—Services at 11 a.m. and 7 p.m. Sunday School and Bible Class 3 p.m. Holy Communion the first Sunday in the month. Seats free in the evening. Rev. J. R. Jones, Incumbent.

St. George's Church, Station Road.—Sunday School and Bible Class at 3 p.m. Service on Thursday evening at 7.30.

Rev. J. R. Jones, Incumbent.

ROMAN CATHOLIC.

St. Michael's, Church Street.—Morning Mass at 7.30 a.m.; High Mass at 10.30 a.m. Vespers at 3.30 p.m.

Rev. James Farrelly, V. G.

METHODIST CHURCH OF CANADA.

Bridge Street Church.—Morning Service, 11 a.m.; Sunday School, 2.30 p.m.; evening service, 7 p.m.; lecture Tuesday evening, 7.30; prayer meeting Friday evening at 7.30. Strangers met at the door and escorted to seats.

West Belleville.—Morning service, 11 a.m.; Sunday School, 2.30 p.m.; evening, 7 p.m. Seats free.

Kingston Road Church.—Sunday School, 10 a.m.; afternoon service, 3 p.m. Seats free. Rev. E. B. Harper, Rev. S. P. Rose, } Ministers.

Bleecker Street Church.—Morning Service 11 a.m. Evening service 7 p.m. Sunday School 2.30 p.m. Seats free.

Rev. T. H. Cullen, Pastor.

EPISCOPAL METHODIST.

Corner of Church and Bridge St. James Sts.—Divine Service—Sabbath, 11.00 a.m., and 7.00 p.m. Prayer Meeting, Wednesday evening. Sabbath School at 2.30 p.m. Rev. B. Lane, B. A., Pastor.

University Church, near Albert College.—Divine Service—Sabbath at 7.00 p.m., and Sabbath School at 9.30 a.m.

Rev. B. Lane, B. A., Pastor.

CANADA PRESBYTERIAN.

John Street.—Morning Service at 11.00 o'clock a.m.; Evening Service at 7.00. Bible Class at 3 p.m. Sabbath School at 3 p.m. Weekly Prayer Meeting, Friday evening at 7.30 p.m. Rev. John Burton, Pastor.

PRESBYTERIAN.

St. Andrew's, Church Street.—Services—Sunday, 11.00 o'clock a.m., and 7.00 p.m. Sabbath School, 3 o'clock p.m. Prayer Meeting on 7.00 p.m. Mission Sabbath School at Northcott's School Hou[se], a

Rev. W. M. McLean, M. A.,

CONGREGATIONAL CHURCH.

Hotel Street.—Services on Sunday at 11 a.m., and 7 p.m. School at 2.30 p.m. Prayer Meeting, Wednesday evening at 7.30

A. O. Cossar,

BIBLE CHRISTIAN.

Services Sunday at 11 a.m., and 6.30 p.m. Sunday School, in the new Bible Christian church, Front Street, west.

Rev. L. W. Wright,

BAPTIST.

Coleman Street, West Belleville.—Morning service at evening service at 7 o'clock. Sunday School at 3 p.m.

Rev. Alex. Turnbull,

The oldest Sunday School in Belleville is that which is held [in connec]tion with the Bridge Street Canada Methodist Church. It was [organized in] 1822.

EDUCATIONAL MATTERS.

It is difficult to get information regarding the early schools of [Belleville]. Indeed, for the most part, they consisted in rude houses, and [the peda]gogues were generally inferior men whose own educational advan[tages had] been few. Besides this, the same rule which governed the farm[er and] mechanic in those old days, in other settlements, worked here [to the] disadvantage of scholastic affairs. The soil had to be tilled, and men and others, who had a limited amount of knowledge themse[lves, hesi]tated about sending their children to school when there was work to do at home and on the farm. Even had there been good schools sixty odd years ago, unless education was made a compulsory [it is] extremely doubtful if they would have been well attended. One [of the] teachers in Belleville, of the old school, was Mr. John Watkins. Mr. Hutton, who taught in a building on the north-west corner [of ...] and Dundas streets. A Scotchman named Bland—an excellent [man in] the way—taught school in a frame house opposite the butchers' [...] Pinnacle street. This school-house perished by fire a few years a[go]. [For a] long time Alex. Burdon, since deceased, taught in the upper s[tory of the] engine-house, on the corner opposite the Merchants' Bank, and a Mr. Burdon taught the grammar school. The common school sys[tem, how]ever, was a great boon. It did not come into operation a moment [too soon,] for to-day Belleville can point to an educational system which [is not sur]passed anywhere. Her High and Public Schools are daily at[tended by] 1800 of the cleverest boys and girls in the Dominion, who vie wit[h each] other in a friendly way, for those grand honors which a thorough [education] confers on man. Belleville has one High School, five Public Scho[ols, a] sixth in prospect, and the Roman Catholic Separate School, whi[ch is in a] high state of efficiency.

Besides the above there is the Albert University which, two [...] ago, was established as the Belleville Seminary by the Methodist [Episcopal] body. This College enjoys a splendid reputation, and its staff of [teachers] is composed of well-known and eminent names, at the head of wh[ich is] Rev. J. R. Jaques, D.D., Ph. D., &c., &c. Alexandra College fo[r ladies] is also a successful hall of learning and a real credit to the City of [Belleville]. It was established about the year 1857.

In 1870 the Government of Ontario established at Belleville th[e] Institution for the Deaf and Dumb. It is under the management [of Dr. W.] J. Palmer, who has been principal from its commencement. It is [second] in importance on the continent, while the attendance averages [...] Ontario Business College is a largely attended and admirable seat [of educa]tion. The system is thorough, and the curriculum embraces all [com]mercial branches, including actual business, penmanship and tele[graphy], the whole under the management of Messrs. S. G. Beatty & [...]. Sisters of Loretto have also founded an excellent educational insti[tution for] girls, which gives promise of being one of the most important in [...]. In addition to these, there are scattered all through the city a n[umber of] well-kept private schools.

THE PRESS.

The first newspaper published in Belleville was *The Anglo-Ca[nadian]* [...] February, 1831, by Alexander T. W. Williamson, editor, and W. A[...] terms $4.00 a year. *The Phœnix* was the second journal, publis[hed by ...] Sheet, editor and proprietor. It died July 3rd, 1832, aged one y[ear]. *Hastings Times* succeeded *The Phœnix*, with Rollin C. Benedict as p[...]. It also died young. In 1834 George Benjamin started *The Int[elligencer]*. In 1847 Mr. Mackenzie Bowell became a partner in the busine[ss ...] year Messrs. Bowell & Moore became proprietors of the paper. [This part]nership lasted for three years and a quarter, when Mr. Moore re[tired and] Mr. Bowell became sole proprietor. In 1875 *The Intelligencer* Pri[nting and] Publishing Co., purchased the paper, retaining Mr. Bowell, who [is a large] stockholder, as editor. In 1867 *The Intelligencer* was merged int[o ...]. *The Sun* was established in 1835 by Mr. Greenleaf. It lived on[ly a short] time. Mr. Hart started in 1836 *The Plain Speaker*. It was frien[dly to the] rebels, and the editor was put in the Kingston Penitentiary for att[empting a] raid on a bank at Cobourg. The soldiers (volunteers) afterwards [entered] the office of the *Plain Speaker*, upset the typefonts and trailed th[em] in the snow and slush. This movement occurred because the paper [appeared] one morning with the British coat of arms turned upside do[wn in its] columns.

The Victoria Chronicle was the next venture. It was started [by ...] Washburn & Sutton. In 1849 E. Miles purchased the paper and [...] cessfully until it became amalgamated with *The Ontario* in 1873. *The Victoria Magazine* was placed on the market by Joseph Wi[lson, ...] Sheriff and Mrs. Moodie as editors, but it only lived one year. In 1848 Mr. Washburn published a monthly magazine of miscellaneo[us ...]. Mr. Wilson then attempted various publications, *The Eclectic*, *Wilson's Experiment*, *Wilson's Canada Casket*, &c. But none of t[hese were] of long or very robust life. Some other papers were started only t[o ...] a few years' life. There were *The Tribune*, *The Independent*, w[hich lived] seven years, and some others. In 1870 Carman & Yeomans establ[ished the] *Daily Ontario*. This journal is still a vigorous newspaper and in [...] the living. The latest venture was *The Free Press*—a small a[...] sheet circulated free, which was established in 1876 by Messrs [...]lough & Hines.

TRADE.

Belleville enjoys an extensive and well managed trade facilit[ies] [...] saw-mills are established, and enormous quantities of lumber are manufactured. During the last few years, however, this importa[nt branch] of industry, through various causes, has visibly declined. Belle[ville does] every year a large business in the exportation of barley, and las[t year the] amount reached 600,000 bushels. The products of the dairy have al[so become] a great feature in the commercial field of this city, the farmers of [the] county receiving annually no less than six hundred thousand dollar[s for] butter and cheese. These articles are regularly shipped to England [and] always bring good prices on account of their magnificent quality and [...]. This statement, showing the growth of Belleville's trade, comp[iled from] Sessional Papers, will be very interesting, as it shows how rapidly [the ex]ports have grown:—

hool, 8 o'clock p.m. Prayer Meeting on Friday at
s School at Northcott's School House, at 3.30 p.m.
Rev. W. M. McLean, M.A., Pastor.

GREGATIONAL CHURCH.

ices on Sunday at 11 a.m., and 7 p.m. Sabbath
er Meeting, Wednesday evening at 7.30 p.m.
A. O. Cossar, Pastor.

BIBLE CHRISTIAN.

1 a.m., and 6.30 p.m. Sunday School, 2.30 p.m.,
n church, Front Street, west.
Rev. L. W. Wickett, Pastor.

BAPTIST.

st Belleville. Morning service at 11 o'clock;
k. Sunday School at 3 p.m.
Rev. Alex. Turnbull, Pastor.

hool in Belleville is that which is held in connec-
est Canada Methodist Church. It was begun in

CATIONAL MATTERS.

formation regarding the early schools of Belleville,
rt, they consisted in rude houses, and the peda-
stier men whose own educational advantages had
the same rule which governed the farmer and the
ge, in other settlements, worked here to the sad
affairs. The soil had to be tilled, and husband-
a limited amount of knowledge themselves, hesi-
children to school when there was work for them
farm. Even had there been good schools fifty and
as education was made a compulsory thing, it is
y would have been well attended. One of the best
the old school, was Mr. John Watkins, and later,
in a building on the north-west corner of Church
Scotchman named Bland an excellent teacher by
a frame house opposite the butchers' market, on
hool-house perished by fire a few years ago. For a
since deceased, taught in the upper story of an
er opposite the Merchants' Bank, and afterwards,
rammar school. The common school system, how-
It did not come into operation a moment too soon,
point to an educational system which is unsur-
High and Public Schools are daily attended by
s and girls in the Dominion, who vie with one an-
or those grand honors which a thorough education
ille has one High School, five Public Schools, with a
e Roman Catholic Separate School, which is in a

ere is the Albert University which, twenty years
e Belleville Seminary by the Methodist Episcopal
es a splendid reputation, and its staff of instructors
n and eminent names, at the head of which is the
Ph. D., &c., &c. Alexandra College for females
learning and a real credit to the City of the Bay.
the year 1857.
nt of Ontario established at Belleville the Ontario
nd Dumb. It is under the management of Dr. W.
principal from its commencement. It is the fifth
tinent, while the attendance averages 225. The
is a largely attended and admirable seat of instruc-
rough, and the curriculum embraces all the com-
ng actual business, penmanship and telegraphing.
anagement of Messrs. S. G. Beatty & Co. The
so founded an excellent educational institution for
of being one of the most important in the land.
re are scattered all through the city a number of

THE PRESS.

ablished in Belleville was *The Anglo-Canadian*, in
nder T. W. Williamson, editor, and W. A. Welles
Phœnix was the second journal, published by T.
on. It died July 3rd, 1832, aged one year. *The*
The Phœnix, with Rollin C. Benedict as proprietor,
1834 George Benjamin started *The Intelligencer*.
Bowell became a partner in the business. Next
oore became proprietors of the paper. The part-
years and a quarter, when Mr. Moore retired and
roprietor. In 1875 *The Intelligencer* Printing and
d the paper, retaining Mr. Bowell, who is a large
n 1867 *The Intelligencer* was merged into a daily.
in 1835 by Mr. Greenleaf. It lived only a short
n 1836 *The Plain Speaker*. It was friendly to the
put in the Kingston Penitentiary for attempting a
s. The soldiers (volunteers) afterwards marched to
aker, upset the type-fonts and trailed the manager
his movement occurred because the paper appeared
British coat of arms turned upside down in its

was the next venture. It was started in 1841 by
1849 E. Miles purchased the paper and ran it suc-
malgamated with *The Ontario* in 1873. In 1841
as placed on the market by Joseph Wilson, with
s editors, but it only lived one year. In 1847 and
shed a monthly magazine of miscellaneous pieces.
ted various publications, *The Eclectic Magazine*,
son's Canada Casket, &c. But none of them were
Some other papers were started only to die after
were *The Tribune*, *The Independent*, which lived
ers. In 1870 Carman & Yeomans established *The*
ual is still a vigorous newspaper and in the land of
enture was *The Free Press*—a small advertising
ich was established in 1876 by Messrs. McCul-

TRADE.

xtensive and well managed trade facility. Many
and enormous quantities of lumber are annually
he last few years, however, this important branch
ous causes, has visibly declined. Belleville does
ss in the exportation of barley, and last year the
ushels. The products of the dairy have also become
mercial field of this city, the farmers of Hastings
no less than six hundred thousand dollars for their
articles are regularly shipped to England, and they
account of their magnificent quality and character.
the growth of Belleville's trade, compiled from
ery interesting, as it shows how rapidly these ex-

Year.	Exports.	Imports.	Duty.
1855	$346,963	$280,310	$32,063.47
1856	342,771	395,843	39,150.05
1857	295,616	283,515	21,701.98
1858	392,239	169,128	15,615.57
1859	289,726	180,794	18,231.32
1860	190,698	173,340	22,260.27
1861	291,391	173,935	
1862	276,225	150,279	
1863	301,799	126,224	
1864*	70,199	33,091	1,531.60
1865	319,730	101,650	7,131.79
1866	550,217	125,479	8,508.82
1867	597,510	142,669	17,119.58
1868	532,120	153,182	21,485.88
1869	575,011	192,592	24,506.38
1870	587,834	155,283	30,891.42
1871	520,526	146,641	35,602.55
1872	544,156	215,211	34,676.09
1873	730,500	247,867	37,782.63
1874	536,869	290,913	39,880.70
1875	598,190	438,162	36,043.63
1876	551,634	273,399	42,166.41

* The returns for 1864 are for six months only, as the close of the fiscal year was at that time changed from 31st December to 30th June.

MANUFACTURES.

Belleville has at present one axe factory, five sash factories, two grist mills, two furniture manufactories, four saw-mills, two cigar manufactories, four carriage manufactories, eight blacksmiths, one organ manufactory, three woollen mills, breweries, four marble factories, one sewing machine manufactory.

RAILWAYS.

In 1852 a charter was obtained to build the Grand Junction Railway. This road was to connect Belleville with Peterborough. Fifty thousand pounds worth of stock was subscribed for by the County Council. Some preliminary work was done, when the enterprise was abandoned for a time. In 1870, however, substantial encouragement was offered by the municipalities interested, and a bold push was made. Mr. A. Brooks, of Brockville, was awarded the contract, and in 1873 the first sod was turned and the work of construction began in earnest. Before long, however, the contractor failed, and this important work received a further check. There is good reason to believe, however, that some time during the present year the work will be completed. The Belleville & Marmora Railway was another project early in the field, but it was abandoned in 1865. In 1875 a charter was procured for the Belleville & North Hastings Railway. This line was to connect Belleville with the Seymour and Moore iron works, near Madoc. In August, 1876, work was begun. In 1877 this line suffered a reverse, and work was postponed after a good deal of money had been expended on it. There is hope that this line will also be advanced towards completion during the present year.

SOCIETIES.

The Masonic bodies, Orange bodies, Oddfellows' organizations, various national, religious, temperance, and charitable societies, have a firm foothold in Belleville. The Y. M. C. A. is a strong body, and they occupy a large and pleasant suite of rooms in Bull's Building, Campbell Street. The Mechanics Institute is the leading literary society, and the library in connection with it contains a goodly number of valuable books. The Murchison Club is a scientific society of much promise. It has an active membership, and is destined to do a great deal of good in the community. Its meetings are held monthly. The large number of private literary clubs which abound in Belleville is a sure index of the culture and artistic taste of the citizens.

THE FIRE DEPARTMENT.

This important branch of public service is one of the best and most thoroughly equipped in Canada. It is composed of two hose companies (Moira, No. 1, and Independent No. 2), a hook and ladder company (Quinté, No. 3), and a company attached to the large chemical engine (Active, No. 4); 150 men "run with the machines." The Department owns two first-class steam fire engines, one hand engine, and two chemical engines, and a hook and ladder apparatus. The efficient chief engineer is Mr. John Taylor; and his able assistant is the popular officer, Mr. W. H. Campbell.

POLICE.

This department is quite efficient and well officered. Hugh McKinnon, a name which strikes terror into the hearts of evil-doers, is chief; and under him are two sergeants and five men. Abraham Diamond, Esq., is the chief magistrate, and he dispenses justice daily to the inhabitants.

MILITARY.

In 1799 the militia of Hastings was organized. John Ferguson, of Kingston was lieutenant-colonel, Alex. Chisholm was major, and William Bell, of Thurlow, was captain and adjutant. In those good old days the militiaman did not indulge much in "child's play." He drilled rigorously, and accordingly every other Saturday this fine body of men were accustomed to assemble at Wallbridge's, in Belleville, for platoon exercise, etc. In 1812 we find the body officered as follows: John Ferguson, colonel; William Bell, lieutenant-colonel; John Thomson, major; Alex. Chisholm, senr., J. McNabb, S. B. Gilbert, Jacob W. Meyers, George Meyers, David Simmons, Gilbert Harris, and John McIntosh, captains. On June 29th, the Hastings militia were ordered to Kingston, but as that stronghold was not attacked, the soldiers returned home in a few weeks. A few names survive the war of 1812, though their owners have long since lain in their graves. These are, Lieutenant Wm. Ketcheson, of Sidney; Captain Thomas Coleman, of Belleville; and A. O. Petrie, of the same place.

In 1837 this *corps* also did efficient duty. In December of the same year a dreadful accident occurred, which deprived the company of a brave officer and a courteous gentleman. Captain James McNab, the organizer of the first volunteer company raised in the town, was out with his men; some excitement occurred in the street, when the gallant Captain was shot dead. A musket carried at the trail by one of his men, accidentally went off, and one of the most dire tragedies which ever happened in Belleville, took place. The occurrence threw the little town into gloom, and for many months afterwards the sad affair continued the topic of discussion. A gallant officer succeeded to the command of Capt. McNab. This was Wellington Murney, who inherited the military capacity of his great Peninsular namesake. His first duty was to organize the 1st Hastings Rifle Company. This corps was ordered to Gananoque to meet the rebels, under the command of General Van Rensselaer. In April 1839, this corps was disbanded. Companies at various times since have been organized by Captains George Bleecker, James Fraser, Donald Mackenzie, Donald McLellan, and Peter O'Reilly. In 1835 the 1st Hastings Cavalry was organized by Captain Perry, with Jacob Fralick, as Lieutenant. The 2nd Hastings Cavalry was raised in 1838, Captain Charles O. Benson, Lieutenant Robert Potts, and Cornet Charles L. Herchmer. John Turnbull received the Lieutenant-Colonelcy of the 1st Regiment, Thomas Barker was Major, Lieut.-Colonel Thomas Coleman was Commandant at Belleville, and Billa Flint was Commissary. In 1837-'38 militiamen were allowed three shillings a-day for their services, but as the Adjutant General, Richard Bullock, refused to certify the pay and allowances of the troops, the pay was not forthcoming. This caused a great outcry against Bullock, and Sir John Colborne was memorialized in consequence.

The present military organizations are:—The 1st Batallion (Argyle Light Infantry), Lieut.-Colonel S. S. Lazier; No. 1 Company of the 49th Battalion Hastings Rifles, Lieut.-Colonel Brown, M.P.P. These men have ever proved equal to emergency, and anxious to do their duty. They performed distinguished services at Amherstburg in 1864, and later at Prescott, Aultsville and Niagara.

GAMES, YACHTING, &c.

The good old English game of Cricket largely obtained here in times past, and some years ago no finer "Eleven" could be found in Canada. Though of late years the game has declined a little, there is still an excellent club, "The Belleville," in existence. The Curling Club boasts of an extensive membership, and a Quoiting Club has also become an institution in the place. Yachting, however, is the prevailing and the most favorite pastime of the people. In Belleville some of the swiftest yachts ever built are to be found, and the Bay of Quinté Club, though young in years, is certainly vigorous.

GAS.—The city is well lighted with gas.

BANKS.—With monetary institutions the city is well supplied, there being no less than three strong financial corporations. These are, branches of the Bank of Montreal, Merchants' Bank of Canada, and the Consolidated Bank of Canada. The business done is of large volume, and grows more profitable every year.

NATURAL ADVANTAGES.—It may be well here to speak briefly of the many natural advantages which Belleville possesses. She has all the advantages which rich, varying scenery bestows. Her situation is most picturesque. She is the centre of an agricultural district, and her mines and minerals are no less abundant and prolific. Her people are thrifty and ambitious, and no one can doubt the great future which awaits this enterprising western city.

TRENTON.

The early history of Trenton is similar to that of many other Canadian towns, and its experience the experience of all early settlements, with all the exciting scenes and deprivations of frontier life, and the gradual unfolding and development of a community complete in its organization, distinctive in its character and expression, and rich in the higher elements of domestic, social, and religious life. The pioneer moves into the forest with his few household goods around him, and rises a king and conqueror. Here he erects his altar, builds his house, levels the forest, calls down the sunlight to thrill with life the sleeping soil and adorn its surface with the bloom of vegetable life, while Nature in her supreme loveliness matures and yields to him the ripening fruit, the richest treasures of her bosom. Here is laid the keystone in the arch of a new social structure, above which are to cluster and unfold all the arts and elements of the highest civilization. Here we see the importance of collecting in successful order all the scenes and events of a community's growth, from the earliest settlement, its first germ, to its full organization and its most recent form: together with the influence, local characteristics, and other combinations that may have modified or directed its development. Thus we are enabled to grasp the science that underlies and governs its life: a science that should be perpetuated, in imperishable records, to our children and our children's children.

We of the present day, who have witnessed the rapid occupation of the western part of our Province, can have but a faint idea of the slow and tedious process of settlement in the closing years of the last century and the beginning of the present, nor appreciate the difficulties and discouragements by which it was attended. Especially is this the case with the early settlement of the river Trent district, an interminable wilderness without roads, and with but indifferent facilities for water communication: together with the scarcity of the necessaries of life and the general poverty of the inhabitants, a condition which they accepted when they abandoned, for their devotion to cherished principles, their homes in the revolted colonies. The early settlers of Trenton endured untold hardships, which will for ever form a theme for the wonder and admiration of posterity. The Trenton of to-day is a standing monument to the wonderful achievements of the founders of Upper Canada, the United Empire Loyalists.

The village is situated at the confluence of the river Trent with the Bay of Quinte, and was incorporated in 1853. It originally formed part of the Township of Murray, in the County of Northumberland, but since its incorporation it has been connected with the County of Hastings for municipal and judicial purposes. Its situation at the head of the Bay is beautiful and picturesque, and the many elevated positions with which it is surrounded present to the observer delightful views of the Bay, the river, and the adjoining country.

Scientifically, much interest attaches to this locality, as here are to be found the upper and lower shaly beds of the Trenton limestone, in the Lower Silurian series of rock, rich in fossils and other evidences of pre-antediluvian existence.

The first permanent settler of Port Trent, as it was then designated, was one James Smith, a U. E. L., from Schoharie Co., New York, who, in 1790, drew land on the west bank of the river and erected, at the base of Bunker Hill, a log house, traces of which still exist. The property was afterwards transferred to Henry Ripson, also a U. E. L., who built a grist mill, the first on the Trent. In 1808, Adam Henry Myers, father of Col. Adam Henry Myers, a native of Bremen, in the Kingdom of Hanover, who came to Canada in 1805, settled at Port Trent and purchased the Ripson property, where he engaged in the lumbering and mercantile business. John Bleeker, a son-in-law of Capt. John Walter Myers, of Sidney, and a U. E. L., also settled here in 1790, securing land on the west side of the river, and built a log house at a point now known as Bleeker's grove. He afterwards erected a frame building on the hill which, after his death, which occurred in 1807, his widow moved to the site of the present garden of J. W. Ryan, where for many years she kept an inn. In 1817, Sheldon Hawley settled at Port Trent, on the east side of the river, and engaged in mercantile pursuits. David Johns, who purchased the remainder of the Smith estate, also came in about this period. John V. Murphy settled here in 1802. The Hon. Robt. Charles Wilkins, of the Carrying Place, also owned property in the village at this date. A. H. Myers, Denis McAulay, William Robertson and James Ford, who is said to have been the first school teacher on the Front of Sidney, on the west side of the river, with their associates on the east side, were the pioneer merchants of the settlement and to them is largely due the prosperity of the present flourishing and prosperous town of Trenton. The Hon. and Rev. Dr. Strachan, first Lord Bishop of Toronto, was prominently associated with the early settlement. He purchased in 1801, the broken front of the gore of Sidney, which he surveyed and laid out in town lots, naming the plot Annwood, after the maiden name of his wife. The name, together with the lots, were subsequently merged with and became a part of Trenton. He generously presented the Church of England authorities with land upon which St. George's now stand, accompanying the gift with a handsome donation of money. Eligible building sites were rapidly taken up by parties who were drawn thither by the splendid advantages offered for

manufacturing, and also on account of the
trade. The lumbering business on the river
sixty years has, undoubtedly, had much to
development of this flourishing village. In
and square timber have been rafted over its
tensive limits in the northern extremities of
Victoria and other points penetrated by its
is either dressed at the large mills at its
Quebec markets. Hundreds of men are annu
in running the rivers and rearranging the
Lawrence run, all of whom are fed and cloth
Trenton merchants. Large quantities of saw
to the United States from this port. We ha
at hand to represent the actual proportions
formed that as much as 20,000,000 feet of saw
season by the steam mills on the Bay shore,

The manufacturing interests of Trenton
follows: two steam saw-mills, two grist-mill
one foundry, three sash and door factories,
manufacturing establishment, one stove and
fulling-mill, one tannery, one tin and copper
tory, one plaster mill; and upon a tributary
There are also several blacksmith's shops,
mercantile business is represented by a nu
hardware, and grocery stores, and also a num
shoe stores, flour and feed, crockery and gla
er trade. There are also several good hotels.

Besides these industries there are five place
Church of England, built of stone, with tow
English style of ecclesiastical architecture;
D.C.L., Senior Canon of St. George's Cath
Roman Catholic church, a large and handsom
therewith a new Separate School, of brick,
Rev. Henry Brittargh, Pastor. Canada Meth
posing spire; Rev. Peter Addison, resident
substantial building; J. L. Stewart, B.A. E
Crosby, Pastor.

The educational interests are represented by
common schools, and one separate school. D
300.

The Town Hall is a large commodious buil
ber on the upper flat, the lower being occup
etc. Adjoining the hall is the drill-shed.

There are also a number of benevolent
Mark's Chapter of Royal Arch Masons, No. 2
& A. M., No. 38, G.R.C., H. W. Day, W.M
Loyal Orange Lodge, No. 100; one lodge o
Andrew's Society.

A large, covered timber bridge across the
by a Board of Commissioners appointed by t
east and west sections of the village.

Trenton is approached from the east and
lines of steamboats on the Bay of Quinte, affor
and the shipment of freights, and is a port of

Since the incorporation of the village in 185
men have respectively officiated in the capa
Alexander McAulay; 1854, E. W. Myers; 18
Thompson; 1856, D. R. Murphy; 1857, E.
don; 1859, J. S. Peterson; 1860, J. Cumming
D. R. Murphy; 1862-1872, William Shea; 18
William Jeffs; 1878, Charles Francis. Jere
Clerk 9th Division Court, and Registrar of V
head, Town Treasurer. Assessed value of re
Number of ratepayers 599. Total population,

Trenton is distant from Belleville, 12 mil
Brighton, 10 miles; and the Carrying Place,

The village of Trenton, its settlement and
an example of the many prosperous corporation
The proverbial log hut of the pioneer settler
appears, and is superseded by blocks of brick
ecclesiastical edifices, substantial schoolhous
other evidences of advanced civilization. To
man Trenton presents many natural and su
require proper development to place this beau
the front rank of manufacturing centres.

SIDNEY.

The townships on the Bay of Quinté, with
township Tyendenaga, were first settled upon
pire Loyalists, and were numbered in the ord
ston being first town, Earnestown, second tow
of Sidney, which was called eighth town. S
are still retained by their inhabitants, especia
the township of Ameliasburgh being still calle
of the first settlers caused these to be chang
upwards, to the names of the sons and dau
those on the Peninsula of Prince Edward
females of that line. These names being ext
ships on the Bay and a kindred settlement on
Elizabethtown downwards, Sidney was name
the time of the revolutionary war, was Secret
ment, and was surveyed and laid out about th
assisted by one McDonald. A map on file in t
has inscribed upon it, "Sidney in the Dis
bounded on the north by the township of Raw
and the city of Belleville, on the south by th
west by the river Trent and the township of M
thumberland, and is at present one of the best
ships in the County of Hastings. The settleme
effected during and from the year 1787, by
settlers, 1st. United Empire Loyalists; 2nd.
Loyalists of older townships on the Bay, who h
by persons of Loyalist origin who came from
and elsewhere, where they had first settled;
quently settled there after the war of 1812, and
Thus, its settlement was progressive in charac
pear, that the entire occupation of the first 6
laid out by Kotte, was not entirely completed
years. The northern section of the township,
veyed and settled later still, and one of the
if not the very first, was that of the Sine Se
24, in the 6th concession.

The first concession of Sidney, from the p
Bay of Quinté, became of necessity the scen
Capt. John Walter Myers, noted during th
bravery and enterprise, and whose career, in
related with varying details in Stokes' "Ame
Canniff's "History of the Settlement of Canad
appears to have been the actual pioneer settler
descent, and came from the vicinity of Albany

'ORICAL SKETCH OF THE

›cturing, and also on account of the centralization of the lumber

The lumbering business on the river Trent during the last fifty or ;ears has, undoubtedly, had much to do with the rapid growth and pment of this flourishing village. Immense quantities of saw-logs uare timber have been rafted over its turbulent waters from the ex- limits in the northern extremities of the counties of Peterboro' and ia and other points penetrated by its tributaries. The raw material er dressed at the large mills at its confluence, or re-rafted for the markets. Hundreds of men are annually employed at the mills and ning the rivers and rearranging the timber in the Bay for the St. nce run, all of whom are fed and clothed by supplies purchased from on merchants. Large quantities of sawed lumber are annually shipped United States from this port. We have not statistical information al to represent the actual proportions of this industry, but are in- d that as much as 20,000,000 feet of sawed lumber has been cut in one by the steam mills on the Bay shore, and on Mill Island.

e manufacturing interests of Trenton and vicinity are represented as s: two steam saw-mills, two grist-mills, one brewery, one paper-mill, undry, three sash and door factories, four carriage houses, one marble facturing establishment, one stove and barrel factory, one carding and g-mill, one tannery, one tin and copper manufactory, one pump fac- one plaster mill; and upon a tributary of the Trent are two grist-mills. are also several blacksmith's shops, and two grain elevators. The atile business is represented by a number of first-class dry-goods, care, and grocery stores, and also a number of general stores, boot and stores, flour and feed, crockery and glass, stationery, and other lines de. There are also several good hotels.

ides these industries there are five places of worship, viz., St. George's, h of England, built of stone, with tower, and in the advanced early sh style of ecclesiastical architecture; Rev. Wm. Bleasdell, M.A., s, Senior Canon of St. George's Cathedral, Kingston, rector. The n Catholic church, a large and handsome building, and in connection with a new Separate School, of brick, of unique architectural design; Henry Brittargh, Pastor. Canada Methodist, frame, with tall and im- g spire; Rev. Peter Addison, resident minister. Presbyterian, a neat antial building; J. L. Stewart, B.A. Episcopal Methodist, Rev. D. O. by, Pastor.

educational interests are represented by one excellent high school, two on schools, and one separate school. Daily average attendance about

e Town Hall is a large commodious building, containing council cham- the upper flat, the lower being occupied by clerk's office, lock-up, Adjoining the hall is the drill-shed.

ere are also a number of benevolent and other societies, viz.: St. 's Chapter of Royal Arch Masons, No. 26, G.R.C.; Trent Lodge A. F. M., No. 38, G.R.C., R. W. Day, W.M.; Wm. Isaac Nelson, Secy.; Orange Lodge, No. 160; one lodge of Good Templars, and one St. ew's Society.

large, covered timber bridge across the Trent, erected many years ago Board of Commissioners appointed by the Government, connects the nd west sections of the village.

enton is approached from the east and west by the G.T.R. and daily of steamboats on the Bay of Quinte, affording ample facilities for trade he shipment of freights, and is a port of entry.

nce the incorporation of the village in 1853, the following named gentle- have respectively officiated in the capacity of Reeve: First Reeve, ander McAulay; 1854, E. W. Myers; 1855, Sheldon Hawley and A. C. pson; 1856, D. R. Murphy; 1857, E. W. Myers; 1858, G. H. Gor- 1859, J. S. Peterson; 1860, J. Cummings; 1861, Robert Francis and Murphy; 1862-1872, William Shea; 1873, Charles Francis; 1874-5, am Jeffs; 1878, Charles Francis. Jeremiah Simmons, Town Clerk, 9th Division Court, and Registrar of Vital Statistics, etc. R. Long- Town Treasurer. Assessed value of real estate for 1878, $400,075, ber of ratepayers 509. Total population, 2,722.

enton is distant from Belleville, 12 miles; from Frankford, 8 miles; ton, 10 miles; and the Carrying Place, about 5 miles.

e village of Trenton, its settlement and gradual development, is but mple of the many prosperous corporations to be found in this Province. roverbial log hut of the pioneer settler of one hundred years ago dis- rs, and is superseded by blocks of brick or stone, handsome public and astical edifices, substantial schoolhouses, palatial residences, and evidences of advanced civilization. To the manufacturer and trades- Trenton presents many natural and superior advantages, which but e proper development to place this beautiful and prosperous village in nt rank of manufacturing centres.

SIDNEY.

e townships on the Bay of Quinte, with the exception of the Indian hip Tyendenaga, were first settled upon both its sides, by United Em- oyalists, and were numbered in the order of their settlement, King- eing first town, Earnestown, second town, and so on to the township ney, which was called eighth town. Some of these numbered names ll retained by their inhabitants, especially in Prince Edward county, wnship of Ameliasburgh being still called seventh town. The loyalty first settlers caused these to be changed from Kingston township ds, to the names of the sons and daughters of King George III., on the Peninsula of Prince Edward Co., being named after the s of that line. Those names being exhausted by the earlier town- n the Bay and a kindred settlement on the River St. Lawrence from ethtown downwards, Sidney was named after Lord Sidney who, in ne of the revolutionary war, was Secretary for the Colonial Depart- and was surveyed and laid out about the year 1787, by Louis Kotte, l by one McDonald. A map on file in the Crown Lands Department scribed upon it, "Sidney in the District of Mecklenburg." It is d on the north by the township of Rawdon, on the east by Thurlow e city of Belleville, on the south by the Bay of Quinte, and on the y the river Trent and the township of Murray in the County of Nor- erland, and is at present one of the best settled and wealthiest town- n the County of Hastings. The settlement of Sidney was permanently l during and from the year 1787, by three classes or grades of s, 1st, United Empire Loyalists; 2nd, Sons and daughters from the sts of older townships on the Bay, who had drawn land there, and also sons of Loyalist origin who came from Nova Scotia, New Brunswick ewhere, where they had first settled; 3rd, by Americans who unless settled there after the war of 1812, and probably at an earlier period. ts settlement was progressive in character. It would, however, ap- hat the entire occupation of the first five concessions surveyed and s by Kotte, was not entirely completed, until after a lapse of several

The northern section of the township, or the Oak Hill range, was sur- and settled later still, and one of the first portions occupied there, the very first, was that of the Sine Settlement, on lots Nos. 23 and he 6th concession.

first concession of Sidney, from the proximity of the waters of the Quinte, became of necessity the scene of the earliest settlement. John Walter Myers, noted during the revolutionary war for his and enterprise, and whose career, in connection with the same, is with varying details in Stokes' "American Border Wars," and Dr. 's "History of the Settlement of Canada," together with his family, to have been the actual pioneer settler. Capt. Myers was of Dutch and came from the vicinity of Albany, on the Hudson, in the State of New York, about the close of the war, and settled upon the front of Sidney, where he drew a large grant of land, a short distance east of the present village of Trenton. His enterprise was here auspicious, not only in clearing land for cultivation, but in erecting a grist mill upon a small stream on his land in Sidney, and thus became a pioneer in mill building as well as trading and sailing Batteaux and other craft on the waters of the Bay of Quinte. Traces of this mill near the Bay shore can still be seen; and the position of the dam, the mill-race, the foundation of the mill and residence of the late owner still exist; and, what is more especially interesting, there is embedded in the surface soil, but still exhibiting its upper side, the lower mill stone, or bed-stone as it is usually designated. The stream proving, eventually, inadequate for the desired end, he subsequently erected another mill on the present river Moira, on a dam constituted for that purpose, and also a log saw-mill on the opposite or east bank of the river, where Belleville now stands. At a much earlier period, on leaving the revolted colonies as a U. E. Loyalist, he sojourned for a short time at Adolphustown, or Fourth town, and also for a time on the front of Thurlow, then unsurveyed, from whence he removed, upon the representation of his eldest son George, to the front of Sidney, as a more eligible place for settlement. Capt. Myers had four sons, George, Tobias, Leonard and Jacob, and at least two daughters, one of whom was married to I. J. Bleeker, the first settler at Trenton, and the other to John Row, an early settler of Sidney. His grant of land consisted of 800 acres, and here descendants of his sons, sons and daughters, reside at the present time. Tobias W. Myers and John G. Myers, sons of his eldest son George, a Major, as he subsequently became, are the present patriarchs of the settlement. Capt. Myers died in the year 1810, and was interred in the old or original burial ground on the front of Sidney.

The original U. E. Loyalists, as far as can be ascertained, settled on the front of first concession of Sidney, in the following order, commencing at Trenton:—Capt. Marsh, Capt. Myers and his four sons, John Scott, George Smith, Abel Gilbert, Chrysdale and Ostrom. George Smith was the first person interred in the old front of Sidney burial ground. To these were added offshoots of U. E. Loyalists from elsewhere, and followed by Americans, who emigrated from the United States. At this period we find the names of Zwick, Vandervoort, White, Bonesteel, Simmons, Kelly, Finkle, Graham, Jones, Laurence, Elijah Ketcheson, and others, in the first concession of the township. In the second concession the early settlers were chiefly composed of the second class descendants of U. E. Loyalists from other townships and provinces of the British Rule, with a few of the third class, and this was also the case with the remaining concessions, including those to the fifth. On the second concession we have the names of Hogle, John Row, from Nova Scotia, Simmons, Gilbert, Ostrom, Vandewater; James Farley, who is said to have come here in 1799, and others. On the third concession there occur the names of John Smith, John Lott, John Stickle—three Johns—hence the name, Johnstown to the west end of the concession. Ira Billings, Bonestell, Percy, Aikens, Crouter, McMullen, Vandervoort, Goldsmith, Ruliff, Pursley, Hagerman, Roblin, Caleb Gilbert, Frulick, and others. On the fourth concession came William Ketcheson with his sons, in 1800, being an U. E. Loyalist family from Nova Scotia, who purchased land and settled here. Other settlers came in about the same time, or it may be a little before this period. The settlement of this concession began towards the eastern boundary, where we find the names of Longwell, Sherard, Hazelton, William Ketcheson sr., William Ketcheson, jr., Youmans, John Ketcheson, Graham, Huffman, Henry Gross, Ackers, Thomas Ketcheson, and others.

The settlement of the rest of the township, as previously mentioned, was still later; many changes have occurred since the days when Capt. Myers with his family took up his residence on the front of the township. Lands have changed hands, and in some cases the names of their owners, together with the original names, to whom the Crown patents were issued, have disappeared from the township records, but in a large majority of cases they still remain. In the old burial ground on lot No. 19, on the front of Sidney, sloping pleasantly down to the bay shore, where the forefathers of the township sleep, where the hardy and energetic pioneers of Sidney are laid in their last resting place, the names recorded are seen now, some without a record, whose vigorous arms felled the forest trees, cleared the land, and raised the first humble dwellings, which have almost entirely given way to more stately edifices of brick, stone and other materials. Around the remains of old John Walter Myers, are gathered his kindred neighbors, and associates; his loyal patriotic friends and opponents are sleeping quietly by his side. The primitive old woods of former times that covered the shore of the [illegible] of Quinte, have almost entirely disappeared, and the Indian hunter [illegible] his wigwam, who fished and hunted, is also gone, and a new order of th[illegible] prevails. The smiling fields ripe with cultivation, the comfortable homesteads, the beautiful and smiling orchards, the well-kept, and tastefully laid out gardens, the neat substantial mansion, with occasionally a rising town, or pleasantly located village, look forth upon the Bay water, and gather pleasure and profit from the situation upon this beautiful arm of Lake Ontario.

On Lot No. 13, on the front of the township and on the road from Trenton to Belleville, there was formerly a tavern, blacksmith shop, store—probably Ferguson & Bell's—and a group of houses which was called Rhinebeck, but the name, with the old features, as a village, have long since disappeared.

Owing to some difficulty arising about the side lines throughout the township, and the general inaccuracy of the division lines, a re-survey was ordered at a later day, the surveyor being one Atkins, whose lines are still considered authority.

In order to show how marriages were performed in those early days, we reproduce a certificate, issued by one of the Sidney Magistrates, in 1819. We withhold the names of the parties directly interested, they belonging to two of the oldest and wealthiest families in the township.

"SIDNEY TOWNSHIP:

"Whereas, and both being of the Township of Sidney, are "desirous of intermarrying with each other, and have presented a written "licence for that purpose. Now, these are to certify that I, Solomon Hazle-"ton, one of His Majesty's Justices of the Peace, have this day married the "said and together in marriage, and they are become contracted to "each other in marriage.

SOLOMON HAZLETON, J.P.

"Sidney, May 19th, 1819."

Through the kindness of F. B. Prior, clerk, we were permitted to make the following excerpts from the Record of the township, which date back to 1790:

"Names of persons who subscribed seven pence halfpenny to pur-"chase this book for a Township Record.

" 1. Caleb Gillbert,
" 2. George Smith,
" 3. Peter Lott,
" 4. Nicholas J. Stickles,
" 5. Aaron Rose,
" 6. Cornelius Lawrence,
" 7. Henry Ketcheson,
" 8. Buliff Ostrom,
" 9. Solomon Hazleton,
" 10. Hugh McMullen,
" 11. James Sharrard,
" 12. John Barnum,
" 13. George Finkle,
" 14. Samuel B. Gilibert,
15. Gilbert Harris,
16. Alex. Chisholm, Jr.,
17. John Hennesy,
18. Cornelius White,
19. William Kelly, Sr.,
20. William Kelly, Jr.,
21. Leonard W. Myers,
22. John Row,
23. Samuel Tompkins,
24. David Marshall,
25. Charles Simmons,
26. Alex. Gillard,
27. Moses Simmons.

The above subscriptions being paid, the book was purchased, upon whose second page appears the following record of the first town meeting ever held in the township of Sidney:

"1790. } UPPER CANADA.
"May 15th. }

"Pursuant to an Act of the Legislature of the Province of Upper "Canada, in such case made and provided, the first annual meeting of the "inhabitants of the Township of Sidney, was held at the dwelling house "of Aaron Rose, in Sidney aforesaid, on May the 15th, 1790, and from "thence adjourned to the dwelling house of Stephen Gilbert, Esq., and to "be held on the first Tuesday of May ensuing.

"May 15th.—The inhabitants of Sidney being assembled as aforesaid, "to act upon town business, have nominated and appointed for town "officers the following persons, viz:—

"MODERATOR—John W. Myers; TOWN CLERK—Leonard Soper; "CONSTABLE—David Simmons; PATHMASTERS TO LAY OUT ROAD—George "Myers, Caleb Gillbert; FENCE VIEWERS—Nathaniel Marsh, William "Lounsbury."

"BY-LAW—It was ordered, the town clerk be entitled to a fee of seven "pence half-penny for entering the ear marks of the inhabitants of Sidney "in the town book."

It appears that three years later, or in 1794, the townships of Sidney and Thurlow were united for municipal purposes, for on Tuesday, the 3rd day of May of that year, at a meeting held in the dwelling house of Caleb Gilbert:—

"It was ordered by a majority of votes, that fences be 4 feet 6 inches "high, in the Township of Sidney, and not to exceed 5 inches between "the rails, Thurlow township not to exceed 6 inches."

At this meeting Archibald Chisholm and George Myers were appointed assessors, the first in these townships.

In the year 1798 "Rams were ordered confined from 1st of Sept. to 10th "of December, under a penalty of 20 shillings; hogs to be free commons "until they done damage."

During the month of May, 1798, the townships of Sidney and Thurlow separated, and the former elected its own officers as follows: William Lounsburry, Town Clerk, Paul Graber and Joseph Rosebush, Town Wardens. In 1799, Henry Smith was elected Town Clerk, and in 1800 John Hagerman, followed by James Farley, James W. Sharrard, Reuben White, Abel Gilbert, Elijah Ketcheson, Jacob W. Myers, Joseph M. Lockwood, Gideon Turner, John S. Hulbaun. The town meetings were held respectively in the Inn of Ketcheson, 5th Concession, Ketcheson's store, 4th Concession, and Ketcheson's school-house. In 1849 John Ketcheson was District Councillor, and Gilbert Bleeker, Township Clerk.

On the erection of the Township of Sidney, in the year 1850, to an independent Municipality, the following persons were elected by a popular vote of its inhabitants to the several positions required by the Act of Parliament made and provided, the returns being made at the dwelling house of Gilbert Bleeker:

Gideon Turner, Reeve; Caleb Gilbert, Deputy Reeve; Robert Bird, and Gilbert Bleeker, Councillors; Thos. D. Farley was appointed Clerk of the first Council.

The chief executive offices of the Council since the above date have been filled respectively by the following named gentlemen: Thomas D. Farley, George Zwick, Baltis Ross, 13 years Reeve—Caleb Gilbert, Ketcheson Graham, Gideon Turner,—many years Clerk—and James A. Chisholm. C. Armstrong is the present Reeve, Frank G. Prior, Clerk.

The town hall is situated about the centre of the township, at the post village of Wallbridge, distance from Belleville, 9 miles. It is a good, substantial, commodious building, with ample shed room for the teams of the ratepayers. The township contains about 68,400 acres of excellent land; its surface is somewhat rolling, well watered with several streams, the principal of which is the Trent. Sidney has about 1,225 ratepayers, with a population of 6,175; assessed value, 1878, $2,588,765, which is one of the best evidences of its prosperity. There is a large cheese manufacturing interest in the township of Sidney, its lands being well adapted for dairy purposes. There are scattered throughout the municipality several large cheese factories, which annually turn out thousands of pounds of a superior quality of cheese, paying their several patrons well for their investments. The River Trent flows through the western part of the township and empties into the Bay of Quinte; it is spanned by two substantial covered bridges, one at Trenton and the other at Frankford, affording to the farmers on either side an excellent means of passage. The Trent river was one of the original routes of Indian and French traders, and is of interest from the fact that Champlain is reported to have entered the Bay of Quinte by this means and discovered Lake Ontario. Its Indian name was *Sennessagee*, and was sometimes called Quinte: unnumbered feet of squared timber and saw logs have been floated over its turbulent waters on their way to the great markets of the old world.

Frankford, once a post village, is situated on the river Trent, in the township of Sidney, distance from Trenton about eight miles, Belleville fourteen miles, and Stirling seven miles. This village has made little progress of late years, its population and business being about the same as in 1870. It is thought that Abel Scott was the original settler and founder of this village. About the year 1807 he built a mill, and the place was known for many years as Scott's Mill, sometimes called Cole Creek, after the creek of that name, a tributary of the Trent, and upon whose water privileges these mills were located. At this date there was no bridge across the Trent, and the inhabitants were compelled to ford the stream, which often was full to its banks, and impassable. During Sir Francis Bond Head's administration he visited Scott's Mills, and named the place Frankford, which name it has retained ever since. We have no data as to the precise time the bridge was built, but it was not until after several years had passed from the visit of the Lieut.-Governor. The gravel road between Trenton and Frankford was completed about 1852. The village has flouring and saw-mills, a woollen factory, tannery, pump factory, and several stores, taverns, etc., good stone Public School building, and three churches—Canada Methodist, Episcopal Methodist, and Roman Catholic. Sills's paper mill is also located at this point; and in 1871 Rabbins erected at considerable expense a dam across the Trent river, with an immense head of water, making it one of the finest mill privileges in the county, suitable for any description of manufacturing purposes. Population about 500.

The incorporated village of Stirling, seven miles from Frankford, in the township of Rawdon, and northern boundary of Sidney, has portions of lots 18, 19, 20, 21, and 22 of this township within its limits.—See Rawdon.

The Belleville and Stirling, and Belleville, Frankford, and Stirling stages, gravel roads pass through the township. The farmers seem to be in a comfortable and prosperous condition, as is evidenced by the character of their several buildings, the well cultivated fields, now covered with promising crops of grain, and the many acres of thriving fruit orchards everywhere to be met with throughout the municipality. School-houses and churches are erected at various points, whose teachings are free as the air we breathe.

Oak Lake is situated upon the Oak Hill range, in the north of the township, is triangular in form, having an area of about 100 acres, and is something of a natural curiosity. The water of the lake is pure and limpid, and abounds with rock bass and other species of fresh water fish. It has no apparent outlet, neither has any visible means of supply ever been discovered.

The Sine family, United Empire Loyalists, came to Sidney at an early date and located upon lots 23 and 24 in the 9th concession, on what is known as the Oak Hill range, and were the pioneer settlers in this section of the township. They are a numerous family, of much intelligence, owning a large quantity of valuable land and occupying respectable sta[...]
in society.

Timothy Soper, son of Leonard Soper, was the first white c[...]
the township of Sidney.

THURLOW.

The Township of Thurlow, or ninth town in the U. E. Loy[...]
Bay numbers, is another of the oldest settled municipalities o[...]
Quinte. It is bounded on the south by the Bay of Quinte, on the[...]
township of Sidney, on the north by Huntingdon, and on the e[...]
dinaga, and is admirably located with reference to railroad an[...]
accommodation and its close proximity to the city of Belleville,[...]
the people have not been slow to avail themselves of as is show[...]
creased prosperity of the agriculturists throughout the town[...]
people are almost entirely descendants of United Empire Loyalis[...]
drawn thither after the close of the Revolutionary war; a people [...]
perous and intelligent, they are justly noted for hospitality an[...]
social virtues, while their broad charity and public spirit find th[...]
in every enterprise conducive to the general welfare. The first p[...]
penetrated the wilds of Thurlow, about the close of the last cent[...]
beginning of the present, found interminable forests, a heavy was[...]
with dense undergrowth, the home of the bear and howling wolf,[...]
very numerous. With their axe and gun and with sturdy arm[...]
they began their work to carve out that grand civilization w[...]
throbs and pulsates in its bosom. When the scattered clearin[...]
admit the sunlight and crops were put in, the people began to re[...]
tility of the soil. Productions of every variety indigenous to t[...]
were certain of rapid growth and large returns, as attested b[...]
wealth that has been drawn from the bosom of the soil durin[...]
quarters of a century that has passed—a wealth that has cove[...]
beautiful homes, drifted to every part of the world, and fed and c[...]
lions. The soil is generally a fine quality of calcareous loam,[...]
siderably with clay. The limestone formation occasionally crops o[...]
sections, particularly on the front, but not sufficiently prominent[...]
with cultivation. The surface of the township is undulating, el[...]
able, and general altitude well adapted to agriculture. It is suffi[...]
watered for farming purposes, with abundant springs of pure cool wa[...]
part thereof. The Moira River, with its several tributaries, runs [...]
Township in a south-westerly direction, and empties itself at Bel[...]
the Bay of Quinte. Apple orchards are beginning to be extensively[...]
and fruit of the finest and hardiest varieties yield abundantly, a[...]
raised in large quantities, while the already large area of orcha[...]
yearly additions. This product alone in a few years will form[...]
principal articles of export during its season. The soil of Thurlow[...]
mentioned, is noted for its productiveness, and wheat has freq[...]
known to yield fifty bushels to the acre, and other cereals in [...]
while it cannot be surpassed for dairy purposes. Large quant[...]
finest quality of cheese are annually manufactured within its li[...]
command the highest market price in the markets of the old worl[...]
it is principally shipped. We have not statistical information at ha[...]
sent the actual extent of this industry, it is, however, sufficient [...]
to-day it forms one of the staple products of this section of the Pr[...]
turning yearly hundreds and thousands of dollars to its patrons—[...]
of Thurlow and the surrounding country. The river Moira, wit[...]
mills and manufactures, has added materially to the development [...]
of Thurlow, and since the days of old John Walter Myers, when he [...]
his first dam across its mouth where Belleville now stands, mar[...]
have taken place. Of the inevitable hardships and deprivations t[...]
attended the pioneer settlements of that day, Thurlow had its sha[...]
sions were not to be obtained nearer than Kingston, and the ne[...]
mill was at Napanee Creek, distance some forty miles, where, in t[...]
of oxen and horses the settlers carried on their backs their g[...]
ground into flour, often inadequate to the wants of a large famil[...]
will be remembered that the settlement of Thurlow was begun afte[...]
when the Government allowed provisions, and being isolated an[...]
the means of sustenance, with but small hopes of speedy help fro[...]
side world, this little band of pioneers suffered hardships and lo[...]
might well have discouraged a less indomitable energy than that w[...]
acterized the first settlers of the front of Thurlow. During the [...]
Louis Kotte, who appears to have been the general and only ava[...]
engineer at that time, surveyed and laid out the front concession o[...]
which was taken up by the families of Capt. John Singleton, Lieut. [...]
an Indian Trader, David Vanderhyden, John and Alex. Chisholm[...]
bably Capt. John Walter Myers, who, it is said, located upon the fro[...]
low previous to his removal to Sidney.

We have not been able to ascertain the origin of the name T[...]
connection with the township, but are informed that the name wa[...]
derived from some titled nobleman who had at one time held an o[...]
the British Government. During the year 1789, Thurlow received[...]
tion to its settlement by the arrival of a large number of Loyalist[...]
had been compelled to leave their several homes in the United St[...]
count of the persecutions by the triumphant insurgents, and seek[...]
and new homes under the British flag. Among these last arrivals[...]
some distance back from the front, we find the names of Russ[...]
Archibald McKenzie, Solomon Hazelton, —— McMichael, Wil[...]
Seth Thrasher, Asa Turner, Stephen and Laurence Badgley, Jo[...]
William Reed and his sons, Samuel William, John and Solomo[...]
Smith, John Longwell, Conelly, Sherard and others. These were[...]
few years afterwards by the families of Richard Cannifl and Robert [...]
In 1790 Capt. Myers built a saw mill and afterwards a grist mill o[...]
bank of the Moira River, which was in all probability the found[...]
present flourishing city of Belleville, for many years after the [...]
known as Myers' Creek. The names of Edward and J. Carscallion[...]
Bidwell, Wm. Johnson, Samuel Sherwood, Coon Frederick, Cr[...]
others are inscribed upon an old map of Thurlow on file in the C[...]
office, as having about 1792-3 settled upon the front and first conc[...]
rear concessions were afterwards rapidly taken up by the offspring[...]
Loyalists, who had settled on other portions of the Bay.

Previous to the year 1798 the townships of Thurlow and S[...]
united for municipal purposes, but in that year we find Sidney [...]
own town meetings, an account of which is given in connectio[...]
history of the early settlement of that municipality.

The records of the township of Thurlow previous to 1862 ar[...]
possession of the clerk, M. E. Thrasher, and as he knows noth[...]
whereabouts, they have probably been destroyed. We however [...]
Canniff's "History of the Settlement of Upper Canada," the re[...]
first town meeting held in Thurlow, which we are kindly permitte[...]
use of.

"At the annual town meeting for the township of Thurlow, h[...]
"fifth day of March 1798, whereat the following persons were el[...]
"officers, viz:—

"John McIntosh, Town Clerk; John Chisholm and Wil[...]
"Assessors; Joseph Walker, Collector; Samuel B. Gilbert, John[...]
"William Johnston, Pathmasters; John Cook and Daniel Lawre[...]
"Wardens; John Taylor, Pound-keeper; John Fairman, Const[...]
McIntosh, Jabez Davis, Caleb Benedict, Roswell Levens, John [...]
Daniel Cannifl, James McDonald, D. B. Sole, Dr. Hayden, who [...]
to have taken sides with the Rebels in 1837 and escaped—his wife [...]
hand over the township records—all held the office of town clerk [...]
to 1850 was one of the most important in the township. It is be[...]
J. J. Farley was the first Reeve of Thurlow, an office he filled w[...]

ole land and occupying respectable social positions

f Leonard Soper, was the first white child born in

THURLOW.

irlow, or ninth town in the U. E. Loyalist order of
of the oldest settled municipalities on the Bay of
n the south by the Bay of Quinte, on the west by the
he north by Huntingdon, and on the east by Tyen-
located with reference to railroad and steamboat
lose proximity to the city of Belleville, advantages
slow to avail themselves of as is shown by the in-
e agriculturists throughout the township. The
descendants of United Empire Loyalists who were
lose of the Revolutionary war; a people to-day pros-
hey are justly noted for hospitality and the many
r broad charity and public spirit find them foremost
ive to the general welfare. The first pioneers who
Thurlow, about the close of the last century and the
found interminable forests, a heavy waste of timber,
the home of the bear and howling wolf, which were
heir axe and gun and with sturdy arms and will,
to carve out that grand civilization which to-day
s bosom. When the scattered clearings began to
rops were put in, the people began to realize the fer-
ctions of every variety indigenous to this latitude
owth and large returns, as attested by the vast
rawn from the bosom of the soil during the three
at has passed—a wealth that has covered it with
to every part of the world, and fed and clothed mil-
rally a fine quality of calcareous loam, mixed con-
limestone formation occasionally crops out in certain
the front, but not sufficiently prominent to interfere
urface of the township is undulating, climate agree-
well adapted to agriculture. It is sufficiently well
ses, with abundant springs of pure cool water in every
River, with its several tributaries, runs through the
terly direction, and empties itself at Belleville into
e orchards are beginning to be extensively cultivated,
d hardiest varieties yield abundantly, and is being
s, while the already large area of orchards receives
product alone in a few years will form one of the
rt during its season. The soil of Thurlow, as already
its productiveness, and wheat has frequently been
shels to the acre, and other cereals in proportion.
assed for dairy purposes. Large quantities of the
are annually manufactured within its limits, which
rket price in the markets of the old world, to where
We have not statistical information at hand to repre-
f this industry, it is, however, sufficient to say that
e staple products of this section of the Province, re-
and thousands of dollars to its patrons—the farmers
ounding country. The river Moira, with its many
has added materially to the development and wealth
days of old John Walter Myers, when he constructed
mouth where Belleville now stands, many changes
e inevitable hardships and deprivations that always
ements of that day, Thurlow had its share. Provi-
ained nearer than Kingston, and the nearest grist
k, distance some forty miles, where, in the absence
settlers carried on their backs their grain to be
inadequate to the wants of a large family. For it
the settlement of Thurlow was begun after the period
lowed provisions, and being isolated and without
with but small hopes of speedy help from the out-
d of pioneers suffered hardships and losses which
ed a less indomitable energy than that which char-
s of the front of Thurlow. During the year 1787,
s to have been the general and only available civil
veyed and laid out the front concession of Thurlow,
families of Capt. John Singleton, Lieut. Ferguson,
Vanderhyden, John and Alex. Chisholm, and pro-
Myers, who, it is said, located upon the front of Thur-
al to Sidney.
de to ascertain the origin of the name Thurlow in
ship, but are informed that the name was probably
ntleman who had at one time held an office under
During the year 1789, Thurlow received an acquisi-
he arrival of a large number of Loyalists who, too,
e their several homes in the United States on ac-
the triumphant insurgents, and seek protection
British flag. Among these last arrivals who settled
the front, we find the names of Russell Pitman,
amos Hazelton, —— McMichael, William Cook,
er, Stephen and Lawrence Badgley, John Taylor,
ms, Samuel William, John and Solomon. Richard
onelly, Sherard and others. These were followed a
e families of Richard Canniff and Robert Thompson.
a saw mill and afterwards a grist mill on the east
which was in all probability the foundation of the
d Belleville, for many years after the above date
The names of Edward and J. Carscallon, Fairman,
Samuel Sherwood, Coon Frederick, Crawford and
an old map of Thurlow on file in the Crown Lands
23 settled upon the front and first concession. The
wards rapidly taken up by the offspring of U. E.
on other portions of the Bay.
1798 the townships of Thurlow and Sidney were
poses, but in that year we find Sidney holding its
count of which is given in connection with the
ent of that municipality.
nship of Thurlow previous to 1862 are not in the
. E. Thrasher, and as he knows nothing of their
obably been destroyed. We however find in Dr.
Settlement of Upper Canada," the record of the
Thurlow, which we are kindly permitted to make

eeting for the township of Thurlow, held on the
whereat the following persons were chosen town

wn Clerk; John Chisholm and William Reid,
er, Collector; Samuel B. Gilbert, John Reid, and
masters; John Cook and Daniel Lawrence, Town
Pound-keeper; John Fairman, Constable; John
Caleb Bouchet, Roswell Levens, John Fredrick,
Donald, D. B. Sole, Dr. Hayden, who is reported
e Rebels in 1837 and escaped—his wife refusing to
ards—all held the office of town clerk, which up
important in the township. It is believed that
Reeve of Thurlow, an office he filled with ability

for many years. From 1866, J. J. Farley, Daniel Clapp, Alex. Wilson, Philip Clapp, S. Chrysler, John Canniff, John Vandwater, John Thompson, Alex. Wilson, Simon Chrysdale, Wm. Stevens, Archibald Ross, Wm. H. Sills, D. D. Chesboro, Wm. McDavitt, John Thompson, F. R. Daley, J. N. Diamond, and S. H. Fairman, have served in the Township Council, and as its officers, Herbert Ashley, Foxboro, at present occupying the chief executive office; William Hudson Roslin, Deputy Reeve; with E. Thrasher, as clerk of the Council. The Town Hall is a fine, substantial brick building, situated in the village of Canniton, with capacity sufficient to accommodate the wants of the ratepayers.

The total assessed value of real and personal property in the township for 1878 amounts to $2,400,000, and embraces within its geographical limits about 65,800 acres of land. There are 1,334 ratepayers with a population of about 6,750, which speaks volumes for the prosperity and enterprise of its inhabitants. An excellent system of Common School Education has been inaugurated throughout the municipality. Comfortable school houses are located at convenient points which are under the supervision of an experienced corps of teachers. One general characteristic of all Canadian pioneer settlements is the early attention which is always paid to schools. Almost as soon as the first acre is cleared and planted, and protection against starvation secured, measures are set on foot to secure means of instruction for their children, and the primitive log huts of 1800, in which the first rudiments of the three "R's" were taught, contrast forcibly with the advanced educational system, and beautiful and pleasantly situated school houses of to-day. The same remarks apply with equal force to spiritual and religious edifices to be met with every few miles on any of the principal roads of Thurlow, some of which would do credit to any city.

The Belleville and Grand Junction R. R., of which mention is made in the General Co. History, and the proposed route of the Moira Valley R. R., to Tweed and Marmora, runs through the township, while macadamized roads lead in every direction to any part of the back country.

The Moira River, named after the Earl of Moira, afterwards Marquis of Hastings, or Sagonaska, its original Indian name, flows at a rapid pace through the township, affording some of the finest water privileges that are to be found in the Province, while large quantities of saw logs are annually rafted from the north woods to the many mills along its banks, and at Belleville, there to be dressed for the market. There are several small villages in the township, the principal of which is Canniton, situated upon the river Moira, about four miles from Belleville, and was first settled about the year 1800 by the family of John Canniff, from whom the village derived its name and who was its founder. Mr. Canniff in 1812 erected a flouring mill upon the fine water privilege on the Moira River at this point. The surrounding country was at that date an interminable wilderness when James Canniff, the father of Dr. Canniff, settled further down the river and commenced to clear up the forest. The village of to-day presents to the traveller a different appearance from what it did when John and James Canniff first settled there. The progress of the village for some years was slow, probably owing to its close proximity to the rising town of Belleville. During the last ten years several new buildings have been added to the village, mills, schools and churches built and enlarged, and other evidences of enterprise are visible on every hand. There are flouring, saw and woollen mills, tanneries, and other manufacturing establishments, while a short distance down the river are situated extensive paper mills, and about a mile and a quarter up the river is located Corby's flouring mill and distillery, one of the largest institutions in the country. Canniton still belongs to the corporation of Thurlow, and has a population of about 1,000.

Plainfield, a small post village of recent date, is situated at a point on the Moira, a short distance above Corby's distillery. There are flouring and saw mills of immense capacity, an hotel, store, &c.

Still further up the river are located Latta's mills, comprising three saw and one flouring mill, all of which are run to their full capacity, except in certain seasons when raw material is not to be obtained.

Thrasher's Corners is situated upon the Roslin and Tweed macadamized road about eight miles from Belleville, and is principally interesting as having been the scene of an extraordinary combat between old Mr. Thrasher, the father of the clerk of the township, and two full grown bears, both of which he killed with a club. There is a tavern, &c., at the corners.

A portion of the post village of Roslin is situated in the north east corner of the township. There are several stores, the extensive carriage manufactory of Wm. Hudson, the 1st Deputy Reeve of Thurlow, an hotel, &c. Considerable business is done here on account of its being on the principal road to Belleville, 15 miles distant, and the fine agricultural country by which it is surrounded.

The post village of Foxboro, formerly Smithville, is situated upon the elbow of the river Moira in the western part of the township of Thurlow, distance from Belleville about eight miles, and was founded by Wm. Ashley, who purchased in 1821 lot No. 2, in the 5th concession, and in 1835 commenced the manufacture of waggons which formed the nucleus of the present thriving village. There are at present Charles Ashley's carriage manufacturing establishment, an extensive cheese factory, other manufactories, several stores, schools, churches, and a number of beautiful private dwelling houses. Population about 150.

TYENDINAGA.

Tyendinaga is a large and comparatively new township, fronting on the northern extremity of the so-called "Long Reach," containing within its geographical limits about 92,700 acres of excellent land. It is bounded on the north by the Township of Hungerford; on the east by Richmond, in the County of Lennox; on the south by the Bay of Quinte; and on the west by Thurlow. It was named after and in honour of the celebrated chief of the Six Nation Indians, better known by his English appellation, Joseph Brant, who came to Canada with remnants of his tribes about 1793-4, from Lewiston, New York, and located upon the northern shore of a portion of the Bay of Quinte. England, on the conclusion of peace with her revolted colonies in 1783, made no provision for her faithful Six Nation allies, and as a conquered people these Indians were left at the mercy of the victorious Continentals, many of whom, smarting under a sense of what they considered a deadly injury, desired the territory of the Indians to be held forfeit, but the influence of Generals Washington and Schuyler prevailed in favour of purchase, and thereby prevented the recurrence of another war. In the early part of 1784, the New York Legislature passed an act appointing a Board of Commissioners on Indian affairs, when a partial arrangement was effected. The first lands purchased of the Indians by the State of New York, included a tract lying between the Chenango and Unadilla Rivers. The treaty, as such transactions have ever been designated, was made June 28th, 1785, whereby the Oneidas and Tuscaroras received $11,500. This cession of land was followed by another on the 22nd of September of the same year, of the remainder of the Oneidas territory, excepting some small reservations. The immense tract of land comprising all the State of New York west of Seneca Lake, was in the possession of the Seneca nation, whose old men were determined to hold it; while, as later appears, the British, who held forts Oswego and Niagara, laid claim to the entire grant. Pending these negotiations, a company was formed in New York City, called the "New York Land Co.," whose plan was to lease from the Six Nations, at a yearly rental of $2,000, for a period of nine hundred and ninety-nine years, their entire land and possessions. A branch company, in connection with the New York company was organized in Canada, and the combined influence of its members over the Six Nations was such that a "lessee contract" was duly signed on Nov. 30th, 1787, by Red Jacket, Little Beard, Corn Planter, Farmer's Brother, and other chiefs, whereby the entire lands of the Nations were ceded to the company. Much has been said by American writers, of the cruelty and bloodthirstiness of the Six Nation Indians under Brant, during the border wars; but when we consider how they were treated in the matter of land negotiations, and by the aggressive policy of the pioneer settler, driven from their hunting-grounds and farms, it is not surprising to find them in arms against their traditional foe, the Americans. As a specimen of the mode of acquiring Indian territory, and as a relic of the times, we here give a synopsis of the contracts above mentioned.

"An agreement made on Nov. 30th, 1787, between the Chiefs or Sachems "of the Six Nation Indians, of the first part, and John Livingstone, Caleb "Benton, Peter Ryckman, John Stevenson, Ezekiel Gilbert, for themselves "and their associates, of the County of Columbia and State of New York, "of the other part," witnessed that the said Chiefs or Sachems of the Six Nations, on certain conditions afterwards mentioned, "leased to the said "John Livingstone and his associates, for a period of nine hundred and "ninety-nine years, all the lands commonly known as the lands of the Six "Nation Indians, in the State of New York, and at the time in the actual "possession of the said Chiefs or Sachems." The Chiefs or Sachems were privileged to make such reservations for themselves or their heirs as they chose, and, "said reservations to revert to the said lessees in case they "should afterwards be relinquished by the Indians." The payments were specified as "a yearly rent of two thousand Spanish milled dollars," payable on the 4th day of July in each year of the nine hundred and ninety-nine, for which the lease was drawn.

It appears that this conveyance was declared illegal and set aside by the State authorities, a matter of little importance so far as the original owners of the soil were concerned, as they never regained possession of their territory; other schemes were introduced by less scrupulous companies and individuals, whereby the red man was gradually deprived of his birthright, the hunting grounds of his fathers, and either driven westward or to Canada.

It was considered no dishonor to defraud an Indian, and the remembrances by these tribes of the conduct of the U. S., and State Governments, private individuals, and companies, present an accumulation of violated and broken pledges which will forever redound to American disgrace.

Captain Joseph Brant, it appears, took no part in these or subsequent transactions of a similar nature, nor does it appear that he was cognizant of what was transpiring, he being absent at the time. On the breaking out of the Revolution, Joseph Brant removed the Mohawk tribe or nation from their homes in the Mohawk and Schoharie valleys, to Lewiston, on the American side of the Niagara river, where he built a log church; a bell which he brought with him from the east was hung upon a crossbar in the fork of a tree, and services were occasionally held by the British Chaplain of the Fort. He afterwards removed his tribe to Canada. *Thayendinaga*, or Joseph Brant, was born in the year 1742, upon the banks of the Ohio River, while his tribe were on a visit to that country. He was, according to one writer, the son of Tehowaghwengaraghkwin, a Mohawk of the Wolf tribe, and was made Chief of the Six Nation Indians after the battle of Lake George, in which the old chief Soi-enga-rah-ta was killed. To follow and detail the eventful life and movements of this renowned chieftain from this date of his appointment as leader of the Six Nations, to the day of his death, would occupy too much space for a work of this description. He was always deeply interested in the welfare of his nation, as was attested by his eloquent appeal to the British Government—who, on the conclusion of peace had forgotten her faithful allies—for a home for his people. His application was listened to, and promises made to grant them in Canada an amount of land equal to what they had lost in the United States. This promise was ratified January 14th, 1793. Brant, on the evacuation of Forts Oswego and Niagara by the British troops, went with his tribe east, returning the following year and selecting a tract upon the north shores of the Bay of Quinté, comprising 92,700 acres, which General Haldimand caused to be purchased from the Mississaugas, and conveyed in fee simple to the Six Nations, who were composed of remnants of the following named tribes: *Mohawks, Oneidas, Cayugas, Onondagas*, and *Senecas*; the sixth nation, the *Tuscaroras* of North Carolina, who, driven from their country in 1725, were adopted by the *Iroquois*, a powerful tribe, and given land between the *Oneidas* and *Cayugas* in fee. He also obtained another grant of land six miles square, on the Grand River, near Brantford, where the larger portion of the Mohawk tribe located, in order to be nearer to their old allies, the *Senecas*, who had settled further down the river. The portion of the tribe who settled on the Bay of Quinté Reservation were under the immediate supervision of Capt. John Deserontyon, a cousin of Brant's, and who was subsequently joined by Capt. Isaac Hill. The deed of the tract to the Mohawks, now Tyendinaga township, bears date 1804. The land being granted to the chiefs, warriors, people, and women of the Six Nations, a portion of whom only settled there, the remainder, as previously remarked, proceeded westward with Brant, and settled upon the Grand River Reservation.

Brant, after he had settled the remainder of his people on the reservation, retired to a grant of land conferred upon him by the Government at where Wellington Square is now located; there he lived in peace until the day of his death, which occurred on the 24th November, 1807. Immediately after the occupation of their reservation, the Mohawks erected a log church at the Indian village of Tyendinaga, the first in the bay region; a school was also established, but shortly afterwards discontinued on account of the indifference of the children to learn, and the general lack of attendance. The old church having become dilapidated, a new and more substantial one was erected in 1844, which was dedicated with imposing ceremonies. During the years 1818 or 1819 the Indians surrendered the first four concessions of the township, which were at once surveyed and put upon the market. The land being of first-class quality, and heavily timbered, was rapidly taken up by actual settlers. In 1840 the remaining north part of the township was surveyed, placed in the market, and also sold by the Government for the benefit of the Indians. The first two and a-half concessions, including the broken front, containing about 20,000 acres, were reserved by the aboriginal owners, and is at present all that remains of the original reservation of the Six Nation Indians. Prominent among the first settlers in the township we find the names of the Portts, Sweeneys, Nealons, McKinneys, Hanleys, Englishs and Kilmurrays from Ireland; the McLarens, McFarlens, Andersons, Fulloughs and Fosters, from Scotland; the Roberts, Palmers, Emmons, Jones, and others from England; the Applebys, Laziers, Osbornes, Mordens, Ross, Trips, and Demills—United Empire Loyalists and their descendants, very many of whom, or their children, still hold front positions in the township or county.

After the first occupation of the reservation lands, in 1820, the township made rapid progress, roads were laid out and improved, school houses and churches sprang up in every direction, and the entire surroundings were changed from Indian woods to one of the finest agricultural municipalities to be found in the county of Hastings. The log shanty of the pioneer settler is now replaced by comfortable looking farm dwellings with substantial out buildings, indicative of the thrift and industry of their respective owners. The whole surface of the township was originally covered with a very heavy growth of timber, of the variety indigenous to this part of Ontario, and a dense undergrowth in many localities, that completely obscured the soil from the sun's rays, making all other vegetation impossible. It was probably these circumstances, and the numerous streams that flow through the township, that induced Brant to select this tract of land as a suitable home for his people. The surface of the land is undulating and inclining toward the south-west, rendering drainage every way possible. The soil is similar to that of Thurlow, is of a calcareous nature; occasionally the limestone formation approaches so near the surface as to render the soil difficult of cultivation, and in some few places the rock is quite prominent. The land, where not too rocky, is of the greatest

fertility, and is second to none in the county in
All cereals are raised abundantly, also fruit;
perhaps in this latitude, is there better encoura
—a soil of peculiar adaptability and a climate
ensure a fair quality of fruit, especially apples
townships in the county the best watered, seve
tributaries run through the whole width of t
westerly direction, emptying themselves into
Sucker creek takes its rise in the township of
Lennox, and runs through the south-east corne
ing into the Bay directly south of the Indian
also takes its rise in Richmond, and runs paral
distance from Sucker creek, emptying into
river, a considerable stream running parallel
which are situated the villages of Shannonvi
Kingsford upon the western boundary, takes it
ship of Kennebec, in the County of Frontenac,
dians as the *Hassigue*; upon this river are loca
factories, &c., and it affords water power fo
vicinity. Fisher creek, a tributary of the Sa
supply upon lot No. 21, in the 6th concession
P. O., and empties into the Salmon on lot 23
unnamed creek also takes its rise immediatel
ning in a south-westerly direction, emptying i
15 and 16 in the township of Thurlow. Pa
tributaries, waters the north-west part of the
into the Moira, at Plainfield, in Thurlow.
across the north-west corner of Tyendinaga, a
saw and grist mill; on the northern boundar
John White, M.P., a small stream, rising upon
empties into the Moira River, in the township
Mud Lake, on the eastern boundary, betwee
is also situated in the township of Tyendinaga,
species of fresh water fish. There are ten P
which are situated as follows:—Shannonvil
Milltown, lot 11, 1st con.; Blessington, lot
32, 2nd con.; Marysville, lot 32, 3rd con.;
Indian Reserve; Melrose, lot 15, 3rd con.; Re
lot 34, 6th con.; with offices on the eastern
Roslin, in the north-west corner of the townsh

There are a number of large cheese factor
township, and considerable money is invested
in this industry. John White, M.P., has a
residence at Roslin, which manufactures la
cheese, as do all the other factories in the tow
particularly well adapted to the manufacture
diet.

The village of Shannonville is situated up
the Indian Reserve, distance from Belleville 9
bay shore. The Grand Trunk Railway passes
ing ample facilities for travel, and the transporta
acres upon which the village stands, together
leased from the Mohawks for 999 years, by W
Keeler, in 1818 or 1819, who built a mill, the
above contract was confirmed by the Departm
bridge, the owner, agreeing to furnish to the
flour as a consideration. The name was
Ports, who came from the vicinity of the Sh
here at an early date, and was by them calle
Shannonville. The Indians named the stream
the large quantities of that species of fish found
extensive saw mills of Mr. Wallbridge, situate
capable of turning out several million feet of
down the stream is the mill of Rathburn & So
mense capacity; there is a large stone flouring
barrels per day; two tanneries, some very fin
smith shops, three churches—Church of Engl
Episcopal Methodist, and a good Common sch
attendance of about 100 pupils. The Division
the County are held at Shannonville. The vil
of late owing principally to the depression of t
a portion of the population chiefly depend.
entry, Richard L. Lazier, being Collector of
formerly was navigable for flat-bottoms up to
such an extent that navigation is rendered im
distance up from the bay. The present popul
700.

Mill Point is situated in the south-east co
second village in point of importance in the
from Napanee, eight from Shannonville, and
Belleville. The village is pleasantly situate
finest views of the bay to be found upon its no
landscape from this point, and the waters of t
shores, combine to make the scene one of sur
with in any other part of the Province. The
the erection in 1849 of the large saw mills b
Lewis E. Carpenter, and H. B. Rathburn, A
Millions of feet of lumber are here sawed for
equally extensive business is done in shingles
to be one of the best equipped in the Provin
dred hands. The raw material is floated do
Salmon and Napanee, Messrs. Rathburn & So
ing large timber limits in North Hastings, an
able attention has of late years been given to s
some of the finest schooners that float upon ou
burn's yard. There is also an excellent pier a
loading of vessels, with lumber, grain, &c.
are neat and comfortable, and the general app
enterprise and prosperity. There is a church,
blacksmith shop, shoemaker's shop, harness s
tavern, &c. The present population is about
the several families being principally employ
upon the lake in connection with the Messrs. R

The village of Milltown is appropriately na
Shannon, about one mile up the river from Sh
and Richmond macadamized road. This is th
township, and like the others is largely enga
waters of the Salmon River at this point have
of dams, which drive numerous mills. The M
and a saw mill, which do a large business. N.
Burdett, carry on the upper mills, manufac
flour; while the saw mill connected furnishes its
of feet of lumber that are annually shipped from
iron foundry and machine shop of R. F. Peg
manufactory, two carriage and blacksmiths' sh
Present population about 250.

Lonsdale is a small post village of recent
Salmon River, and lot No. 32 in the 2nd co
Tyendinaga. It has a population of about 2
done at this point, drawn principally from the

TORICAL SKETCH OF THE

y, and is second to none in the county in its natural productiveness. eals are raised abundantly, also fruit; nowhere in the Province, or in this latitude, is there better encouragement for the fruit-grower, l of peculiar adaptability and a climate modified by the Bay breezes a fair quality of fruit, especially apples. Tyendinaga is of all the ips in the county the best watered, seven parallel streams with their ries run through the whole width of the municipality in a south- ly direction, emptying themselves into the Bay and Moira River. creek takes its rise in the township of Richmond, in the County of x, and runs through the south-east corner of the Reservation, empty- the Bay directly south of the Indian Council house. Mud creek kes its rise in Richmond, and runs parallel with and about two miles from Sucker creek, emptying into Hungary Bay. The Salmon considerable stream running parallel with Mud creek, and upon are situated the villages of Shannonville, Milltown, Lonsdale, and ord upon the eastern boundary, takes its rise in Crow Lake, town- Kennebec, in the County of Frontenac, and was known by the In- s the *Ossipsa*; upon this river are located flouring and saw mills, es, &c., and it affords water power for all the requirements of the y. Fisher creek, a tributary of the Salmon river, receives its first upon lot No. 21, in the 6th concession of the township, near Read and empties into the Salmon on lot 23, in the 2nd concession; an ed creek also takes its rise immediately south of Read P. O., run- a south-westerly direction, emptying into the Bay upon lots No. 16 in the township of Thurlow. Parker creek, with its several ries, waters the north-west part of the township, and empties itself he Moira, at Plainfield, in Thurlow. The Moira river also runs the north-west corner of Tyendinaga, and upon which is situated a d grist mill; on the northern boundary, near the cheese factory of White, M.P., a small stream, rising upon lot 27, in the 9th concession, s into the Moira River, in the township of Hungerford. A portion of Lake, on the eastern boundary, between the 7th and 8th concessions situated in the township of Tyendinaga, which abounds with several s of fresh water fish. There are ten Post-offices in the township, are situated as follows:—Shannonville, lot 9, Indian Reserve; wn, lot 11, 1st con.; Blessington, lot 5, 4th con.; Lonsdale, lot l con.; Marysville, lot 32, 3rd con.; Millpoint, lot 40, 1st con., Reserve; Melrose, lot 15, 3rd con.; Read, lot 21, 6th con.; Albert, , 6th con.; with offices on the eastern boundary at Kingsford and , in the north-west corner of the township.

ere are a number of large cheese factories scattered throughout the hip, and considerable money is invested by the farmers of Tyendinaga s industry. John White, M.P., has a large establishment near his ce at Roslin, which manufactures large quantities of first-class , as do all the other factories in the township, the Bay region being ularly well adapted to the manufacture of this very useful article of

e village of Shannonville is situated upon the Salmon River, and in lian Reserve, distance from Belleville 9 miles, and about 1 from the ore. The Grand Trunk Railway passes through the village, afford- ple facilities for travel, and the transportation of freight. The land, 200 upon which the village stands, together with the mill privilege, was from the Mohawks for 999 years, by Warren Noble and Frederick , in 1818 or 1819, who built a mill, the first in the township. The contract was confirmed by the Department of the Interior, F. Wall- , the owner, agreeing to furnish to the Indians yearly 30 barrels of as a consideration. The name was originally derived from the who came from the vicinity of the Shannon, Ireland, and settled at an early date, and was by them called Shannon, and afterwards onville. The Indians named the stream Salmon River, on account of ge quantities of that species of fish found in its waters. There are the ive saw mills of Mr. Wallbridge, situated on the river at this point, e of turning out several million feet of lumber yearly; while farther the stream is the mill of Rathburn & Son, of Mill Point, with an im- capacity; there is a large stone flouring-mill, with a capacity of 200 s per day; two tanneries, some very fine stores, carriage and black- shops, three churches—Church of England, Canada Methodist, and pal Methodist, and a good Common school, having a daily average ance of about 100 pupils. The Division Court sittings for this part of unty are held at Shannonville. The village has not improved much owing principally to the depression of the lumber trade, upon which ion of the population chiefly depend. Shannonville is also a port of Richard L. Lazier, being Collector of Customs. The Salmon River ly was navigable for flat-bottoms up to the village, but has fallen to n extent that navigation is rendered impracticable, except for a short e up from the bay. The present population of the village is 650, or

ll Point is situated in the south-east corner of Tyendinaga, and is the village in point of importance in the township. It is seven miles Napanee, eight from Shannonville, and seventeen from the city of ille. The village is pleasantly situated, and commands one of the views of the bay to be found upon its northern shores. The varying ape from this point, and the waters of the long reach, with indented , combine to make the scene one of surpassing grandeur seldom met any other part of the Province. The village received its impetus by ection in 1849 of the large saw mills by Messrs. Thomas Y. Howe, E. Carpenter, and H. B. Rathburn, Americans from Albany, N.Y. s of feet of lumber are here sawed for the American market, and an extensive business is done in shingles and laths. The mill is said one of the best equipped in the Province, employing about one hun- ands. The raw material is floated down the rivers Trent, Moira, and Napanee, Messrs. Rathburn & Son, the present owners, hav- ge timber limits in North Hastings, and other counties. Consider- tention has of late years been given to ship-building at this point, and f the finest schooners that float upon our waters were built at Rath- yard. There is also an excellent pier affording every facility for the g of vessels, with lumber, grain, &c. The cottages of the workmen t and comfortable, and the general appearance of the village is one of ise and prosperity. There is a church, school, stores, waggon and ith shop, shoemaker's shop, harness shop, telegraph office, and a &c. The present population is about 1,000, the representatives of eral families being principally employed in and about the mills or e lake in connection with the Messrs. Rathburn.

village of Milltown is appropriately named, and is situated on the n, about one mile up the river from Shannonville, on the Belleville hmond macadamized road. This is the third largest village in the ip, and like the others is largely engaged in manufacturing. The of the Salmon River at this point have been utilized by the erection , which drive numerous mills. The Messrs. Lazier have a flouring w mill, which do a large business. N. S. Appleby, M.P.P., and Mr. , carry on the upper mills, manufacture an excellent brand of hile the saw mill connected furnishes its quota to the many thousands f lumber that are annually shipped from this point. There is also the ndry and machine shop of R. F. Pegan, and a cabinet and chair tory, two carriage and blacksmiths' shops, shoemakers' shops, etc. population about 250.

dale is a small post village of recent date, also situated upon the River, and lot No. 32 in the 2nd concession of the township of aga. It has a population of about 200. Considerable business is this point, drawn principally from the fine agricultural country by which it is surrounded. The village is growing rapidly, and promises fair to become one of importance in the township.

Melrose P.O. and village is located upon lot No. 13 in the 3rd conces- sion of Tyendinaga, and is the seat of municipal government. The Town Hall, a most substantial brick building, with every convenience for the accommodation of the council and ratepayers, is situated in the village. There is a blacksmith's shop, tavern, etc., two churches and a cheese fac- tory. A school is situated a short distance west of the village. Popula- tion, 100.

A portion of the village of Roslin is situated in the north-west corner of the township, a description of which is given in connection with the history of the township of Thurlow. A short distance east of Roslin, on the bound- ary line between Hungerford and Tyendinaga, and on the Moira River, are the mills of J. Wilson, and the cheese factory of John White, M.P.

Kingsford is another post village on the Salmon River and eastern boundary of the township, principally situated in the township of Richmond, in the County of Lennox. The council-house of the Mohawk Indians is located upon lot 24 in the 2nd concession of the reservation. The reserved land of the Indians is far behind the rest of the township in point of culti- vation. Some few clearances are to be met with, the majority of the tribe preferring to earn their living by light handicraft and other work than to clearing and cultivating their land, which is of a superior quality and com- paratively free from limestone. There are at present about 400 Indians upon the reservation.

The early townships records of Tyendinaga, as in other municipalities in the county, have either been lost or destroyed: Mr. E. Hollingsworth, town- ship clerk, &c., has nothing in his possession further back than 1830, in that year a town meeting was held in the house of Richard Lazier, Shan- nonville, but beyond the appointment of John Portt, as Town Clerk, nothing further is stated in the minutes. In 1831, John Portt was again appointed clerk, and held the office until 1832, when Thomas D. Appleby was elected to fill the position, an office he held until 1836, when Michael Nealon was appointed for one year. In 1837-8, Thomas D. Appleby again held the office, and in 1839 John H. Blecker. From 1840 to 1850, Michael Nealon officiated.

John Portt and John Sweeney for many years represented the town- ship in the old district Council. On the passage of the Municipal Act in 1849, establishing the present system of township councils, the following gentlemen were returned as Councilmen:—

"TOWN REEVE.—Wellington Frizzell; DEPUTY REEVE.—Alexander Mc- "Laren; COUNCILLORS.—Michael Nealon, Hugh Keys, John Hanley; "TOWNSHIP CLERK.—Thomas McKinny; TREASURER.—Francis English; "ASSESSORS.—Reuben Gorseline, Henry Rutland, James Anderson; COL- "LECTOR.—John Shaughnessy."

From 1850 up to the present date the following named gentlemen have been identified with municipal affairs, having held the principal offices in the township during that period:—Nathaniel S. Appleby, M.P.P., Reeve for twenty consecutive years; Thomas D. Appleby, Clerk, under the old and new Councils, for eleven years; Alexander McLaren, Reeve and Deputy-Reeve for ten years; Wellington Frizzell, Reeve and Deputy- Reeve for six years; Michael Nealon filled the position of Township Clerk, previous to 1850, and subsequently upwards of twenty-five years, dying on the 1st of February, 1870, in the 73rd year of his age. Michael Sweeney was Deputy-Reeve for many years; Frederick Warwick, was clerk during 1854; Francis English, was Treasurer for about twenty years, an office he filled with credit to himself and the municipality; Alexander McLaren is the present treasurer, having succeeded Mr. English. John R. Hoffman, Donald Anderson, Thos Deasey, Charles Hudson, and John White, M.P., Roslin, William McLaren, Michael McCullough, Samuel Osborne, and Thos. Casey, have all respectively filled offices of trust.

The principal Township officers for 1878, are John White, M.P.P., Reeve; Donald Anderson, and Thomas Casey, Deputy-Reeves; E. Hollingsworth is the present Township Clerk, having succeeded the late Michael Nealon to that office. The assessed value of all real and personal property for 1878 is

No. of acres assessed, 79,132, exclusive of the Indian Reserve; No. of Ratepayers, 1,303; assessed value of property, $1,306,082; No. of inhabi- tants, white, 4,550; No. of cattle, 2,835; No. of sheep, 3,974; No. of hogs, 1,251; No. of horses, 1,482.

HUNGERFORD.

This is the largest township in the county, having an area of 108,200 acres of land, the greater part of which is well adapted to agricultural pur- poses. From time immemorial it has been the hunting ground of the Mississaugas and other tribes of Indians, who found game and fish of every variety indigenous to this country and latitude, in great abundance in the beautiful lakes and streams within its limits. With the advent of the pioneer settlers the red man plunged farther into the wilderness, and his existence is but a remembrance. The tribes that then roamed over the lands now covered with well tilled fields of waving grain and comfortable homes, have for ever disappeared, and another race are the undisputed possessors of the soil.

The township of Hungerford is situated in the north riding of the county of Hastings, and is bounded as follows, that is to say, on the north by Elzevir and a portion of Madoc; on the east by Sheffield, in the county of Lennox; on the south by Tyendinaga, and on the west by the township of Huntingdon.

The early history of Hungerford is similar to that of the adjoining town- ships; especially is this the case with those facts that are matters of record. It is quite probable that during this period it was united with some other town- ship for municipal purposes. Settlements ten and fifteen miles apart were regarded as not very distant neighbours, and met together at the same town meetings for local organization, and to elect the same officers and co-operate in all matters of public improvements, for laying out new roads, improving old ones, building bridges, establishing schools, and, in fact to inaugurate any new enterprise essential to the prosperity of the settlements or conducive to the general welfare. The earliest recorded settlement in the township is that of Sugar Island, on the south side of Stoco Lake, so called from the fact that the Mississaugas were accustomed to make every spring large quantities of maple sugar, which they paddled down the Moira to Myer's Creek Village and the settlements on the front to be traded for different articles. It is stated that Owen Dirkin and Martin Donohue located upon the island in 1826 and were shortly followed by Philip Huffman and Nichol Conlin. In 1828 the Woodcock family came in and settled near the present village of Tweed. Robt. McCammon, J.P., Feix Gabourie, J.P., George Host, John R. Way, J.P., Thomas Close, J.P., James Martin, William Caton, J. D. Roblin, J.P., Henry Mattes, Mattaniah Kerr, and a number of others were also pioneer settlers.

It is impossible to fully realize the hardships, privations and sufferings of these first settlers of Hungerford. In the midst of an immense forest, without society, far removed from villages where anything could be purchased, and oftentimes destitute of the means to purchase; with twenty miles of almost impassable roads to travel before a grist mill or a store could be reached, a journey, with ox teams, occupying six or seven days. In a sickly country, where fever and ague was the common lot of nearly everyone, and no physicians near, the wolf without and sometimes the wolf of hunger within, all conspired to try the stoutest heart. The little produce raised could find no market, as there were no transportation facilities, and each settler supplied his own wants. As a result, little money was in circulation, all groceries were paid for in produce at extremely low rates, as the store keeper must find a market over nearly impassable roads. Such were among the trials and ex- periences of the early settlers of this township, and for years the permanent hindrance to its increase in population, values and property. But happily these difficulties have all passed away. The country is nearly all improved, perfectly healthy, as productive and well cultivated as the surrounding municipalities, and considerably ahead of some. Societies organized, churches built, stores convenient, schools excellent, mills abundant and flourishing, and every convenience at hand, make all a prosperous and happy rural people.

The records of the first town meetings are, as in other municipalities, lost or destroyed, but we are informed that the first meeting in Hungerford was held at the dwelling house of John R. Way, lot No. 7, in the 7th concession, in the year 1830. John Johnston was appointed town clerk; commissioners, John R. Way, Robert McCamnon and Philip Huffman. The first Court of Requests was held in a little log shanty near Manes' Bridge, John R. Way, J.P., Robert McCamnon, J.P., Owen Dirkin, J.P., and James Morton, J.P., were the commissioners. The town meetings were subsequently held in Caton's school house at Chapman's Corners. Robert McCamnon and Owen Dirkin were the first appointed magistrates in the township.

In 1848 a meeting was held on the 3rd day of January—the first recorded—when George Hart was elected District Councillor; Benjamin Read, Town Clerk; and James Willson, Assessor and Collector.

When the Municipal Act came into force in 1850 the township was divided into five wards and the following named gentlemen were elected councillors: George Benjamin, John Johnston, George Hart, James Allen, and Hugh McDonald. George Benjamin was chosen Reeve. At a meeting held at Hungerford Mills in 1850, a by-law was passed to build a town hall at Georgetown, Felix Gabourie, J.P., generously presenting the municipality with a sufficient quantity of land whereon to build the same.

Previous to 1830 the following names occur as having been identified with the affairs of the township:—

John Wilson, James Barker, Isaiah Clarke, Joseph W. Whittaker, J.P., Charles McGinnis, Archibald Parks, Philip Huffman, James Pree, William Thompson, Allen Mills, Archibald Benedict, William Bayley, J.P., James Kaner, Charles Lee, Daniel Vatley, J. D. Koddin, Sidney Way, John Wilson, John Hewett, James Corbett, Thomas Close, Henry Cook, George Cannill, Noah Ratney, Richard Woodcock, Samuel Secord, Sylvester Parsons, Jeremiah Prince, William McFall, William Potts, Peter Woodcock, Alexander Bonner, Patrick Vatley, James Rogers, Robert Kincaid, Isaac Dean, John Bonn, Henry Fice, John West, Uriah West, James Corkey, Samuel Darens, Peter Casey, William Rath, Gillespie Sayers, Phelix Collins, Henry Manes, William Caton, Cremmins Parks, William Alexander, William Barshaw, Hugh O'Donnell, Patrick Kincaid, John Gibson, James Kenney, Nathaniel Dunn, Andrew Huyke, William Loesing, James Alexander, Silas Cannill, Joseph Gartland, John Adams, Phelix Gabourie, J.P., James Carlton, Alex. Exter, Joseph Hirsch, Mattaniah Kerr, Alan Morrow, Arthur Youmans, and several others.

In 1850, and subsequently, we find the following as occupying the principal township offices:—

Felix Gabourie, Robert Sanderson, Robert McCamnon, Abram L. Bogart, Windsor W. Janes, John Graham, Isaiah Clarke, James Galkin, Robt. Morton, Paul Durkin, Joseph Elliott, Samuel Durkin, George Stocks, Robert Rudolph, John McGill, Dr. Tar, Thomas Close, George Howell, John Johnson, J. G. Osborne, John Castleton, John Newton, Thomas Maltooney, John Allen, John Graham, Wm. Rath, and Richard Gabourie.

The Municipal Council for 1878 is constituted as follows:—

Robert Gordon, Reeve.
Patrick Murphy, Deputy Reeve.
S. C. Johnston, do.
Francis Murphy, W. H. White, Councillors.
William Weir, Township Clerk.
Joseph Elliott, Treasurer.
John Thompson, James Finley, Assessors.
George Morton, Thomas Graham, Collectors.

The old Town Hall at Georgetown having been destroyed by fire, a handsome brick structure was recently erected at a cost of \$2,500, in which the Council at present meets for the transaction of public business. This is by far the finest town hall in the county, and is built of brick. Dimensions of body of hall, 30 by 28; clerk's office, with iron fire-proof vault in rear, 20 by 18; lock-up, with cells, ditto; all of which are conveniently arranged, both as regards seating capacity, clerk's office, safety of public documents, records, criminals, etc., the whole being surrounded by a neat picket fence. The building is pleasantly situated upon the main street of the village, and reflects credit alike upon the contractors, Messrs. John Foley and W. H. Gabourie, and the municipality of Hungerford, and at the same time adds to the appearance of the village.

The township of Hungerford was settled principally by Irish and people of Irish descent, together with the descendants of United Empire Loyalists, who first located in this and other front townships. Its progress, since the occupation of Sugar Island by Durkin & Dandine in 1824 has, as previously stated, been rapid, and to-day it occupies a position second to none in the county in its natural resources and productiveness. The surface of the township is somewhat rolling, rendering drainage everywhere practicable. The soil is of that calcareous nature common to north Hastings, and is well watered. Stoco Lake, situated in the north centre of the township, derived its name from the famous Mississauga chief, Stongoung, who, with his tribe, yearly ascended the Saganaska (Moira) River to fish and hunt, and has an area of about 1,200 acres. Bass, Maskenonge, Pickerel and other species of fresh water fish abound within its waters. Sugar Island, embracing within its limits about 2,000 acres of land, is situated immediately south of the lake, is formed by two branches of the Moira, which leave the lake about one mile apart and run parallel to each other for about two miles, when the east branch winds around to the west and forms a connection with the west branch at a point where steam saw-mills are located.

The Moira River enters the township at its north-west extremity, and flows in a south-easterly direction until it reaches Stoco Lake, leaving the lake by two branches which unite after running parallel to each other for about two miles, thence in a south-westerly direction to the boundary. The characteristics of the Moira in Hungerford are identical with other localities through which it flows, except that the current is not quite so rapid, affording less facilities for mill privileges than in Huntingdon.

The Sagamamata Creek, a stream of considerable size, with its tributaries, flows through the south-east portion of the township.

A tributary of Parks Creek takes its rise on lot No. 30, in the 9th concession, and flows in a southerly direction towards the Tyendinaga Boundary, in which township it forms a connection with the main branch.

Black Creek takes its rise on the south eastern boundary of the township, and empties at the mouth of a large cranberry marsh, into Stoco Lake, at its southern extremity.

Clare River also takes its rise in divisions upon the eastern boundary, upon which is situated Bogart's mills, and empties itself into Stoco Lake at its eastern extremity.

It is related that this stream derived its name from the late Thomas Clare, then quite a youth, whose occupation was chain-bearer to the surveying party engaged in laying out the township, and who one night encamped upon the banks of an unnamed stream. Young Clare accidentally fell into the river and disappeared underneath the ice; it being frozen, all thought him lost, when he was discovered, in an exhausted condition, by one of the party some distance down the stream, having come to the surface through an air hole. The surveyor, Samuel Benson, named the stream Clare River in commemoration of the event.

John R. Way, father of Sidney Way, owned the first team of horses ever used in Hungerford.

Hungerford is not behind the neighbouring municipalities in educational matters, as is evident from the many neat, substantial school houses every-where to be met with throughout the township. Able teachers ar[e] ... and a proper interest in their general efficiency generally maintain[ed] ... are a large number of well equipped cheese factories located in th[e] ... the many living springs in different parts of the township, and ... interest is manifested by the farmers in this enterprise. The pro[duct of these] establishments finds a ready market at Belleville, which has of ... one of the principal cheese centres in the province.

Gold, as well as other minerals, has been discovered in conside[rable quan]tities in the eastern part of the township. A crusher was erected, ... the extraction of small quantities of the precious metal, the res[ult ...] satisfactory. The Village of Tweed is pleasantly and favorably ... the Moira, adjacent to Stoco Lake, and in the north-west part o[f the town]ship, and is connected with the City of Belleville, 25 miles distan[t ...] macadamized road. The early settlement of the village is contem[poraneous with] that of the surrounding country, and was in all probability first e[stablished with] a view to utilising the fine water privileges of the Moira at this p[oint ...] an eligible site whereon to build a village. At first the growth ... large town was slow, owing chiefly to the close proximity of ... Notwithstanding these disadvantages, it kept pace with the dema[nds and re]quirements of the surrounding country, and at present affords amp[le ...] for a flourishing local trade. Valuable and substantial improveme[nts have of] late been made, principally in connection with the manufacturing ... the village. Many fine private residences have also been built, ... opened, and on every hand evidences of thrift and enterprise are ... ding fair, in the course of time, to make Tweed an important inla[nd ...]

The village contains four churches, viz.:—Church of Engla[nd ...] Methodist, Episcopal Methodist and Bible Christian. There is al[so a] Catholic church on Sugar Island, adjacent to Tweed; an excelle[nt] school, built of stone, with a daily attendance of about 200 ... Masonic lodge, who have recently erected a new hall, 18x38, ... Worshipful Master, A. H. Gilbert; Secretary, H. H. Warren. The ... Orange Hall, No. 717, and a lodge of Good Templars, who hold ... ings in Dr. T. E. Pomeroy's Public Hall, over his drug store. ... facturing interests of Tweed is represented by the extensive flouri[ng mill of] Mr. Eastbrook, on the Moira, containing three run of stone ... mill of J. H. Jordan, containing large circular, edging, cutting ... cheese box machinery and planer; Gossett's Iron foundry and ... two carriage manufactories, extensive concerns; a woollen factory ... industries. There are also a number of first-class general and ... three good hotels, telegraph office, &c. Population about 1,000.

Georgetown, a small village about one mile south of Tweed, upon the banks of Stoco Lake. This was formerly the seat of government, the first town-hall having been erected at this point, at that time the principal village in the township. Felix Gab[ourie] owned the land, settled here at an early date, and was the origi[nator] of the place. There is a good common school, etc. Population 100.

Stoco is a small post village opposite Tweed, on the east ban[k of the] Moira River and Lake, distance from Belleville 25 miles. It conta[ins] Murphy's general store, and other places of business, the carriag[e shop of] Thos. Maltooney, blacksmiths' shops, etc., and a couple of hotels. ... about 100.

Bogart's Mills and post-office are situate upon Clare River, in ... part of the township. This is a comparatively new place, and ... by Abram L. Bogart, Esq., who owns considerable land in this township.

Thomasburgh, a pleasant post village, is situated on lot No. 1, ... concession of Hungerford, on the macadamized road leading from ... Tweed, and was founded by Thomas Clare, Esq., at an early date. It ... name from three Thomas's, viz., Thomas Clare, Thomas Graham, an[d ...] Nichols, all pioneer settlers. A post-office was established in 18.. Clare being the first P.M. There are two general stores, carriage ... smith's shops, a tannery, large cheese factory and potash w[orks,] common school, etc., etc. There are four churches: Church of ... Bible Christian, Canada Methodist, and Episcopal Methodist, and ... Hall. Population about 200.

HUNTINGDON.

The early settlement of the township of Huntingdon is so ... with the history of that section of the county, already described ... tory of the settlement of the adjoining township, that a further ... would be only a repetition. The same difficulties were experien[ced by the] pioneer settlers of Huntingdon that for years retarded the advan[cement and] development of older municipalities. There were no roads, ... plenty of wolves; no stores nor mills nearer than Myers Creek; ... except of a very primitive character; no places of worship, ex[cept the] houses of the pioneers on the front, which for years was but spar[sely settled.] These and other experiences of a similar character were what the ... of Huntingdon had to contend with in the days of its infancy. ... quence of the above mentioned and other drawbacks, arising fro[m its geo]graphical position, remote from business centres, markets, &c. ... regular thirst of enterprise, it was not permanently settled and bro[ught under] general cultivation until more than twenty years later than the ... to the front of it.

Huntingdon is rectangular in shape and is bounded on the no[rth by the] township of Madoc, on the east by Hungerford, on the south by ... and on the west by Rawdon, and contains within its geograph[ical limits] 56,000 acres of land. It was first settled about the year 18.. during the latter year it is recorded that there were five perman[ent settlers] in the township, with a total population of twenty-six inhabitan[ts ...] were probably the families of the Ketchesons and Ostrom from ... Anthony Demke, Philip Lake and Foster, who settled in the ... of what is now Moira P.O. A contemporary settlement was ... point to the west, now known as West Huntingdon P.O., ... Ashley family, closely followed by others.

Subsequently the following names occur as being among the ... on the first three concessions, viz., Ketchesons, Vanhuntens, ... Spencers, Vandusells, Cronks, Thompson, Morton, Emerson, Clare, Utman, Snider, Thrasher, Aikenbrack, Green, Merrill, Post, Sills, Sexton, Lake, Bashnell, Lloyd, Hamilton, Fitchet, Marsh, Hodges, Stanfield, Way, Newton, Coulter, Dafoe, Hagg... Jeffrey, Read, Mitts, Clapp, Cook, Sarles, Hawkins, Detlor, Wilso[n,] Ross, Vandusen, Rupert, Sharp, Rutledge, Dorland, Anders[on,] Nicholson, Hollinger, Collins, McAvoy, Herity, McKim, Haigh, Post, Murray, and others. A large number of the above n[ames were] United Empire Loyalists, from the older settlements on the front ... land in the second and third tier of townships. As usual with to[wnships at] that early period, the town meetings were held in the dwelling ... ratepayers, and beyond the simple recording of the proceedings ... business was transacted throughout the year. The "Minutes" ... year were "taken down" upon a piece of paper, which eventua[lly was] mislaid or destroyed, hence the difficulty experienced in obtain[ing] information in relation to the early meetings and affairs of the m[unicipality.]

The earliest township records extant, in the possession of th[e ...] G. Foster, only date back to 1848. We have, however, through ... ness of John McCamnon, of Hungerford, been furnished with a ... first town meeting held in Huntington, which we reproduce, ... meeting being omitted.

"The annual township meeting of Huntingdon for the ... The following resolutions were made and resolved:—That Ph...

oughout the township. Able teachers are employed, heir general efficiency generally maintained. There ll equipped cheese factories located in the vicinity of in different parts of the township, and considerable the farmers in this enterprise. The product of these ady market at Belleville, which has of late become e centres in the province.

minerals, has been discovered in considerable quan- d the township. A crusher was erected, but beyond uantities of the precious metal, the result was not e of Tweed is pleasantly and favorably situated on oro Lake, and in the north-west part of the town- h the City of Belleville, 25 miles distant, by a good early settlement of the village is contemporary with ountry, and was in all probability first effected with e water privileges of the Moira at this point, and as o build a village. At first the growth of this em- ing chiefly to the close proximity of Georgetown, advantages, it kept pace with the demands and re- ding country, and at present affords ample facilities s. Valuable and substantial improvements have at ly in connection with the manufacturing interests of private residences have also been built, new stores al evidences of thrift and enterprise are visible, bid- time, to make Tweed an important inland town.

our churches, viz.:—Church of England, Canada thodist and Bible Christian. There is also a Roman Island, adjacent to Tweed; an excellent common with a daily attendance of about 200 pupils; a e recently erected a new hall, 18x30, two stories. Gilbert; Secretary, H. H. Warren. There is also an d a Lodge of Good Templars, who hold their meet- y's Public Hall, over his drug store. The manu- ed is represented by the extensive flouring mills of Moira, containing three run of stone; the saw- aining large circular, edging, cutting and lath saws, l planer; Gossett's Iron foundry and stove works; es, extensive concerns; a woollen factory and other o a number of first-class general and other stores, ph office, &c. Population about 1,000.

village about one mile south of Tweed, is situated Lake. This was formerly the seat of municipal n-hall having been erected at this point, which was al village in the township. Felix Gilbourne, who ere at an early date, and was the original founder good common school, etc. Population from 75 to

village opposite Tweed, on the east branch of the istance from Belleville 25 miles. It contains Messrs. and other places of business, the carriage-works of ths' shops, etc., and a couple of hotels. Population

st-office are situate upon Clare River, in the eastern his is a comparatively new place, and was founded q., who owns considerable land in this part of the

ant post village, is situated on lot No. 1, in the 9th , on the macadamized road leading from Belleville to by Thomas Clare, Esq., at an early date. It derived its 's, viz., Thomas Clare, Thomas Graham, and Thomas ts. A post-office was established in 1853, Thomas . There are two general stores, carriage and black- , large cheese factory and potash works, good . There are four churches. Church of England, Methodist, and Episcopal Methodist, and an Orange 200.

HUNTINGDON.

of the township of Huntingdon is so interwoven section of the county, already described in the his- the adjoining township, that a farther description n. The same difficulties were experienced by the ngdon that for years retarded the advancement and unicipalities. There were no roads, so to speak, ss nor mills nearer than Myers Creek; no schools, e character; no places of worship, except at the the front, which for years was but sparsely settled es of a similar character were what the first settlers tend with in the days of its infancy. In conse- tioned and other drawbacks, arising from its geo- te from business centres, markets, &c., and the at was not permanently settled and brought under more than twenty years later than the townships

gular in shape and is bounded on the north by the e east by Hungerford, on the south by Thurlow, don, and contains within the geographical limits t was first settled about the year 1816 or '17, for s recorded that there were five permanent settlers otal population of twenty-six inhabitants. These ies of the Ketchesons and Ostroms from the front, Lake and Foster, who settled in the neighborhood 3. A contemporary settlement was effected at a own as West Huntingdon P. O., probably by the lowed by others.

owing names occur as being among the early settlers sions, viz., Ketchesons, Vanderwaters, Hagermans, ranks, Thompson, Morton, Emerson, Salisbury, Thrasher, Aikenbrock, Geros, Merrill, Hoskins, e, Rashnell, Lloyd, Hamilton, Fitchett, Maynes, , Way, Newton, Coulter, Dafoe, Haggerty, Ryan, p, Cook, Sarles, Hawkins, Detlor, Wilson, Streeter, , Sharp, Ruthedge, Dorland, Anderson, Woods, llins, McAvay, Hefty, McKim, Haight, Darling, rs. A large number of the above were sons of from the older settlements on the front, who drew d tier of townships. As usual with townships at n meetings were held in the dwelling houses of the e simple recording of the proceedings no farther roughout the year. The "Minutes" from year to upon a piece of paper, which eventually became e the difficulty experienced in obtaining correct the early meetings and affairs of the municipality. records extant, in the possession of the Clerk, J. to 1848. We have, however, through the kind- of Hungerford, been furnished with a copy of the in Huntington, which we reproduce, the place of

ip meeting of Huntingdon for the year 1836, were made and resolved:—That Philip Lake, serve as Town Clerk; Robert Reid, Samuel T. Darling, as Commissioners; John Detlor, as Collector; Henry Ketcheson, as Assessor; James Haggerty, as Town Constable; Robert Wilson, Merrell Streeter, Daniel Ross, Robert Vandusen, John Anderson, John McConnell, Robt. Rutledge, Phillip Ketche- son, Henry Newton, Jonathan Haggerty, William Coulter, James McClusky, Elisha Phillips, Hugh McMullen, Andrew Sharp, John Wood, Conrad Rupert, as Overseers of Highways; Westell Streeter, David Nicholson, as Poundkeepers; John Ketcheson, John Andrews, and John McKay, for rest of Township.

"Resolved, That all hogs be free commoners over and above the weight "of 60 pounds. That all cattle be free commoners for the present year. "That rams be not free commoners, from 1st Sept. to 18th November. "That boars be free commoners. That fences to be lawful, should be five "feet high, and no intervening space over six inches. That the next annual "township meeting be held at John McKay's Inn.

"Huntingdon, Jan. 5th, 1836, RICHARD NEWTON, Chairman."

At a meeting held on the 16th January, 1836, the following rules were allowed and established by the Board of Commissioners for pound-keepers in the township of Huntingdon:—

"7½d per head for each horse, or each head of horned cattle for 24 hours, "and to furnish them with 14 lbs. of hay each; 1d per head for sheep and "calves for 24 hours, and to feed them at the rate of 14 lbs. of hay for six "sheep or six calves; 1d per head for swine, each 24 hours, to feed them "with a pint of peas or corn, each head. Poundkeeper's fees:—Each horse, "cow, or ox or bull 2d; each sheep, calf or hog, 1d.

"Commissioners, { NESBIT REID, SAMUEL T. DORLAND, SIMON ASHLEY."

It is related that much excitement prevailed at these early town meetings, on account of the "west towners," who came in force armed with clubs, &c., who wanted to have things all their own way; matters were however amicably arranged, and the meetings thereafter conducted in an orderly manner. From the date of this first meeting up to 1849-50, the business of the township was transacted at the usual town meetings, and in a manner similar to the above. The several officers were appointed for the year next ensuing, bills and ac- counts audited, and except the "assessing" and collecting of rates and taxes, the whole work of the municipality was completed in one day.

In 1848, the earliest records extant, we find that at a meeting held in the public school-house on lot No. 13 in the 3rd concession, a short distance of the present village of Moira, Thomas Baker was appointed Clerk, Anthony Denike, Commissioner, and Henry Ostrom, Collector.

In 1849-50, under the Municipal Act, Anthony Denike, George Graham, Samuel Baker, Archibald Pringle, and Henry Ostrom were elected by a popu- lar vote of the township, Councillors for their respective wards. Anthony Denike was chosen Reeve, Thomas Baker, township Clerk, Henry Ketcheson, Treasurer, Phillip Lake and George Mowatt, Auditors, Daniel Ross, Col- lector, and Joseph Foster, Surveyor of roads.

From 1850 to date, the following named gentlemen have represented the township in the County Council, in the capacity of Reeve.—George Graham, 1851-2 and 3; John Maynes, 1855; Phillip Lake, 1856-7; Thomas Emo, 1858, and who is the present representative, having held the position for eighteen years; James Haggerty and Henry Ostrom also held the Reeveship for one year each.

The township Council for 1878 is constituted as follows:—

Thos. Emo, Reeve; Mathew Robinson, Deputy Reeve; James Foster, John Gordon, and Henry Morton, Councillors; J. G. Foster, township Clerk and Division Registrar of vital statistics, &c.; Elijah Ketcheson, Treasurer; Alex. Harvey, Collector; A. R. Ketcheson, Surveyor.

The total assessed value of the real and personal property for the year 1878, is $421,362. Total number of ratepayers 610. Total population, about 3,200. The township Council meet in a substantial and commodious town hall, in the village of Moira, situated upon lots Nos. 11 and 12 in and between the second and third concessions. This village was established at an early date in the history of the township. It contains, besides the town hall, one general store, two blacksmith shops, one cheese factory, J. G. Foster, President; one church, one general store and post office, the first in the township, previously situated further east, Henry Ostrom, P. O.; and several private residences. It is pleasantly situated at the base of a large hill which shelters it from the west winds, and is the most important village in the township.

There are three Canada Methodist churches in the township,—one at Moira, one on lot No. 8 in the 8th concession, and another at West Hunting- don; Revs. Philp and McCamon, pastors. There is also an Episcopal Metho- dist church at West Huntingdon, Revs. Finn and Pomroy, ministers. On lot No. 12, in the 6th concession, there is a Bible Christian church, Rev. R. Baker, minister; and on lot No. 7, in the 7th concession, is the Presbyterian, presided over by Rev. M. Gray, of Stirling. At the village of Roslin are two other churches, upon lots 19 and 20, in the first concession. In 1845 there were only two frame and a few log schoolhouses in the township, with but an indifferent staff of teachers; to-day there are two large stone, two large brick, two frame, and two good log schoolhouses, situated respectively upon lots Nos. 6 and 11 in the 2nd concession; upon lot No. 2 in the 4th concession; upon lot No. 2 in the 5th concession; upon lot No. 7 in the 6th concession; upon lot No. 8 in the 8th concession; upon lot No. 13 in the 9th concession; upon lot No. 5 in the 10th concession; and upon lot No. 5 in the 13th con- cession. The schools of Huntingdon of to-day are of the highest character and efficiency, and have always been a matter of paramount interest with the people.

There are five post-offices in the township: West Huntingdon, James Gay, P.M.; Moira, Henry Ostrom, P.M.; Fuller, lot 13, con. 8, Mrs. John Fuller, P.M.; Ivanhoe, Thos. Emo, P.M.

The 7th Division Court sittings are held in the Town Hall, Moira; Thomas Emo, J.P., and Commissioner in B.R., License Commissioner for North Hastings, clerk, etc.

There are three cheese factories in Huntingdon. One is situated upon a branch of a creek that takes its rise upon the farm of H. Ostrom, at Moira. Another is situated upon lot No. 3, near West Huntingdon; and another upon lot No. 9, in the 6th concession. These factories have a large manufac- turing capacity, and are fitted throughout with all latest modern appliances, which enable them to compete favourably with the factories in the front. Their stock is of an excellent quality, and commands the highest price in the Belleville markets.

There are three saw mills in the township,—one upon Rawdon creek, lot No. 3 in the 2nd concession; another, stream, on the north side of the narrows of Moira Lake; and another upon a small stream known as Deer creek, which empties itself into Moira Lake. These mills are of modern construction, and are capable of turning out many thousand feet of lumber daily.

There are eight blacksmith shops situated at different points of the town- ship, all doing a fair country business. The township is well watered by Lake Moira and the Moira River, Rawdon creek and its tributaries, a branch of the Scoutamatta, and Beaver creek.

Moira Lake, better known as Hog Lake, is one of the largest and finest of the back country chain of fresh water lakes, and is situated in the northern part of the township and on the Moira river. It embraces an area of some- thing over 2,000 acres, and its waters contain a variety of fish indigenous to this section of the country. It is also noted for its unusually large and sonorous bull frogs, immense turtles, &c., large numbers of the former being annually caught and shipped to the Montreal and New York markets.

The Moira river in Huntingdon is deep, and is in places completely sur- rounded by impenetrable marshes and swamps; it is a favourite haunt of the wild duck and other game, and also of sportsmen. Rawdon creek takes its rise from a small natural lake upon lot No. 12 in the 12th concession, and traverses the township in a south-westerly direction; a tributary of Rawdon creek has its source upon lot No. 4 in the 5th concession, connecting with the main branch at a point upon lot No. 8 in the 4th concession.

The Belleville and North Hastings railway runs through the township, entering upon lot No. 1 in the 2nd concession, and running in a north-easterly direction through the township, crossing the narrows of Moira Lake, thence to Madoc. The road will be completed this fall.

The Huntingdon macadamized road runs through the whole length of the township, commencing at the south-western extremity and crossing Moira at the Narrows by an extensive and well-built bridge, and striking the northern boundary at a point upon lot No. 12, which forms a part of the corporation of the village of Madoc. There are a number of other excellent roads in the township affording every facility for travel, and by which any point in the township can be reached with any description of vehicle.

The surface of the township is rolling, and in some places hilly, being originally covered with a heavy growth of hardwood timber, which, since the advent of the pioneer settler, has almost entirely disappeared. The soil is similar to that of Hungerford on the east, and Rawdon on the west,—a clay loam mixed with shale, the limestone formation occasionally assuming a degree of prominence which renders it unfit for cultivation. With the exception of the localities where these geological specimens exist, the township is in an advanced state of cultivation, as is attested by the many fields of heavy and nearly matured grain to be met with on every hand. The people are apparently prosperous and happy, neat comfortable dwelling houses and well appointed outbuildings have superseded the log shanty and lean-to of the early settlers, modern and luxurious furniture has replaced that of antiquated design. Light traps and carriages, &c., have become a necessity, the ox cart of the pioneer being of no use to the present generation. Costly fabrics, gotten up in the latest and most approved fashion, are worn instead of the homespun of our pioneer mothers. The fathers of Huntingdon belong to another age, and may in justice exclaim, "The times are changed and we are changed with them." A truism that applies with equal force, not only to the present generation of the township of Huntingdon, but to the country at large.

MADOC.

The township of Madoc derived its name from Lord Madoc of Wales, England, and is situated in the North Riding of the county. Its settlement has been rapid, and commenced about the year 1830. It is bounded on the south by Huntingdon, on the east by Elzevir, on the west by Marmora, and on the north by Tudor. Madoc, aside from its mineral wealth, is excellent for agricultural purposes and is in a prosperous condition, as is evidenced by the numerous comfortable and home-like farm houses, and well tilled lands to be met with throughout the township. Cyrus Riggs, Barnabas Vankleek, James O'Hara, Donald McKenzie, Thomas Allan, John R. Ketcheson, Jacob Hazzard, Uriah Seymour, Louis Empey, William Allen, and Robert Cooper were among the first settlers of Madoc township; Donald McKenzie built a grist and saw mill and opened a store, the first in the township, on Deer creek, a tributary of the Moira river, which formed the nucleus of what is now known as the village of Madoc. The discovery of gold in the quartz rocks of Madoc during the years 1866-7 caused an immense excitement, and the little village of Hastings, now Madoc, received an impetus that was readily taken advantage of by its inhabitants. Houses and stores sprang up in every direction, and business of all kinds flourished. Large and valuable deposits of iron, copper, lead, marble, and lithographic stone are found within a short distance of the village, which only awaits the completion of the B. & N. H. R. R. to develop the wealth of these products. Uriah Seymour, of Madoc village, erected at considerable expense a smelting furnace, and commenced the manufacture of iron, bringing the ore from his mines lot No. 11 in the 5th concession of Madoc township. The works were carried on for some years, but owing to the great inconvenience and expense attending the transportation of the pig iron to the front, he was compelled to abandon the enterprise, and the furnace was closed. The ore is reputed to be valuable, yielding about 90 of pure metal of a very superior quality. Gold exists in small, and in some places paying, quantities, in nearly every part of the township, but owing to a want of proper mechanical appliances much difficulty has been experienced in separating the precious metal from the quartz rocks in which it is found, consequently very little money has as yet been made in this direction; a full and complete description of the famous Richardson and other mines is given, under its proper head, in the general county history. The settlement of Madoc from 1830 up to 1855 continued, and to-day good substantial dwelling houses and buildings have taken the places of log shanties and dilapidated old sheds. The people are industrious, as is evidenced on every hand by the well tilled fields producing abundantly every description of cereals. The surface of the township is rolling, and contains within its limits about 70,400 acres of land, some of which is extremely rocky but of good quality. The river Moira takes its rise on the northern boundary of the township, and in Tudor, affording excellent water power, and upon which is erected grist and saw mills. Deer creek also takes its rise within the limits of the township and flows through the village of Madoc. The Hastings macadamized road runs through the entire length of the township, affording to the farmer superior advantages for transporting the products of his farm to the front and market.

In the absence of the township records which were unfortunately destroyed by fire in 1873, we can only give the names of the several parties who were connected with the township Council from 1850 to the present time. Formerly Madoc, Tudor and Elzevir were united for municipal purposes, but since the erection of the two latter into separate and independent municipalities, Madoc has elected her own representatives. On the dissolution of the old Midland District, a change of the form of government was established in rural municipalities, and the first election by popular vote, in 1850, resulted in the return of the following named gentlemen:—

"REEVE.—John R. Ketcheson; DEPUTY REEVE.—Joseph Bradshaw; COUNCILLORS.—Henry Cook, James O'Hara, Mathew Herlheson; ASSESSORS.—P. Olmstead, W. H. Bristol; CLERK.—John McDonald; COLLECTOR.—Sylvanus Bond."

During December, 1850, a By-law was passed, ordering the erection of a town hall, which was built the following year, and since burned. From 1850 and up to the present year, the following gentlemen have represented the township in its several Councils,—Wm. Blair, Daniel Thompson, Henry Cook, Jeptha Bradshaw, John R. Ketcheson, A. F. Wood (Reeve for 22 years), Wm. Tumilty, Barnabas Vankleek, ——— Carrigan, W. Findlay, Mark Rennings, J. McCoy, J. N. Moore, E. D. O'Flynn, Thomas Allen, D. C. Brown, John Dale, S. Read, John Fassett, William Ward, Thomas Cross, F. D. Ross, Peter Vankleek, ——— Coskeys, Charles English, J. W. Allen, Joshua Broad. Peter Vankleek is the present Reeve and John R. Ketcheson, Clerk.

The population of the township is about 2,473, of which 1,323 are Methodists, 596 Presbyterians, 324 Church of England, and 230 Roman Catholic. Total assessed value, 1878, $512,550. With increased capital and railroad facilities, the mineral wealth of Madoc will be rapidly developed, a desideratum anxiously looked for by the inhabitants, several of whom have invested large sums of money in experimental work.

The village of Madoc, on the south boundary line, is now incorporated, having assumed that position, January, 1878. The elective officers for the present year are Thomas Cross, Reeve; E. D. O'Flynn, Dr. Loomis, Dr. Sutton, and S. D. Ross, Councillors. The population is about 1,000, and promises to be, ere many years pass away, double that number. There are five churches, viz:—Presbyterian, Church of England, Canada Methodist, Episcopal Methodist, and Roman Catholic. The new Presbyterian church,

an engraving of which appears in this work,
There are a number of fine residences, with ta
stores, of which there are a large number, are
ness than those of a back country village. A
the course of erection, and when completed w
an excellent gravel road connects the village
miles, and whenever the North Riding of the c
municipality, Madoc will be the county town
Hastings, and the proposed route of the To
through the village, the former of which will
doubt will add largely to its population and
its reverses; in 1873 a general conflagration to
pal part of the village was destroyed. Solid
place of the wooden ones burned, and the Mado
one of importance in the county of Hastings.

In 1870 an effort was made to connect Mad
railway, which was christened and known as t
way. After considerable labour and money ha
up, the Kingston people changed their minds.
Pembroke road. In 1873 the Belleville & Nort
jected, and in 1874 a charter obtained. The pr
Lewis Wallbridge, A. F. Wood, H. C. Floyd,
Brown, M.P., McKenzie Bowell, M.P., Thom
N. B. Falkiner, S. B. Burdett, Thomas Kerr, E
kleek. In the same year the company was o
Esq., elected provisional president. In 1876
Beard, and A. Foster, Esq., of Pennsylvania, w
money in the road and iron mines in Madoc,
H. C. Floyd, A. F. Wood, McKenzie Bowell,
and A. B. Foster, directors. Bonuses of $85,00
and Madoc township, and the Government gav
The road commences at a point on the Grand Ju
ship of Sidney, about fifteen miles north of B
through the townships of Rawdon and Hunting
iron mines, in all 22 miles. The road is about
distributed, and it is confidently expected the
this year (1878). The scheme is intended to dev
projectors expect to extend the road into the in

Bannockburn is a rising post village on the M
of the township, and is also becoming a place
flouring and saw mills, and considerable business
the year. Gold was discovered, and the Tow
whose lands are adjacent to the village, erected
small quantities of gold have been taken.

Eldorado is another gold mining and post villa
It is at this point the Richardson mine is situate
mills, but none of them appear to be working ex
and since the reaction that took place some years
have set, for the village has not improved much
or business since that date.

The old Presbyterian church in the village of
in 1873, and the congregation, which was formed
expense of about $25,000, an imposing new struct
seen for miles in any direction from the village.
long, 44 feet wide, and 58 feet from the basement
feet high from the ground, its seating capacity, w
finished inside will be one of the finest places of
Province. The present pastor, the Rev. Mr. Wi
and is a native of Culross, graduated at St. Andr
ties, and studied theology in the F. C. Divinity
After preaching the gospel for two years in the
cardine and Duncan Temple, he was sent out
Committee of the Free Church, and was settled as
then a small hamlet on Deer creek. Perhaps the
village of Madoc is more due to Dr. Wishart's
vidual. His private residence was the first stone
has built two other churches in the township of M
in the erection of a church in Huntingdon Towns
the Hastings road. The erection of St. Peter's, M
his exertions, and when entirely finished will be
memory. The Rev. Dr. Wishart has always bee
for the improvement of the people. Of the most
ouraged and respected by all who have the pleasure

RAWDON TOWNSH

The township of Rawdon is bounded on the
Marmora, on the east by Huntingdon, on the so
west by the township of Seymour, in the count
contains within its limits an area of about 67,2
its name from Lord Rawdon, an English noblem
with the Colonial Department of the British Gover
settlement dates back to about the year 1820, w
cessions were occupied by sons and descendants o
families, who lived in the older settled townships
who, on the occupation of the available land in th
drew lots in the rear tier of townships, upon
wards settled. The vicinity of Stirling was, owin
the soil and the close proximity to Sidney, toget
vileges afforded by Rawdon creek, we are informed
settlement in the township.

Amongst the pioneers of Rawdon, descenda
and others, we find the names of Chad, We
Hogle, Mayhee, Johnson, Potts, Vandervoort, H
Rupert, Hagerman, Frelick, Seeley, Sarles, H
were subsequently followed by families of Ameri
located the many eligible lots upon the fron
amongst whom we are able to trace the name
Wilson, Ketcheson, Kyle, Montgomery, Rogers,
lip, White, Fox, Green, Merrick, Brooks, A
Walker, Parks, Todd, Jeffs, Heard, Carlisle, B
Reed, Glass—for many years township clerk—Ell
Coats, Muller, Ralph, Morton, Heimer, Caverl
Fletcher, Snore, James Cook—the present Re
Bateman, Stiles, Haslett, Buskell, Livingstone, Th
Haggerty, Orser, Tompkins, Austin, Hurst, Volli
Patterson, Scrimshaw, Eastman, McWilliam,
Bedford, Jordison, Dethu, Gurnsey, Dusk, Fink
Peterson, Shearman, Thrasher, Welman, McQuin
Bailey, McKeown, Ward and others of more recen
their energies to the cultivation of their respectiv
ances, they have been well repaid for their industr

The experience of the pioneers of Rawdon was
that of the early settlement of other townships, a de
be found in other portions of this history. In 1839
thriving village of Stirling, now incorporated, only
habitants, whose wants were supplied by one grist
stores in the village; 84 horses, 103 oxen, and 23
total of the live stock, (with perhaps the excep
township for that year. Since that date its progr

ORICAL SKETCH OF THE

aving of which appears in this work, would be a credit to any city. re a number of fine residences, with tastefully laid out grounds; the of which there are a large number, are more like city places of busi- an those of a back country village. An elegant Model School is in rse of erection, and when completed will accommodate 250 pupils; llent gravel road connects the village with Belleville, distance 24 nd whenever the North Riding of the county is erected into a separate ality, Madoc will be the county town. The Belleville and North gs, and the proposed route of the Toronto & Ottawa Railroads, run the village, the former of which will be completed this fall, and no will add largely to its population and business. Madoc too, has had res; in 1873 a general conflagration took place, in which the princi- of the village was destroyed. Solid brick blocks have taken the the wooden ones burned, and the Madoc of to-day bids fair to become mportance in the county of Hastings.

870 an effort was made to connect Madoc village with Kingston by a which was christened and known as the Kingston and Madoc Rail- After considerable labour and money had been expended in working it Kingston people changed their minds, and took up the Kingston & ke road. In 1870 the Belleville & North Hastings Railway was pro- and in 1874 a charter obtained. The provisional directors were Hon. Wallbridge, A. F. Wood, H. C. Flovd, Dr. Boulter, M.P.P., James M.P., McKenzie Bowell, M.P., Thomas Kelson, R. S. Patterson, alkiner, S. B. Burdett, Thomas Kerr, E. D. O'Flynn, and Peter Van- In the same year the company was organized, and A. F. Wood, ected provisional president. In 1876 the stockholders organized a and A. Pardee, Esq., of Pennsylvania, who had invested considerable n the road and iron mines in Madoc, was elected president, with loyd, A. F. Wood, McKenzie Bowell, R. S. Patterson, W. Mershon, R. Foster, directors. Bonuses of $60,000 were granted by the county he township, and the Government gave a subsidy of $60,000 more. t commences at a point on the Grand Junction Railway, in the town- Sidney, about fifteen miles north of Belleville, and runs northerly the townships of Rawdon and Huntingdon into Madoc to the Moose es, in all 32 miles. The road is about all graded, bridges built, ties ted, and it is confidently expected the road will be fully completed (1878). The scheme is intended to develop the iron mines, and the rs expect to extend the road into the interior as a colonization road. ockburn is a rising post village on the Moira River, in the north part ownship, and is also becoming a place of importance. There are and saw mills, and considerable business is done at certain seasons of . Gold was discovered, and the Toronto & Whitby Mining Co., ands are adjacent to the village, erected a quartz mill from which antities of gold have been taken.

rado is another gold-mining and post village in the township of Madoc. his point the Richardson mine is situated. There are several quartz t none of them appear to be working except, perhaps, at intervals, e the reaction that took place some years ago Eldorado's star seems to , for the village has not improved much either in point of population ss since that date.

ld Presbyterian church in the village of Madoc was destroyed by fire and the congregation, which was formed in 1858, have erected, at an of about $20,000, an imposing new structure of stone, which may be miles in any direction from the village. Its dimensions are 80 feet feet wide, and 58 feet from the basement to the ridge, the spire is 118 from the ground, its seating capacity, with gallery, 600, and when inside will be one of the finest places of worship to be found in the . The present pastor, the Rev. Mr. Wishart, who settled in 1857, native of Culross, graduated at St. Andrew's and Glasgow Universi- studied theology in the F. C. Divinity Hall, Edinburgh, Scotland. eaching the gospel for two years in the capacity of assistant at Kin- and Duncan Temple, he was sent out to Canada by the Colonial e of the Free Church, and was settled as above-mentioned at Madoc, small hamlet on Deer creek. Perhaps the prosperity of the thriving Madoc is more due to Dr. Wishart's energy than any other indi- His private residence was the first stone building in the place. He two other churches in the township of Madoc, and was instrumental ection of a church in Huntingdon Township, and also two others on ings road. The erection of St. Peter's, Madoc, is also chiefly due to ons, and when entirely finished will be a lasting monument to his The Rev. Dr. Wishart has always been identified in every effort provement of the people. Of the most exalted integrity, he is hon- respected by all who have the pleasure of his acquaintance.

RAWDON TOWNSHIP.

wnship of Rawdon is bounded on the north by the township of on the east by Huntingdon, on the south by Sidney, and on the e township of Seymour, in the county of Northumberland, and ithin its limits an area of about 67,289 acres of land. It derived rom Lord Rawdon, an English nobleman, at one time connected olonial Department of the British Government. Its first permanent dates back to about the year 1820, when the front range of con- ere occupied by sons and descendants of United Empire Loyalists ho lived in the older settled townships on the Bay of Quinte, and e occupation of the available land in their respective corporations, in the rear tier of townships, upon which their families after- ed. The vicinity of Stirling was, owing to the superior quality to the close proximity to Sidney, together with the fine water pri- nded by Rawdon creek, we are informed, the scenes of the earliest in the township.

st the pioneers of Rawdon, descendants of the U. E. Loyalists , we find the names of Chad, Weese, Weaver, Huddle, Sine, yhes, Johnson, Potts, Vandervoort, Huffman, Denike, Caverley, agerman, Frelick, Seeley, Sarles, Horton, and Dafoe. These quently followed by families of American and other origin, who many eligible lots upon the front and centre concessions, hom we are able to trace the names of Westcott, Kingston, etchesan, Kyle, Montgomery, Rogers, Cumming, Faning, Has- , Fox, Green, Merrick, Brooks, Allard, Williams, McKim, arks, Todd, Jeffs, Heard, Carlist, Bryant, Woodbeck, Sharp, —for many years township clerk—Elliott, Tucker, McCann, Scott, ler, Rolph, Morton, Heimer, Caverley, Crier, Barges, Mack, more, James Cook—the present Reeve—McMullen, Danforth, iles, Haslett, Buskell, Livingstone, Thompson, Falkner, Courtney, neer, Tompkins, Austin, Hurst, Vollick, Outwater, Micklejohn, Scrimshow, Eastman, McWilliam, McQuaig, Berger, Duncan, rdison, Dethes, Gurnsey, Doak, Finkle, Montgomery, McKer, eatman, Thrasher, Wellman, McQuinsh, Wright, Moster, Kelly, Keown, Ward and others of more recent date, all of whom applied s to the cultivation of their respective farms, and, from appear- have been well repaid for their industry.

rience of the pioneers of Rawdon was similar in every respect to rly settlement of other townships, a detailed account of which will ther portions of this history. In 1830 the township, including the age of Stirling, now incorporated, only contained about 1,050 in- hese wants were supplied by one grist and saw mill and two small village; 81 horses, 103 oxen, and 231 cows, furnished the sum live stock, (with perhaps the exception of a few hogs) of the that year. Since that date its progress has been rapid, and on

the development of the mineral wealth of Marmora, the township lying im- mediately to the north, affording a ready and available market for her pro- duce, Rawdon suddenly became a populous and thriving township. The Stirling and Marmora macadamized road which runs through the centre of the township, and other principally travelled roads, are in good repair, and almost any point in the township is easy of access. The dwelling houses of the farmers are comfortable and home-like, some exhibiting a degree of archi- tectural beauty and refinement not often to be met with except in old and wealthy municipalities. The outbuildings are of a substantial character, well calculated to protect the dumb animals from the inclement weather of our Canadian winters. The surface of the township is undulating and in some places rather precipitous. The soil, like that of the adjoining [illegible] is on the blue limestone formation, and, with the exception [illegible] sections where the rock crops out to the surface, is excellent for agricultural purposes, and is apparently in a high state of cultivation. No one can drive through the township of Rawdon and look upon its well tilled farms, now covered with waving fields of grain ready to be harvested; with the numerous comfortable homesteads, where fifty years ago the forest stood in all its grandeur and wildness, but must confess that real and substantial progress has marked its course. Such lands as are at present unimproved are owned either by pri- vate individuals, the Government, or the Canada Company, and may be ob- tained from either at reasonable terms.

Compare the old log school house of early days, with indifferent pedagogue, to the dozen neat, commodious, and conveniently situated buildings with well educated teachers of to-day, with our excellent free school system, where every child may obtain a thorough knowledge of all the English branches, and if necessary, can, through grammar schools, acquire a good commer- cial and classical education.

There are also located at various points in the township, comfortable places of public worship, presided over by regularly ordained ministers of the gospel who administer to the spiritual and religious wants of the inhabitants. There are five post offices, by which the people of Rawdon communicate with the outside world through Her Majesty's mails, viz:

STIRLING VILLAGE,—Mrs. A. Judd, postmistress.

BONNSWORTH, situated upon lot 22, in the 2nd concession,—Allen Wil- liams, postmaster.

WELLMAN'S CORNERS, situated upon lot 18, in the 6th concession,—D. S. Ostrander, postmaster.

HAROLD, situated upon lot 12, in the 8th concession,—A. McWilliams, postmaster; and

SPRING BROOK, upon lot 13, in the 9th concession,—Peter Welch, post- master.

Rawdon is well watered by several rivers, creeks, and rivulets, which pursue their tortuous course in a south-westerly direction through the township, affording at different points excellent facilities for almost every description of manufactures. Rawdon creek takes its rise in the township of Huntingdon, and upon lot No. 1, in the 9th concession of Rawdon, tra- versing the township in a south-westerly direction, passing through the vil- lage of Stirling, and emptying itself into the river Trent, in the township of Sidney.

A stream of considerable importance takes its rise upon lot No. 9, in the 5th concession, and runs parallel to Rawdon creek, also emptying into the Trent in Sidney.

Squires' creek most effectually drains the north and east and west centres of the township. It takes its rise upon lots 6, 8 and 21, in the 6th con- cession, and upon lot 13 in the 12th concession, with tributaries at the south- western extremity of the township, in which direction it flows until it meets the Trent in the township of Seymour, in the county of Northumberland. Upon this stream a number of grist and saw mills and cheese factories are erected, with ample supply of water to drive all the necessary machinery.

The Moira river, with a tributary called Mud creek, crosses the north- eastern corner of the township, and upon which are a couple of saw mills, one of which has, we are informed, fallen into disuse. Crow River, a large tributary of the Trent, traverses the north-western extremity of the town- ship, widening to the proportions of a small lake. There are no mills upon this river in Rawdon. The manufacturing interests of Rawdon are confined to grist and saw mills, brick yards, and cheese factories. Upon lot No. 19, in the 5th concession, upon Squire's creek, directly south of Wellman's Cor- ners, is a grist and saw mill, and blacksmith's shop. On lot No. 5, in the 2nd concession, on a tributary of Rawdon creek, is a saw mill, and further down the stream, on lot No. 7, is situated another, both in running order, but owing to the scarcity of raw material, only a local business is done at certain seasons of the year. Upon lots Nos. 8 and 9, in the 8th concession, and upon a branch of Squires' creek, Mr. A. Ketcheson owns and carries on a saw mill; and upon lot No. 2, in the 13th concession, on the Moira, there is another saw mill.

There are several first-class cheese factories located upon the many living springs and streams which abound throughout the township, all of which are fitted out with the latest approved apparatus and mechanical appliances to facilitate the manufacture of this very important branch of agricultural in- dustry. They are situated as follows:—Upon lots Nos. 14 and 17, in the 1st concession; upon a tributary of the Trent, directly west of the village of Stirling at Wellman's Corners; upon lot No. 19, and a branch of Squires' creek, in the 6th concession, is situated another factory with a large manu- facturing capacity. Upon lot No. 19, in the 9th concession, and upon an- other branch of Squires' creek, is situated the Batton factory. Another factory is located upon the farm of Mr. S. S. Vandervoort, lot No. 13, in the 6th concession, and Squires' creek, having a large patronage; while another is situated upon lot No. 8, in 9th concession, immediately north of Ketche- son's saw mill. Upon lot No. 14, in the 12th concession, is the Read fac- tory, also upon Squires' creek. Two other factories are situated respectively upon the properties of Mr. J. Gay and Mr. Wright, upon lot No. 9, in the 4th and 5th concessions, making a total of nine establishments which, dur- ing the season, are run to their utmost capacity, netting annually thousands of dollars to their patrons, the farmers of Rawdon.

The Grand Junction Railway runs across the front of the township. The road is graded, and iron laid from Belleville to the village of Stirling. A daily Stage is run between the two corporations, consisting of an open hand car, the motive power being two stalwart and enterprising citizens of Belle- ville.

The Belleville and North Hastings Railway leaves the Junction track upon lot No. 1, in the 1st concession, and passes through the township of Huntingdon to Madoc.

We have been unable to procure any statistical information whatever in connection with the township's municipality or otherwise, hence our inability to reproduce the records of the early town meetings, the names of the sev- eral municipal officers, and other information necessary to show the gradual progress and development of the township from its first settlement to the present time. Mr. James Cook is at present, and has been for many years, the reeve of the township. Mr. Joseph Bogle is the township clerk, &c., and Mr. Philip Brown, treasurer. The municipal council meet monthly in the Town Hall, a substantial and commodious brick building, situated in the centre of the township, at Harold P. O. There is a tavern adjoining, store, blacksmith's shop, &c., affording every facility for the accommoda- tion of the ratepayers.

About the year 1820 a primitive grist mill was erected upon Rawdon creek, which formed the nucleus of the present flourishing village of Stir- ling. The principal portion of the village is situated in the township of Rawdon, to which it formerly belonged, until its incorporation in 1858, when the limits were extended to the south, embracing several lots in the town- ship of Sidney. It is surrounded by a fine agricultural country, and, being centrally located, commands a large trade. The excellent water privileges of Rawdon creek, which flows through the centre of the village, have been utilized to advantage, and several large mills and factories have been erected, adding largely to its commercial prosperity. There are a large number of first-class dry goods, grocery, and general stores; boot and shoe, harness, and other shops, all of which seemingly do a safe and prosperous business. Its educational facilities are exceptionally good, comprising an excellent grammar and common school, with efficient teachers, under the direct supervision of a united board of county, grammar and common schools. Its churches are a credit to its inhabitants, and are in point of architectural beauty second to none in the county. They consist of the Presbyterian, Church of England, Canada Methodist, Episcopal Methodist, and Baptist Churches. The village also possesses two good hotels. Distance from Belleville, 15 miles; Trenton, 16; Frankford, 7; Marmora, 16; Madoc, 19; and Campbellford, 16 miles. Daily stages connect the village with Belleville and other points, and when the Grand Junction Railway is completed to its destination, Stirling will, from its geographical position and abundant resources, become an important inland town. The present Reeve of the village is Charles Craige; Treasurer, James Milne; Clerk, &c., F. B. Parker. Population 1000.

BIOGRAPHIES.

CAPTAIN JUSTUS SHERWOOD, a Pennsylvanian of English descent, was an officer in the Royal Forces during the Revolutionary war. He was wounded and made prisoner, with the surrender at Saratoga, of the British army, under Burgoyne. For his espousal of the Royal Cause he was obliged to fly the country, and came to Canada shortly after the above-mentioned event, settling at St. John's, Lower Canada, and being among the very first pioneers of what was then an almost unexplored wilderness. There his son was born, who afterwards became Hon. Lewis P. Sherwood, Judge of the Court of Queen's Bench of Upper Canada. He married Miss Charlotte Jones, daughter of Ephraim Jones, Esq., another U. E. Loyalist, who was one of the earliest settlers in the township of Augusta. One of the children of this marriage was the Hon. George Sherwood, Q.C., now senior Judge of the County Court of the County of Hastings. He was born in the County of Leeds, Upper Canada, in May, 1811, and educated at the Johnstown District Grammar School. He was called to the Bar in Michaelmas Term, 1833, and commenced practice at Brockville, where he soon took position as one of the ablest lawyers of the country. He continued to do a very large, successful and lucrative business here, till his appointment to his present position. He was a short time in partnership with the late Hon. Henry Sherwood, Q.C., and for thirteen years a law partner of the late Judge Steele.

From a very early age he seems to have identified himself with public affairs. His first public services began by filling for a number of years the position of Town Councillor of Brockville, he afterwards became Mayor, and subsequently filled the Warden's chair in the County Council.

He was elected to a seat in the Legislative Assembly of Canada, in 1841, a position which he held uninterruptedly for ten years. In 1851 he was defeated for the same position, and again in 1854; but recovered the lost seat in 1858, and continued in it till 1863.

His eminent ability and undoubted integrity received a deserved acknowledgment in his assignment, at various times, to a great number of public positions of the highest honor and importance. He was appointed Sept. 5th, 1845, a Commissioner to inquire into the management of the Board of Works; he was elected a Bencher of the Law Society in 1846, and created a Q.C. in 1856. He was a member of the Executive Council of Canada from Aug. 6th, 1858, to May 23rd, 1862, during which time he filled the offices of Receiver-General, and Commissioner of Crown Lands, in succession. He received the appointment to his present position, September 2nd, 1865.

During a long and busy life he has always found time to devote to the encouragement and forwarding of all beneficial public enterprises—many of which to-day bear witness to his wisdom, energy, and devotion to the public welfare. He was one of the influential and active promoters of the Brockville & Ottawa Railway, of which he was President for many years. He is married to Marianne, daughter of Dr. Keegan, of Halifax, Nova Scotia, but has no children.

The Sherwood family have all and always been strongly Conservative in politics, and among the most influential men in the country of that party; a party to which their allegiance dates back to the time when their ancestors for the love of the King and Constitution—abandoned home, and possessions, and luxurious surroundings, for privations and hardships in a Canadian wilderness.

THOMAS APPLEBY LAZIER, Junior Judge of the County Court of Hastings county, is descended from U. E. Loyalists on both his father's and mother's side.

The Lazier family was of Dutch extraction, their ancestors having come from Holland, near the French border, at a very early day in the history of the western continent, and settled above Manhattan Island, on the Hudson. Jacobus Lazier, the great-great-grandfather of the Judge, was born near Yonkers, New York, in 1707, and died there in 1792. His descendants lived in that vicinity till the Revolution divided the colonies into hostile factions. The Laziers espoused the Royal cause, and to escape the persecution of the Americans, and at the same time enjoy freedom of their political opinions, many of them left their homes after the war, some going to New Brunswick and Nova Scotia, while Nicholas, the eldest son of Jacobus, born in 1758, came to Canada in 1791, and settled in Prince Edward county, where he remained till his death. His wife's maiden name was Charity Conklin, and their whole family of seven sons and two daughters lived to be very old. Mary, the eldest child—afterwards Mrs. Bogart—died at Adolphustown, as recently as 1874, in the 102nd year of her age. Nicholas, the third in order of age of this family; and of his family, Richard, born in Prince Edward, in 1805, was the eldest son. He married Anna Bookhout Appleby, and removed in 1828, to Tyendinaga, being among the very first settlers in that neighborhood, near Shannonville.

The Applebys were Americans, from Duchess county, New York, but originally of English extraction. They left their old home in Duchess county, in 1797, and came to Canada, as did the Laziers, to enjoy freedom of political opinion.

Richard Lazier became first Clerk of the old Court of Requests of Hastings county, and subsequently—and until the Court was abolished—one of the Commissioners of the same. He was a Captain of Volunteers in active service in 1837-38, and was afterwards promoted to a Lieut.-Colonelcy. He was a man who deserved well of the country, and the Government recognized his ability and public services by appointing him to the position of Collector of Customs for Shannonville, a position held by him up to the time of his death, June 29th, 1874.

Of his family of five sons and two daughters, Thomas Appleby Lazier, born in 1826, was the eldest. He was educated at Victoria College, and studied law chiefly with Hon. Lewis Wallbridge, of Belleville, finishing his studies in the office of the Hon. Judge Adam Wilson, of Toronto, and was called to the Bar in Hilary Term, 1851.

He commenced the practice of his profession soon after, at Belleville, where his talents and application soon won for him a large and lucrative practice, and marked him as the man to fill the Junior Judgeship of the County Court of Hastings county, to which position he was appointed, July 11th, 1873.

The Judge has always lent his valuable assistance to the development of all public or private enterprises which promised for the benefit of Belleville, of Hastings county, or of the Province. He was a very active advocate of the

Grand Junction Railway, the organization of which was in a
due to him, and a few more like him; and he became the firs
that Corporation. His younger brother, S. S. Lazier, who stu
Judge, and afterwards practised in partnership with him f
years, is now Master-in-Chancery for the counties of Hastings
ward, Lennox, and Addington. His uncle, S. N. Appleby, re
Hastings in the Ontario Legislature. He himself never s
Laziers are a very numerous family throughout the Bay of
trict, and their number, intelligence, and high social position
munity enable them to exert a very strong influence in all affai
or political nature.

The HON. EDMUND MURNEY, M.L.C., born at Kingston,
day of November, 1812, was the second son of Henry Murney,
He was educated at Upper Canada College, acquired the k
law in the office of Marshall S. Bidwell, at Kingston, an
to the Bar at Osgoode Hall, in 1834. He commenced the pr
in Belleville, and for many years dedicated himself to his p
which he established a brilliant reputation, ranging first among
sional associates. Mr. Murney entered Parliament in 1836, the
Canada. The first election after the union of the Provinces, he
by the Hon. Robt. Baldwin, by 37 votes. In 1842 Mr. Murn
tested the Riding in the Conservative interests and was return
majority over his former opponent, the Hon. Robt. Baldwin,
defect in the returns another election was ordered by the House,
defeated the Reform candidate, Joseph Cannifl, Esq. On the
Parliament in 1844, new elections were held, and Mr. W. H
brought out to oppose Mr. Murney, but again his political
the ascendant. In 1848 he was defeated by the Hon. Billa
1851 Mr. Murney was again elected, defeating his former oppo
division of the county Mr. Murney defeated Mr. George Benja
North Riding. In 1857 he became a candidate for the Trent
signing his seat in the Lower House, he was elected to the Uppe
feating Mr. Shorts, of Peterboro, and was a member of this
Legislature at the time of his death, which occurred on the 15th
1861, leaving his wife and one son, Edmund H. Murney—both si
and nine daughters surviving him, viz.: Mrs. Ridley, wife of H
Esq., Mrs. D. Hamilton, Mrs. McLeod, wife of Henry A. Ma
Engineer Prince Edward Island R. R., Mrs. Falkner, wife of N.
Barrister-at-Law, Belleville, Mrs. Baldwin, wife of the Rev. A.
M.A., Rector of All Saints Church, Toronto, Mrs. Ridley, wife
C. Ridley, Esq., Belleville, and four unmarried. The following
the *Intelligencer* of August 13th, 1861, will be read with interest

"As a politician he took a leading part from the moment
"upon his public career; he assisted materially in reforming ma
"rendered them suitable to the wants of the community, and up
"sions he gave his support to the enactment of such new laws a
"and prosperity of his country required.

"Although a firm and faithful supporter of his own Church,
"his political career he assented to the passing of a law fully
"other denominations from what was considered the powers of
"clergy in the Province.

"He was an excellent lawyer, and his clear perception an
"rendered him an opponent of no small magnitude at the Bar.
"tainly one of the most successful practitioners in what was k
"old Midland District. Some of his cotemporaries in the Provin
"sibly have gained in some passages of their career, more brill
"tions, but for discretion, judgment, and honesty of purpose, as
"sistency in the whole of his political life he was unsurpassed by

"No client of his ever knew what it was to be persecuted or
"payment, none were ever asked even to meet claims strictly ju
"who would thereby embarrass themselves or distress their famili
"ferred the inconvenience of a non-compliance of engagements on
"others, to bring distress and trouble upon any one.

"He was never violent or extreme in his language, yet with t
"integrity and firmness he ever maintained his opinions, and wa
"most consistent politicians of the age.

"In all social relations of life Mr. Murney was greatly gifte
"husband—a doting father—a sincere friend—loved, revered an
"he went down to his grave—

"Like the stars, by day
"Withdrawn from mortal eye,
"But not extinct—they hold their way
"In glory through the sky."

ALEXANDER ROBERTSON, the subject of this sketch, was born
in the county of Hastings, on the 5th of December, 1839, but has
the city of Belleville, of which he is now its first Mayor, since he
years of age. He was called to the Bar in 1864, and commenced t
of law. He was a member of the Town Council for a number of
Mayor in 1870. In 1871 he was offered by acclamation the positi
ber for East Hastings in the Commons, on the elevation of the Hon.
the Senate, but declined its acceptance at the request of his fa
Robertson is identified with every enterprise calculated to prom
terests of the city in which he lives, and last year erected a beautif
buildings on Front street, which are an ornament to the city. Mr.
is a widower, having lost his wife, the eldest daughter of Dr. R.
1874; he has one daughter. His father, the late William Robert
native of Glenelg, Inverness-shire, Scotland, and came to Canada in
married in 1857, the daughter of an old U. E. Loyalist family
and embarked heavily in the lumber business, on the Trent and M
which business he continued for many years, and died in 1861.
descendant of the Robertsons of Strowan, who were noted for th
and fine physique; when James I. was cruelly murdered in Bl
Monastery, Perth, in presence of the Queen and ladies, by Sir Rober
having made his escape to the Highlands, hiding in the Braes o
he was captured by Robert—grandson of Robert de Athole—fou
Clan Robertson. For the arrest of Graham and taking him to th
Stirling, he was rewarded with a Crown Charter, dated 1451, e
whole lands into a free Barony. He also received the honourabl
to his coat of arms, a naked man, manacled, under the achievemen
motto *virtutis, gloria, merces.*

MR. GEORGE RITCHIE.—The firm of Messrs. Ritchie is
oldest business houses in this part of the Province. It was establi
year 1858, by Mr. George Ritchie, the late senior partner, and has
with uninterrupted success, so that notwithstanding having passe
several periods of great commercial depression, we believe it hol
position to-day than at any previous time. Few firms in this new
can show a clearer record, for though originating in comparati
beginnings, there is not a single instance of reflection upon its cre
mains throughout the twenty years of its existence, while now it ha
rank among the dry goods houses of the Dominion.

This firm, who are really the only *bonâ fide* importers of a gener
dry goods between Kingston and Cobourg, are connected with an
shipping house in the old country, which connection gives them an
advantage, enabling them to purchase their goods at lower rates
otherwise be done; for in this way they get access to the leadin
turers in Great Britain and the continent, who will not sell direct
tail trade on account of arrangements with wholesale houses at h
apart from this advantage their system of doing business is well

, the organization of which was in a great measure
ete like him; and he became the first Secretary of
ounger brother, S. S. Lazier, who studied with the
practised in partnership with him for about ten
Chancery for the counties of Hastings, Prince Ed-
lington. His uncle, S. S. Appleby, represents East
the Legislature. He himself never married. The
erous family throughout the Bay of Quinte Dis-
intelligence, and high social position in the com-
xert a very strong influence in all affairs of a public

MURNEY, M.L.C., born at Kingston, on the 11th
was the second son of Henry Murney, of Kingston.
pper Canada College, acquired the knowledge of
Marshall S. Bidwell, at Kingston, and was called
Hall, in 1834. He commenced the practice of law
many years dedicated himself to his profession, in
rilliant reputation, ranging first amongst his profes-
Murney entered Parliament in 1836, the last in Upper
ion after the union of the Provinces, he was defeated
win, by 37 votes. In 1842 Mr. Murney again con-
Conservative interests and was returned by a large
opponent, the Hon. Robt. Baldwin, but owing to
ther election was ordered by the House. In 1843 he
ndidate, Joseph Cannill, Esq. On the dissolution of
w elections were held, and Mr. W. H. Yager, was
Mr. Murney, but again his political star was in
8 he was defeated by the Hon. Billa Flint, and in
gain elected, defeating his former opponent. On the
Mr. Murney defeated Mr. George Benjamin, for the
he became a candidate for the Trent Division; tr-
ower House, he was elected to the Upper House, de-
Peterboro, and was a member of this branch of the
of his death, which occurred on the 15th of August,
d one son, Edmund H. Murney, both since deceased,
iving him, viz.: Mrs. Ridley, wife of Henry Ridley,
n, Mrs. McLeod, wife of Henry A. MacLeod, Chief
t Island R. R., Mrs. Falkner, wife of N. B. Falkner,
ille, Mrs. Baldwin, wife of the Rev. A. H. Baldwin,
nts Church, Toronto, Mrs. Roller, wife of Frederick
lle, and four unmarried. The following extract from
st 13th, 1861, will be read with interest.
took a leading part from the moment he entered
; he assisted materially in reforming many laws, and
to the wants of the community, and upon all occa-
ort to the enactment of such new laws as the welfare
country required.
st faithful supporter of his own Church, very early in
assented to the passing of a law fully relieving all
rom what was considered the powers of a dominant

st lawyer, and his clear perception and judgment
nent of no small magnitude at the Bar. He was cer-
successful practitioners in what was known as the
Some of his contemporaries in the Province may pos-
me passages of their career, more brilliant reputa-
n, judgment, and honesty of purpose, as well as con-
of his political life he was unsurpassed by any.
r knew what it was to be persecuted or pressed for
ver asked even to meet claims strictly just and due,
barrass themselves or distress their families. He pre-
er of a non-compliance of engagements on the part of
s and trouble upon any one.
nt or extreme in his language, yet with true sterling
he ever maintained his opinions, and was one of the
hans of the age.
ts of life Mr. Murney was greatly gifted—a loving
er—a sincere friend—loved, revered and respected,
ave—

"Like the stars, by day
Withdrawn from mortal eye,
ut not extinct—they hold their way
n glory through the sky."

son, the subject of this sketch, was born at Trenton,
, on the 5th of December, 1839, but has resided in
which he is now its first Mayor, since he was seven
led to the Bar in 1861, and commenced the practice
er of the Town Council for a number of years, and
he was offered by acclamation the position of mem-
he Commons, on the elevation of the Hon. R. Read to
its acceptance at the request of his family. Mr.
th every enterprise calculated to promote the in-
h he lives, and last year erected a beautiful block of
which are an ornament to the city. Mr. Robertson
his wife, the eldest daughter of Dr. E. Stewart, in
er. His father, the late William Robertson, was a
s-shire, Scotland, and came to Canada in 1827, and
ghter of an old U. E. Loyalist family in Murray.
the lumber business, on the Trent and Moira rivers,
ed for many years, and died in 1861. He was a
ons of Strowan, who were noted for their bravery
James I. was cruelly murdered in Black Friar's
nce of the Queen and ladies, by Sir Robert Graham,
the Highlands, hiding in the Braes o'Mar, until
rt grandson of Robert de Athole founder of the
arrest of Graham and taking him to the Queen at
with a Crown Charter, dated 1451, erecting his
rony. He also received the honourable addition
d man, manacled, under the achievement, with the
ers.

.—The firm of Messrs. Ritchie is one of the
his part of the Province. It was established in the
Ritchie, the late senior partner, and has since met
s, so that notwithstanding having passed through
ommercial depression, we believe it holds a firmer
y previous time. Few firms in this new Dominion
l, for though originating in comparatively small
single instance of reflection upon its credit or good
ty years of its existence, while now it holds a front
houses of the Dominion.
ly the only *bona fide* importers of a general stock of
n and Colonial, are connected with an extensive
ountry, which connection gives them an undoubted
to purchase their goods at lower rates than could
this way they get access to the leading manufac-
the continent, who will not sell direct to the re-
rangements with wholesale houses at home. But
their system of doing business is well calculated

to secure the popular favor, it being one of the most reliable places to deal in anywhere to be found.

Mr. George Ritchie, the founder of the business, died on the 8th of May, 1878, respected by all who knew him, and among business men was one whose word was regarded as good as his bond. He was a man thoroughly honorable and upright in all his dealings, and generous to a fault. He was a son of the late Robert Ritchie, Esq., formerly of the Ordnance Department, at Ottawa, who removed to Belleville with the rest of the family, in the fall of 1858, having previously retired from her Majesty's service in which he had served for nearly fifty years. Mr. Thomas Ritchie succeeds his brother in the management of the business, and conducts it under the same auspices. We present elsewhere in this work an interior view of this magnificent store, showing the grand staircase leading to the millinery and mantle show rooms on the second floor.

P. V. DORLAND, M.D., was born in Adolphustown, on the Bay of Quinte, and is a son of the late Col. Samuel Dorland of the First Lennox and Addington Militia, which for many years had been in active service, and during the colonial times were mustered several times during a year for drill. Col. Dorland was also one of the three Commissioners appointed to hold the Court of Requests every fortnight alternated, one week in the Court House and the second one at Fredericksburg. Dr. Dorland's two grandfathers were Capt. Thomas Dorland and Capt. Edward Huyck, both of whom were in General Burgoyne's army when he surrendered to General Gates, at Saratoga, and were among the earliest settlers on the Bay of Quinte, bringing with them three families of slaves, and material to assist in clearing up the country. Capt. Dorland was appointed Commissariat officer for the whole Bay District. He represented this section of Canada for eighteen years in Parliament, retiring from political life at an advanced age. He located one thousand acres of land presented him by the Government for services rendered in Adolphustown, commencing at Dorland's ferry opposite the steam mills, within a short distance of the village, a quantity in Prince Edward County, and the rest in the County of Hastings. Capt. Huyck entered into a general mercantile business in addition to that of farming in the 4th concession of Adolphustown, on what is called Hog Bay, where he did a large business. Dr. Dorland received his preliminary education at Knox and Queen's Colleges, Toronto and Kingston. He studied at Trinity and the Toronto Schools of Medicine, and took his degrees at the College of Physicians, Edinburgh, viz.: L.R.C.P., and M.R.C. Surgeons. He graduated at the New York Eye and Ear Institute, and at Philadelphia University of Medicine. He has resided *for many years in Savannah, Georgia, and was in that city when it was de*-populated by cholera, and has written a treatise on the theory of that dreaded disease, also a valuable paper on inhalation that have withstood the criticisms of many able writers of medical jurisprudence. He owns one of the finest residences in Belleville, and recently built a block on Front street, near the upper bridge, which adds much to the appearance of that part of the city. Dr. Dorland has been an extensive traveller, and a description of his travels would fill a volume; he has visited the Pacific slope, *via* Cuba, he crossed over and under the Alps, *via* the Mount Cenis tunnel, he travelled through Great Britain, Hungary, Italy, Germany, France, Spain, and Central America, and is one of the best read men of the times. Of a genial and generous disposition, with fine conversational powers, his society is highly enjoyable. He is devoted to the science of medicine, in which profession he has ever been a successful practitioner, and is regarded as one of Belleville's most public spirited and valued citizens.

W. D. FULLER was born in Washington County, New York, and came with his family to Canada in 1846, settling in the township of Huntingdon, where he engaged, in connection with his father and brothers, in the lumbering and other lines of business in which he was eminently successful. In 1868 he removed to Belleville and commenced business as a grain merchant, since which time he has been the heaviest operator in the Bay district. He was appointed United States Consular agent for Belleville, a position he still holds. *He enjoys the reputation of being a gentleman of unblemished char*-acter, and is highly respected by a large circle of friends in Belleville and throughout the County of Hastings.

WILLIAM HUDSON, carriage manufacturer, Roslin, was born in the year 1842, within the limits of the township of Thurlow, which municipality he has for some years represented in the County Council, having been repeatedly elected by acclamation to the position of Deputy-Reeve, an office he fills with ability. He is a son of Charles Hudson, who came to Canada in 1820 and settled upon lot No. 1, in the 9th concession of Tyendinaga, which he cleared, the surrounding country being at that period an unbroken wilderness. Mr. Hudson commenced the manufacture of carriages at the village of Roslin, then known as Wilson's Corners, in the year 1869, the annual business being but small compared with that of the present day. The success of this enterprise is due in a great measure to the fact that Mr. Hudson is a practical mechanic, and gives the business, in all its details, his personal supervision, turning out a class of work that for finish and durability cannot be excelled in the country. His business is steadily increasing, and the establishment annually turns out upwards of one hundred vehicles, and employs from fourteen to sixteen hands. Mr. Hudson by strict business habits has gained the confidence of his fellow men; his goods have a wide-spread reputation for general excellence, unsurpassed and seldom equalled in the County of Hastings.

JOHN C. VERMILYEA, is a Canadian by birth, and is of Dutch and English descent. His father, the late Solomon Vermilyea, emigrated with his parents from the Catskill Mountains, Green County, N.Y., about the year 1800, and settled upon what is known as Beach Ridge, north of Montreal, where the family remained for a short time; not being suited they removed westward, and purchased lot No. 1, in the 4th concession of the township of Thurlow, which he converted from an interminable forest into one of the finest farms in Ontario. John C. Vermilyea, owns an improved farm of 150 acres adjacent to the city of Belleville, having an excellent dwelling house and good substantial out-buildings, the whole being surrounded by an orchard of superior fruit, of which he makes a specialty. He is a prominent member of the Society of Orthodox Friends, and has officiated in that body, in the capacity of local preacher, for upwards of ten years. The Vermilyea family have always been prominently identified with that religious organization, the father of the subject of this sketch having erected the first meeting house in this section of the county, at the Corners, near his farm, where he enjoyed the quiet of a home life until the day of his death. John C. is highly respected by all his neighbours, is a man of unblemished character, and a useful member of society.

The parents of the late JOHN J. BRADSHAW were among the early pioneers of the township of Thurlow, where the subject of this notice was born, in the year 1803. He learned the house carpenter's trade with William Yager, and *in 1825 married Nancy McMullen. During his early life he became very suc*-cessful as a builder, and by dint of industry, honesty, and sound judgment, acquired a competency which enabled him to secure a farm in Sidney, which he afterwards sold and purchased lot No. 8, in the 4th concession of Thurlow, near the place where he was born, upon which he lived until the day of his death, which occurred on the 8th day of September, 1877. Mr. Bradshaw early identified himself with the Church, in which body he was a useful and consistent member. While his memory remains fresh, his virtues will be remembered; and whether active in public life, at home upon his farm, or in the family circle, his example and influence were potent for general and individual good.

WILLIAM ASHLEY, the subject of this sketch, was born at Fredericksburg, then called 4th town in the order of U. E. L. settlements, in the year 1808.

He is of English descent, his parents emigrating from the State of Massachusetts about the year 1800, preferring the then inhospitable wilds of Canada, under British rule, to the persecutions of the victorious rebels. The family first settled and commenced business at Canniff's Mills, on the Moira river, and in the year 1824 removed to lot No. 2, in the 5th concession of Thurlow. In 1835 William Ashley erected at this point a shop, and commenced the manufacture of waggons, etc., which was undoubtedly the origin of the present flourishing village of Foxboro', for many years called Smithville. A post office was established at an early date, Mr. Ashley being the first Postmaster. On his retirement he left the carriage manufacturing, together with its appliances, to his son Charles, who at present carries on the business. Mr. Ashley owns two farms and a beautiful private residence, and by his energy and enterprise has added much to the development and improvement of the village in which he resides.

CHARLES ASHLEY is a son of William Ashley, and was born in the village of Foxboro'. He is at present extensively engaged in the manufacture of carriages, etc., having succeeded his father on his retirement, in the business. This establishment is fitted with all the modern appliances necessary to the manufacture of every description of first-class vehicles, and is run to its full capacity in order to supply the increased demand for its products. Mr. Ashley has extended his business in every direction, and at present employs fifteen skilled workmen, a force often inadequate to fill the many orders that come from all parts of the country. His work gives unusual satisfaction, as is attested by the increase of business, and he justly merits the enviable reputation he has attained.

The JOSE family came from Cornwall, England, to Canada, in the year 1859, and rented lots 1 and 2, in the 8th concession of the township of Thurlow, where they have resided ever since. They are engaged in farming and raising thorough-bred Durham and grade stock, Leicester sheep, and Berkshire hogs, and have one of the finest herds to be found in the County of Hastings. Their stock is known throughout the Province, for its general superior excellence and purity of breed. They have at present in their herd, two thorough-bred Durhams, with authenticated pedigrees, fine specimen of the Short-horn, while their Durham and grade cows are choice, excelling those of ordinary breeders. Their Leicester sheep are pure, and their thoroughbred Berkshire hogs have done much towards improving the breeds, throughout the townships, of this very useful and necessary animal. The Jose brothers by industry and hard work have gathered together a herd they may feel justly proud of, and which is a credit to the municipality in which they reside.

GIDEON SPENCER, the subject of this sketch, was born in the year 1838 upon lot No 2, in the 8th concession of Thurlow, which property was granted by the Government, under proclamation, to his grandfather Corey Spencer, a United Empire Loyalist, who emigrated from Rhode Island during the Revolutionary War, and settled in the neighbourhood of East Lake, about the year 1784. Mr. Spencer is an industrious and successful farmer, and has served in sev[illegible] [illegible]tions of public trust, discharging the many arduous duties connec[illegible] [illegible]erewith in a manner alike satisfactory to the municipality and the ra[te]payers, and creditable to himself. He is a member of the Society of Friends, and is highly respected by the community in which he resides.

COL. WILLIAM KETCHESON was born in the town of Bedford, New York, in the year 1772, and was of English descent. He came to Adolphustown from Nova Scotia, and married Nancy Robbin in the year 1800, removing to Sidney shortly afterwards, and locating and settling upon lot 26, in the 5th concession, which he cleared. He reared a family of fifteen children, nine sons and six daughters, all of whom eventually became united with the Church, and these all married and settled before death entered the family. There were living at his death—which occurred March 15th, 1848—eighty-four grand-children, one hundred and thirty-seven great grand-children, and five great, great grand-children. Total, two hundred and forty-one descendants.

The ancestors of Col. William Ketcheson saw service in Lord Cornwallis' army during the Revolutionary war, and were engaged in the many battles fought under that general. In the reign of George III. William Ketcheson held the commission of Acting Ensign under Lieut. Sturgeon, who commanded the Hastings Militia, and in 1809 received from Sir Francis Gore a commission as Full Ensign, in 1812 a Lieutenancy, in 1815, under Gen. Brock, he was commissioned Captain, and in 1832 Major from Sir John Colborne's Government. Serving under Col. Thomas Coleman in 1834, during Sir Francis Bond Head's administration, he became joint commissioner of the 11th division of the Court of Requests. In the first year of Queen Victoria's reign he was honoured with a Coloneley, a title which he wore with honour until the day of his death. In 1839 he received his Commission as a Justice of the Peace in the 2nd division of the Court of Requests. In all of these positions he discharged their several duties with credit to himself and advantage to his country. Mr. Ketcheson and his amiable wife were consistent members of the Canada Methodist Church, having been converted at the first camp meeting ever held in Canada, at Adolphustown. Of commanding presence, with a vigorous mind, his judgment was good—possessing quick perceptions, he was self-reliant. He possessed the confidence of his neighbours and of the community at large. As a citizen he was loyal to the core, and a staunch supporter of the Government. Possessed of that modesty that ever commands respect, he never forced himself upon the people as a claimant for public honours, though he served the township in several capacities. He was ever governed by a conscientious regard for his word and his obligations, possessing such a love for his family and regard for his friends that he ever laboured for their enjoyment.

ELIJAH KETCHESON was born in Fredericksburg, in the county of Lennox, in the year 1795, and is a son of the late William Ketcheson, who came from Yorkshire, England, and settled in Virginia. On the breaking out of the rebellion he joined the British forces, and served until the close of the war. The subject of this sketch (Elijah Ketcheson), while a trifle over sixteen years of age, in the year 1812, in company with three brothers, started for Kingston and helped to build the fort. In 1819 he was united in marriage to Minerva Ostram, long since deceased. During the rebellion of 1837 he held a Captain's commission, and did service at the Trent, and now holds the rank of Colonel. He receives a pension of $20 per year for services rendered in the war of 1812, and is at present the oldest Magistrate in the County of Hastings. In his eighty-third year, with all his faculties perfect, he has prospectively many years before him yet of happy green old age.

The RANDEWATER family originally came, with other Loyalists, from Duchess county, New York, about the year 1806, and tarried for a short period at what was then known as Hay Bay, afterwards purchasing and settling upon lots 28, 29, and E ½ 30, in the 6th concession of the township of Sidney. Daniel Randewater, the present owner of the property—a view of whose residence appears elsewhere—is largely engaged in the dairy business, having one of the most productive stock farms in Hastings county. He was elected to the Township Council during the years 1872-3, and is regarded as one of the most energetic and progressive men of the period.

MOSES BOARDMAN is a descendant of New England stock, his father coming from Boston, Mass., to Canada at an early period of our country's history. He purchased the property—lot 27, in the 7th concession of Sidney—upon which he resides, during the year 1842, from George Fairman,

the original grantee of the Crown. Mr. Boa
300 acres, well stocked with choice fruit trees,
house, and well-appointed buildings. He e
munity, and is looked upon as one of the l
Sidney.

The MASSEY family are of English descent
of the subject of this notice, was born in E
sachusetts some time previous to the Re
annually $80 pension for services rendered d
After the close of the war, or about the begin
he came to Canada and settled in the tow
county of Northumberland, where he lived t
years. Levi Massey was born in Haldima
McClutchie, a school teacher of Scotch desce
He came to Sidney in 1869, when he purcha
and 30, in the 5th con., where he now reside
of the best wheat-growing sections in the tow
is an active and energetic farmer, and is lo
minent men of the township.

SAMUEL T. WILMOTT is a Canadian by
Clarke, in the county of Durham. He cam
years of age, and resided with his grandfathe
U. E. Loyalist, for many years connected wit
lot No. 29, in the 5th concession of Sidney.
father Mr. Wilmott inherited the homestea
Wilmotts are of English descent and came fr
about the year 1800, and were engaged in th
finally settled in the township of Clarke, wh
reside, amongst whom is Samuel Wilmott
Fisheries.

RICHARD DAVIS, the subject of this bio
Duchess County, N.Y., in the year 1795, a
father, who first settled in Fredericksburg,
During the following year he came to Sidney
lot No. 31, on the front of the township, wh
He has four children living—two sons and tw
followed the occupation of a tiller of the soil.
he held a Captain's commission in the Light
while stationed at Toronto, and was appointe
Peace as a Magistrate by Sir George Cartier.
fort at Kingston. He owns 180 acres of excel
the Canada Methodist Church, with which h
relations of life; and, at the ripe age of ei
vigorous. Mr. Davis is of Welsh descent, his
Wales, England, to the United States, previo
was killed by the rebels at the battle of Whit
ing for his country.

COL. SHELDON HAWLEY, whose portrait is
of this work, sprang directly from U. E. L
revolutionary period, leaving their home in
route for a short time in Lower Canada,
Earnestown, sometimes designated 2nd town.
about the year 1795, he was educated and spe
the pastoral supervision of the Rev. John Lang
teachings he showed to the latest years of his
did garrison duty as a volunteer at Kingston d
this period until 1817 he remained chiefly in K
Trenton, then known as River Trent, where h
lumbering pursuits, and in which he was consp
placed on the Commission of the Peace as a M
for the improving of the navigation of the Tr
large timber bridge across the same. He was
the Murray Battalion of the Northumberland
the military post at Trenton during the rebelli
ment of the commissariat was also entrusted to h
Hon. and Right Rev. John Strachan, D.D., first
was co-founder of St. George's Church, Trent
fested a strong interest in the same, pecuniar
day of his death, which occurred on the 25th
year of his age. Thus closed the loyal and en
earliest and most highly respected pioneers in t

ADAM HENRY MEYERS, late of Trenton,
Henry Meyers, of the same place, and Mary H
Wallbridge, a U. E. Loyalist, and one of th
influential citizens of Ameliasburg. His fathe
known as Colonel Meyers, was a British subje
and Kingdom of Hanover, in 1780, when tha
appendix to the British Crown. He was a ma
the land of his birth, having been highly edu
on his advent to Canada, in 1805, a large stock
up the first store at Belleville, where he locat
him, Meyer's Creek—a name which it retaine
remained in "Meyers' Creek" about three ye
1808, where he engaged in the same pursuits a
at "The Creek." During his short residence
and made other improvements, which formed
prosperous and beautiful city of Belleville.
well as of means, he did not relax his energies
went to work building more mills, making roa
bourhood soon emerged for its primeval state
munity. He was one of young Canada's most
men—in public life strong in his efforts to adva
vate, hospitable and charitable—letting not
was doing; in all things honourable and above
successively from the ranks of the 1st Northu
tions of Ensign, Lieutenant, and Captain, an
Coloneley of the 2nd Northumberland, which
the 9th May, 1832, most deeply mourned.

His son, ADAM HENRY MEYERS, who w
Murray in 1812, had the advantage of a liber
treal and partly in New York—and studied l
Thomas Kirkpatrick, Q.C., and afterwards
Having been called to the Bar in 1834, he co
in Trenton, and continued to practise there
time of his death in 1876.

From early life Mr. Meyers took an activ
affairs. He was a staunch and honest Tory o
exerted great influence in favour of John A.
leader, whose warm personal friend he was,
ally. He unsuccessfully contested Northum
interest for the Assembly in 1842, against G
Judge of the County Court of Northumberla
sequently more successful, and represented th
for a number of years. To the many positio
dence which he filled, was added that of L
Northumberland Militia. He was a man of s
had been many years successfully exerted t

ORICAL SKETCH OF THE

inal grantee of the Crown. Mr. Boardman has an improved farm of s, well stocked with choice fruit trees, with good substantial dwelling- nd well-appointed buildings. He enjoys the respect of the com- and is looked upon as one of the leading men in the township of

Massey family are of English descent. Merrill Massey, grandfather ubject of this notice, was born in England, and emigrated to Mas- tts some time previous to the Revolutionary War. He drew y $80 pension for services rendered during those troublesome times. ie close of the war, or about the beginning of the present century, to Canada and settled in the township of Haldimand, in the of Northumberland, where he lived to the good age of ninety-eight Levi Massey was born in Haldimand, and married Miss A. E. chie, a school teacher of Scotch descent, from Huntington, Quebec. e to Sidney in 1869, when he purchased the S. halfs of lots Nos. 29 in the 5th con., where he now resides. His farm is located in one est wheat-growing sections in the township of Sidney. Mr. Massey tive and energetic farmer, and is looked upon as one of the pro- men of the township.

rel T. Wilmott is a Canadian by birth, from the township of in the county of Durham. He came to Sidney when about two f age, and resided with his grandfather the late Gideon Turner a oyalist, for many years connected with township affairs, who owned 29, in the 5th concession of Sidney. At the demise of his grand- Mr. Wilmott inherited the homestead where he now resides. The ts are of English descent and came from New Brunswick to Canada he year 1800, and were engaged in the wars of that period. They settled in the township of Clarke, where several descendants still amongst whom is Samuel Wilmott, Government Inspector of es.

hard Davis, the subject of 'this biography, was born in Clinton, s County, N.Y., in the year 1795, and came to Canada with his who first settled in Fredericksburg, County of Lennox, in 1800. the following year he came to Sidney and purchased and cleared up 34, on the front of the township, where he has resided ever since. four children living—two sons and two daughters—and has always d the occupation of a tiller of the soil. During the rebellion of 1837 l a Captain's commission in the Light Horse service, and was for a tationed at Toronto, and was appointed upon the Commission of the as a Magistrate by Sir George Cartier. He also helped to build the Kingston. He owns 180 acres of excellent land, and is a member of nada Methodist Church, with which he has been consistent in all as of life; and, at the ripe age of eighty-three, is still hale and s. Mr. Davis is of Welsh descent, his grandfather emigrating from England, to the United States, previous to the revolution, and who lled by the rebels at the battle of White Plains while gallantly fight- his country.

. Sheldon Hawley, whose portrait is represented in another part work, sprang directly from U. E. Loyalist stock, who, after the ionary period, leaving their home in Vermont, after sojourning for a short time in Lower Canada, settled in the township of town, sometimes designated 2nd town. Born in Ernestown in or the year 1795, he was educated and spent his early days there under storal supervision of the Rev. John Langhorne, the effects of whose ngs he showed to the latest years of his life. When but a youth he rison duty as a volunteer at Kingston during the war of 1812. From riod until 1817 he remained chiefly in Kingston, when he removed to n, then known as River Trent, where he engaged in mercantile and ing pursuits, and in which he was conspicuously successful. He was on the Commission of the Peace as a Magistrate; also a commission improving of the navigation of the Trent, and the erection of the mber bridge across the same. He was commissioned a Colonel in rray Battalion of the Northumberland Militia, and had command of itary post at Trenton during the rebellion of 1837-8. The manage- f the commissariat was also entrusted to him. In connection with the nd Right Rev. John Strachan, D.D., first Lord Bishop of Toronto, he founder of St. George's Church, Trenton, built in 1845, and mani- strong interest in the same, pecuniarily and otherwise, until the his death, which occurred on the 25th day of April, 1868, in the 73rd his age. Thus closed the loyal and energetic career of one of the and most highly respected pioneers in the settlement of Trenton.

u Henry Meyers, late of Trenton, was the son of Col. Adam Meyers, of the same place, and Mary Halloway, daughter of Elijah dge, a U. E. Loyalist, and one of the earliest settlers and most ial citizens of Ameliasburg. His father, so widely and popularly as Colonel Meyers, was a British subject born in the town of Hollen, ngdom of Hanover, in 1780, when that part of Germany was an x to the British Crown. He was a man of wealth and influence in l of his birth, having been highly educated, and brought with him dvent to Canada, in 1805, a large stock of goods, with which he set first store at Belleville, where he located, and which was called after eyer's Creek—a name which it retained for many years. He only d in "Meyers' Creek" about three years, removing to Trenton in here he engaged in the same pursuits as he had previously followed e Creek." During his short residence at the Creek he built a mill le other improvements, which formed the nucleus of the present ous and beautiful city of Belleville. Being a man of enterprise as of means, he did not relax his energies on removal to the Trent, but work building more mills, making roads, etc., etc., till the neigh- d soon emerged for its primeval state to that of a civilized com- He was one of young Canada's most distinguished representative a public life strong in his efforts to advance the general good; in pri- spitable and charitable—letting not his right know what his left g; in all things honourable and above suspicion. He was promoted vely from the ranks of the 1st Northumberland Militia to the posi- Ensign, Lieutenant, and Captain, and finally to the Lieutenant- y of the 2nd Northumberland, which rank he held till he died, on May, 1832, most deeply mourned.

son, Adam Henry Meyers, who was born in the township of in 1812, had the advantage of a liberal education—partly in Mon- l partly in New York—and studied law in the offices of the late Kirkpatrick, Q.C., and afterwards with the Hon. Mr. Baldwin. been called to the Bar in 1834, he commenced the practice of law on, and continued to practise there with enviable success till the his death in 1876.

n early life Mr. Meyers took an active part in public and political He was a staunch and honest Tory of the old school, and as such great influence in favour of John A. Macdonald, the Conservative whose warm personal friend he was, as well as zealous political e unsuccessfully contested Northumberland in the Conservative for the Assembly in 1842, against George M. Boswell, now Senior the County Court of Northumberland and Durham. He was sub- y more successful, and represented that constituency in Parliament mber of years. To the many positions of public trust and confi- hich he filled, was added that of Lieutenant-Colonel of the 5th berland Militia. He was a man of superior parts, whose influence many years successfully exerted to the benefit of the Midland District, and his memory will have a green spot in the hearts of the whole Bay Quinte region for many a long day to come.

Charles Francis, Barrister and Attorney-at-Law, was born in the vil lage of Trenton on the 22nd of February, 1837, and is of Irish extraction. He matriculated before the Law Society in Hilary term, on the 10th of February, 1857, and entered immediately on his studies, and was called to the Bar in Hilary term, on the 11th of March, 1862. He has since prac- tised law at Trenton, and is at present the senior member of the popular law firm of Francis & Forbes. He was elected to represent his native vil- lage in the County Council of the County of Hastings in 1878, and is the present representative of Trenton in that deliberative body, a position which his intellectual capacities and educational attainments eminently qualify him to fill with credit to himself and constituents. He is also an active member of the Church of England. Mr. Francis has a lucrative practice, and his fine talents and correct business habits have placed him among the leading members of the Bar of Hastings County.

Selah Sarle is one of the pioneer settlers in the rear of Sidney, and is a descendant of a United Empire Loyalist family who came to Canada after the close of the Revolutionary War. He owns a well cultivated farm and a beautiful residence, built of stone, the whole of which is surrounded by an elaborately finished and rustic verandah, adding much to its appearance and comfort. The Sarle family are highly respected by the community in which they reside, and throughout the township of Sidney.

J. O'Connor, the subject of this sketch, was born in the United States, and is of Irish extraction. He has by his industry acquired a valuable pro- perty and comfortable home, where he now resides, a view of which is represented elsewhere in this work. Mr. O'Connor is held in high esteem by his acquaintances, and is regarded as one of the most successful and pro- gressive young men in the neighbourhood—an example well worthy of emulation by the rising generation of the township of Sidney.

Baltis Rose is one of Sidney's representative men, and has for many years held the position of Reeve of the Township, an office he fills with ability, discharging the many duties connected therewith in a manner entirely satisfactory to his constituents, as is attested by his repeated re- election. He possesses an excellent farm in a high state of cultivation, and a comfortable and substantial dwelling house, with well appointed outbuild- ings. The Rose family are well and favourably known throughout the township of Sidney, and were among its first settlers, and own a number of farms of very fine land. Baltis Rose, the present Reeve, is a cultivated and public spirited citizen, and is always prominently identified with any movement or enterprise calculated to improve the condition of the people.

Thomas D. Appleby, J. P., the father of the subject of this sketch, was born in Duchess County, New York, in the year 1778, and came to Canada in 1780 with a party of emigrants to Kingston. He shortly afterwards located in Sophiasburgh, in Prince Edward Co., where he married a daughter of Nathaniel Solmes, and settled down to the life of a farmer, subsequently removing to Tyendinaga, and was one of the pioneer settlers of that township. He held for several years many positions of public trust in the township. He was a Justice of the Peace, Commissioner of the Court of Requests, Boundary Line Commissioner, &c., and died in 1865, in the 86th year of his age. Anna, his eldest and only living daughter, married Richard Lazier, in 1824, and is mother of Thomas Appleby Lazier, Junior Judge of the Co. of Hastings; of Samuel S. Lazier, Master in Chan- cery for the Counties of Hastings, Prince Edward, Lennox and Addington, and of Richard L. Lazier, J. P. and Collector of Customs at the Port of Shannonville.

Nathaniel S. Appleby, the eldest and only living son of Thomas D. Appleby, was born in the County of Prince Edward in the year 1820, and came with his father to Tyendinaga in 1828, and was elected Reeve of the township for twenty consecutive years, and was appointed a justice of the peace in 1854, Census Commissioner for Hastings in 1861 and 1871; Warden of the county for three years, and elected M.P.P. for East Hastings, in 1875. Mr. Appleby is at present largely engaged in milling and farming, and is the only active magistrate in that part of the county. The Appleby family are originally from Westmoreland County, England, and came to the United States previous to the Revolution. After the close of the war the late Thos. D. Appleby emigrated to Canada with a number of other U. E. Loyalists, in consequence of their firm attachment to British laws and rule, preferring the wild and rugged forests of Canada with laws and institutions they loved, to the fair and familiar homes on the banks of the Hudson and Mohawk with a form of government they detested, and they have never regretted the choice they have made. N. S. Appleby, M.P.P., is one of the most prominent men in the history of this section, in every effort for the improvement of its people or the advancement of its material interests; of unblemished integrity and unassuming manners, he is honored and respected by a host of friends.

Richard L. Lazier, Collector of Customs, was born in the village of Shannonville, in Tyendinaga, and is a son of the late Richard Lazier who came from Prince Edward County to the Shannon in 1828, and was one of the pioneer settlers of the township, his mother being a daughter of the late Thos. D. Appleby, a U. E. Loyalist, also originally from Prince Edward Co., and an early settler in Tyendinaga. The Lazier and Appleby families after the close of the war of the Revolution, in order to escape the persecu- tions of the arrogant continentals and enjoy the freedom of British rule and British institutions, came to Canada and settled in Prince Edward Co., their sons afterwards settling, as already stated, in Tyendinaga. The Lazier family are of Dutch descent, their forefathers having originally come from Flanders, near the French border, at an early date. Jacobus Lazier, the great great grandfather of Richard L. Lazier, was born in Yonkers, New York, in 1707, and died there in 1792. His descendants continued to re- side in that vicinity until the breaking out of the Revolutionary war when the family, as above mentioned, emigrated to Canada. Mr. Lazier was ap- pointed Collector of Customs at the Port of Shannonville in the year 1871, a position he still retains. He is also engaged in the milling business, owning the large flouring mills at Milltown. Mr. Lazier is highly esteemed by all who know him, and is distinguished for his many virtues and digni- fied propriety of conduct.

John White, M.P. for the East Riding of the County of Hastings, is a native of the County of Donegal, Ireland, and came to Canada when but a boy, in the year 1850. His first introduction to the people of Hastings was in the capacity of chore boy amongst the farmers of Thurlow, with whom he lived for some years. He is an eminent example of the many self-made men of our times. Removing to Tyendinaga when still a youth, with but a limited education, yet possessing an unflinching determination and indomit- able will, coupled with integrity and native tact, he soon attained to a lead- ing position in society and politics. He was elected Reeve of Tyendinaga in 1869, a position, with the exception of 1873-4, he has ever since occu- pied. In 1871 he was elected, in the Conservative interest, to the Do- minion Parliament; re-elected in 1872 and 1874, and at present represents this constituency in the House of Commons of the Dominion. Mr. White has always been prominently identified with the Orange Society, and is at present Right Worthy Grand Master of Ontario East, and has filled many subordinate positions in the same order. In addition to his many Parlia- mentary and municipal duties Mr. White is engaged in building and farm- ing. He has erected two large cheese factories, one on lot No. 4 in the 9th, near his residence, and another on lot 20 in the 8th concession of Tyendinaga, and is also interested in an iron foundry at Madoc. He is regarded as one of the leading and most influential men in the county.

JAMES FORRESTER is one of the pioneer settlers of the township. He emigrated with his father, D. Forrester, from Forfarshire, Scotland, when quite a youth, and purchased from the Government lot No. 20 in the 3rd concession of Tyendinaga, which he cleared, and where he now, with his family, resides. Mr. Forrester at an early age became converted to Christ, and being of a retiring nature has never striven to make himself popular but has performed his several duties to God and man unostentatiously and quietly. For upwards of forty years he has been connected with the Presbyterian Sabbath school, and was the first superintendent of that very beneficial and instructive adjunct of the Church of which he is an active and consistent member. An example of untiring industry and integrity in all the relations of life, he is held in high estimation by the community in which he has lived for so many years.

THOMAS CLARE.—The subject of this sketch was born in the county of Wicklow, Ireland, on the 8th of January, 1802, and came to Canada in 1815, and first settled in Sidney. He assisted Samuel Benson to survey and lay out the township of Hungerford. On the completion of the survey he purchased and built upon lot No 1, in the 4th concession—the site of the present village of Thomasburg—which he cleared and where he resided until his death, which occurred in May, 1869. Mr. Clare was one of the pioneers of Hungerford, there being only six other settlers in the township at the time he located and settled upon his land. His house was ever open to the early settlers, who often came in without means and with but an imperfect knowledge of the country or the whereabouts of their grants. He was accustomed, without remuneration, to pilot them through the forest to their several new homes. Mr. Clare was always an active man in the affairs of his township; all measures and movements contributing to its prosperity or the welfare of his fellowmen always received his hearty support, and found in him a most efficient aid. Of the strictest integrity, and with an uncompromising sense of right, he was called to many positions of trust, which he faithfully administered, and retained through all the confidence of his neighbours.

ROBERT GORDON, a son of Robert and Jane Gordon, was born at Maguire's Bridge, in the county of Fermanagh, Ireland, on the 18th day of August, 1828. He came out to Canada in 1846 and purchased a portion of lot No. 4, in the 11th concession of Hungerford, containing 160 acres, which he cleared and which is at present in an excellent state of cultivation. He was elected Reeve in 1877 and re-elected for 1878, a position he fills with ability. He is also President of the Victoria Cheese Factory, and county master of the Orange order for Hastings. Early in life he allied himself with the B. C. Church, of which he is a consistent member, and has for many years officiated in that body in the capacity of local preacher. Liberal, public spirited, and always awake to the needs of society, no measure for the public good or movement for the advance of moral or religious interests fails to receive his support or the aid of his best efforts, and an appreciative community have rewarded him with positions of trust.

C. POMEROY, M.A., M.D., of Tweed, is a son of the late M. Pomeroy, a U. E. Loyalist from Vermont, who distinguished himself during the war of 1812, and who settled in Addington county at an early period in the history of our country. His mother, as the daughter of a U. E. Loyalist, drew lot No. 10 in the 7th concession, adjoining the present village of Tweed. Dr. Pomeroy removed from Addington to Tweed in 1851, when this part of the country was in its infancy. He is a finished scholar, being a University graduate and gold medalist, and has been coroner for the county of Hastings for upwards of 22 years, and was appointed in 1857 surgeon of the 4th battalion of Hastings militia, a position he still holds.

In order to acquire a more thorough knowledge of surgery, Dr. Pomeroy entered the U. S. army as surgeon, and was present on the field at many of the principal battles during the late civil war. He has a large and extensive practice, and is looked upon both in medical circles and throughout this section of the country as one of the most skilful practioners in this part of the Province. He owns a farm and a beautiful residence with well appointed out-buildings, surrounded by a tastefully arranged lawn and garden. Dr. Pomeroy is one of the most uncompromising temperance men of his day, both by precept and example, and is a recognized leader of all temperance movements, never refusing aid to its earnest workers. Well-informed and highly cultured, with rare conversational powers, his hospitality is richly enjoyed by any who may be so fortunate as to have the pleasure of his acquaintance.

SIDNEY WAY.—The subject of this sketch was born in Prince Edward Co., in the year 1825. He is a son of the late John R. Way, one of those devoted adherents to his king and the British constitution, who preferred hardships and poverty in the wilds of Canada to riches and plenty in the
land which had revolted from its allegiance to its rightful rul
land in Prince Edward Co. where he lived with his family fo
years. He afterwards removed to Hungerford and purchased
the 7th concession, upon which he resided until his death, wh
in 1837, and which was a loss to the public; a vacancy to be
created in his family, the Church, and the community. Sid
posed of the homestead to the Clare family, and purchased fr
Kerr part of lot No 7, in the 6th concession, where he with
present reside. Mr. Way owns a pleasant home, delightf
and 150 acres of choice land in the highest state of cultivatio
for location and natural productiveness, cannot be equalled in t
He is also, in addition to farming, largely interested in the m
cheese, and is one of the principal stockholders and patrons
Factory. Liberal, large-hearted, and generous, and of a genia
he is universally respected by his neighbours and acquaintance

D. ROBLIN was born in the township of Sidney, in the y
settled in Hungerford in 1838. Here he purchased part of lot
7th concession, which he cleared and where he at present
Roblin is of Welsh extraction, his ancestors emigrating fr
the United States early in the eighteenth century. On the re
provinces and the commencement of active hostilities in 1775
allied themselves with the British forces and were actively e
many sanguinary conflicts of that period. At the close of
branches of the family came from Jersey and Vermont to
settled originally in Adolphustown; the father of the subject
moir, afterwards removed to Sidney, and as a U. E. Loyalist
in the 3rd concession. He also took an active part in the
having held a captain's commission in the Hastings militia.
owns a comfortable dwelling house and an unusually producti
which is situated the noted "Roblin" cheese factory. An inv
spring supplies the factory with an unlimited quantity of pu
an important requisite in connection with the manufacture of
Roblin looks well after his farm and factory, and is regarded
trious, honourable, and highly intelligent citizen, enjoying t
respect and esteem of his neighbours and acquaintances.

ABRAHAM L. BOGART is a descendant of one of the oldest D
on this continent. As early as 1689 we find one Jacob Va
with others, petitioning the good people of Albany, and later,
name of Bogart appears in connection with the census returns
city. The Bogarts came from Jappan town, near New York.
the conclusion of the war, with a number of other U. E. L
settled on the 4th concession of Adolphustown, where many d
the family still reside, his mother living to the great age o
Abraham L. Bogart, the subject of this sketch, was born in A
and has resided in Belleville for a number of years, removin
ago to Hungerford, where he owns a large tract of valuable
principally engaged in building and farming, and erected som
considerable expense a saw mill on the splendid water priv
creek, where large quantities of raw material have been pre
market. Bogart P. O. is situated at this point, lot 20, in the
sion, and was named in honour of the founder of the place.
has, within this last twenty years, converted this part of the t
an uncultivated wilderness into what seems destined to beco
tant day one of the most flourishing and productive sections o
The public have on several occasions shown their just appreci
Bogart's abilities by electing him to important and responsi
which he has ever filled to the satisfaction of his constitue
credit to himself.

JAMES JAMIESON.—The subject of the following sketch was
wick on the Tweed, August 13th, 1808. At an early age he ca
with his parents and settled in Belleville, where he resided pri
within a few years of his death, when he removed with his
village of Tweed, where he died at the age of 62. Fond of m
of an enterprising disposition, he built the foundry in Bellevi
the Victoria Foundry, which he subsequently traded for lan
power in the township of Hungerford, where he greatly impro
power and laid out the present village of Tweed, which he nam
of his birth place. He also built a grist mill in the township
where he shortly resided, which mill was subsequently burned
years he acted in the capacity of county surveyor, but steadil
municipal honours tendered him. . . . Of a quie
disposition, his death was deeply regretted by a large circle of a
who valued his sterling qualities as a man and a Christian.

BARNABAS VANKLEECK was born in 1803, in the townshi
bury, in the then County of Prescott, and is of Dutch descent
nal ancestors were forced, on account of their loyalty to the B

GEOGRAPHICAL DESCRIPTION.

Prince Edward is one of the forty-two counties of Upper Canada, and is comprised of an irregular shaped peninsula lying to the south of the counties of Northumberland, Hastings, and Lennox on the mainland, connected with the former by the narrow isthmus called the Carrying-place; and separated from the two latter by the Bay Quinte, which, with its numerous small bays and inlets, form its northern boundary; the waters of Lake Ontario, with the various bays which are a part thereof, surrounding the peninsula on the other three sides. Its area is 241,500 acres, valued at $6,855,0[illegible], and inhabited by 18,[illegible] souls.

TOPOGRAPHICAL AND GEOLOGICAL FORMATION.

The topographical and geological formation of Prince Edward will be found noted with sufficient fulness under head of the various minor municipalities of which it is composed, to render anything further in that connection here unnecessary, except merely to remark that the former is singularly *regular*, without being anywhere, or to any great extent, very *even*; while the latter is chiefly of a calcareous nature—in places rocky to the surface, in others possessing a rich deep soil, and interspersed here and there with scattered evidences of the glacial period, which prove it at one time to have been covered with water, and to have received these reminders of a northern region from the droppings of melting icebergs.

LAKES AND STREAMS.

Of lakes and rivers Prince Edward has but few, and those of comparatively insignificant size. It were not possible otherwise from its geographical shape and dimensions; for though the extreme length of the peninsula is about sixty miles, it is in no place one-third of the distance in breadth, often but five to ten miles, and very considerable parts of it not more than two from water to water.

Three of the chief lakes, *viz.*, Lake of the Mountain, Fish Lake, and Roblin's Lake, in the townships of Marysburg, Sophiasburg, and Ameliasburg respectively, will be found more fully described in the sket
several townships. East and West Lakes, the former in Athol
in Hallowell, are also noted in their proper connection, tho
more accurately but parts of Lake Ontario, as Weller's Bay,
the Carrying-place, and Pleasant Bay in the township of Hilli

The largest lake in the county is Consecon Lake, whose na
from an Indian word meaning pickerel, a variety of fish which
fairly thickened its waters when the first white settlers cam
This body of water is several miles in extent, north-easterly a
terly, and is situated between the townships of Ameliasburg
and emptied by the river of the same name—the only consid
within the limits of the county—at the mouth of which, upon
the flourishing village of Consecon is situated.

Mills of various capacity are driven by the streams whi
other above-mentioned lakes; also at Bloomfield, in Hallowell
running into West Lake, and by the Black River, at Milford.
days of the county, and until the greater portion of the land w
became cleared, these would all seem to have been streams of
time, though they are now of very inconsiderable magnitude.

PECULIARITIES.

The above-mentioned peculiarities in regard to lakes and st
no means the only ones resulting from the singular and ex
graphical characteristics of Prince Edward,—possessing mis
nowhere else in the country exist. For instance, the coast line
great in proportion to the area,—and there being so many spl
harbours all along the shore, have been the means of giving t
this strictly farming community, greater advantages as to
marketing their crops, than can be found in any other part of
America. There is probably not a single farmer's barn in the
at a greater distance than seven or eight miles from a good wh
house; while the great majority have a half dozen such to
within less than half that distance.

Again, the crops raised throughout the district are of a cla
find market during the season of navigation. On account of

ed from its allegiance to its rightful ruler. He drew
Co. where he lived with his family for a number of
removed to Hungerford and purchased lot No. 7, in
on which he resided until his death, which occurred
s a loss to the public ; a vacancy to be long felt was
he Church, and the community. Sidney Way dis-
l to the Clare family, and purchased from Mattaniah
in the 6th concession, where he with his family at
Way owns a pleasant home, delightfully situated,
land in the highest state of cultivation, and which,
d productiveness, cannot be equalled in the township.
to farming, largely interested in the manufacture of
he principal stockholders and patrons of the Roblin
ge-hearted, and generous, and of a genial disposition,
cted by his neighbours and acquaintances.

n in the township of Sidney, in the year 1814, and
in 1838. Here he purchased part of lot No. 7, in the
he cleared and where he at present resides. Mr.
xtraction, his ancestors emigrating from Wales to
y in the eighteenth century. On the revolting of the
mencement of active hostilities in 1775, the Roblins
the British forces and were actively engaged in the
flicts of that period. At the close of this war four
ly came from Jersey and Vermont to Canada and
Adolphustown: the father of the subject of this me-
oved to Sidney, and as a U. E. Loyalist drew lot 33,
He also took an active part in the war of 1812,
a's commission in the Hastings militia. Mr. Roblin
velling house and an unusually productive farm, upon
noted "Roblin" cheese factory. An invaluable living
tory with an unlimited quantity of pure cold water,
in connection with the manufacture of cheese. Mr.
his farm and factory, and is regarded as an indus-
d highly intelligent citizen, enjoying the universal
his neighbours and acquaintances.

rr is a descendant of one of the oldest Dutch families
s early as 1689 we find one Jacob Vaulen Bogart,
ng the good people of Albany, and later, 1750-60, the
s in connection with the census returns of New York
me from Jappan town, near New York, to Canada on
war, with a number of other U. E. Loyalists, and
cession of Adolphustown, where many descendants of
, his mother living to the great age of 102 years.
he subject of this sketch, was born in Adolphustown,
elleville for a number of years, removing some years
here he owns a large tract of valuable land. He is
building and farming, and erected some time ago at
a saw mill on the splendid water privilege of ——
antities of raw material have been prepared for the
is situated at this point, lot 20, in the 10th conces-
in honour of the founder of the place. Mr. Bogart
wenty years, converted this part of the township from
rness into what seems destined to become at no dis-
ost flourishing and productive sections of the county.
veral occasions shown their just appreciation of Mr.
lecting him to important and responsible positions,
ed to the satisfaction of his constituents and with

-The subject of the following sketch was born in Ber-
ugust 13th, 1808. At an early age he came to Canada
ettled in Belleville, where he resided principally until
his death, when he removed with his family to the
re he died at the age of 62. Fond of mechanics and
osition, he built the foundry in Belleville known as
which he subsequently traded for lands and water
of Hungerford, where he greatly improved the water
present village of Tweed, which he named in memory
e also built a grist mill in the township of Thurlow
ed, which mill was subsequently burned. For many
capacity of county surveyor, but steadily refused all
ffered him. . . . . Of a quiet and retiring
was deeply regretted by a large circle of acquaintances,
g qualities as a man and a Christian.

eck was born in 1803, in the township of Hawks-
nty of Prescott, and is of Dutch descent. His pater-
ced, on account of their loyalty to the British crown,
to leave their homes in Duchess County, New York, joining the Royalist party, and taking an active part in all the stirring times of that period. At the close of the revolutionary war the Vankleecks followed the British army to Nova Scotia, and from thence removed to Upper Canada about the year 1800. The place where the Vankleecks first settled in Hawksbury, still bears the name of Vankleeck's Hill. Simon Vankleeck, the father of the subject of this sketch, was one of the earliest pioneers of this country. He held a commission in the militia signed by Lord Dorchester, then Governor of Quebec, and was in the Commission of the Peace. He was well and favourably known throughout the county, dying at his son's residence, Madoc, in his 98th year, respected by all who knew him. Barnabas Vankleeck came to Madoc in 1841, and cleared the farm upon which he resides. He lives with his son, Peter Vankleeck, the present Reeve of Madoc. He held a captaincy in the Hastings County Militia, and is a Justice of the Peace, in which capacity he has always acted without fee or remuneration, and enjoys the reputation of being the best read man in North Hastings. In politics he is an active and staunch reformer. A respected old gentleman of seventy-five years, he commands the universal esteem of a wide circle of friends.

A. F. Wood, J.P., is a son of Thos. H. Wood, Esq., who came to Canada from Saratoga County, New York, in 1810, and took part in the war of 1812. He is still living, is 88 years of age, and draws a pension for his services. Mr. A. F. Wood, whose portrait appears on another page of this work, is of English descent on his father's side, and Scotch on his mother's, and was born in 1828. The early part of his life was spent in the township of Fredericksburg, on the Bay of Quinté. He came with his father, in 1843, to this county, and to Madoc village in 1853, where he has since resided. In 1862 he built a large flouring-mill in the village, on Deer Creek ; and was elected Reeve of Madoc in 1858, filling that office with ability for nineteen years. He was elected Warden of the County of Hastings in 1864, and filled that position for ten years. He is one of the board of directors of the B. & N. H. R. R., and was also prominently connected with the Grand Junction R.R. Mr. Wood's family are noted for their activity and intelligence in public matters, the Hon. S. C. Wood, Provincial Treasurer, being a brother ; also Dr. Wood, of Ottawa. His intelligence in municipal affairs of the township and county, has placed him in the front rank of municipal legislators, and he is considered one of the best authorities of municipal law in the county. As chairman of the Public School Board, and being a public-spirited citizen, he takes a deep interest in the cause of liberal education ; believing, that on the intelligence and virtue of the people, the safety and welfare of the Dominion rests.

E. D. O'Flynn, J.P., a view of whose residence appears in this work, was born in the County of Lennox in 1831, and is of Irish descent on his father's side, who came from the County of Waterford, Ireland, in 1818, and Scotch on his mother's. Mr. O'Flynn came to Madoc in 1842, then a wilderness, and has seen every building go up in the village. He commenced the dry-goods trade in 1851 without capital, and has built up the largest business in Madoc. Being a prominent and energetic citizen, he has, by careful management, added largely to the improvement of the village. He is in the village council and School Board, and is at present contesting the North Riding of Hastings in the Reform interest for the Commons. He has been an active and consistent member of the Methodist Church for upwards of twenty-five years, and is respected by all his neighbours and acquaintances.

G. D. Rawe is a native of the town of Portsmouth, England, and came to Canada in 1858 and settled in Madoc. Here he commenced the jewellery and watch-making business, which he has successfully carried on for a period of twenty years. His stock of jewellery and silver-ware would do credit to any city establishment. In 1866 he organized No. 4 Company, 49th Regiment, retaining the command of it until 1875, when he retired with the rank of major. He was appointed, in 1866, Clerk of the 6th Division Court, a position which he still holds. Mr. Rawe is well and favourably known in Madoc and throughout the county, and is regarded as a man of character and a useful member of society.

John R. Ketcheson, J.P., Township Clerk, Registrar of Vital Statistics etc., was born in Sidney, and is a son of Col. Wm. Ketcheson, one of the pioneer settlers of the township, a biographical sketch of whom will be found on another page of this work. John R. Ketcheson, when a young man, moved back into Madoc, and settled upon lot No. 4 in the 7th concession, which he converted from an uncultivated wilderness into one of the finest and most productive farms in the county. He was early appointed on the Commission of the Peace ; he has always been prominently identified with public affairs and every enterprise calculated to advance the interests of his municipality, or improve the condition of the people. He has been honoured with many positions of public trust, and on the erection of Madoc into a separate and independent municipality he was chosen its first Reeve. He has held the position of Township Clerk for many years, an office he discharges with credit to himself and advantage to the municipality.

E EDWARD COUNTY.

be found more fully described in the sketches of those
ast and West Lakes, the former in Athol and the latter
noted in their proper connection, though they are
arts of Lake Ontario, as Weller's Bay, south-west of
d Pleasant Bay in the township of Hillier.
the county is Consecon Lake, whose name is derived
meaning pickerel, a variety of fish which is said to have
aters when the first white settlers came to its shores.
several miles in extent, north-easterly and south-wes-
between the townships of Ameliasburg and Hillier,
ver of the same name—the only considerable stream
e county—at the mouth of which, upon Weller's Bay,
of Consecon is situated.
apacity are driven by the streams which empty the
l lakes ; also at Bloomfield, in Hallowell, by the creek
ke, and by the Black River, at Milford. In the early
d until the greater portion of the land within its limits
would all seem to have been streams of no mean vol-
now of very inconsiderable magnitude.

PECULIARITIES.

ed peculiarities in regard to lakes and streams, are by
es resulting from the singular and exceptional geo-
es of Prince Edward,—possessing many such which
untry exist. For instance, the coast line being so very
the area,—and there being so many splendid natural
shore, have been the means of giving the farmers in
community, greater advantages as to shipping and
than can be found in any other part of Canada or of
obably not a single farmer's barn in the whole county
han seven or eight miles from a good wharf and store-
at majority have a half dozen such to choose from
hat distance.
ised throughout the district are of a class that chiefly
e season of navigation. On account of the chemical
composition of the soil, and the limestone foundation which underlies its surface, the land about Bay Quinte seems more peculiarly adapted to the growth of barley than any other section of America. Crops of that staple are produced which yield fair returns as to quantity, and for quality defy competition ; the "Bay barley" always being, without exception, the highest quotation in the American barley centres.

Another peculiarity is, that although one of the first settled counties of Upper Canada, and one which has made most rapid strides in very many directions which tend toward the higher civilization of the age, yet they have never had a railroad, the one thing of all others now considered the most necessary adjunct to *settlement*, to say nothing of the requirements and attributes of a highly enlightened state. This, of course, is but another natural result of its position, isolated, probably, but still giving its inhabitants in turn many compensating benefits in other directions.

THE HIGHWAYS.

It has been asserted, and wisely so, that the avenues of communication are an undoubted evidence of the state of society. The history of this planet from its earliest days furnishes indisputable proof of this now universally admitted truth. As civilization progresses, intercommunication increases, and the channels of trade are improved ; while the conveyance of products and the movements of armies require an unobstructed highway.

Of the eastern nations who comprehended the truth of this great principle, the chief were the Romans, whose broad stone-ways and ruined arches still survive—if a ruin can be said to survive—to remind us of the former power and greatness of those ancient masters of the world ; while in the Western Hemisphere, Mexican causeways and Peruvian stone roads attest the vigor of a national life for centuries departed ; and whatever remains is upon a scale immense and enduring, indicative of indefinite periods of construction and the employment of masses of population.

But the trail across the Carrying-place, and the birch-bark canoe upon the Bay—ample for the aborigines of Prince Edward, and withal equal to their capacity,—have given place in turn to a network of highways which, if not comparable to the military roads of the Romans or ancient Mexicans, are at least equal to the requirements of a highly civilized people, to whom the arts and sciences are as familiar as were the shield and javelin to the ancient warriors for whose benefit those stone-ways were built; or the rude stone tomahawk and flint-pointed barb to the painted savage who traversed from time immemorial the forest trails of the peninsula. And although its inhabitants have yet no railway, they have a greater proportionate length of excellent carriage roads, and a smaller proportionate length of poor or indifferent ones, than any other territory of similar extent in the Dominion.

Governor Simcoe's celebrated military road, connecting the east and the west, runs through almost its entire length. A full description of this, commonly known as the Danforth Road, is elsewhere given.

To be brief,—a drive through the county, in any or all directions, is to be delighted with it; not simply the excellent carriage roads and pleasantly shaded avenues, but the comfortable homes, beautiful groves, romantic lakes, well kept farms, fine schools, handsome churches, and general air of thrift and prosperity, and of a refined and highly moral sentiment of the people which pleases even the infidel, and fills with joy the christian traveller. In short, we doubt—notwithstanding the comparative absence of manufactures, and the entire want of such public works as generally go hand in hand with a community's wealth and greatness,—whether a section can be pointed to within this broad Dominion, bounded by three oceans, embracing half a continent, and stretching across two zones,—where an area containing an equal population, or a population contained within an equal area, can be found possessing in a greater degree the elements of

MATERIAL PROSPERITY

And genuine rural felicity than can here be seen. As is well known, Prince Edward is pre-eminently an agricultural county; and, compared in size and population, there are among its inhabitants a greater number of comfortably situated *owners* of the soil they till, than in any other section we have ever visited. It is not claimed for it that its progress has been as rapid as some other parts of the country in population or wealth, but a perusal of official statistics proves that in educational advantages and acquirements—which is the true foundation of a nation's greatness,—this county is second to none, if not, indeed, the very first, in Ontario, the banner Province of the Dominion. Figures to substantiate this assertion may be found in another place.

On all hands is the most convincing evidence of the existence of high moral principles and a sense of religious duty on the part of the people. The eye can at almost any time rest upon one or more church steeples. This was almost the first great field of labour of the Methodists in Canada; and these devoted servants of the Master were the first to find their way to the log-cabins of the pioneers, and preach the everlasting truths of the Gospel to the early inhabitants.

Being what might be termed an "aggressive" church, the Methodists have let go no hold, but pushed their conquests further and further as pioneer settlements became thriving communities, and straggling hamlets grew into prosperous villages and busy towns. This religious denomination is now by far the most numerous in the county, as well as throughout the Bay Quinté District, and have literally covered the land with beautiful churches, while the other Protestant denominations follow close behind. The Roman Catholics are not so numerous, though their comparative increase has been rapid and satisfactory.

To sum up its material advantages in a very few words, we may truthfully say that as to varied and delightful scenery, magnificent roads, pleasant drives, interesting natural and historical landmarks, and an intelligent, refined, and hospitable people, Prince Edward beyond question claims a foremost position; while in everything which tends to make a country prosperous, its people contented with their lot, and others contented with them, it occupies no second place.

EARLY HISTORY.

Age has succeeded age, and centuries have ripened into cycles of time, since first the highlands of Prince Edward arose from out the depths. This seems as sure as the eternal truths of science itself.

That Prince Edward was also inhabited by a pre-historic race, possessed even of many of the attributes of what we now term a high degree of enlightenment, is also quite probable; as it is certainly true of many other portions of the North American continent, which must have been brought out of chaos at the same time and under the same force of circumstances.

Who were those strange people? Whence came they? and whither did they go? These questions must remain to form a melancholy interest in the wondrous past, and a mystery which time, nor circumstance, nor science, nor the more wondrous future, may unravel.

But since their time another race has come and gone—gone from their ancient homes and hunting-grounds, though not yet quite extinct. Whether these latest aborigines (if the term be allowable), came by migration from the north-west, across Behring's Straits, as some scientists assert is proven by tradition, legend, and geography combined; or whether, as is stoutly affirmed by others, who bring forward quite similar arguments to substantiate their theory, they are descended from the Norsemen, whose inclination or necessities brought them to the south-west, is matter which no amount of research can to a certainty establish, and will forever be a subject of deepest conjecture. Nor is it matter of serious concern, so far as affects our local history, how comes it that the Indians, with whom more recent discoveries prove the bulk of the North American continent, and more particularly Canada, to have been peopled, are of a different race, with characteristics in direct antagonism to those of the more southern tribes described by Columbus, Cortez, and Pizarro. The former found the West Indies populated by a peaceful, pleasant people, ruled over by Caciques, enjoying existence, and knowing naught of war-cry, stake, or crimson trophy. Cortez found the Aztecs residents of cities, advanced in the arts of civilization, builders of causeways, dwellings and temples, and tillers of the soil; and it is a question time ne'er can solve, whether, left to themselves, the Mexican and Peruvian were not types of civilization which in time would have emulated Eastern stages of progression, had not a higher order of intellect crushed out the rising national instinct, and implanted its germs upon the ruins.

Undoubtedly the first white man who planted his foot upon the Prince Edward peninsula was Samuel Champlain, one of a company of French traders who set out for the New World in 1603, with the primary object of exploring the St. Lawrence with a view to establishing a depôt for the fur trade upon its banks.

At that time the territory south of the St. Lawrence and Lake Ontario was occupied by the Iroquois, so called by the French, though afterwards by the English the "Five Nations," a confederacy of the Cayugas, Mohawks, Onondagas, Oneidas, and Senecas, who were subsequently joined by the Tuscarowas, whom they adopted, and were thenceforward and to this day known as the "Six Nations." This confederation, then and for many succeeding generations, formed the most powerful savage nation on the American continent. At the same time the country to the north of the waters named, was peopled by the Algonquins, Hurons, and Ottawas, so called by themselves, but by the French named Northern Iroquois; between whom and the Five Nations of the south there were evidently common points of ancestry at no very remote period, as well as an eternal enmity, which, but for the existence of the chain of waters separating them, would have long since resulted in the extermination of the one or the other.

Students of character have denominated these Iroquois the "Romans of the western world," and—considered either from the extent of their conquests, the wisdom and eloquence of their chiefs, their impatience of control, their

treatment of the vanquished, or their passio
taken.

The advent of the French under Champla
the opportunity of securing a most useful all
ern foes. Anxious to please them, and amb
French arms throughout the continent, he
leaving part of his force at the present site of
panied the warriors of these tribes with the
River, to the lake whose name still recalls re
coverer. On its shores the Northern and S
battle, in which the former, through the aid
ious, after which they returned to their home

Subsequently to this, Champlain erected
Montreal now stands, and from thence plann
in as many directions, partly of a commercia
ter. One of these was up the Ottawa River
thence up that stream towards its source, and
head-waters from the head-waters of Lake
lake and down French River to the shore of
southward to the mouth of the Severn, up w
and rapids; thence through Lake Conchiching
and up the Talbot River to the Balsam Lak
Balsam Lake, and through the numerous cha
past the sites of the present towns of Fenelo
Keene, Campbellford, and down the Trent
waters he gazed in admiration,—the first of
from here was *via* Picton Bay and across the E
his party crossed Lake Ontario in their cano
were severely beaten in that locality by the F
escaping, twice wounded, across the lake wit
allies, to find refuge for the winter of 1615-16 o

It should be in this connection that Cham
first instance, to explore; but subsequently
occasion, as the ally of the Northern Iroquois
Thus was this French adventurer, at this
manner, the discoverer of the Bay Quinté an
his race who set foot upon the soil which divid

Subsequently it was no doubt frequently t
and fur-traders; but it was not till 1783 that
was made by Mr. Weese, on the north shore o
well as the successive settlements, in their reg
the county, are sufficiently detailed in the sk
themselves, together with the time and manne
bered, named, and surveyed. We will therefo

POLITICAL HIST

Which may be said to have commenced with
Simcoe, dated at Kingston, July 16th, 1792, w
Upper Canada into nineteen counties for elec
comprised the counties of Glengarry, Stormo
Frontenac, Ontario, Addington, Lennox, Prin
berland, Durham, York, Lincoln, Norfolk, Su
order named. Ontario, as then constituted, c
south of the mainland, and between Prince Ed
The proclamation commenced with the usual pr
sought and the causes therefor; and contained
clauses, and was signed by William Jarvis, Gov
signed by the Governor with his initials only,
of the instrument precludes the advisability of
give the part referring to this county, as follows

"That the tenth of the said counties is to b
of the County of Prince Edward; which County
by Lake Ontario; on the west by the Carrying
Presque Isle de Quinté; on the north by the B
from Point Pleasant to Point Traverse, by its s
ing the late townships of Marysburg, Sophias
said County of Prince Edward to comprehend
Ontario and Bay of Quinté nearest to the said c
part fronting the same."

The Proclamation also gave an approximate
Glengarry had two members, a part of Lincoln
of it, grouped with counties on either side, ret
counties returned one each; while in a number
or parts of counties, were grouped together to se
The township of Adolphustown, in Lennox, wa
Edward, the clause referring thereto being as fo

"That the County of Prince Edward, bound
together with the district of the late Township o
Lennox, shall together send one representative
Prince Edward, together with the said district, la
shall and may be represented together in the sa
member."

The whole number provided for from the n
members; which number composed the Leg
Canada for many years.

The first election held under this proclamatio
of Hallowell, to which township the first elec
Besides Philip Dorland and Peter Van Alstine th
was early represented by Simeon Washburne,
Wilson. The latter served his county in that ca
a century, though not consecutively.

The next political change to note in the hist
ting-off of Prince Edward as a separate distri
justice, by Act of Parliament, in 1831. Mr.
county in Parliament, and introduced the Bill
desired change into effect. Heretofore the peo
Kingston for the transaction of all judicial b
building of the jail and court-house for the ne
thereof, may be found under the head of Picton.

Since that time Prince Edward has been a sepa
for judicial purposes; for many years a separate co
purposes in the Legislative Assembly; and, sinc
division returning one member to the House of C

Lt.-Col. Walter Ross now represents the co
Lt.-Col. Gideon Striker is the present sitting m
lature.

MUNICIPAL HIST

The oldest county or district record now in
the Surveyor of Highways." The subject matte
the connection indicated by its title. It is signed
of Highways," and endorsed as follows: "Appro
July, 1811. A. McLean, Clerk of the Peace."
uminous, and of no peculiar or general interest.

The earliest record of the deliberative procee
of the district, are minutes of the first District Co
following are extracts from the same:—

"Journal of the Municipal Council of Prince
8th, 1842.

ORICAL SKETCH OF THE

nt of the vanquished, or their passion for war—the comparison is well

advent of the French under Champlain afforded the Northern Iroquois portunity of securing a most useful ally to make war upon their south-. Anxious to please them, and ambitious to spread the terror of the arms throughout the continent, he acceded to their demands; and part of his force at the present site of Quebec to build a fort, he accom- the warriors of those tribes with the balance of his party, *vià* the Sorel to the lake whose name still recalls remembrances of its European dis-. On its shores the Northern and Southern Iroquois fought a severe in which the former, through the aid of their white allies, were victor- fter which they returned to their homes north of the St. Lawrence.

sequently to this, Champlain erected a fort and trading-post where al now stands, and from thence planned and executed many expeditions any directions, partly of a commercial and partly of a military charac- ne of these was up the Ottawa River to the mouth of the Matawan, up that stream towards its source, and across the portage separating its aters from the head-waters of Lake Nipissing; thence through that l down French River to the shore of Georgian Bay, which he followed ard to the mouth of the Severn, up which he passed, avoiding its falls ds; thence through Lake Couchiching, the Narrows, and Lake Simcoe, the Talbot River to the Balsam Lake portage; thence over that to Lake, and through the numerous chain of lakes and connecting rivers e sites of the present towns of Fenelon Falls, Bobcaygeon, Peterboro', Campbellford, and down the Trent to the Bay Quinté, upon whose he gazed in admiration,—the first of his race. His supposed course ere was *vià* Picton Bay and across the East Lake portage. At all events ty crossed Lake Ontario in their canoes to a point near Oswego, and verely beaten in that locality by the Five Nations, Champlain himself g, twice wounded, across the lake with the remnant of his northern to find refuge for the winter of 1615-16 on the Prince Edward peninsula. doubt be in this connection that Champlain ascended the Ottawa in the stance, to explore; but subsequently turned south, as on a previous n, as the ally of the Northern Iroquois against their southern enemy. was this French adventurer, at this early day, and in this peculiar r, the discoverer of the Bay Quinté and Lake Ontario, and the first of e who set foot upon the soil which divides the two.

sequently it was no doubt frequently traversed by the French Jesuits -traders; but it was not till 1783 that the first permanent settlement ade by Mr. Weese, on the north shore of Ameliasburg. That event, as the successive settlements, in their regular order, in the other parts of unty, are sufficiently detailed in the sketches of the several townships lves, together with the time and manner in which the same were nam- named, and surveyed. We will therefore proceed directly with its

POLITICAL HISTORY.

may be said to have commenced with the proclamation of Governor , dated at Kingston, July 16th, 1792, which divided the Province of Canada into nineteen counties for electoral purposes. This division sed the counties of Glengarry, Stormont, Dundas, Grenville, Leeds, nac, Ontario, Addington, Lennox, Prince Edward, Hastings, Northum- d, Durham, York, Lincoln, Norfolk, Suffolk, Essex, and Kent, in the named. Ontario, as then constituted, consisted of the various islands of the mainland, and between Prince Edward and the foot of the lake. oclamation commenced with the usual preamble, setting forth the object and the causes therefor; and contained in addition nineteen distinct , and was signed by William Jarvis, Governor's Secretary, and counter- by the Governor with his initials only, "J. G. S." The great volume instrument precludes the advisability of a repetition *in extenso*, but we e part referring to this county, as follows:—

That the tenth of the said counties is to be hereafter called by the name County of Prince Edward; which County is to be bounded on the south e Ontario; *on the west by the Carrying-place on the Isthmus, of the* e Isle de Quinté; on the north by the Bay of Quinté; and on the east, oint Pleasant to Point Traverse, by its several shores and bays, includ- e late townships of Marysburg, Sophiasburg, and Ameliasburg. The unty of Prince Edward to comprehend all the islands in said Lake and Bay of Quinté nearest to the said county, in the whole, or greater nting the same."

Proclamation also gave an approximate representation by population. rry had two members, a part of Lincoln returned two, and other parts rouped with counties on either side, returned one each. A number of s returned one each; while in a number of cases two or more counties, of counties, were grouped together to send one member in conjunction. nship of Adolphustown, in Lennox, was thus grouped with Prince , the clause referring thereto being as follows:—

at the County of Prince Edward, bounded as hereinbefore described, with the district of the late Township of Adolphus, in the County of shall together send one representative; that is, the said County of Edward, together with the said district, late the township of Adolphus, l may be represented together in the said House of Assembly, by one "

whole number provided for from the nineteen counties, was sixteen s: which number composed the Legislative Assembly of Upper for many years.

first election held under this proclamation is fully noted in the sketch well, to which township the first elected representative belonged. Philip Dorland and Peter Van Alstine there mentioned, Prince Edward ly represented by Simeon Washburne, James Cottle, and James The latter served his county in that capacity for nearly a quarter of y, though not consecutively.

next political change to note in the history of the county, is the set- of Prince Edward as a separate district for the administration of by Act of Parliament, in 1831. Mr. Roblin then represented the n Parliament, and introduced the Bill which brought that much- hange into effect. Heretofore the people had been obliged to go to n for the transaction of all judicial business. Particulars of the of the jail and court-house for the new district, and the location may be found under the head of Picton.

that time Prince Edward has been a separate and independent district al purposes; for many years a separate constituency for representative in the Legislative Assembly; and, since confederation, an electoral returning one member to the House of Commons.

ol. Walter Ross now represents the county in the Commons; and Gideon Striker is the present sitting member in the Ontario Legis-

MUNICIPAL HISTORY.

dest county or district record now in existence, is the "Report of yor of Highways." The subject matter of the report is entered in tion indicated by its title. It is signed, "James Young, Surveyor ys," and endorsed as follows: "Approved in Quarter Sessions, 9th . A. McLean, Clerk of the Peace." The contents are quite vol- and of no peculiar or general interest.

rliest record of the deliberative proceedings of the governing body rict, are minutes of the first District Council meeting of 1842. The re extracts from the same:—

nal of the Municipal Council of Prince Edward District, February

"Court House, Picton.

"This day, at 2 o'clock, the Warden being present, the following members presented themselves, and giving satisfactory evidence to the Warden of their having taken the necessary oaths prescribed by law, took their seats in the Council.

"For Hallowell—Benjamin Hubbs and David B. Stephenson, Esquires.

"For Marysburg—Henry Van Dusen and John Rose, Esquires.

"For Sophiasburg—Henry Dunning and John Howell, Esquires.

"For Ameliasburg—William Dempsey and Owen Roblin, Esquires.

"For Hillier—James T. Lane and Arnoldi Dorland, Esquires.

"For Athol—Wilson Bentley, Esquire.

"Roger B. Conger was appointed clerk, *pro tem.*, by the Warden.

"The Clerk then read at the table the commission appointing John P. Roblin, Esquire, Warden of the District."

A Select Committee was then moved for and nominated, to "draft and report rules for the guidance of this Council at its several sittings," which shows that this was the first meeting under the District Council system, which was inaugurated by the Act of Parliament of the previous year. These rules number twenty-two, the last of which we quote:—

"22. *That should any case occur for which no special rule is made*, reference shall be had to parliamentary proceedings, so far as such may apply."

Several matters of public interest were deliberated upon at that session, among which was the "motion on the address to the Governor-General." This gave rise to a very animated discussion, and on the third reading, the yeas and nays being called, the vote stood seven for and three against. The address was on the occasion of the arrival in the country of a newly-appointed Governor-General, and commenced thus:—

"To His Excellency the Right Honourable Sir Charles Bagot, Knight Grand Cross of the Most Honourable Military Order of the Bath, one of Her Majesty's Most Honourable Privy Council, Governor-General of British North America, and Captain-General and Governor-in-Chief in and over the Provinces of Canada, Nova Scotia, New Brunswick, and the Island of Prince Edward, and Vice-Admiral of the same," etc., etc.

Since then there have been no municipal changes to note, beyond the introduction of the elective system in choosing the warden; and later in electing the representatives to the County Council by the *popular* vote of the several townships, instead of from the five township councillors previously chosen from the various wards.

The incumbents of the county municipal offices, for the year 1878, are as below:

WARDEN—Samuel N. Smith, Reeve of Sophiasburg.

COUNCILLORS—*Josiah Pinson, Deputy Reeve of Sophiasburg*; T. W. Flagler, Reeve of Wellington; Benjamin A. Hubbs, Reeve of Athol; Henry Hubbs, Deputy Reeve of Hallowell; Nelson Hudgins, Deputy Reeve of South Marysburg; J. A. Johnson, Deputy Reeve of Ameliasburg; R. S. Jones, Deputy Reeve of Hillier; J. H. Knox, Reeve of South Marysburg; Joseph R. Nightengale, Reeve of Ameliasburg; J. Platt Nash, M. D., Deputy Reeve of Picton; Wm. P. Reynolds, Reeve of Picton; Lewis B. Stinson, Reeve of Hallowell; Levi Williams, Reeve of North Marysburg; John Young, Reeve of Hillier.

COUNTY CLERK—Robert Boyle.

COUNTY TREASURER—R. J. Chapman.

The principal business transacted at the June session of the County Council this year (1878) was in connexion with the railroad agitation. "To have a railroad, or not to have a railroad—that's the question" which agitated the assembled wisdom of the county.

Briefly stated, the history of the railway question is this:—For thirty years past the subject has been agitated, and a number of years since a company was organized and chartered, which succeeded in obtaining a grant, by way of bonus, from the county, of $60,000, on condition of its being built within a certain time. Meantime, the commercial panic struck the country, and the prosecution of all public works—the Prince Edward County Railway included—was either suspended, or entirely abandoned. The time being about to expire within which the company were to complete the work, in order to receive the bonus, a movement was inaugurated by the friends of the enterprise, to have the time extended. The opponents of the scheme fought it persistently and consistently; but it at last carried by a respectable majority of the County Council; and the chances now look, to an outsider, as if the people might soon have a railroad. As for themselves, they are confident of having cars running to Picton this season—which is perhaps too sanguine a view. In any case they will have it eventually, and ought to have had it long ago; while too much praise cannot be awarded to the promoters, who have worked long and earnestly, and without rest, till the result they nobly strove for seems assured.

The chief officers of the company are Lt.-Col. Ross, M. P., President, and S. P. Niles, of Hillier, Vice-President. That the enterprise may flourish is the wish of all fair minded people.

MILITARY HISTORY.

Prince Edward was settled by a military race. A few were discharged soldiers of the regular army. For the most part, however, they were not a military people in the strictly professional sense of the word; but their military qualities were the result, as well as the adjunct, of that patriotism which preferred the fatigues of the campaign, the sufferings of the winter bivouac, and the dangers of the battle field, to the abandonment of cherished political principles, or the acceptance of those which were hostile to their sentiments as devoted subjects of their king, and loyal adherents to the British Constitution and the laws of their fathers. In such a patriotism, and in such a loyalty, Prince Edward and Upper Canada had their origin.

Under such circumstances we are not surprised to learn that the war of 1812-13 found this noble county to the front with its best men—and probably more in proportion to population than many other localities whose exposed position rendered them far less safe from invasion than itself, whose only invasion was that of American prisoners of war, as they were marched across the Carrying Place, or over the old Danforth military road, under escort, on their way to Kingston and Quebec, from Detroit, Queenston, and other places in the west, where the fortunes of war, aided by the superior bravery of the Upper Canada militia, had placed them in our hands.

We do not consider the Rebellion of '37-'38, or the action taken by our ancestors on either side, as necessarily the true criterion of real patriotism, or the reverse; for it is now universally admitted that the wrongs of the "Patriots"—as they were called by the Loyalists in derision—were long since arrived at a stage, under the misrule and tyranny of the Family Compact, when forbearance ceased to be a virtue; and moreover, we all know that the very principles for the advocacy of which Lount and Matthews, and many other good men, were ignominiously hung as *traitors*, are now the foundation and copestone of our political structure.

We are of those who believe in the truth and justice of that time-worn, but not less honored proverb, that "The truly loyal subject will neither advise nor submit to arbitrary measures." And so, too, no doubt thought many of our people here in those troublous times when brother was arrayed against brother, and father against son—though the vast majority were in favor of peaceful and constitutional means of obtaining redress—doubtful as it is whether it could ever have been thus obtained—while very many went for sustaining the Government at any and all hazards; and we thus find Prince Edward furnishing a large quota of men for the military service at that time, some of the officers from this county (notably the Hon. Colonel Wilkins, of the Carrying Place) taking most prominent and honorable positions, and performing most arduous and important services.

Some of the volunteer companies of 1812-'13, and later, of '37-'38, remained in nominal existence till the "Trent" affair, which roused the military spirit of the nation; and nine splendid infantry companies were then filled, and being accepted by the government, were consolidated into the famous 16th Regiment, of which Walter Ross, M.P., was appointed Lt.-Col. in 1863. Other companies were also raised, but their services declined—the crisis having passed, by the action of the American government in delivering up Mason and Slidell, the Confederate Ambassadors to Great Britain and France, whose capture by the "San Jacinto," American man-of-war, under command of Captain Wilkes, was the original cause of the excitement which for a time threatened the peace of the Empire.

The 16th Regiment—as also a fine troop of cavalry—are now in the best of condition, and have their head-quarters at Picton; and a more detailed notice of them may be found in the historical sketch of that place.

PRESENT RESOURCES.

Prince Edward being pre-eminently an agricultural constituency, its resources lie chiefly in its soil, and the products of the field. The chief products of the various sections may be found described under their respective heads. In regard to the amount of production, the distribution of the chief staples—notably barley and rye—is so scattered among the many storehouses which line the shores on all sides, that any accurate figures are not to be arrived at, and an approximate estimate would be simply conjectural. The principal depôt, and in fact the only considerable one in the county, for the collecting and distributing of some particular branches of farming industry, is Picton; and under that head we have given, as fully as circumstances would admit, the average yearly results.

Official statistics, where obtainable, are the only true estimate of a country's material resources; and we give below some figures compiled from the records in the County Clerk's office, which bear more particularly on the point:

Total number of acres in the County	229,241
Assessed value of Real Estate	$6,862,670
do. Personal Estate	457,168
do. Taxable Income	31,910
Number of Horses owned in the County	8,371
do. Cattle do. do.	14,409
do. Sheep	12,746
do. Swine	3,971
Total number of Ratepayers	5,218
Total Population	18,933
Amount of Taxes collected last year	28,343
Received from government on account of administration of Justice	1,114 83
Expended on account of same	4,634 00

Financially, the county, whether taken as a whole, or as separate minor municipalities, is in a most prosperous state. The only liabilities of the county are the $60,000 grant to the Prince Edward Railway. Picton is the only minor municipality having any debt: while some of the townships have very considerable sums invested in the "School Trust"—notably Ameliasburg, where the amount exceeds $15,000

The indebtedness of Picton is merely nominal compared with that of other towns of similar size and importance; being but $6,500, of which $5,000 are in debentures, bearing interest at six per cent.—besides $20,000 lately granted to the P. E. Railway. These railway grants will not become liabilities till the road is built; so the people run no risk of paying for what they fail to receive.

There was raised in Picton last year by taxation	$ 9,040
Received from other sources	6,047
Total revenue	15,087
There was paid on account to Schools	3,600
do. do. do. Salaries	1,088
do. do. for local improvements	4,000
do. do. Interest on Debentures	300

The total rate of taxation, including county and all local rates, was one and a half cents in the dollar.

The rates of taxation throughout the townships are very light compared with some sections of the central and western parts of the Province, averaging two to three mills in the dollar. The *local* rate in Picton this year is eight mills.

The following table shows the result of the labors of the "Equalization of Assessment Committee," of the County Council, at the June session, 1878:

Municipality.	No. acres.	Val. $ ac.	Real Pr'ty.	Personal.	Total.
Athol,	22,891	$20	$ 457,820	$25,000	$ 482,820
Ameliasburg,	43,092	25	1,077,300	75,000	1,152,300
Hallowell,	42,280	30	1,268,400	80,000	1,348,400
Hillier,	31,311	25	782,525	50,000	832,775
Marysburg, North,	29,935	24	550,000	25,000	575,000
Marysburg, South,	26,632	19	449,008	25,000	474,008
Sophiasburg,	43,100	25	1,077,500	75,000	1,152,300
Wellington (capitalized)					130,000
Picton do.					87,000
Grand total					$6,535,243

A tax was levied at the rate of one and five eighths mills in the dollar, which makes the proportion of each municipality as follows:

Ameliasburg, $1,872.48; Athol, $784.58; Hallowell, $2,191.15; Hillier, $1,358.25; Marysburg, North, $935.09; Marysburg, South, $770.26; Sophiasburg, $1,872.81; Wellington, $141 37; Picton, $698.75.

In addition to the above, a by-law was passed to raise the further sum of $2,621 for school purposes. And this reminds us that there are a few general facts concerning

EDUCATIONAL MATTERS,

Which it would be as well to refer to in this connection. As is well known, one of the greatest disadvantages experienced in all the early settled portions of the county, arose from the entire absence of facilities for educating the children of the pioneers. This has been the case even in those counties which, compared to Prince Edward in point of age, are yet young. How much more so must it have been in the early days of Prince Edward?

We of the present day, with our universities and colleges, normal and model schools, high and public schools, almost without number, can scarce realize the vast difference, when even the scenes depicted in that popular and much-perused work "The Hoosier Schoolmaster," would have been looked upon as a wonderful advancement towards what we might term the extravagances of a higher order of civilization. For years and years after the advance guard of that army of Loyalists who made Upper Canada what it is first came into Prince Edward, schools nor intellectual training were [illegible] thought of, except as an adjunct of that civilization which they had [illegible] hand—a thing to be ardently desired, but not seriously hoped [illegible] for years simply a question of keeping body and soul together [illegible] laborious toil; and the hardships endured in procuring the [illegible] necessaries of life precluded the idea of looking for intellectual improvement. But as the clearings gradually broadened, the little settlements grew more numerous, and the toil of the inhabitants had, by the favor of a kind Providence,

placed them beyond immediate want, they bethought themselv
duty to their little ones, and it was only then that the real diffic
situation was fully realized.

Many of the early Loyalists had been men of wealth, influence
cation in their old homes. These, so far as more necessary du
permit, instructed their own children. Even when this opportun
wanting, the difficulty of want of books presented itself, most
settlers having abandoned everything in their flight and escape
For many years after neighbourhoods commenced to be form
gathering together of five or six families, within a radius of as m
the only pretence to regular school teaching was by some of t
opening a private school in the winter season, and instructing th
the "settlement" in the "three R's." The more advanced pu
lessons in the New Testament and "English Reader;" and thos
so fortunate as to be able to spend their winter evenings in poris
pages of those text-books by the light of the blazing back-log on
were happy indeed, and the envied of all the settlement. Bu
standing these great disadvantages, some of our most able men we
of the earliest settlers of the Bay Quinté District, who gaine
education in the manner described.

The names and location of the first teachers in the county wi
in the various township sketches, wherever it has been possibl
reliable information of the same; which, in the absence of a
records, and from the great length of time since elapsed, it has be
and in some cases impossible, to do.

Generally speaking, the progress of the school system here,
elsewhere throughout the country, has been of a most remarkabl
factory character. The Legislature seems early to have recogniz
which centuries of history had already proven, that upon the ex
and efficiency of our educational institutions must be laid the gre
our country's success and national greatness. Parliament, theref
the first Common School Act in 1816, which provided for the electi
trustees in each township, whose duties were to hire teachers,
books, and make general regulations such as are now provided for
law, or attended to by the Department of Education. This Act
panied by a grant for school purposes, which was supplemented
to time, and continued to be spent without system and to com
little advantage.

Finally, by the Hon. S. B. Harrison's bill of 1841, which prov
annual grant of $200,000 to the various counties of the Province,
tion to their *school population*, and conditional upon the said coun
menting said respective sums by like amounts for similar purpo
Francis Hincks's amendment thereto in 1843, the scheme of divi
ships into school sections was initiated, and other reforms in
which have been extending and improving till impartial, disinte
well-qualified judges are not slow to assert that Ontario now p
finest school system in the world.

This is the fruit, in a very great measure, of the superior a
untiring exertions of the Rev. Dr. Egerton Ryerson, whose nan
than half a century has been a familiar household word in every
country. He was of U. E. Loyalist stock, a self-made man—o
experienced in his youth and early manhood the difficulties of
pioneers, and was therefore especially qualified to deal with them
for many years one of the ablest and most popular Methodist pre
ever graced that or any other Christian church. Being appointe
Superintendent of Education in 1844 (the Provincial Secretary b
time, and long afterwards, *ex officio*, the nominal Chief Super
Dr. Ryerson made several trips to the eastern hemisphere and
parts of the United States, to study the educational systems of fo
tries, and on his return prepared a School Bill, very comprehen
character, and embodying the best points in the various syst
different States and European nations. The chief features of t
still in force.

But the difficulty of bringing order and symmetry out of the c
then existing, was a matter of more time and labour than simpl
Bill or passing an Act of Parliament; and to comprehend thi
necessary to understand the existing state of affairs previous to
taking the matter in hand. What school-houses existed were o
description. Rough log shanties covered with bark, or thatched
hay, were quite common; while in many the well-known chara
the proverbial "Irishman's shanty" celebrated in song, were
found, in an aperture which served the treble purpose of door, w
chimney. Interior arrangements were in keeping with the struc
selves—rived slabs, with sticks stuck in for legs, forming seat
The sparsely settled districts, with intervening swamps and fores
of good roads, made the location of the school a matter of no smal
and more than all, the objection on the part of those who had rea
without education, or educated them themselves, was very str
any system which distributed the expense of a school among
owners, whether with or without children requiring its use.

But all these, and many more difficulties of a similar characte
gradually overcome. A few primitive, ill-ventilated, and un
shanties have given place to fine, commodious frame-houses,
brick and stone structures. The teachers, too—of whom very
possessed of but indifferent scholastic attainments, and would
sidered far from competent,—have been supplanted, through t
the splendid Normal and Model Schools, which the liberal-mi
our legislators has given us, with those who are a credit to
system and the county which supports it; and, in a word, "
School system of Ontario, from a condition of perfect infancy,
within the compass of a generation into one which will compa
with any in the world," and conclusive proof of this statem
sought no further than in a comparison of the state of circumsta
previous to the passage of the Harrison-Hincks Acts above not
facts set forth in a few statistics given below, which we have g
the very exhaustive, voluminous, and able report of G. D. Platt
the County Superintendent of Public Schools, for the present sc

This report refers simply to the schools in the county proper,
the corporate limits of the Town of Picton being referred to in t
that place.

We find the number of schools at present in operation thr
County, as follows:—Ameliasburg, 15; Athol, 8; Hallowell, 15;
North Marysburg, 9; South Marysburg, 9; Sophiasburg, 13;
1; or a total of 81 schools, having each one teacher, except the
School, those at Milford, Demorestville, and the "Union," n
Mills, each of which has two, making a total of 85 teachers in
There are 36 brick schoolhouses in the county, 18 of stone, a
buildings.

During the past year seven new schoolhouses have been e
frame and five of brick. There is not a single log schoolhouse
the county.

The amounts of school taxes raised last year by the several m
were as below:—Ameliasburg, $4,121; Athol, $1,587; Hallow
Hillier, $3,712; North Marysburg, $1,911; South Marysbu
Sophiasburg, $5,820; Wellington, $767; or a total of $24,6
Total amount received from all sources, $40,077.

Highest salary paid to any teacher, $507; lowest do., $1
salary to male teachers, $493; ditto female, $211; total numb
1,789; daily average attendance, 45 per cent.

Compared with other counties of Ontario, the school expens
to be greater in proportion to the school population than in an

mediate want, they bethought themselves of their and it was only then that the real difficulty of the ed.

yalists had been men of wealth, influence, and edu-s. These, so far as more necessary duties would own children. Even when this opportunity was not f want of books presented itself, most of the first d everything in their flight and escape to Canada. neighbourhoods commenced to be formed by the e or six families, within a radius of as many miles, gular school teaching was by some of the settlers in the winter season, and instructing the youth of "three R's." The more advanced probably took ment and "English Reader;" and those who were e to spend their winter evenings in poring over the s by the light of the blazing back-log on the hearth. the envied of all the settlement. But notwith-dvantages, some of our most able men were the sons of the Bay Quinté District, who gained all their described.

on of the first teachers in the county will be found sketches, wherever it has been possible to obtain the same; which, in the absence of any positive eat length of time since elapsed, it has been difficult, sible, to do.

the progress of the school system here, as well as e country, has been of a most remarkable and satis-Legislature seems early to have recognized the fact ry had already proven, that upon the completeness cational institutions must be laid the groundwork of l national greatness. Parliament, therefore, passed Act in 1816, which provided for the election of three ip, whose duties were to hire teachers, select text-l regulations such as are now provided for by general e Department of Education. This Act was accom-hool purposes, which was supplemented from time to be spent without system and to comparatively

S. B. Harrison's bill of 1841, which provided for an) to the various counties of the Province, in propor-*lation*, and conditional upon the said counties supple-sums by like amounts for similar purposes, and Sir ment thereto in 1843, the scheme of dividing town-ns was initiated, and other reforms inaugurated, ing and improving till impartial, disinterested, and not slow to assert that Ontario now possesses the he world.

a very great measure, of the superior abilities and e Rev. Dr. Egerton Ryerson, whose name for more been a familiar household word in every home of our E. Loyalist stock, a self-made man—one who had h and early manhood the difficulties of the early are especially qualified to deal with them. He was e ablest and most popular Methodist preachers who other Christian church. Being appointed Assistant-ation in 1844 (the Provincial Secretary being at that rds, *ex officio*, the nominal Chief Superintendent), ral trips to the eastern hemisphere and to various es, to study the educational systems of foreign coun-prepared a School Bill, very comprehensive in its ing the best points in the various systems of the opean nations. The chief features of this Bill are

bringing order and symmetry out of the chaotic state tter of more time and labour than simply framing a of Parliament; and to comprehend this it is only the existing state of affairs previous to Dr Ryerson ad. What school-houses existed were of the rudest shanties covered with bark, or thatched with marsh ; while in many the well-known characteristics of an's shanty" celebrated in song, were there to be hich served the treble purpose of door, window, and ngements were in keeping with the structures them-h sticks stuck in for legs, forming seats and desks. tricts, with intervening swamps and forests, and want location of the school a matter of no small difficulty; objection on the part of those who had reared families educated them themselves, was very strong against ributed the expense of a school among all property without children requiring its use.

many more difficulties of a similar character, have been A few primitive, ill-ventilated, and unh[illegible] log ace to fine, commodious frame-houses, or splendid ures. The teachers, too—of whom very many were tent scholastic attainments, and would now be con-tent,—have been supplanted, through the means of nd Model Schools, which the liberal-mindedness of en us, with those who are a credit to the present which supports it; and, in a word, "the Public rio, from a condition of perfect infancy, has ripened generation into one which will compare favorably l," and conclusive proof of this statement need be in a comparison of the state of circumstances existing of the Harrison-Hincks Acts above noted, with the statistics given below, which we have gleaned from luminous and able report of G. D. Platt, of Picton, lent of Public Schools, for the present school year. ply to the schools in the county proper, those within the Town of Picton being referred to in the sketch of

r of schools at present in operation throughout the meliasburg, 15; Athol, 8; Hallowell, 15; Hillier, 11; South Marysburg, 9; Sophiasburg, 13; Wellington, ols, having each one teacher, except the Wellington rd, Demorestville, and the "Union," near Roblin's s two, making a total of 85 teachers in the county. oolhouses in the county, 18 of stone, and 26 frame

r seven new schoolhouses have been erected—two There is not a single log schoolhouse remaining in

nd taxes raised last year by the several municipalities iasburg, $4,121; Athol, $1,587; Hallowell, $5,089; n Marysburg, $1,911; South Marysburg, $2,414; Wellington, $767; or a total of $24,626 in taxes. from all sources, $40,077.

l to any teacher, $507; lowest do., $180; average , $103; ditto female, $241; total number of pupils, ttendance, 45 per cent.

er counties of Ontario, the school expenses are found tion to the school population than in any other; or, in other words, the people of Prince Edward tax themselves the highest of all the counties of the Province for educational purposes. Comparing the last year's report of the various County Superintendents, we find the figures as below for the six counties standing first on the list: Prince Edward, $5.40; Wentworth, $4.73; Oxford, $4.44; Elgin, 4.25; Durham, $4.21; Waterloo, $4.13. This represents the total cost of each pupil for the school year 1877, including salaries to teachers, building and repairing of schoolhouses, and all expenses for school purposes.

THE PUBLIC CIVIL SERVICE.

HER MAJESTY'S CUSTOMS.—The officers of this department of the Civil Service for Prince Edward District, are as follows:—

Walter T. Ross, collector, Picton; F. W. Mandeville, landing waiter, Wellington; W. H. McLean, landing-waiter, Port Milford; John Prinyer, preventive officer, Prinyer and Cressy; Joshua M. Cadman, preventive officer, Consecon. The latter two belonged till recently under the supervision of the Kingston and Brighton offices respectively, but have lately been added to the jurisdiction of Picton to which all the above now belong.

John S. Clute, the late collector at Picton, was recently superannuated, after twenty-seven years' service.

INLAND REVENUE.—Robert Boyle is Inspector of Weights and Measures for the district, which includes the whole county. There are no regularly stationed excise officers in the county, those duties being attended to by officers from the Belleville station.

MARINE AND FISHERIES.—The following are the names and addresses of the "Inspectors of Fisheries" for the various districts into which Prince Edward is divided:—David Conger, Wellington; Peter Huff, Salmon Point; Eli Ketchum, Cherry Valley; Wesley Hicks, South Bay (Port Milford); Abram Welbanks, South Bay (Port Milford); William Plews, Prinyer.

There are seven lighthouses within the Prince Edward District. Their locations are apparent from their names, which are given below, with the names of their keepers:—"Blue Bonnet," Robert Pye; "Carrying-place," Reuben Young; "Salmon Point," Peter Huff; "Point Petre," James Burlingham; "False Ducks," William Sweetman; "Indian Point," John Prinyer; "Telegraph Island," — Mason.

ADMINISTRATION OF JUSTICE.—Prince Edward has long been proverbial for the lightness of its criminal calendar. Various causes have been assigned for it, but whatever the true one the fact remains the same. This is attested by the exceptionally small expenditures charged to the "Administration of Justice" account year after year. This amount for the last judicial year was $1,364; while the amount received from Government on account of the same was $1,114.83.

Following are the County Court officials:—Judge of the County Court, Robert Patterson Jellett; County Crown Attorney and Clerk of the Peace, Philip Low, Q.C.; Clerk of the County Court, Clerk of the Surrogate Court, Deputy-Clerk of the Crown and Pleas, John Twigg; Master-in-Chancery, S. S. Lazier; Registrar, Walter McKenzie; Deputy-Registrar, Major Bog; Sheriff, James Gillespie; Sheriff's Clerk, J. F. Gillespie; High Constable, Ellis Grimmon; Jailor, W. E. Patterson; Turnkey, Ezekiel Harris.

The following table, giving the number and location of the various Division Courts in the county, with their clerks and bailiffs and their post-office addresses, will be found useful for reference.

No. and Name.	Clerk.	Bailiff.	P.O. Address.
1. Picton	J. P. Downes	Andrew Buchanan	Picton
2. Milford	H. H. Haight	Richard Ladd	Milford
3. Demorestville	Israel Hamilton	Edward Nixon	Demorestville
4. Roblin's Mills	Edward Roblin	J. S. Tice	Ameliasburg
5. Wellington	William Young	Thomas Jackson	Wellington
6. Bloomfield	Obadiah Cooper	John Thompsett	Bloomfield
7. Consecon	J. D. Cadman	D. H. Weeks	Consecon
8. North Marysburg	Ed. Harrison	James Rose	Waupoos

A full description of the minor municipalities of Prince Edward, and of the principal objects of local or general interest connected therewith, may be found in the respective town and township historical sketches.

PICTON AND HALLOWELL.

PICTON, which is, as to date of incorporation, a comparatively new place, was in point of settlement one of the oldest in the district. Previous to incorporation it was part and parcel of Hallowell, by which name it was also called. Thus it happens that the history of town and township is so intimately interwoven as to be in a great degree identical, the one with the other; which fact we judge sufficient cause to treat them, for the purposes of this history, still practically as one—which they were in reality, until a comparatively recent date. And although Hallowell as a township, was neither the first organized, the first surveyed, nor the first settled, still its central location and superior natural advantages very soon gave it the foremost position among the townships of the district—a position which it still maintains; and this circumstance, coupled with the fact of Picton (still within its limits) being the County Town, and consequently the judicial, as well as the commercial, financial, educational, and social centre of the county, gives the palm to this ancient township, which therefore deserves the first place in the history of the various municipalities of the county.

The township of Hallowell was formed by virtue of an Act of Parliament, passed July 3rd, 1797, which provides "that a township be struck off from the southernmost parts of Marysburg, and Sophiasburg." The Act also sets forth the reasons therefor, viz.: That "the inhabitants of the said townships experience many difficulties from the uncommon length of such townships." It strikes us as a little singular, however, that "Surveyor-General Smith was the person employed to lay out the new township;" as we are led to believe from every account we have seen of the original three townships, that the surveys of the same had been completed, and all remaining to be done was simply to describe the metes and bounds. There is another curious thing about it, *i.e.*, the Act specified that the new township should be taken "from Marysburg and Sophiasburg," notwithstanding which, it contains so much of the original township of Ameliasburg as lies west of a line produced southward from the eastern boundary of the present townships of Ameliasburg and Hillier. Again, the Act states that it is to be taken "from the southernmost parts," &c.—whereas, as regards Marysburg, the relative positions of the townships, as they at present stand, show that no such instructions were obeyed. From all of which we conclude that the real point aimed at by the Act was simply to ameliorate "the many difficulties experienced by the inhabitants," &c., and that surveyor-General Smith was detailed with liberal discretionary powers to relieve the said inhabitants from the difficulties therein complained of. He apparently succeeded in executing the duty allotted to him, and succeeded at the same time in laying out a township, the metes and bounds of which it exceeds our simple powers to geographically describe. If it was the object of the surveyor-General to confuse and puzzle the enterprising Directory and Atlas man, we are constrained to admit that he was in an eminent degree successful. Strictly speaking, however, the new township was surveyed in a most extraordinary manner; and, marked out on paper, its gogs, gores, triangles, and *double*-triangles, coupled with its curiously irregular outline, formed in part by the indentations of its natural boundaries—Picton Bay and West Lake—it presents a picture bearing a striking similarity to a Chinese puzzle, or Mark Twain's celebrated map of the scene of the Franco-Prussian war.

It may, however, be described as being bounded northerly by Sophiasburg, Hillier, and Picton Bay; easterly by north and south Marysburg; southerly by south Marysburg and Athol, and westerly by Hillier and Lake

Ontario. In very many respects, it is the finest township of Prince Edward county—all that can be said in praise of the "fifth" "sixth" and "seventh towns" being true in respect to Hallowell. It excels them all in wealth and population, though only third in size—a fact which may be taken as a general indication of its splendid natural attributes, and their judicious and prosperous development by a class of citizens whose general intelligence, morality, hospitality, and superior social qualities mark them the worthy descendants of a race, who, animated by the purest principles of loyalty, and with a devotion to the cause of their King, which moved the highest admiration even of their enemies—preferred the loss of all of this world's goods; and war with its entailing sufferings and dangers; and even death itself, to a servile submission to principles of political government, their hatred and contempt for which could only be measured by their love and reverence for the glorious British Constitution, and the laws of their fathers. If the sacrifices involved in an enforced or voluntary exile from once-loved homes, to an unknown land, the home only of savage wild beasts and wilder human savages, can be repaid, after the toils and trials of a century, by the possession of magnificent estates, palatial residences, and a country universally admired by travellers from all parts of the world, then the present citizens of Prince Edward County, and more particularly those of Picton and Hallowell, are now enjoying the well-earned reward of lives devoted to the most noble principles.

The land of the township bears a greater proportion of improved to unimproved, than is to be found in any other section of the district. The surface is nearly all of an even and gently undulating character, while its soil is for the most part of a depth and richness which yields a most generous return to the toiling husbandman. The "equalized" assessment of the county shows Hallowell to be a 20 per cent. higher per acre valuation than the next highest township.

The following account of the origin of the township's name is gleaned from Sabine:—Benjamin Hallowell, of Boston, Commissioner of Customs, while passing through Cambridge in his chaise, in the year 1774, was pursued toward Boston by a detachment of one hundred and sixty Continental cavalry, at full gallop, but eluded them. To escape persecution at the hands of the Americans, he sailed to England on the evacuation of Boston by the British. He seems to have been an object of special animosity to the Americans, as he was proscribed and "banished" in 1778, if any one could be said to be banished who had already been two years in England. Prior to the Revolution he owned extensive landed estates on the Kennebec River, in Maine. These were confiscated by the "Conspiring Act" of 1779, in which Mr. Hallowell was included. He returned from England to America in 1796, with his daughter and son-in-law, Mrs. and Mr. Elmsley, the latter of whom had just been appointed Chief Justice of Upper Canada. His estates all being confiscated and himself proscribed, he could not remain there, but came to Canada, and took up his residence in Toronto, where he died in 1799, at the age of seventy years—the last surviving member of the Board of Commissioners. To compensate for the losses sustained by his devotion to the King's cause, "the British Government granted him lands in Manchester, and two other towns in Nova Scotia, and *a township in Upper Canada which bears his name*." It appears quite plain, however, that his land-grant near the head of Picton Bay, gave the name of "Hallowell Bridge," first to the bridge itself, which was built over the stream emptying into the head of the bay, and, as a natural consequence, to the village which subsequently sprang up there. This being the chief place within the limits of the new township, and the place where public business was transacted, the name "Hallowell" was adopted for it. Thus, although it appears pretty certain that Benjamin Hallowell owned land "in a township which bears his name," yet it is equally certain that the "township which bears his name" derived that name directly from the village (now Picton), and *not* the name of the village from the township, as is popularly supposed.

In our sketch of Athol, we notice the settlement formed by the Youngs, at East Lake, in 1783, or 1784. This tract of land originally belonged to Marysburg, and was of part taken off by the Act of July 3rd, 1797. The Youngs were, beyond dispute, the oldest settlers in the limits of the original township of Hallowell. The next settlers came in and located at the head of Picton bay, but at what precise date is not quite certain. Dr. Canniff, an admitted authority on such matters, states that probably the first settlers took up their residence at the head of Picton bay in 1788. It is pretty certain however, that there was at least *one* settler at "Hallowell Bridge" before that date, as we have it from Mr. Lewis B. Stinson, that his great-grandfather came in and settled on the present Stinson homestead (Lot 3, north of West Lake), in 1774, or certainly not later than 1775, "when his nearest neighbor on the east was at the present town of Picton." It thus appears that there was at least *one* settler at Picton Bay in 1775, and that after him, John Stinson and family were the first settlers within the present limits of Hallowell. Ebenezer Washburne, father of the Hon. Samuel Washburne, is said to have been one of the very first to settle at Hallowell Bridge, though we have been unable to learn the precise date of his advent; and about the same time, or shortly after, came Captain Richardson, Richard Hare, James Dougall, and Henry Ferguson. It seems that before 1788, the Congers, Peterson, Cory Spencer, and Henry Johnston, had also taken up their residences at the head of the bay; and they were followed very shortly by Van Dusen, and Abraham Barker. The place gradually acquired accessions to its population. It soon became "Hallowell Bridge," and took the lead in all township, and even county matters, and has succeeded in holding to the present day the leading position thus early acquired. In the neighborhood of West Lake, the McFauls, McDonalds, Congers, Whites, Coopers, and Balls, were very early settlers. There were several families of the name of Conger—all relatives, who settled in different parts of the township. The following Obituary is taken from one of the newspapers of the day. "DIED.—At West Lake, Hallowell, on the 27th May, 1825, *D. E. Conger*, in the sixtieth year of his age. He held a commission in the Prince Edward Militia during twenty-three years. He resided in Hallowell forty years, and lived an exemplary life, dying regretted by all who knew him." From this it will be seen that the West Lake Congers came in about 1785; with David Conger came also Thomas Goldsmith, a U. E. Loyalist, in 1786. He had previously owned extensive estates in Ulster County, New York; and had performed important and dangerous services to the Royal cause, meeting with many thrilling adventures and hair-breadth escapes. On first coming to the country he settled temporarily in Bath, but soon removed to Hallowell, where the Government granted him a tract of land as partial compensation for his loyal services, and heavy losses he thereby entailed. The Garretts, Ellises, Dorlands, Haights, and Hubbs, were also among the earliest, and further on towards the spot where Bloomfield now stands, the Bowermans, Bulls, and Leavens'. Philip Dorland, one of the above, was the first member elected by Prince Edward, to represent the county in the first Parliament of Upper Canada. The Proclamation of Governor Simcoe, calling upon the people to elect members to this Parliament, was dated at Kingston, July 16th, 1792. William Jarvis, as Governor's Secretary, signed the document, which established nineteen counties; giving some two members, some one each, and in some instances, a number combined to elect one member, according to the population. "Fourth," or Adolphustown, then as now, in the county of Lennox, was annexed to Prince Edward County for the purpose of forming one electoral division. There were in all sixteen members elected by the nineteen counties. The elections were held in the early autumn of 1792, and the Assembly met at Niagara (then called Newark), on the 18th September of the same year, and continued in session till October 15th ensuing. "Nor marble domes, nor gilded spires" cast their classic shade upon the assembled wisdom of the first Parliament of Upper Canada, who met in a common army-tent to give laws to the nation. John McDonell,

of Glengarry, was elected Speaker, and the
pointed viz.:—

CLERK.—Angus McDonell. SERGEANT-AT-
LAIN,—Rev. Mr. Addison.

The oaths of office were administered by
Secretary; and from the fact that the Assembl
Dorland incompetent to serve in that body, on
victions (he was a Quaker) forbidding him t
proves that our early legislators had lost that
toleration, the denial of which had driven m
new world, in the "Mayflower," nearly tw
populous and important county of Prince Edw
the first session of the first Parliament. A W
new election, it resulted in the return of Pete
the second member elected, and the first who s
Prince Edward county, in the Parliament of U

Hallowell shows an official record of its p
date than any of the three townships from wh
seen that it was created a township by the Act
inspected the records of the first "Town me
antiquity imparts to it an interest which call
we herewith give, *verbatim et literatim*.

"TOWN LAWS, 1798.—At the annual meeti
"township of Hallowell, held on Monday, th
"held by virtue of an Act of the Legislatur
"Augustus Spencer, and John Stinson, jr., tw
"of the Peace,—the following persons were
ensuing year:—

"TOWN CLERK.—Rozel Ferguson. ASSESSO
"D. Conger. TOWN OR CHURCH WARDENS.—Ja
"smith. OVERSEERS OF HIGHWAYS AND FENCE V
"John Miller, Owen Richards, Henry Zufelt,
"White, Corey Spencer, George Baker. POUN
"Isaac Bedall, Samson Striker, Henry Johns
"Williams, Isaac Garrett.

"It is enacted at the said meeting this fifth
"is to be lawful in the township of Hallowell,
"eight inches high, sufficiently made.

"Horses and horned cattle are to run a
"except stags, which are not to be a free comm
"any such shall be found at large, it shall and
"habitant to impound the same.

"All hogs are to run at large, except boars,
"commoner from the first of March, to the 15th
"such shall be found on the commons of the tow
"the two periods above-mentioned, it shall be
"takes him up, shall hold and keep him as his
"to run at large in the said township under the

"All sheep is to be a free commoner the
"which are not to have the liberty of the comm
tember to the tenth of November ensuing.

"It is further enacted, that if any freeholder
"any Canadian thistle to go to seed on his own f
"the sum of twenty shillings, H. Cy., (meaning

"If any person in this township shall put
"stubble, or whatever, whereby his neighbors,
"danger of receiving damage thereby, without p
"his neighbors, so that he may have help to
"damage, he shall forfeit and pay the sum of fo
"benefit of the 'highways.'"

It appears from the above that even at th
Thistle" had made its appearance there; and it
the yeomanry of this part of the country, to be
nection, that the land throughout Hallowell and
County, is for the most part very free from this p
so, in fact, in comparison with some of the town
where the settlements are of a much more recent
justly boast of the superiority of their lands. A
the ins and outs of the agricultural profession as
the above speaks volumes, not only for the *pres*
lightful agricultural township, but for those w
cultivation of the soil since the place has had a h

Following the course of municipal events: af
ng," we find that Rozel Ferguson continued to
James B. Armstrong was chosen to the position,
being succeeded by Arva Ferguson, who held the
followed, in succession, Simeon Washburne, E
Ferguson (three years), Robert Scott (two years
Washburne, Arva Ferguson (seven years), Willia
brings us down to 1832, when James McDonald w
first "special" meeting was held on March 24th
the meetings gradually became more frequent, a
ceedings more voluminous, till they finally gre
Up to 1812 the following names appear in the li
Wardens"—corresponding with our Councillors
Thomas Goldsmith, Barret Dyre, James Fulton
Johnston, Peter D. Conger, Samson Striker, F
Garrett, Asa Worden. Two of the above gentle
Some of them served quite a number of times. M
served the greatest number of terms.

There has been no record kept, during that e
consider *necessary* statistics, so we are kept in ig
population, &c., from year to year. The last a
following figures:—

Total population of township, 3,515; total n
976; total number of acres improved, 31,157; va
950; value of personal property $80,000; amoun
total number of horses, 1,085; cattle, 2,456; sh

The present incumbents of the various munic
REEVE.—Lewis B. Stinson; DEPUTY REEVE.
LORS.—A. Huyck, G. I. Lazier, Wm. H. Johns
Morgan; ASSESSOR.—Albert Clarke; COLLECTOR.
SEER.—Obadiah Cooper.

BLOOMFIELD, the only village of any size wit
Hallowell, is near the head of West Lake, and is
the township, on the old Danforth Road, and ab
and Wellington. It is comparatively an old pla
having been formed here in the very early days
the first to make the place their home were the L
Jonathan Clarke, Cornelius White, James Barke
mans were both large families. Jonathan Bow
mill here, and John Bull the first saw-mill. T
runs through the village and empties into the
splendid opportunities for the milling business;
early utilized by the enterprizing citizens, thus
centre of trade. The village extends over a mil
with a considerable number of cross-streets. Th
gular, as is also the ground on which it is bu
tends to add to its picturesque appearance, a
pleasing variety extremely enjoyable to the tra
of a class far exceeding those of the average town

]RICAL SKETCH OF THE

garry, was elected Speaker, and the following officers were ap- viz. :—

ĸ.—Angus McDonell. Sergeant-at-Arms.—George Law. Chap- Rev. Mr. Addison.

oaths of office were administered by William Jarvis, Governor's ry ; and from the fact that the Assembly *unanimously* voted Philip l incompetent to serve in that body, on account of his religious con- (he was a Quaker) forbidding him to take an oath, conclusively that our early legislators had lost that Christian spirit of religious m, the denial of which had driven many of their ancestors to the rld, in the "Mayflower," nearly two centuries before. Thus the s and important county of Prince Edward was disfranchised during session of the first Parliament. A Writ having been issued for a ction, it resulted in the return of Peter Van Alstine, who was thus nd member elected, and the first who served Hallowell, and with it Edward county, in the Parliament of Upper Canada.

lowell shows an official record of its public Acts of a more ancient an any of the three townships from which it was taken. We have t it was created a township by the Act of July 3rd, 1797. We have d the records of the first "Town meeting," and its comparative ty imparts to it an interest which calls for its reproduction, which with give, *verbatim et literatim*.

wn Laws, 1798.—At the annual meeting of the inhabitants of the hip of Hallowell, held on Monday, the fifth day of March, 1798, by virtue of an Act of the Legislature of Upper Canada, before stus Spencer, and John Stinson, jr., two of His Majesty's Justices Peace,—the following persons were chosen town officers for the year :—

wn Clerk.—Rozel Ferguson. Assessors.—Caleb Ellsworth, Peter nger. Town or Church Wardens.—James Blakeley, Thomas Gold- Overseers of Highways and Fence Viewers.—Benjamin Waite, Miller, Owen Richards, Henry Zufelt, Ichabod Bowerman, Aaron , Corey Spencer, George Baker. Pound Keepers.—Daniel Young, Bedall, Samson Striker, Henry Johnston. Constables.—Samuel ms, Isaac Garrett.

is enacted at the said meeting this fifth day of March, that no fence be lawful in the township of Hallowell, under the height of four feet inches high, sufficiently made.

orses and horned cattle are to run at large, in this township, t stags, which are not to be a free commoner the ensuing year. If ch shall be found at large, it shall and may be lawful for any in- nt to impound the same.

l hogs are to run at large, except boars, which are not to be a free oner from the first of March, to the 15th December ensuing. If any shall be found on the commons of the township of Hallowell, between o periods above-mentioned, it shall be forfeit, and the person who him up, shall hold and keep him as his lawful property. No pig is at large in the said township under the age of four months.

l sheep is to be a free commoner the ensuing year, except rams, are not to have the liberty of the commons, from the tenth of Sep- o the tenth of November ensuing.

is further enacted, that if any freeholder in this township shall suffer nadian thistle to go to seed on his own farm, he shall forfeit and pay m of twenty shillings, H. Cy., (meaning Halifax currency).

any person in this township shall put fire to any rubbish, brush, e, or whatever, whereby his neighbors, or the publick may be in r of receiving damage thereby, without previously acquainting two of ighbors, so that he may have help to suppress the fire in case of e, he shall forfeit and pay the sum of forty shillings, H. Cy., to the of the 'highways.'"

ppears from the above that even at that early day the "Canada had made its appearance there ; and it is extremely creditable to nry of this part of the country, to be able to state, in this con- that the land throughout Hallowell and the whole of Prince Edward is for the most part very free from this pestiferous weed—extremely t, in comparison with some of the townships of Western Ontario, e settlements are of a much more recent date, and where the owners ast of the superiority of their lands. And to those acquainted with nd outs of the agricultural profession as carried on in this county, e speaks volumes, not only for the *present* inhabitants of this de- agricultural township, but for those who preceded them in the on of the soil since the place has had a history.

wing the course of municipal events : after the first "Town Meet- find that Rozel Ferguson continued to be clerk till 1810, when Armstrong was chosen to the position, which he held two years, ceeded by Arva Ferguson, who held the office three years. Then in succession, Simeon Washburne, Ebenezer Washburne, Arva (three years), Robert Scott (two years), Arva Ferguson, Simeon ne, Arva Ferguson (seven years), William Rorke (two years). This down to 1832, when James McDonald was appointed clerk, and the cial" meeting was held on March 24th of that year ; after which ings gradually became more frequent, and the reports of the pro- more voluminous, till they finally grew into the present system. 12 the following names appear in the lists of "Town and Church "—corresponding with our Councillors of to-day, James Blakeley, Goldsmith, Barret Dyre, James Fulton, Samuel Gardiner, Henry Peter D. Conger, Samson Striker, Farrington Ferguson, Isaac Asa Worden. Two of the above gentlemen were elected each year. hem served quite a number of times. Messrs. Dyre and Goldsmith greatest number of terms.

has been no record kept, during that early day, of what we now *necessary* statistics, so we are kept in ignorance as to the wealth, n, &c., from year to year. The last assessment returns give the figures :—

population of township, 3,515 ; total number of acres of land 42,- l number of acres improved, 31,157 ; value of real property, $1,417,- e of personal property $80,000 ; amount of taxable income, $1,200 ; ber of horses, 1,685 ; cattle, 2,456 ; sheep, 1,853 ; hogs, 752.

resent incumbents of the various municipal offices are as follows :— .—Lewis B. Stinson ; Deputy Reeve.—Henry Hubbs ; Counci- . Huyck, G. I. Lazier, Wm. H. Johnston ; Clerk.—Thomas H. Assessor.—Albert Clarke ; Collector.—Oliver Dingman ; Trea- Obadiah Cooper.

field, the only village of any size within the corporate limits of , is near the head of West Lake, and is situated near the centre of hip, on the old Danforth Road, and about midway between Picton ington. It is comparatively an old place, quite a large settlement en formed here in the very early days of the township. Among make the place their home were the Leavens', Bulls, Bowermans, Clarke, Cornelius White, James Barker. The Bulls and Bower- e both large families. Jonathan Bowerman built the first grist and John Bull the first sawmill. The beautiful stream which gh the village and empties into the head of West Lake, affords pportunities for the milling business ; and the advantages were zed by the enterprizing citizens, thus making Bloomfield quite a trade. The village extends over a mile along the Danforth Road nsiderable number of cross-streets. The plan of the place is irre- is also the ground on which it is built ; all of which, however, dd to its picturesque appearance, and give it an ensemble of ariety extremely enjoyable to the traveller. The residences are far exceeding those of the average town of equal size, and have been seemingly laid out and built with a more than usual view to the health, comfort, and convenience of their respective occupants. Beautifully shaded avenues, luxuriant flower-gardens, well-kept grounds,—these and other attributes of a competency of this world's goods, coupled with refined taste and praiseworthy liberality, are to be seen on every hand.

The business portion of the place contains one boot and shoe store, two general stores, two water grist-mills, two steam saw-mills, steam shingle and stave factories, planing mills, one tannery, one woollen factory, one cheese factory, two brick-yards ; also a post office, with daily mails both ways, to and from Picton and Trenton per stage ; two telegraph offices and four churches—some of them very fine ones,—viz : one Methodist, one Universalist, and two Society of Friends.

The Town Hall is also here situated ; it was built in 1857, by John Thompsett, contractor, at a cost of $2,000 ; and with ground and sheds attached is worth about $3,000. For a long time—probably till the incorporation of Picton—the town meetings were held at Hallowell village ; generally at some private house at first, and later, in 1807, at the "Inn of Thomas Eyre," where they continued to be held for many years. It was customary in those days to hold town meetings "in the presence of certain of His Majesty's Justices of the Peace." As will be seen from the minutes, Augustus Spencer and John Stinson, junior, presided in that capacity at the first town meeting ever held in the place. The next year John Peters was added to the list ; and under the supervision of those three, who were without doubt the first magistrates in the township, the town meetings continued to be held for many years.

Bloomfield is the seat of the sixth Division Court of Prince Edward County. Obadiah Cooper is Clerk of the Court, and John Thompsett is Bailiff.

There is another post village in Hallowell, viz: West Lake, a pleasant hamlet located about eight miles south-east of Picton, and on the south shore of that body of water from which its name is derived.

A short distance further on and about ten miles from Picton, we come to "The Sand Banks." These are high hills, or ridges, running in a general northerly and southerly direction, and composed of pure white sand, which has been driven hence from time immemorial, by the winds and waters of Lake Ontario, and by the raging south-west storms, until, by degrees, these "Sand Banks" have acquired their present form and size. They extend respectively across the mouths of East and West Lake ;—in the former case, being more broad, and of less altitude, and consisting of two points, or "tongues," overlapping each other to a long distance, and divided by a narrow channel, which is the outlet of East Lake. The "Wellington Banks," or those across the mouth of West Lake, are so called from their proximity to the village of that name, though within the municipal bounds of Hallowell. The ridge here is very high and quite narrow, and of a length (approximately) of three miles ; divided only near the north end, which joins the south-east portion of the township, by a narrow channel fordable for men or horses, which connects West Lake with Lake Ontario. The Sand Banks are, beyond dispute, curiosities of no mean order. The northern, or "Wellington" Banks, are the most celebrated, from their greater height. They are covered in spots by groves of beautiful trees, which form pleasant picnicing grounds for the numerous parties of tourists and pleasure seekers during the summer season. In other places, the bare white sand glittering in the sunlight, presents a curious and agreeable contrast to the variegated foliage of its many natural parks ; while these, with the blue waters of West Lake at their feet, and the sea-green of Lake Ontario, dotted with snow-white sails, and covered with the ethereal blue of a cloudless sky, combine to form a panorama of nature which it delights the eye of the artist to rest upon, and commends the Banks as one of the chief points of attraction and interest in this extremely attractive and interesting locality. There is a magnificent wide and sandy beach stretching from the Banks into the Lake on either side, forming a most delightful place for bathing ; and it is in the contemplation of some of the enterprising citizens of Picton to build a large summer hotel here, in which case "The Banks" will undoubtedly become a highly popular summer resort, and fashionable watering-place.

In the eastern part of the township, we have not yet brought the history of the locality up to the period when it became a town, so we may be allowed to revert for a short time to "Hallowell Bridge," the facts connected with the early settlement of which have already been noted. There was a public-house built here at an early day, known as "Hovington House," from the name of the man who built and kept it. By this time the settlement of the Youngs, at East Lake, and that of the Stinsons and Congers, at West Lake, had received considerable accessions to their numbers and importance ; and as they were all obliged to go to Kingston mills, (and after 1796, to the "stone mills"), their routes across the camping places terminated at Hallowell Bridge. This made quite a traffic through the Bridge, and Hovington House met with a liberal degree of patronage, and became the centre of all scenes of sport and gaiety, not only for those living at East and West Lakes, but it is said to have been frequented by settlers from along the bay shores, to the distance of forty miles either way. This made things lively at the old Hovington, until one by one, both mills and public houses were built in various other localities. Still, after that the local trade increased, and the Hovington continued for many years to deal out food to the hungry, drink to those who thirsted, and music to the disciples of Terpsichore, for many years. The Hovington still occupies a place in the memory of the oldest inhabitants, as one of the oldest land-marks of the place, about which—though but a tavern—many pleasant reminiscences of the early days of Hallowell Bridge are entwined.

The village grew apace ; not rapidly, 'tis true—but surely, still. Meantime another village also sprang up at the head of the Bay, across the bridge, on the east side of the creek. It was founded by the Rev. W. Macaulay, who came into Hallowell at an early day, and by grant or purchase, or both, came into possession of a large tract of land. This estate, and the collection of houses which he built upon it, near the bridge, he named Picton, in honor of Major General Picton, of the British Army, a distinguished officer, who was a friend or relative of the Macaulay family, and who fell at Waterloo a short time previously. Great rivalry existed between Picton and Hallowell villages, from the time the former took its start—each striving to outstrip the other, and secure to itself the advantages accruing to the place of greater size.

When Prince was set off as a separate district, in 1831, and it became necessary to build a Jail and Court House, a bitter strife arose between the two villages as to which should have them. Rev. Mr. Macaulay being a man of public spirit and wealth combined, succeeded in securing those public buildings to his village, by giving a free grant of land and a present of two hundred dollars in cash, to assist in their erection. The two villages continued to be called by their above respective names till their incorporation into a town, neither having ever occupied the middle position of an incorporated village. Then the inhabitants of Hallowell wanted the town thus named ; those of the other village were indifferent, while Mr. Macaulay was determined that Picton should be the name, and through his influence at Court, he again succeeded in carrying his point against the wishes of the majority, and had the name of Picton inserted in the Act of Parliament which incorporated the new town.

As Mr. Macaulay was the one who named the present Town of Picton, and the founder of the village of the same name, now a part of the town, a short sketch of the rev. gentleman might not be uninteresting. He was one of a family of several sons, whose father settled in New York previous to the Revolution, and being a man of considerable prominence among the English, their abandonment of that city was the signal for his departure for Canada. Here he engaged in mercantile pursuits, and for many years after taking up his residence in this country, he was the principal contractor for supplies to the troops in and about Carleton Island and Kingston. He amassed a great fortune in trade at Kingston, where he died in 1800, and his business was subsequently carried on by his family with great success. The sons were among those exerting the highest influence of any family, or any individuals in the country, in public and political affairs. One of them was the Hon. John Macaulay, a man whose name was closely identified with all public questions of the day. Among the many honourable and highly important positions filled by one or other of the brothers was that of Member of the Legislative Assembly, Chaplain to the Legislative Assembly, Provincial Secretary, Surveyor-General, Inspector-General, and various other positions of almost equal importance. At the time of the incorporation of Hallowell, one of the Rev. Mr. Macaulay's brothers was Chaplain to the Canadian Houses of Parliament. The result of all this was that in spite of the protests of the people, the place received the official name of Picton, as above stated; and although Mr. Macaulay evinced in this matter a very determined disregard for the wishes of his parishoners, still it must be admitted that in all matters of public concern—whether of Church or State, or municipal affairs, or public or private undertakings for the advancement either of his country or his own particular parish, he was beyond question a most enterprising and public-spirited citizen, and one to whom the present town of Picton undoubtedly owes much of its material prosperity, as well as its name.

In regard to the incorporation of Picton, we have been unable to arrive at the exact date, as there are no preserved records of the public acts of the governing body of the town previous to the commencement of the operation of the Municipal Act. We have the authority of a gentleman who had been identified with the affairs of the place, however, for many years previously, that the incorporation took place about three years—or certainly not more than five—previous to the passing of the Municipal Act, which would probably be in 1847, or possibly as early as 1845. From this time up to 1850 the town was governed by what was called a "Board of Police," also called "Commissioners of Police." The commissioners who comprised the first Board were,

Philip Low, chairman; Alexander McAllister, now of the Customs Department, Kingston; and the late Sheriff McDonald. Mr. Low was Chairman of the Board for some years, and at the time the Municipal Act came into force.

We find by reference to the oldest preserved minutes, that the first meeting of the Town Council under the said Act, was held at the Court House, on the 21st January, 1850, when the following-named gentlemen handed in their respective certificates of election and qualification as town councillors, and took the oath of office, viz.:—

For Hallowell Ward—William Hale, James McDonald, Calvin Pier.

For Brock Ward—James Porter, David Barker Stevenson, Walter Ross.

For Tecumseth Ward—John Proctor, John Miller, Philip Low.

On motion of Mr. Hale, seconded by Mr. Ross, Mr. Low was *unanimously* chosen the first mayor of the town. Mr. Lanprière was then chosen Town Clerk and Treasurer, and the following officers were subsequently chosen at the same meeting in the usual manner, viz.:—

John G. Reynolds, Chief Constable, Assessor, Collector, and Census Enumerator.

Cecil Mortimer (by Mayor), James Cook (by Council), Auditors.

David Barker Stevenson, Town Reeve.

James Shannon, Weigh-master.

Thomas Donnelly, Superintendent of the Public School.

The Council and town officers for the year 1878 are as follows, viz.:—

John R. Allen, Mayor; Wm. P. Reynolds, Reeve; J. Platt Nash, M.D., Deputy-reeve; John Twigg, Clerk; R. J. Chapman, Treasurer; George H. Hatt, Assessor; John Gibson, Collector; Dr. Platt, Superintendent of Public Schools; Ellis Grimmon, Chief of Police; Wm. P. Reynolds, Chief of Fire Brigade; Jabez Bennett, William Carter, William Coughlin, Jacob J. Fralick, Richard A. Foster, Charles Garner, Arthur W. Hepburn, John Perrill, George C. Van Blaricom, George E. Van Dusen, J. F. Van Horn, Isaac N. Wait, Councillors.

The Picton of to-day is a delightful town, possessing very many advantages of location as to the rest of the county, and of position as to water communication, which have always given it an advantage over other parts of the district, and resulted in making it the general centre of all commercial and financial business, as well as the county seat. It is composed of two parts, one on each side of the creek which empties into the bay, and formerly known as Hallowell and Picton respectively. The east side contains quite a large number of stores, the jail, the court-house, two churches, two hotels, and some of the finest residences, without exception, which are to be seen in any town in the country. This can also be said of the residences in all parts of the town, for proof of which we refer our readers to some views herein contained, many of which are fine specimens of architectural beauty rarely met with in edifices of like class.

The west side, or old Hallowell, is still the principal part of the town. Here are situated the Banks, Insurance Offices, County Offices, Registry Office, Post Office, Telegraph Offices, Express Office, the finest Churches, the leading Hotels, the Driving Park, the splendid grounds of the County Agricultural Society, the High School, the Town Hall, the Newspaper Offices, the finest stores, and by far the greatest number of them, most of the Iron-working and Carriage Shops, &c., &c., and the larger portion of the inhabitants.

The two parts of the town are connected by bridges over the creek—that originally known as "Hallowell Bridge" being now a very handsome arched stone structure, as broad as the street on which it is built; and one side separated from the extreme head of the bay, only by the stone wall forming the usual and necessary protection to pedestrians on bridges of similar character.

The Bay here runs to a narrow point, opening gradually towards the mouth, with shores of comparatively regular outline, and of the general shape of the letter "Y," with the angle very acute. Along these shores on either side, and extending for a considerable distance from the angle of the "Y" there are wharves, docks, and store-houses, sufficient to accommodate a very large amount of marine commerce.

The western part of the town is built upon a level table land of the probable height of one hundred feet above the waters of the Bay. This dips down somewhat suddenly to the creek which divides the two portions of the town, while along the Bay it runs on a level almost to the water's edge, which is reached by descending a steep and almost perpendicular precipice which extends nearly to the water level, then breaks abruptly and slopes gradually to the water, leaving a street, as it were, of sufficient width which has been utilized for the purposes of commerce, and is capable of improvement in the same direction, with comparatively light cost, to a practically unlimited extent. The ground on the east side rises more gradually from both creek and Bay, but to a much greater height, forming in fact a mountain, while the west side is simply a part of the "High Shore," so called, and as such known from the earliest days, which extends all along that side of the Bay to the farthest point of Sophiasburg. Taken together, the beautifully shaded streets, well-built business blocks, fine public buildings, and splendid private residences, with the magnificent grounds which in a great majority of cases surround them, make Picton one of the pleasantest places imaginable. But a more comprehensive idea of the place than can be derived from any general description, may be formed, if we mention in detail some of those features which particularly strike the outside observer, under distinct heads.

First then we might call attention to the large amount of trade transacted, which is proven by the large number of business establishments of the place, a careful estimate of which places the number as follows:—

2 Newspapers, 2 Banks, 6 Lawyers, 7 Doctors, 4 Drug Sto
Stores, 3 Jewellery Stores, 2 Photographers, 4 Hardware Stor
Goods Stores, 5 Merchant Tailors, 7 Boot and Shoe Stores, 1 H
Store, 3 Butcher Shops, 4 Bakeries, 31 Groceries, 4 Confectioneri
Fancy Stores, 3 Liquor Stores, 7 Hotels, 1 Temperance Hotel, 2
1 Brewery, Several Tobacconists, 3 Barber Shops, 3 Livery Stable
age Shops, 1 Waggon Shop, 7 Blacksmith Shops.

There are many very fine private business blocks, principally
Upper Main street. The principal ones are Striker's block, t
block, Mottashed block, Ayres block, Oddfellow's Hall, and
Office block.

Among the finest business blocks in the place are THE BANK—
there are two, viz.: Bank of Montreal (R. J. B. Crombie, Agt.)
Standard Bank (George P Reid, Agt.)

THE JAIL AND COURT HOUSE were built in 1831, at a cost, inc
ditions and improvements since effected, of about $24,000. The
substantial and moderately handsome stone buildings, beautifull
on the east side.

THE TOWN HALL is a handsome building, well calculated to
wants and requirements of the citizens in matters relating to t
business of the place. It was built in 1866, at a cost of $6,000.

The old HIGH SCHOOL is a very large and handsome brick buil
together with the wings added subsequently to the erection of
building, now forms the CENTRAL PUBLIC SCHOOL. The citizen
building a new HIGH SCHOOL at a cost of $5,500.

The building called THE COUNTY OFFICES is a fine brick stru
ated on Lower Main street, and on the brink of the "High Shore,
erected in 1876, at a cost of $9,000, and contains the offices of t
County Crown Attorney, and Clerk of the Peace, Registrar of the
Court, Deputy Clerk of the Crown and Pleas, County Judge's
Clerk of the First Division Court of the County, Clerk of the Cou
Clerk of the County Council, and the Chamber of the County Co
the head office of the Prince Edward County Mutual Fire Insura
pany. The various offices are laid out too with due regard to th
ments of the respective officials occupying the same, and all are
with the best fire-proof vaults. The County Council chamber is
best we have anywhere seen. For the size of that Municipal Bod
kind of business transacted at their meetings, no more complete
venient hall could be wished for, whether for convenience of the
themselves, or their constituents, or others who may choose to
meetings or witness their deliberations.

THE COUNTY REGISTRY OFFICE is situated immediately beside t
offices. It is of brick, and built on the usual plan approved by th
Government. There are now so many fine buildings of this class t
the country, exactly similar to it, that a description would be supe

THE CHURCHES are five in number, some of them being quite
of considerable pretentions as to architectural beauty. The grou
of these, viz.: the English and Roman Catholic, was presented by
Macaulay; and he also contributed a great share of the cost of th
church, which was built in 1825, the first church in the limits o
sent town of Picton. The first Roman Catholic church was built
by Rev. Father Brennan. It is now used as a separate school.
stone church—Saint Gregory's,—a view of which we give in anoth
this work was built in 1839, by the Rev. Father Lalor. The
Church of Canada, and the new Episcopal Methodist church are
did and handsome structures. They cost about $15,000 and $16,0
tively. The first Methodist church built in Prince Edward County,
one of the very oldest in the Province of Upper Canada, was the c
near Picton. It was erected on land given by Stephen Conger, in
the first trustees were Benson Conger, Dougal, German, Van Blari
Dusen, and Wilson, all long since gone the way of earth. As l
beginning of 1818, but one other place of worship—a Quaker meet
—had been erected in Hallowell, though preparations were then
to build a Presbyterian church.

A PUBLIC CEMETERY was opened in 1873, under the conditi
Cemetery Act. Of an area of twenty-five acres—in a delightful si
it is a most appropriate place for the burial of the dead. Much
done to enhance the natural beauty of the place, which is one of
possible spots for the purposes intended. It already contains a l
ber of beautiful and costly monuments to the memory of loved ones
fore.

THE NEWSPAPERS of Picton compare favorably with those of
of equal size in this country or the United States, both for neatne
pearance, general interest of their contents-matter, and the manner
they are conducted. They are two in number; *The Gazette*, Cons
and *New Nation*, Reform. The former was established as long ag
30th, 1830. It was at that time called the *Hallowell Free Press*,
the first newspaper published between York and Kingston. It ha
fitted office; the machinery is run by steam, and the *bona fide* circ
the paper is twelve hundred. The *New Nation* was started in 1868,
ing *The North American*, which had been published here by the
Brothers, since 1861. In January, 1877, it was merged with *Th
Times*, which had been established by Messrs. Gillespie and Boyle,
Mr. Boyle continuing its management, till it merged with *The Nati*
the present name of *The New Nation and Times*. It prints over tw
dred copies. Another paper, *The Picton Sun*, was established in
Mr. J. Douglas, but removed to Cobourg, in 1853.

EDUCATIONAL.—There are in Picton a High School, a Centra
School, two Primary Public Schools, and one Roman Catholic
School. The principal officials of the schools department are as fol
J. M. Platt, M.D., Inspector of Public Schools; J. A. Clark
Principal of High School; N. Gordon, Head Master of Public Sch
C. Curry, Chairman, A Bristol, Secretary of the Public School Boa
L. Fralick, Chairman, C. D. Morden, Secretary of the High Scho
There are two teachers in the High School, eight in the Centra
each of the Primaries, and two in the Separate School; a total of
teachers employed in the High and Public Schools of the town.

ONTARIO COLLEGE.—This was the name of an institution desi
all the higher branches of education, which was chartered by Act
ment and brought into successful operation in May, 1867. It
especially in connection with the Established Church, though p
freedom from all sectarian forms. Those who were chiefly instrum
principally active in getting Ontario College into operation, were th
of England Clergymen of the Bay of Quinte District, and the leadin
of that religious body. Philip Low, Q.C., was one of the most ac
efficient organizers. Bishop Lewis, of Kingston, was also a ver
worker, and the nominal head of the College. Quite a handsome en
fund was subscribed for its support, and it entered the list with a fai
ance, a fine staff of Professors, and every prospect of a brilliant fu
was found, however, that the great delay and inconvenience in co
with visiting the place, caused by lack of travelling facilities—pa
in the winter season—soon caused the attendance to drop off, and p
fresh accessions to the ranks of the students; so it was closed aft
five years' operation. The building, which is now occupied by A
Hepburn, Esq., was a delightful place in every respect, for the pur
which it was intended. It was built by Col. Ryland, of Montreal,
present one of the finest private residences in the country.

MILITARY.—Picton is the head quarters of the 16th Regiment
teers, a body of men so well and favorably known in the military
the country, that they need no praise at our hands. It formerly
of nine companies, but the parsimonious action taken by the Gover

s, 6 Lawyers, 7 Doctors, 4 Drug Stores, 3 Book s, 2 Photographers, 4 Hardware Stores, 12 Dry Tailors, 7 Boot and Shoe Stores, 1 Hat and Fur Bakeries, 34 Groceries, 4 Confectioneries, Several stes, 7 Hotels, 1 Temperance Hotel, 2 Tanneries, onists, 3 Barber Shops, 3 Livery Stables, 4 Carri- s, 7 Blacksmith Shops.

ne private business blocks, principally located on principal ones are Striker's block, the Conger Ayres block, Oddfellow's Hall, and the Post

ess blocks in the place are THE BANKS, of which of Montreal (R. J. B. Crombie, Agt.), and the Reid, Agt.)

HOUSE were built in 1831, at a cost, including al- since effected, of about $21,000. They are very handsome stone buildings, beautifully situated

andsome building, well calculated to meet the f the citizens in matters relating to the public was built in 1868, at a cost of $6,000.

is a very large and handsome brick building, and, added subsequently to the erection of the main CENTRAL PUBLIC SCHOOL. The citizens are now ot, at a cost of $5,500.

E COUNTY OFFICES is a fine brick structure, situ- , and on the brink of the "High Shore." It was of $9,000, and contains the offices of the Sheriff, and Clerk of the Peace, Registrar of the Surrogate he Crown and Pleas, County Judge's Chambers, Court of the County, Clerk of the County Court, il, and the Chamber of the County Council, also e Edward County Mutual Fire Insurance Com- are laid out too with due regard to the require- ficials occupying the same, and all are provided ults. The County Council chamber is one of the n. For the size of that Municipal Body, and the l at their meetings, no more complete and con- hed for, whether for convenience of the members tituents, or others who may choose to visit their deliberations.

OFFICE is situated immediately beside the county built on the usual plan approved by the Ontario now so many fine buildings of this class throughout ar to it, that a description would be superfluous.

e in number, some of them being quite costly, and s as to architectural beauty. The ground for two and Roman Catholic, was presented by Rev. Mr. ntributed a great share of the cost of the English in 1825, the first church in the limits of the pre- first Roman Catholic church was built in 1830. It is now used as a separate school. The new ary's,—a view of which we give in another part of 839, by the Rev. Father Lalor. The Methodist e new Episcopal Methodist church are both splen- res. They cost about $15,000 and $16,000 respec- st church built in Prince Edward County, as well as the Province of Upper Canada, was the old chapel ed on land given by Stephen Conger, in 1809, and son Conger, Dougal, German, Van Blaricom, Van long since gone the way of earth. As late as the other place of worship—a Quaker meeting house llowell, though preparations were then going on urch.

was opened in 1873, under the conditions of the t of twenty-five acres—in a delightful situation— lace for the burial of the dead. Much has been al beauty of the place, which is one of the finest oses intended. It already contains a large num- monuments to the memory of loved ones gone be-

Picton compare favorably with those of any town ty or the United States, both for neatness of ap- d their content-matter, and the manner in which are two in number; *The Gazette*, Conservative; The former was established as long ago as the. at time called the *Hallowell Free Press*, and was ed between York and Kingston. It has a well- s is run by steam, and the *bona fide* circulation of d. The *New Nation* was started in 1865, succeed- which had been published here by the McMullen January, 1877, it was merged with *The Picton* ublished by Messrs. Gillespie and Boyle, in 1854, anagement, till it merged with *The Nation*, under *ew Nation and Times*. It prints over twelve hun- s, *The Picton Sun*, was established in 1841, by ed to Cobourg, in 1858.

re in Picton a High School, a Central Public lic Schools, and one Roman Catholic Separate ials of the schools department are as below:— spector of Public Schools; J. A. Clarke, M.A., N. Gordon, Head Master of Public Schools; G. sted, Secretary of the Public School Board; Jas. D. Morden, Secretary of the High School Board. in the High School, eight in the Central, one in two in the Separate School; a total of fourteen igh and Public Schools of the town.

his was the name of an institution designed for lication, which was chartered by Act of Parlia- ccessful operation in May, 1867. It was more with the Established Church, though professing tions. Those who were chiefly instrumental and g Ontario College into operation, were the Church he Bay of Quinte District, and the leading laymen ilip Low, Q.C., was one of the most active and ge Lewis, of Kingston, was also a very active nd of the College. Quite a handsome endowment support, and it entered the list with a fair attend- ers, and every prospect of a brilliant future. It he great delay and inconvenience in connection sed by lack of travelling facilities—particularly caused the attendance to drop off, and prevented ks of the students; so it was closed after about building, which is now occupied by Arthur W. htful place in every respect, for the purposes for was built by Col. Ryland, of Montreal, and is at vate residences in the country.

he head quarters of the 16th Regiment of Volun- and favorably known in the military circles of d no praise at our hands. It formerly consisted pardonable action taken by the Government in military matters during the last few years has had its effect here, as well as elsewhere throughout the country, and three very fine companies were dis- banded and "gazetted out." As the Regiment now stands, the Field, Staff and Line Officers are as follows:—

Lieut.-Col. Walter Ross, M.P.; Major, Bvt. Lt.-Col. Bog; Paymaster, Captain Walter T. Ross; Quartermaster, Captain Donald Ross; Surgeon, Dr. [illegible]; Assistant Surgeon, Dr. Ingersoll.

No. 1 Company, *Picton*; Capt. J. J. Fralick, Ensign Aylesworth.

No. 2 Company, *Milford*; Capt. Ostrander, Lieut. Ackerman, Ensign Grimmon.

No. 3 Company, *Milford*; Capt. Van Dusen, Lieut Craig.

No. 4 Company, *Picton*; Capt. Johnston, Lieut. Foster.

No. 5 Company, *Roblinsville*; Capt. Dempsey, Lieut. Anderson, Ensign Cunningham.

No. 6 Company, *Robbin's Mills*; Capt. Peterson, Lieut. Rothwell.

CAVALRY.—Picton is also head-quarters of "D" Troop, 4th Regiment Cavalry, Lt.-Col. Duff, with head quarters at Kingston. The officers of the Troop are, Captain, Major White; 1st Lieut. H. McCullough; 2nd Lieut. Van Vatten.

Lieut.-Col Ross, of the 16th Regt. is now the senior Colonel of this Military District

During the rebellion of 1837-8, Prince Edward raised and sent to the front two companies of infantry, and one of cavalry, besides detachments for the Artillery and Engineers. One of these companies was put into Col. Taylor's Regiment, and did service in the Western Peninsula, at Sandwich, and Fort Malden. The other marched to Toronto, York, and was put into Col. Kings- mill's Regiment, which did duty there while the troublous times lasted. This company was commanded by Capt. Flagler—Wm. Dempsey, Lieut.—and named themselves "The Queen's Own," from the fact that when marching over the camping place, Capt. Wilkins witnessed their "march past," and was so pleased therewith that he involuntarily exclaimed, "Look! the Queen's Own!!" On arriving at York, the question of a name for the regiment came up, and Lieut. Dempsey suggested his company's self-assumed name, which was adopted; and in this way that splendid regiment "The Queen's Own," of Toronto, came by its title.

The cavalry troop was commanded by Capt., afterwards Lt.-Col. the Hon. Robt. Charles Wilkins, and were stationed in detachments along the Danforth road, between Carrying Place and the "Old Marsh Tavern," 5 miles above Port Hope. Prince Edward is embraced in the third Military District, the head-quarters of which are at Kingston.

NAVIGATION.—The splendid natural facilities of Picton, as a shipping point, have already been referred to. The county is well supplied with the best of shipping facilities on every side, but the advantages by Picton in other lines of business, which outside points do not possess, draw the great bulk of the produce of the county to its warehouses. The amount of all kinds of produce, including stock, annually passing over Picton Bay is im- mense. The only figures we have at hand are those of the U. S. Consular Agent at this Post, John Quincy Sullivan, Esq., and include *direct* shipment to the U. S. during 1877, and up to 30th Nov., of that year; 162,066 bush- els barley, 10,605 bushels peas, 11,675 bushels rye, 48 bushels potatoes, 81 barrels apples, 4,851 dozen eggs, 295 hides, 31 cows, 2,084 sheep, represent- ing a total value of $111,570.38. The smallness of these figures is accounted for by the fact that most of the Picton buyers deal in Kingston; and the steamers hence to the U. S. calling there, they take out Consul's certificates at the latter place. A leading produce dealer estimates that about a quarter million of bushels of grain pass over Picton harbor annually, and that an average of at least 10,000 barrels of apples is annually shipped from this port to Montreal and the Maritime Provinces.

INSURANCE.—Almost every Insurance Company operating in Canada— whether fire, life, marine, accident, or guarantee—has an agency in Picton. The reason is to be found in the fact, that being central, and easy of access from all parts of the District, and the only town in the county, the farmers all centre in to that place to do their trading, and consequently their insur- ance business also. There are two companies having their head offices here. "The Prince Edward County Mutual Fire Insurance Company" is the name of the one, which is represented by the following officers:—

President, Lewis B. Stinson, who is also Manager and Inspector; Vice- President, John Prinyer; Secretary, C. D. Morden; Treasurer and Solicitor, Philip Low, Q.C.; Directors, Lewis B. Stinson, John Van Alstine, A. H. Taylor, T. B. Hubbs, John Prinyer, George Martin, Carleton McCartney.

The "Prince Edward" does a large and prosperous business; careful man- agement, economy, and square dealing having succeeded in reducing the ex- penses of the insurer as nearly as possible to the long desired basis of "In- surance at Cost," which so many stock companies claim with more assurance than honesty.

The other is the "Bay Quinte Mutual Fire Insurance Co." The officers are as below:—

President, Archelaus Southard; Vice-President, Andrew Wycott; Secre- tary, Frederick Dodge; Treasurer and Solicitor, R. S. Roblin; Directors, Robert Carson, John Murney, R. R. Burlingham, William Blakeley, William G. Stafford, David B. Spencer, Thomas Welbanks.

This is a comparatively young company, but it has already succeeded in securing a fair share of the Insurance business of those sections of the country in which it operates.

MASONIC.—Prince Edward Chapter of Royal Arch Masons, No. 31, G. R. C., is held at Picton. Following are the officers:—

Z., Wm. Smeaton; H., D. W. Allison; J., T. N. Van Blaricom; S. E., A. J. Corkindale; S. N., Wellington Boulter; P. S., James Tennent; S. S., William Lawson; J. S., W. H. Orchard; Janitor, P. McFadden.

There is a Blue Lodge held here, viz.: Prince Edward Lodge, A. F. and A. M., No. 18, G. R. C., of which the following are the present officers:—

W. M., William Lawson; S. W., N. B. Rose; J. W., E. Blakeley; Sec., Wm. Smeaton; Treas., Wm. P. Reynolds; S. D., Lucius Hart; J. D., Samuel Minaker; I. G., James W. Barker; Tyler, Patrick McFadden.

The masonic district includes Frontenac, Lennox and Addington, Hast- ings, and Prince Edward. Donald Ross, of Picton, previous to the present year, held the position of D. D. G. M. for three successive years.

METEOROLOGICAL AND CLIMATIC.—We owe it to the interest in science of Levi Varney, Esq., of Hallowell, that we are acquainted with the climatic changes and rainfall for a period of nearly a decade. The figures were ob- tained by careful daily observations at 6 o'clock a.m., and at 2 and 6 p.m., with the thermometer in the shade on the north side of the house.

YEAR.	MEAN.	HIGHEST.	LOWEST.	RANGE.	
1857	42.25°	86°	—25°	111°	☞The lowest tem-
1858	45.29°	88°	—7°	95°	perature is below zero,
1859	45.15°	91°	—19°	110°	indicated by the sign
1860	46.06°	86°	—14°	100°	(—) *minus*.
1861	45.44°	87°	—29°	116°	☞ The observation
1862	45.81°	90°	—7°	97°	in 1865 extends through
1863	46.10°	91°	—20°	111°	only 11 months. All
1864	46.43°	92°	—16°	108°	but mean temperature
1865	48.19°	....	......	....	is consequently omitted.

The annual rainfall for six years, commencing with 1860, is as follows :—

YEAR.	INCHES.	YEAR.	INCHES.
1860	27.12	1863	27.59
1861	36.93	1864	35.32
1862	24.13	1865	22.53—(11 months)

It will be seen from above that the highest temperature in eight years was 92°, and the lowest 29°, or a range in eight years of 121°. The greatest range during any single year was 116°, and the lowest 95°. The highest mean temperatrue of any single year was 46.43°, and the lowest 44.25°, or a range of 2.18° of mean temperature in eight years.

The difference in the yearly rainfall is proportionately greater than the variation of the temperature. The lowest depth falling during a single year was 24.13 inches, and the highest 36.93 inches, a range of 12.80 inches—the greatest fall being 50.17 per cent. more than the least, over a period of five years.

AGRICULTURAL.—The Prince Edward County Agricultural Society is in a most flourishing and prosperous condition, owning one of the finest fair grounds of any county society in the Province. It boasts of a delightful situation, and for convenience of location, and excellence of appointments, is all that could be suggested or desired. The main building, for the display of manufactures and arts, is a fine commodious structure, containing plenty of room and all facilities for the purposes for which it was designed; while the grounds are supplied with sheds and all conveniences for the proper care of stock, &c. There is a house on the premises, built expressly for, and occupied by, a caretaker, kept constantly on permanent salary, to attend to the grounds and look after the general interests of the society. The value of the property approaches very near $6000. Below are the officers for the present year :—

President, Wellington Boulter, Demorestville; 1st Vice-Prest., D. Spencer, Rose Hall; 2nd Vice-Prest., Isaac Minaker, Picton; Sec. and Treas., R. S. Roblin, Picton; Andrew Davidson, G. A. Wellbanks, R. G. Davis, W. G. Stafford, Stephen B. Hubbs, W. B. Cooper, Robert McCartney, H. H. Huyck, and Allen Caven, Directors.

The receipts last year were over $1,900, of which upwards of $400 were paid out to township societies. Membership last year 249, which is about the yearly average.

We make a note of the various cereals, &c., which are chiefly grown in the different townships. There is one agricultural interest, however, which all centres to Picton to find market, and for this reason we propose to give our readers an idea of the great cheese product of the county. As is very well known, this industry has grown from very small beginnings within a few years, to be one of paramount importance. Every farmer is directly interested in it, and the English cheese reports are as eagerly watched by the agriculturists here as are the New York and Chicago wheat markets by the millers and speculators of Western Ontario. With the probable exception of Belleville and Ingersoll, Picton is the greatest cheese market in Canada. There are twenty-six factories in the county, twenty-three of which are now in operation and running full. Some of these are very valuable establishments, and all are furnished with modern and most approved appliances for the manufacture of a first-class article of that staple, in the production of which so much capital has been invested, and for the quality of which Prince Edward is justly regarded as the choicest brand in the English market. One of the principal buyers (Mr. W. Crandall) estimates the cash value of these factories at over $45,000, many of them having cost over $3,000 each to build. A fair average number of cows to each would be two hundred and fifty (this is under rather than over the mark). Compared with the stock of other parts of the country they are good. A fair estimate of the value of all those animals (in the spring season) would be, without fractions, $208,000. Then there must be at least $12,000 represented in various traps connected with the business not before counted—such as waggons, &c., making a total, approximately, of $245,000 invested. The number of boxes average 25,000 per season, varying in value, of course, according to the fluctuations of the markets. Some years $200,000 worth of cheese has been made. Mr. Crandall has handled over $75,000 himself in one season at Picton. He is the principal buyer, though there are a number of others.

The Prince Edward horses have been a household word among stockmen all over the country for many years past; and where people have good stock they like to show their good qualities, particularly if those qualities run in the direction of speed. Consequently we are not surprised to find in Picton one of the finest specimens of a driving park, which we remember to have seen anywhere. The course is admitted by sporting men everywhere to be the best half-mile track in the Dominion. Everything which could be improvised for the best display of the speeding qualities of the stock, and the comfort and convenience of visitors has been provided with a lavish hand, and horsemen from all parts are loud in their praises of the judgment and liberality of the Picton Driving Park Association.

Having now briefly touched upon the main points of interest of Hallowell and Picton—those points which strike the outside observer as most worthy of note—we will take leave of it for a season with a few retrospective and commonplace remarks.

The present town of Picton, dating from its *first settlement*, is now nearly a century old. Within its time many and great changes have occurred—not in our own country alone, but throughout the world. The great majority of mankind, who knew of the events transpiring in their several parts of the universe, when Picton was the embryo town of Hallowell Bridge, have long since done their work among their fellows and passed off the stage of action, and the places that once knew them will know them no more forever. Still there are some few—and they are very few indeed—still living, who lived before the foot of the white man ever trod the shores of the Picton Bay. What a strange and wonderful experience has been theirs! Born in a day when this—the finest province of the "Crown Jewel" of the first nation in the world had yet no existence, they have seen an unbroken wilderness reclaimed from a primeval state of nature and give place to fertile fields of waving corn; and the spots which, at their advent, resounded but to the scream of the eagle, the cry of the panther, or the war-whoop of the savage, are now busy centres of life and industry. Under their own eye they have seen this country have its origin in the wrestling from the British Crown, another, which has since become a leading nation in population, riches, power, trade and commerce, and all the mechanical arts. They have seen their own country grow from a "great unknown Northland" to be second only to the one whose revolt from Britain was the origin of this. They have suffered the miseries of war and enjoyed the blessings of peace in their own home; and while their countrymen have never yielded in the tug of war, still their greatest victories have been the victories of peace—the victory of human industry over rugged nature—the victory of mind over matter. Forest and field have been conquered by the pioneer's axe and ploughshare of the husbandman; while steam and electricity have annihilated time and space. The birch-bark canoe and cumbrous batteaux have given place to magnificent floating palaces propelled by steam; land journeys, instead of being made on foot or horseback, are accomplished in luxurious palace cars; and the antipodes are visited in less time and with greater facilities than those pioneers of olden time could make the distance from New York to Picton in the century which gave them birth.

This, in our own country! while in the old world mighty empires have arisen and mightier have crumbled to ruin. Napoleon, an unknown private soldier, has given laws to the world—adorned his capital with the spoils
and treasures of nations conquered, and mad
kingdom and empire worth governing, save H

This also has been a century of the declin
"State" churches, connected with differ
founded by the blood of the conqueror's swo
Saviour. We have seen the "States of th
where they properly belong, and the "Et
capital of a mighty empire. Nearer home w
Church disestablished, and the ban remove
pised religious sect, one of whom now rules
and is to-day the admitted foremost man of
the present hour the disintegration of an em
ticism, which has ruled by the sword for n
Apostles and the Holy Sepulchre, and gras
science sake, the finest Christian provinces o
of all these signs that the nations are gra
period of religious toleration, which is the ne
to the Christian millennium, Britain still ru
red cross—the "flag that braved a thous
breeze," still waves triumphant over half th
an empire on whose shores the sun never sets

When we contemplate the vast amount o
sion of tremendous events which have crowd
first was "Hallowell Bridge," we are lost in
that Supreme Ruler who has made us what w
it to-day. The thought inspires with awe, a
that He may so direct our affairs as to make
mise under His blessing to become—a land w
foundation and cope-stone of a greatness and
brass, and whose glory and grandeur shall
future ages.

MARYSBURG, S

The original "Ten Towns" of Upper Ca
number of the order in which they were surv
instructions were issued from time to time
survey one township after another, as each
"located" by grant or actual settlement;
pushed the furthest settlements, as was so
territory last laid out. Up to the time of the
above referred to, they were officially known
"First Town," "Second Town," etc., and fo
tinued to be so called by the settlers. A
surveyed, however, the authorities adopted t
numbering them as they were laid out, and
to the "Ten Towns" as follows :—The first,
Great Britain, was called King's Town, which
to Kingston; the second was named after t
Ernest Town, afterwards Ernestown; the thi
King's second son; the fourth Adolphustown,
ber of the Royal Family. Then came M
Ameliasburg, in the order named, and call
Amelia, and Sophia, Princesses of the Royal
twelfth, and fifteenth children respectively
eighth, ninth, and tenth towns were not in
they may be appropriately referred to in t
named in the following order—Sidney, Thurlow
in honour of Lord Sidney, the British Colonia
lutionary War; the others after Lord Thurlow
It will be seen from the map of the Province t
are in Hastings County, and the latter in Len
lies a stretch of land apparently missed by the
"Ten Towns." This, however, simply obtain
and was granted in a block to the Six Nati
their Chief, Thayendenagea, which was in time
A full description of these townships appears

The fifth, sixth, and seventh towns above
the whole of Prince Edward County, which w
the Prince of that name, son of George III.,
Duke of Kent and father of Queen Victoria.

We have no accurate data as to the bounds
Marysburg, as parts of what are now Hallow
in the Fifth Town—the northern portion of th
off to form a separate municipality quite recen
constituted, Marysburg is far from being the
the county, it claims notice in this order from
—as to the survey—of any in the county.
Col. Henry C. Young settled at East Lake—
Hallowell, and now Athol—in 1783, the first
(See ATHOL.) This, however, is disputed on
the Weeses, and many others of Ameliasburg.

We will here remark that Marysburg at pr
municipalities—North and South—the form
corporate existence since January 1st, 1871.
of what was originally considerably more t
Peninsula, is a tract of 23,741 acres of lan
shape, called South Marysburg—the main po
the north by Prince Edward, or South Bay;
North Marysburg, Hallowell, and Athol; on
township of Athol; and on the south by L
known as "Long Point" extends in an
Ontario, terminating at Point Traverse. The
from the winds and waves of Lake Ontario con
finest harbours of refuge, in every respect, of
can boast; while the geographical position of
harbour easy of approach in any weather.
with others of the county, is admittedly the p
an average of but nineteen dollars per acre, by
ment" Committee of the County Council—the
four dollars per acre. Notwithstanding this, t
ally fine farms and many wealthy farmers in
ally, the county may be described as princ
occasional and considerable breaks of a some
which a stronger term than "undulating" mi
logically, its composition is in a high degre
coming to the surface with such prominence
places incapable of sustaining vegetable gro
tinual drought. Other parts are less rocky
sections with rocky foundation have still enou
land less susceptible to lack of moisture, and
almost always of excellent quality, and respon
of the husbandman. To repeat a much-used
his land with the plough, and it laughs with
face of the fact that South Marysburg, ta
average of excellence, yet the industry and
have covered it with comfortable and, in
commodious schools, fine churches, and every
a prosperous and thriving community. Th
careful cultivation, proper rotation of crops,
ness, for which the inhabitants of this section
justly been noted.

easures of nations conquered, and made emperors and kings for every om and empire worth governing, save Britain alone.

is also has been a century of the decline of what has been known as e" churches, connected with different religious faiths—whether ed by the ideal of the conqueror's sword, or upon that of Christ, the r. We have seen the "States of the Church" gravitate to Italy, they properly belong, and the "Eternal City" become again the l of a mighty empire. Nearer home we have seen the "Established" h disestablished, and the ban removed from an oppressed and des- religious sect, one of whom now rules the destiny of Great Britain, to-day the admitted foremost man of all the world. And we see at esent hour the disintegration of an empire built up by religious fana- , which has ruled by the sword for many centuries the land of the ies and the Holy Sepulchre, and ground to the dust, for their con- o sake, the finest Christian provinces of Europe. And while we read these signs that the nations are gradually drawing closer to that of religious toleration, which is the nearest approach yet experienced Christian millennium, Britain still rules the wave, as of old, and the ross—the "flag that braved a thousand years the battle and the ," still waves triumphant over half the world—the proud emblem of pire on whose shores the sun never sets.

hen we contemplate the vast amount of history and the rapid succes- f tremendous events which have crowded into the time when Picton as "Hallowell Bridge," we are lost in "wonder, love and praise" of upreme Ruler who has made us what we are and the world as we see ay. The thought inspires with awe, and fills us with a patriotic wish fe may so direct our affairs as to make Canada what it bids fair pro- nder His blessing to become—a land whose Christian virtues are the ation and cope-stone of a greatness and prosperity more durable than and whose glory and grandeur shall remain for the admiration of ages.

MARYSBURG, SOUTH.

e original "Ten Towns" of Upper Canada were so called from the r of the order in which they were surveyed, and from the fact that ctions were issued from time to time by the British Government to one township after another, as each in succession became either ted" by grant or actual settlement; or, as the tide of emigration d the furthest settlements, as was sometimes the case, beyond the ry last laid out. Up to the time of the survey of the last of the ten referred to, they were officially known by their respective numbers, t Town," "Second Town," etc., and for many years after they con- l to be so called by the settlers. After the "Tenth Town" was ed, however, the authorities adopted the plan of *naming* instead of ering them as they were laid out, and at the same time gave names "Ten Towns" as follows:—The first, in honour of the Monarch of Britain, was called King's Town, which afterwards was abbreviated gston; the second was named after the eighth child of the King t Town, afterwards Ernestown; the third Fredericksburg, after the s second son; the fourth Adolphustown, in honour of the tenth mem- f the Royal Family. Then came Marysburg, Sophiasburg, and asburg, in the order named, and called thus in honour of Mary, a, and Sophia, Princesses of the Royal Household, and the eleventh, h, and fifteenth children respectively of the King. Though the , ninth, and tenth towns were not in Prince Edward County, yet may be appropriately referred to in this connection. They were l in the following order—Sidney, Thurlow, and Richmond—the former our of Lord Sidney, the British Colonial Secretary during the Revo- ary War; the others after Lord Thurlow and the Duke of Richmond. be seen from the map of the Province that the two former townships Hastings County, and the latter in Lennox, and that between them stretch of land apparently missed by the surveyors in laying out the Towns." This, however, simply obtained its "metes and bounds," as granted in a block to the Six Nations, and called in honour of Chief, Thayendenagea, which was in time abbreviated to Tyendenaga. description of these townships appears under the proper head.

e fifth, sixth, and seventh towns above-named originally comprised ole of Prince Edward County, which was thus called in honour of ince of that name, son of George III., who afterwards became the of Kent and father of Queen Victoria.

e have no accurate data as to the bounds of the original township of burg, as parts of what are now Hallowell and Athol were included Fifth Town—the northern portion of the remainder being also taken orm a separate municipality quite recently. Though, as at present uted, Marysburg is far from being the most important township in unty, it claims notice in this order from the fact that it is the oldest o the survey—of any in the county. It is also claimed for it that enry C. Young settled at East Lake—then Marysburg, afterwards ell, and now Athol—in 1784, the first actual settler in the county. ruct.) This, however, is disputed on apparently good authority, by eses, and many others of Ameliasburg. (See AMELIASBURG.)

will here remark that Marysburg at present consists of two distinct palities—North and South—the former having had an independent te existence since January 1st, 1871. All that is now therefore left t was originally considerably more than one-third of the whole ula, is a tract of 23,741 acres of land of an extremely irregular called South Marysburg—the main portion of which is bounded on th by Prince Edward, or South Bay; on the north-east by parts of Marysburg, Hallowell, and Athol; on the south-east, also by the ip of Athol; and on the south by Lake Ontario. That portion as "Long Point" extends in an easterly direction into Lake , terminating at Point Traverse. The shelter formed by Long Point e winds and waves of Lake Ontario constitute South Bay, one of the arbours of refuge, in every respect, of which the great inland lakes st; while the geographical position of Point Traverse renders the easy of approach in any weather. This township, as compared hers of the county, is admittedly the poorest of all, being valued at age of but nineteen dollars per acre, by the "Equalization of Assess. Committee of the County Council—the next lowest being twenty- lars per acre. Notwithstanding this, there are numerous exception. farms and many wealthy farmers in the township. Topographic- e county may be described as principally of an even face, with al and considerable breaks of a somewhat irregular character, to stronger term than "undulating" might justly be applied; geo- y, its composition is in a high degree calcareous, the limestone to the surface with such prominence as to render the soil in many incapable of sustaining vegetable growth through seasons of con- rought. Other parts are less rocky and more sandy, while some with rocky foundation have still enough of soil on top to make the susceptible to lack of moisture, and where it is thus, the soil is lways of excellent quality, and responds most liberally to the touch usbandman. To repeat a much-used expression, the farmer tickles with the plough, and it laughs with a bountiful harvest. But, in the fact that South Marysburg, taken as a whole, is below the of excellence, yet the industry and enterprise of its inhabitants vered it with comfortable and, in many cases, beautiful homes, lious schools, fine churches, and everything tending to constitute it erous and thriving community. This end has been attained by ultivation, proper rotation of crops, and a strict attention to busi- which the inhabitants of this section of the country have long and een noted.

Barley, rye, oats, buckwheat, hops, and corn are the chief agricultural products. There are also large quantities of fruit raised, chiefly apples, though all kinds grow here in as great luxuriance and to as great perfection as in any country of similar latitude and corresponding climate.

As can be seen by a glance at the map, the facilities for shipping are very great; the wharves and store-houses being so numerous along the shores that the farmer is obliged to go but a very short distance to market his grain. This very fact, which equally applies to the other townships of the county, renders it difficult to obtain anything like a correct account of the exports, except in case of such articles as gravitate to a common centre—notably cheese, which is a paramount in this as in other parts of the county. This staple is all marketed at Picton, and in the sketch of that place will be found some reliable information in regard to the cheese product.

The first settlers in what is now South Marysburg were the Moncks, Hicks, Colliers, Loneys, Martins, and Ostranders, along the south shore; the Hubbs, Ellis, Minakers, and Ackermans, along South Bay; and the Clapps, Garrisons, Van Dusens, Van Vlacks, Van Alstines, Wellbanks, Palens, Heads, and Dulmages, in the vicinity of Black River and Milford. Most of the above were U. E. Loyalists or their families. Some were discharged British soldiers.

James Gerollamy was a very old resident, one of the first in the township. He was a U. E. Loyalist and quite young when he came to Canada. He performed important service for the British during the Revolutionary War, and the Government granted him one thousand acres of land. The war of 1812-13 again found him under the Royal Standard, together with his two sons, all of whom distinguished themselves by performing arduous and important services of the most daring and dangerous character.

The Hicks above-mentioned were descended from Edward Hicks, who, with his father, was condemned as a spy by the Americans, who had captured them while acting in that capacity. The father was taken from prison at Boston and hung before his son's eyes, which aroused his fury to such a pitch that he broke from his confinement, slew the armed guard while yet handcuffed; and though closely pursued and unable to release his hands from the manacles, he succeeded in eluding his enemies, and after nine days of fasting and untold suffering he reached the British lines.

The first school in the township was taught in a log shanty, long since in ruins, where Milford now stands; and the first religious services were performed for a long time by travelling Methodist ministers, who periodically visited the locality, and held meetings at the houses, and sometimes in the barns, of the different settlers in turns.

Of the early municipal history of Marysburg we remain in comparative ignorance; the first official records in the possession of the Township Clerk being after the passage of the Municipal Act in 1850, from which we learn that the following were the town officers for that year:—Andrew Wycott, Reeve; John G. Hicks, James Clapp, and Alexander Ghaunon, councillors; Richard Lobb, clerk.

In 1851, Palen Clarke was Reeve, and Richard Lobb again Clerk, whose duties he continued to perform in 1852, when E. W. Wright was reeve. In 1853, the first Deputy-Reeve was sent to the County Council from Marysburg. The following gentlemen held the positions respectively credited to them during that year:—Palen Clarke, Reeve; Lewis Hudgins, Deputy-Reeve; Robert Turnbull, Clerk. Mr. Turnbull has performed the duties appertaining to the above position ever since.

From thence forward the following gentlemen filled the offices of Reeve and Deputy-Reeve respectively:—In 1854, Messrs. Clarke and Hudgins; in 1855, Messrs. Lobb and Nelson Dodge; in 1856 and 1857, Nelson Dodge and Lewis Hudgins; in 1858, Messrs. Cavan and Thompson; in 1859 and 1860, John G. Hicks and Robert Clapp; in 1861, James Cavan and Nelson Dodge; in 1862, Nelson Dodge and William Lane; in 1863, William Kerr and John Cavan; in 1864, John Cavan and John Prinyer; in 1866, John Prinyer and James Wilson; from 1867 to 1870 inclusive, Robert Clapp and Andrew Wycott.

In January, 1871, North Marysburg was incorporated as a separate municipality. South Marysburg continued to be represented as follows:—In 1871 and 1872, by Messrs. Dodge and Collier; in 1873 and 1874, by Messrs. Clapp and Van Alstine; in 1875 and 1876, by James Henry Knox and Nelson Hudgins; and in 1877, by J. H. Knox and Solomon Collier.

Below we give a complete list of town officers for the year 1878:—James Henry Knox, Reeve; Nelson Hudgins, Deputy-Reeve; Benjamin Hubbs, John Walters, and Carleton McCartney, Councillors; Robert Turnbull, Clerk; Garrett Dingman, Treasurer and Poor Commissioner; Marshall Palen, Road Surveyor; George A. Wellbanks, Assessor; Samuel Ostrander and George P. Farrington, Collectors.

There was a Town Hall (already built), purchased at Milford in 1864, at a cost of $400.

There are four post villages in the township, viz., Milford, Port Milford, Cardwell, and South Bay. At the latter three places, notably at Port Milford (which contains an office of the Montreal Telegraph Co.), quite a large amount of shipping is done.

MILFORD, however, is the most important point in the township. It takes its name from a number of mills here, erected at an early day, on the banks of Black River; the first of which was built by Mr. Clapp, a U. E. Loyalist, and the second settler in the place. Some interesting facts in regard to its early settlement will be found in our sketch of Robert Clapp, Esq., a grandson of the above-named gentleman, who is now one of the leading men of the place.

The military spirit of their ancestors still fills the breasts of the inhabitants, there being two full companies of the 16th Regiment having their headquarters here. The second division court of the county is also held here. H. H. Haight is Clerk of the Court, and Richard Lobb, Bailiff.

Milford was in some respects the leading place in the county for many years. It was the depot of the lumber trade for the whole county, the southern part of which was covered with a dense forest of as fine pine and oak timber as ever grew. It was also a great sporting centre at one time; but with the failure of the lumber trade Milford lost its former prestige, though it is still a pleasant village, containing three or four stores, carding, grist, and saw-mills, a graded school, two churches, telegraph office, two hotels, boot and shoe shops, etc., etc., and a number of private residences which would be extremely creditable to a town of much larger dimensions and more ambitious pretensions.

NORTH MARYSBURG.

This township was called into existence as an independent municipality by an Act of Parliament which took effect January 1st, 1871. Geographically, it may be described as an arm of the main body of Prince Edward County, of the average width of one and a half to three miles, extending the distance of some sixteen or eighteen miles in a north-easterly direction into Lake Ontario, lying in a generally parallel position to the mainland shore, from which it is separated by the Eastern Branch (a name sometimes applied to that part of Bay Quinte), at an average distance of two to three miles; being surrounded on its eastern end and south-eastern side by the waters of Lake Ontario; and joined to the body, so to speak, of the county by an imaginary line commencing on the south shore of Bay Quinte about a mile and a half miles west of the lake on the Mountain, and drawn southerly along the eastern boundary of the township of Hallowell till it intersects the northern boundary of South Marysburg, thence easterly, terminating at the head of Smith's Bay. The area comprised within these bounds is 26,[illegible] acres, of which the very large proportion of 23,300 is improved. The value of this land is assessed at $324,142, and that of the personal property within the township at $46,410, while the latest enumeration gives the population at 1,524.

COUNTY OF PRINCE ED

In shape this territory is perhaps the most singular of any part of a county which for variety of outline and singularity of contour is almost without comparison within our geographical knowledge. The original surveyor, a Mr. Collins, who was instructed by the authorities to lay out a "Fifth Town," evidently recognized the exceptional configuration of what was afterwards called Marysburg, as can be seen by a glance at the map, and the manner in which he proceeded to execute his orders.

The present township of North Marysburg, then and long afterwards familiarly known as Indian Point, was the first surveyed in the Prince Edward district. It was also the first locality settled in the original "Fifth Town" (including the two Marysburgs, Athol, and part of Hallowell), with the probable exception of the Youngs at Eastlake, the Weeses on the Bay Shore of Ameliasburg, and probably a single settler at Carrying Place.

Mr. Collins entered upon his duties by landing in what was afterwards called McDonald's (now Prinyer's) Cove; and, ascending a little stream running into the head of the Cove, he built a log house for the use of his party on the south part of lot 32, Bay range, now owned by William Plews, and this is believed to have been the first habitation erected by a white man in Prince Edward County. After having obtained a correct outline of Indian Point by the usual means of "traversing," he divided it into two ranges or concessions, by running a somewhat irregular line through the approximate centre of that peninsula, and from this laid off lots of two hundred acres on either side; one range having a lake frontage to the south, the other a bay frontage to the north. These concessions are still designated as "Bay Range" and "Lake Range." The number of acres in a lot was put at two hundred, according to the custom previously followed in laying out the first "four towns," that being the allowance of each U.E. Loyalist or discharged private soldier for whom the said lots were designed. They numbered from a place called "The Rock," on the south shore, both ways, first westward to the head of Smith's Bay, then eastwardly to "The Point." The lots in this range are designated "Lot So-and-so, east (or west) of the Rock," as the case may be. Arriving at the Point, they double back therefrom, numbering consecutively westward along the Bay shore to a point opposite the head of Smith's Bay.

Scientifically, great interest attaches to the formation of North Marysburg. The peninsula, from the Hallowell boundary eastward to what is known as "The Rock," is of rocky formation, the limestone cropping to the surface on all hands. The shores are bold, and of a height of over one hundred feet, and in places nearly two hundred, and almost perpendicular. This high, rocky formation suddenly breaks away, across the whole peninsula, at the Rock, from which place to the Point, a distance of five miles or thereabouts, the country presents an even and unbroken surface, ascending gradually from both shores to the centre; possessing a depth and quality of soil admirably fitted to any and every branch of agriculture, and covered with magnificent schools, churches, and farm residences, giving the traveller a forcible idea of the highest state of prosperity, together with the very best educational facilities, and praiseworthy religious and moral attributes of the citizens; while the farms throughout this section, whether for the general excellence of the soil, the quantity, variety, and quality of the products, or the neat and husbandlike manner in which land, fences, buildings, and everything about them, is kept up, are scarcely to be excelled in the Dominion.

It is the theory of scientific men that this part of the peninsula was covered with water at a comparatively recent date; but that "The Mountain" is a part upheaved in the grand upheaval of Nature centuries ago, when the principal portion of the western continent was raised from the depths. That this also was originally covered by the ocean, however, is asserted by geologists, who offer as proof that here and there are to be found boulders on the surface (some of them of immense size), which are indigenous only to the glacial regions, and could only have been placed there by becoming frozen in with icebergs, and carried by them in their southern course to the various places where the temperature of a more genial clime overcame these northern monsters of the deep, which, in their decay and dissolution, of course dropped the boulders previously embedded in them on the spots where they may now be seen.

In this township is to be found a natural phenomenon which has been the subject of much scientific investigation, and which for nearly a century has been the wonder and delight of the traveller, and a curiosity of the most interesting description. We refer to the "Lake of the Mountain," a body of water over three miles in circumference, and situated literally *on a mountain*, as its name indicates, within five miles of Picton, on the north or bay shore, and separated from the bay by a ledge of solid limestone, terminating in a rocky precipice two hundred feet in height. The most extraordinary stories used to be told about this wonderful lake, it being at one time popularly believed that it was practically fathomless, and that it derived its supply of water from a subterranean passage connecting it with Lake Erie. The latter assertion would seem to be verified by the fact that although it had very little visible means of supply to draw from, there was still a steady and abundant discharge of water from it; also, that a careful calculation had decided that, with few extra influences, such as severe rains during the wet season, etc., it maintained a relative height almost identical with the level of Lake Erie, which is known to vary somewhat from time to time; and finally, that its great depth (for it was at one time supposed to be deeper than the bottom of Bay Quinte, two hundred feet below it), gave reasonable grounds for such a theory. Of late years, however, careful experiments have rather negatived this idea; though it is not claimed by even those who deny it, that the quantity of water which we can actually trace as coming into the lake, is at all equal to that going out, to say nothing of the immense loss by evaporation. The generally accepted theory now is, that it is fed from springs of a local character, which burst forth into the lake at points below its surface. To supply the outflow and waste, however, these springs must be numerous, and of great volume; and probably omit the great bulk of the water which falls within a radius of some miles; the layers or ledges of rock being apparently of such formation as to bring the water gradually downwards and in the direction of the lake, into whose depths it finally bursts forth as above surmised. It has been satisfactorily proven that the lake bottom is of uncommon regularity, and at a depth of about one hundred feet from the mean water level. A celebrated newspaper writer thus describes it: ". . . We then paddled upon its waters upon a small scow. For a considerable distance we can both easily perceive the bottom, which consists of dissolved, or rather corroded, lime, so loose and light that with little or no exertion one may push the whole length of his oar into it. Continuing to look downwards upon the beautiful white bottom as we sail along, we start instinctively upon finding that we all at once lose sight of it, and that we gaze into a deep and frightful abyss, which is formed . . . by the sudden appearance of a precipitous ridge running right across the lake. Nothing can exceed the amazement—terror, I had almost called it—which some people express on suddenly finding themselves surrounded by lofty, dark woods, and floating upon the surface of water as black as ink, over an abyss which they have been told is quite unfathomable."

The waters of this beautiful and wonderful lake are utilized to drive the machinery of the "Little Giant Turbine Water Wheel" works, a large grist mill, and a plaster mill. The locality is known by the name of "The Stone Mills." Here the first mill in the county, and the third in the Bay Quinte district, was erected by Major Van Alstine, in 1796, at the foot of the cliff, and run by water from the lake. The mill contained two run of stones, and three bolts of different degrees of fineness.

There was also a ferry crossing the Bay between this point and Adolphustown, the shores being here about one mile distant from each other. It was by this ferry that the old York and Kingston road lay. It passed close to the water's edge, under the "High Shore," to Hallowell (Picton), thence, *via* Bloomfield, Wellington, and Consecon, over the Carrying-place, whe[nce] it followed near the lake shore all the way to Toronto (York).

In this "land of the mountain and the flood," whose scenery ha[s for] years the admiration of travellers from all parts of the world, there [is no more] beautiful landscape scene to gaze upon than can be witnessed from [the top of] the mountain between the lake and the bay. On the one hand is th[e] [il]legible sheet of water, skirted by a precipitous rocky ridge, covered by the original forest growth. On the other, at a distance of tw[o hundred] feet below, the waters of Bay Quinte ripple and sparkle almost di[rectly be]neath your feet. No more appropriate site could have been select[ed for the] handsome church which adorns the shore of both lake and bay. H[ere is a] collection of well built and commodious houses, including a store, [forming a] small, but very pretty, country village. Down a precipice, so stee[p] as to almost daze one's vision, stand the Stone Mills, on the edge of [the bay,] while the great altitude of the beholder spreads out before him, lik[e a pano]rama, a most magnificent and lovely prospect. To the north, east, [and west,] stretches for many miles Bay Quinte, in some directions as far as [the eye] of vision can extend. Numerous bays or inlets of greater or less siz[e, and of] variety of shapes, indent the land on either side. The islands, like[wise, lend] additional interest and variety to the view; while the ever-changi[ng] intermingled land and water, forest and clearing, impart a surprisi[ng degree] of beauty to the prospect: and, to complete the picture, the fore[st-clad] hills of the mainland, with their vari-colored shades and tints, fo[rm in the] far distance a back-ground which adds immensely to the captivating [charm of] a scene which requires a more able pen to adequately portray, an[d leaves a] painting on one's mind's eye which time nor circumstance will not [efface]. Besides, there seems such an "eternal fitness of things" everywher[e,] whether upon the quiet church beside the deep unfathomable water[s, impres]sing one with the omnipotence of the Creator, or the neat litt[le village,] instinctively recalling the comforts and pleasures of our own loved [homes; or] the busy mills beneath, whose ponderous machinery, driven by t[he waters] from above, is the symbol of that trade, commerce, and prosperity [of which] navigation is always a necessary adjunct; or the distant landsca[pe, rich in] romance and history, within whose primeval forests the loyal refu[gees, the] fathers of our country, endured untold privations and hardships fo[r the sake] of principles they cherished, upon whose shores these pioneers c[ame, and] with sturdy arms and British will, conquered an inhospitable wild[erness and] made it to blossom as the rose, and within whose confines and und[er whose] sight were born many of our most illustrious men, including Rt. [Hon. Sir] John A. Macdonald, Hon. Chief Justice Hagerman, Hon. Edward [...] Hon. Lewis Wallbridge, Hon. Robert Charles Wilkins, Hon. Ric[hard] Cartwright, Hon. Billa Flint, Hon. Robert Read, Hon. Samuel W[...] Judge Lazier, Adam Henry Meyers, the Roblins, Dorlands, Clapps, Ketchesons, and many others, whose history is the history of Ca[nada, and] whose names are particularly identified with all public affairs of th[e country] from its earliest settlement to the present day. Taken altogether, t[he scene] spread out before the observer from the Lake of the Mountain w[ill, whether he] be a lover of the beautiful in Nature, or an admirer of unexample[d] fortitude, and patriotism, or one who delights in tradition and ro[mance, be] one of the grandest imaginable, and well worthy the tributes of [praise be]stowed upon it by all writers since the country has had a history.

We have intimated that Marysburg was surveyed more esp[ecially for] settlement by discharged British soldiers. The first actual settler [within the] limits of North Marysburg was Alexander McDonald, afterwards [fami]liarly known as Col. McDonald, who was a sergeant in the 78th Hi[ghlanders]. He was present with his regiment at the capture of Louisburg an[d Quebec,] and was in the military service of Great Britain for nearly sixty y[ears. He] landed in McDonald's Cove (now Prinyer's) in 1784, and settled on [a lot close] hard by, where he remained till his death in 1815. He had but one [daughter,] who afterwards married a Mr. Prinyer, of French descent, whose s[on, John] Prinyer, still lives there, and is one of the leading men of the town[ship. It] is still related of the Colonel as a fact of the greatest importance [in those] days, that he "owned the first horse below the rock."

In 1796, Major Van Alstine, then of Adolphustown, received some five hundred acres of land alongside the Lake of the Mounta[in, settled] there, and built the Stone Mills. Van Alstine was a man of co[nsiderable] means, who had come in from the Mohawk Valley with a party of se[ttlers and] refugees some years previous, and settled in Adolphustown, wh[ere he] lived till 1796. Many of his descendants are now among the most [respected] citizens of the Prince Edward District.

Capt. Archibald McDonell, Sergt. Harrison, "Squire" Wright, [...] Carson, Daniel McIntosh, the Ross's, Powers' and McCrimmons, w[ere among] the first settlers. There were also about forty discharged Hessians, [who came] in and settled in the autumn of 1784. The circumstances and su[rroundings] of the country were so utterly at variance with what they had herei[ofore been] accustomed to, that many of them sold or gave away their land and [whatever] else they possessed, and left the country, some to settle in the Unit[ed States,] others returning to Europe; many others would gladly have left [had they] been possessed of the necessary means. Among those who rema[ined were] Tobias Schneider, Ernest Noedling, and Hendrick Schmidtz. T[he ortho]graphy of these names has been Anglicised since then, and we find Y[...] a son of the former, now living on Lot 1 east of the Rock, on wh[ich he] settled, and where he himself was born, hard on to four score yea[rs ago, and] has ever since resided. The latter-named had quite a family of son[s, all of] whom settled along the north side of Smith's Bay, which was sub[sequently] so named after them.

Near the point where Col. McDonald settled, Sir Frederick Ha[ldimand, at one] time Surveyor-General, had large grants of land. His son, now an [old man,] hale and hearty, still lives on part of the old estate, Lot 31, Bay Ra[nge. He] is an advanced agriculturist, and has a most beautiful and pleas[ant home.] His wife was the granddaughter of Col. McDonald.

Among the earliest settlers on the bay shore were the Bongards, and Pringles. Old Mr. Bongard was an assistant on the survey in [...] the party running the original lines, and received large grants of lan[d at the] place now known as Bongard's Corners. A large number of his d[escendants] still reside in the neighbourhood, well-to-do and enterprising citiz[ens.]

From the recent date at which North Marysburg was set off as [a separate] township, its municipal history is soon told. A glance at the map [will con]vince the most casual observer that for the people of the Point to att[end to] business at Milford (where the Town Hall of the old township was [situated)] was a matter of very great inconvenience. Being in a minority, they worked for a long time unsuccessfully to effect a separation. T[he] desired end was finally accomplished in 1870, chiefly through the perseverance of Mr. John Prinyer, who had devoted his time and en[ergy to the] object for years past.

The first town officers elected were for the year 1871, and the foll[owing were] their names and respective positions:—John Prinyer, Reeve; Andre[w ...,] John Burley, William Williams, Robert Miller, Councillors; Levi [...,] Clerk; F. W. Wright, Assessor; James Brown, Lewis Ketchum, [...] Bagster, Collectors; Conrad J. Bongard, Treasurer.

After this, Mr. Prinyer was Reeve for a number of years, the[n ...] Kerr followed for two years, and he in turn was succeeded by Levi [...] ney filling the position the second time. Following is a full list of officers for the present year:—Levi Williams, Reeve; A. W. Brown[, ...] Wycott, John Pierce, Allen Cavon, Councillors; Henry A. Powe[rs, ...] James Hart, William W. Bongard, Assessors; Peter A. Minak[er, ...] Hughes, Peter B. Collier, Collectors; Conrad J. Bongard, Treasure[r.]

A Town Hall was built some years ago near Waupoos P.O., a[...] east of the Rock. It is large, convenient, and of modern style, and [with] sheds attached, very near $2,000.

d Conseron, over the Carrying-place, west of which
hore all the way to Toronto (York).
ountain and the flood," whose scenery has been for
vellers from all parts of the world, there is no more
o gaze upon than can be witnessed from the top of
lake and the bay. On the one hand is the magni-
ed by a precipitous rocky ridge, covered in places
rth. On the other, at a distance of two hundred
Bay Quinte ripple and sparkle almost directly be-
e appropriate site could have been selected for the
joins the shore of both lake and bay. Hard by is a
l commodious houses, including a store, forming a
intry village. Down a precipice, so steep and deep
ion, stand the Stone Mills, on the edge of the Bay ;
the beholder spreads out before him, like a pano-
and lovely prospect. To the north, east, and west
Bay Quinte, in some directions as far as the power
ments bays or inlets of greater or less size, and all
the land on either side. The islands, likewise, lend
riety to the view ; while the ever-changing scene of
ter, forest and clearing, impart a surprising degree
; and, to complete the picture, the forest-covered
h their varicolored shades and tints, form in the
which adds immensely to the captivating beauty of
more able pen to adequately portray, and forms a
ye which time nor circumstance will not eradicate.
an "eternal fitness of things" everywhere we gaze,
urch beside the deep unfathomable waters, impres-
otence of the Creator, or the neat little village
comforts and pleasures of our own loved home ; or
chase ponderous machinery, driven by the waters
of that trade, commerce, and prosperity to which
cessary adjunct ; or the distant landscape, rich in
in whose primeval forests the loyal refugees,—the
dured untold privations and hardships for the sake
ed, upon whose shores these pioneers of Canada,
tish will, conquered an inhospitable wilderness and
rose, and within whose confines and under our very
our most illustrious men, including Rt. Hon. Sir
Chief Justice Hagerman, Hon. Edward Murney,
Hon. Robert Charles Wilkins, Hon. Richard John
ilit, Hon. Robert Reed, Hon. Samuel Washburne,
ry Meyers, the Roblins, Dorlands, Clapps, Allisons,
hers, whose history is the history of Canada, and
rly identified with all public affairs of the country
t to the present day. Taken altogether, the picture
iver from the Lake of the Mountain—whether he
in Nature, or an admirer of unexampled courage,
or one who delights in tradition and romance,—is
mable, and well worthy the tributes of praise be-
ets since the country has had a history.
hat Marysburg was surveyed more especially for
British soldiers. The first actual settler within the
g was Alexander McDonald, afterwards and popu-
nald, who was a sergeant in the 78th Highlanders,
egiment at the capture of Louisburg and Quebec,
rvice of Great Britain for nearly sixty years. He
e (now Prinyer's) in 1784, and settled on the shore
d till his death in 1815. He had but one daughter,
Mr. Prinyer, of French descent, whose son John
and is one of the leading men of the township. It
and as a fact of the greatest importance in those
first horse below the rock."
Alstine, then of Adolphustown, received a grant of
land alongside the Lake of the Mountain, settled
e Mills. Van Alstine was a man of considerable
om the Mohawk Valley with a party of several loyal
ons, and settled in Adolphustown, where he lived
his descendants are now among the most respected
arit District.
ell, Sergt. Harrison, "Squire" Wright, William
the Ross's, Powers' and McCrimans, were among
ere also about forty discharged Hessians, who came
mn of 1784. The circumstances and surroundings
rely at variance with what they had heretofore been
of them sold or gave away their land and what little
it the country, some to settle in the United States,
e ; many others would gladly have left had the
essary means. Among those who remained were
Nodding, and Hendrick Schmidtz. The author
been Anglicised since then, and we find Mr. Spicer,
living on Lot 1 east of the Rock, on which Colonel
self was born, hand on to four score years ago, and
he latter named had quite a family of sons, some of
orth side of Smith's Bay, which was subsequently

ol. McDonald settled, Sir Frederick Daute, at one
d large grants of land. His son, now an aged man
on part of the old estate, Lot 31, Bay Range. He
st, and lies a most beautiful and pleasant place.
ghter of Col. McDonald.
lers on the bay shore were the Bongards, Minakers,
ongard was an assistant on the surveying staff of
inal lines, and received large grants of land near the
ard's Corners. A large number of his descendants
hood, well-to-do and enterprising citizens.
which North Marysburg was set off as a separate
ory is soon told. A glance at the map will con-
ever that for the people of the Point to attend public
the Town Hall of the old township was situated, is
t inconvenience. Being in a minority, however,
e unsuccessfully to effect a separation. This much-
complished in 1870, chiefly through the untiring
Prinyer, who had devoted his time and energy to the

lected were for the year 1871, and the following are
positions :—John Prinyer, Reeve ; Andrew Wycott,
liams, Robert Miller, Councillors ; Levi Williams,
sessor ; James Brown, Lewis Ketchum, John H.
el J. Bongard, Treasurer.
r was Reeve for a number of years, then William
s, and he in turn was succeeded by Levi Williams,
second time. Following is a full list of township
r :—Levi Williams, Reeve ; A. W. Brown, Andrew
n Caven, Councillors ; Henry A. Powers, Clerk ;
Bongard, Assessors ; Peter A. Minaker, Calvin
Collectors ; Conrad J. Bongard, Treasurer.
some years ago near Waupoos P.O., on Lot 11,
ge, convenient, and of modern style, and cost, with
$2,000.

Waupoos, an island lying off the south shore, opposite the mouth of Smith's Bay, contains about 1,000 acres mostly of good land, and forms a part of the township of North Marysburg.

The oldest church in the township was built on Lot 16, south part, west of the Rock. Its ruins are still visible.

The oldest burying-ground in the township—and, indeed, in the county—where many of the oldest settlers are buried, is called the "Ross Burying-ground," and is situated on the lake front between the Rock and the Point.

There is no village in the township except Prinyer, situated on Lot 35, Bay Range. The P.O. was named after John Prinyer, Esq., and is kept by G. B. German, who also has the telegraph office, a wharf, warehouse, and general store. In fact he runs the town, and "runs it well" his neighbours say.

Besides the above, there are post-offices at Bongard's Corners, Lot 57, Bay Range; Cressy, Lot 8, east of the Rock: and Waupoos, Lot 7, west of the Rock.

Waupoos is the seat of the Eighth Division Court of Prince Edward County, of which Ed. Harrison, Esq., is clerk, and James Ross, bailiff.

There are a number of wharves and store-houses on both water-fronts of the township ; and the facilities offered the farmers for the shipment of produce is nowhere else equalled except in other parts of Prince Edward County.

SOPHIASBURG.

So-called in honour of the twelfth child of George III., the original "Sixth Town" is bounded on the north by the west branch of Bay Quinte, on the east by the "Long Reach" and Picton Bay, on the south by the township of Hallowell, and on the west by Ameliasburg and Hillier. Commencing at a point on Picton Bay, about two miles down the left shore, we strike the boundary of Sophiasburg ; turning at right angles to the shore, we follow the line dividing it from Hallowell, running thence in a zig-zag direction—alternately to north-west and south-west—till it intersects the south-east corner of the township of Hillier, thence along the eastern boundary of that township, in a northerly direction to the Bay, at a point opposite Huff's Island, which, however, is a part of Ameliasburg, while Big Island, a little further to the east, belongs to Sophiasburg.

As the townships of Prince Edward are now constituted, Sophiasburg is the largest in point of size, containing 43,100 acres, while, in valuation of real property, it is second only to Hallowell, the said value being assessed at $1,077,500.

In point of population it is the third township in the county, the latest returns giving the number of souls at 2,128.

It was the second surveyed and the last settled of the three original townships—the first settlements which we can trace with any degree of certainty as to date having been made in 1777 or 1778.

The surveying had been done three years previously by Louis Kotte, under instructions from Mr. Collins, who had lately been promoted to the position of Deputy Surveyor-General. The survey is a great improvement upon that of Marysburg, the general form of the territory offering facilities for a more regular and symmetrical division.

The settlement mentioned above as having been commenced about 1777 was near the head of Picton Bay, and not in the present limits of Sophiasburg, though it was a part of the "Sixth Town" at that time.

The great bulk of the land around the water fronts is said to have been "located" by U. E. Loyalists of Fredericksburg and Adolphustown, who "drew" it under Governor Simcoe's proclamation ; but having already established themselves they held this land without settling upon it, and sold it as opportunity occurred.

Many of the first actual settlers were what the "U. E.'s" called "Late Loyalists" by way of contempt. These were American colonists, who, though sympathizing with the British, were not driven out of their old homes, as many had previously been, by the victorious Continentals ; but, tired of the new order of things after the return of peace, or more likely, induced by the Proclamation of Governor Simcoe, followed in the steps of the original Loyalists. This Proclamation, issued 7th February, 1792, was prompted by the Governor's opinion that there still remained in the then United Colonies a large number of loyal British subjects who would yet leave the newer Government and older country for the older Government and newer country, did they but receive sufficient encouragement to do so ; and it provided, among other things, that each should receive two hundred acres of land. Whether the Governor's opinion on the above subject was correct or otherwise, the object aimed at was accomplished, as his proclamation had the effect of inducing many to leave their homes in the United States and settle in Canada.

We have no data from which to give an exact statement as to either the time *the first* settlement of the township was made, or the individual who made it. It is known, however, that John Parcells settled on Lot 10, Marsh front, and Nathaniel Solmes and Guillaume Demorest somewhat further west, on the same front, in the beginning of the last decade of the eighteenth century. Parcells had served the Royal cause in Major Van Alstine's company, and Solmes was a U. E. Loyalist from Duchess County, New York.

About the same time Philip Roblin, together with the family of John Roblin, his brother, came in from Adolphustown, where John had died just previously from the effect of wounds received some years before from a foraging party sent out by Washington, whose army was then encamped in the vicinity of their home in New Jersey. These two brothers, together with two other Roblins—relatives—who also came into Canada at the same time and for the same cause, were the ancestors of a very numerous progeny, who have scattered throughout the country—notably the Bay of Quinte region—and taken such a leading, honourable, and intelligent part in all public questions of the day, as to entitle them to the distinction of one of the most influential families of our country. The remark is specially applicable to the Bay District, as any of its citizens can bear witness.

James Morden was also one of the very earliest settlers, in 1791, and Isaac Demill settled at nearly the same time.

James Cotter, one of the earliest settlers along the marsh front, was a man of superior parts and admitted influence among the pioneers. He was the first Justice of the Peace of the township, and one of the first Members of Parliament for the District. Two of his sons still live here, fine specimens of our best class of yeomanry.

The Laziers, Crooks, Spragues, Goslines, and Ways, all U. E. Loyalists, settled among the first ; also Jacobus Peck and his son James, who soon after removed to Ameliasburg, where their descendants still reside. All the above last named families have numerous descendants now living in the district. The Lazier family, particularly, have among them a large number of representative men, and many of them hold distinguished positions of public trust in various branches of the public service.

Among the first to push the settlements into the interior were Ruttan, Dorland, Parliament, Hill, the Howells, Burdettes, and Samuel H. Barton.

GRASSY POINT, in the eastern extremity of the peninsula of Sophiasburg, was at those early days a place of considerable importance. It was this point to and from which all residents from the adjacent parts of Prince Edward and the mainland came back and forth on business or pleasure. John Trumpour was the first settler here, and kept a public house—the first in the township—and for many years, and into the beginning of the present century, the militiamen of the whole district were wont to assemble here annually on "training day."

BIG ISLAND, a very important adjunct of Sophiasburg, lies to the north, being separated by a deep morass from the main body of the county, with which it is connected by a very substantial roadway or bridge of solid stone, over a quarter of a mile in length. The Island is about five miles in length, containing 4,938 acres of land.

Samuel Peck and Samuel Shaw are said to have been the two first to settle on the Island, and after them came the Spragues, De Longs, and Allisons, all U. E. Loyalists.

The Morans, also descended from Virginian U. E. Loyalists, and originally of Irish extraction, were among the early settlers. They are now a numerous and influential family—as are also the Boulters—descendants of George Boulter, who occupy conspicuous positions in public and municipal affairs here and in the County of Hastings.

The surface of the Island is comparatively even, the soil of excellent quality, and on all sides are to be seen farms which will compare with the best in any section of the Province—the beautiful residences and commodious buildings with which they are furnished, bearing testimony to the general prosperity of the owners. The same may be said of a large portion of the township, particularly that part lying along the marsh front, though there are some sections where the rock comes too prominently to the surface to make the land productive, except in the most favourable seasons.

Many of the farmers of Sophiasburg make hop-raising a specialty, this branch of agricultural industry being carried on here to a greater degree probably than in any other section of equal extent in the country. In some parts of the township large hop-yards are to be seen on every farm, and we have passed some farms on which were hop-fields containing fifty acres. The principal hop growers are Ira A. Coolidge and George Dunning, both wealthy and enterprising farmers. Mr. Coolidge has grown as high as fifteen tons of hops per season, and is said to be the largest hop-grower east of Toronto. All other branches of agriculture are carried on with an abundant degree of success. We have seen many magnificent fields of barley which is here, as elsewhere along Bay Quinte, the staple crop. There are a number of stock-breeders in the township also, whose short-horns and roadsters are second to none in a country universally acknowledged as one of the finest stock-producing regions in the world.

There is a very prosperous and flourishing Agricultural Society in the township with the following list of office-bearers:—D. W. Ruttan, President; Matthew Benson, Vice-President; Wellington Boulter, Secretary; Ira A. Coolidge, Treasurer.

Of the three post villages in Sophiasburg, the smallest, Green Point, is a pleasant little hamlet on Lot 31, High Shore.

Northport, the next larger, is so-called from its geographical position. It is situated on lots 25 and 26, on the extreme north point of the marsh front of the township, just below Big Island, and the first bay port at which steamers call east of Belleville, from which it is distant twelve miles, and from Picton thirteen. It was settled by James Morden and Isaac De Mill—the former built the first house in the place in 1791. It is a pleasant village, containing two general stores, two blacksmith shops, waggon shop, two hotels, cheese factory, etc. The Montreal Telegraph Company have an office here. There are two wharves and steamboat accommodation for a large grain trade. A great quantity of produce is annually handled here. Six steamboats call daily to and from Oswego, Kingston, Ogdensburg, Montreal, and all bay ports.

Demorestville is called after the founder of the place, Guillaume Demorest, of French extraction, but a native of Duchess County, New York, who served in the Commissary Department of the Royal forces during the Revolutionary War, being at that time quite young. He came to Canada and located in Sophiasburg in 1790. He built a mill at a very early day on Lot 38, marsh front, though it was about a mile inland from the marsh, and a village soon afterwards sprang up at the place. The mill was built upon the creek running north-west from Fish Lake into Bay Quinte.

This beautiful body of water, whose attributes are very similar to the Lake of the Mountain, is situated near the centre of the township, at a height of over one hundred feet above the Bay level. It is somewhat circular in form, and covers an area of about six hundred acres. Its name was derived from the immense quantities and superior quality of those species of the finny tribe which filled its waters in the early days of the settlement.

Guillaume Demorest was the first postmaster of Demorestville, a very old Justice of the Peace, and one of the oldest officers of Militia. He was a relative of the celebrated Madame Demorest, now and for many years the acknowledged leader of New York fashions. He removed to Consecon at a later period, and died there in 1848, when seventy-nine years of age.

The village was once the chief place of the county, aside from Picton, but does not appear to be in as flourishing a condition now as formerly. The old mill built by M. Demorest is long years ago in ruins, but the place still contains large mills erected at a more recent date. There are also two general stores, three blacksmith shops, three waggon shops, two shoe shops, one cabinet shop, and an agency of the Montreal Telegraph Company. It also contains four churches—some of them quite handsome—and the Town Hall, a very fine brick building, with exceptionally commodious interior arrangements, which was built by James Spurr, contractor, in the year 1876, at a cost of $4,000; and which, with the exception of Ameliasburg, is the best building owned by any township for public purposes, which we have seen in Canada.

The Third Division Court of the County of Prince Edward is held here. Israel Hamilton, Esq., is Clerk of the Court, and Mr. Edward Nixon Bailiff.

From an old memorandum of Mrs. Barton, who came with her husband from Albany and settled at Demorestville in 1815, it appears that there were only some half dozen houses here at that time, and only two settlers between here and Picton, from which it is distant nearly ten miles. The distance to Belleville by land is sixteen miles; in winter, when the Bay is frozen over, it is less than ten.

Just about the same time of the advent of the Barton family, the Howards, Rightmeyers, Eatons, and Thompsons came in from the United States and settled in the neighbourhood.

The roads hereabouts, and indeed in all parts of the township, are of the most excellent description. The character of the soil renders the finest quality of "road metal" easily accessible; and in all directions we drive over mile after mile of such highways as the inhabitants of an alluvial, vegetable, or clay soil country can never realize. But it would seem that the roads hereabouts were not always as we now find them, for we see that as late as 1830 Parliament granted the very liberal sum of £13,650 "for the improvement of roads and bridges." Of this sum £1,900 was apportioned to the Midland district, and it was specified in the Act that a part of it should be spent as follows:—"On the road leading from Wessel's Ferry, in Sophiasburg, to Demorest's Mill, the sum of one hundred pounds; and that Abram Van Blaricom, Daniel B. Way, and Guillaume Demorest be appointed commissioners for expending the same."

The first school in the township was taught by a Mr. Salisbury, on the High Shore, about 1803. The first on the Marsh front was kept by one John James near Grassy Point.

The first public religious services—and the only—held in the township up to the time when Mr. and Mrs. Barton came to the country were conducted by Rev. Thomas Madden, a W. M. minister, who visited the place periodically for that purpose, and held meetings in Demorest's mill.

The first white child born in the township (and it is believed also the first in the county) was Andrew Pringle, afterwards a W. M. minister, and the first native-born Canadian who ever preached Christ and Him crucified.

The oldest burying-ground in the township—and one of the oldest in the district, is on the Cronk farm, between Northport and Grassy Point.

Excepting the township of Hallowell, S[illegible]
oldest public records of any municipality in
connecting link of the past to the present,
observable, make it quite interesting—so we
Though the orthography and syntax differ i[illegible]
recognized rules, the report is written in a m[illegible]

"Passed at Sophiasburg, at a regular to[illegible]
day of March, 1800.

"For the better ascertaining astrays, [illegible]
horses and neat cattle, sheep or swine.

"Be it understood by this town meetin[illegible]
householder shall, within six weeks from the
marks and brands recorded according to law [illegible]

"And be it further enacted by the author[illegible]
horses, neat cattle, sheep or swine that shall
proved lands from the 20th November to the
year, the owner or owners of such improvem[illegible]
in their natural marks, or artificial marks, an[illegible]
possible to the Town Clark, who is hereby o[illegible]
book to be kept for that purpose, for which s[illegible]
shilling for each such horse or neat cattle, an[illegible]
swine.

"Provided always, and be it understood, [illegible]
tioned is not one of his near neighbours, whic[illegible]
Clark to decide; and the Clark shall send w[illegible]
he knows them by mark or brands, and if m[illegible]
is hereby ordered to advertise them in three d[illegible]
for which he shall be entitled to receive f[illegible]
followeth, viz: For sending word, or writing
any way, one shilling and three pence; if adv[illegible]
pence for each horse or neat cattle, and for ea[illegible]
head.

"And be it further enacted by the auth[illegible]
habitant or householder who shall leave any
on his or her cleared lands for eight days from
1st of April, and neglect to give notice ther[illegible]
tioned, shall lose the reward for finding or fee[illegible]
owner one shilling for each horse or neat [illegible]
sheep or swine.

"And be it further enacted by the same [illegible]
owners shall appear by the first Monday in A[illegible]
then, and in that case, the Town Clark sha[illegible]
astrays in three townships, viz: Ameliasburg
for the space of twenty days, describing the
age, as near as possible, and if no owner or o[illegible]
their property, then the Town Clark shall
astrays, by appointing the day of sale, to t[illegible]
deducting the expenses, to be adjudged by
pointed from each parish in this town, the ov[illegible]
the hands of a treasurer, hereafter to be appo[illegible]

"And be it observed: That all well-regul[illegible]
parishes. Be it enacted by the majority of [illegible]
divided into parishes and described as follows
lot No. 45, west of Green Point, to lot No.
name of St. John's; and by the authority afore[illegible]
No. 6, in the Crown lands, west of Green P[illegible]
name of St. Matthew. And be it further enact[illegible]
that including the tenants on the Crown la[illegible]
shall be a parish by the name of St. Giles, a[illegible]
Hallowell shall be a parish by the name of Mo[illegible]

"Whereas, all the fines and forfeitures tha[illegible]
shall be appropriated to charitable uses; we t[illegible]
in our town meeting, on the 3rd day of March
to appoint our treasurer in this town out of th[illegible]
bitants, to be treasurer to this town, to receiv[illegible]
of money that is or shall be ordered to be ap[illegible]
which treasurer is hereby ordered to serve in
behaviour, or until he shall wish a successor.

"And be it enacted by the authority afor[illegible]
nominate Peter Vallen, who is appointed tre[illegible]
and who is to receive all moneys coming into h[illegible]
received, and for what fined; and when a suc[illegible]
up all the moneys he has belonging to the sa[illegible]
receipts, to his successor, and deliver the same
each parish shall nominate one good and re[illegible]
gether with the overseer of the poor, shall be
that all fines and forfeitures of this town is re[illegible]
to the treasurer.

"And if any person who comes and proves
within one year and a day, then the treasurer a[illegible]
the overseers of the poor, shall refund such m[illegible]
treasurer, deducting two shillings on the pound

"And be it enacted by the authority afore[illegible]
money in the hands of the treasurer, and a n[illegible]
charitable use, this body corporate shall have [illegible]
posing of, who is to receipt to the treasurer for [illegible]
his book, and the use they had applied the s[illegible]
seers of the poor, and the parish inspector ma[illegible]
the same when, and as often as they or th[illegible]
choose to do so, and shall be a body corporate
that may appertain to their several offices."

From all of which we conclude that it was n[illegible]
troubled the grangers of those days, but that
question of the hour with our worthy friends i[illegible]
at law-making seems to have "settled it" wit[illegible]
we find nothing further "on the authority afor[illegible]
ing is a copy of the proceedings that year:—

"An Act passed at the town meeting, Sop[illegible]
for the relief of the poor of the township of Sop[illegible]

"Report of the committee on the subject.

"We the committee appointed, who have
said township, have agreed to report that one
each man's ratable property be paid for the p[illegible]
of the committee that when any person is ag[illegible]
poor, that they endeavour to get them to tak[illegible]
fraying said poor.

"Signed on behalf of

Sophiasburg, 3rd January, 1820."

And again in 1821 we find the following in [illegible]

"The report of the committee to the car[illegible]
year is, that one farthing in the pound of each
be sufficient for the present year.

"

At the same meeting, which was held Ja[illegible]
the following town officers were appointed:—

John Short, Town Clerk; John R. Way, M[illegible]
John Shorts, Thomas D. Appleby, Assessors;
Henry W. Fox, Joseph Hazard, Town Warden[illegible]

'ORICAL SKETCH OF THE

cepting the township of Hallowell, Sophiasburg has preserved the public records of any municipality in the county. The age of this ating link of the past to the present, as well as some other points vable, make it quite interesting—so we give it *verbatim et literatim.* gh the orthography and syntax differ in some cases from the present ized rules, the report is written in a neat, legible hand. We quote :—

Passed at Sophiasburg, at a regular town meeting, held on the 3rd f March, 1800.

For the better ascertaining astrays, and knowing and describing s and neat cattle, sheep or swine.

Be it understood by this town meeting that every inhabitant and holder shall, within six weeks from the passing of this Act, have their and brands recorded according to law by the Town Clark.

And be it further enacted by the authority aforesaid, that any astrays, s, neat cattle, sheep or swine that shall be found in any open or im- d lands from the 20th November to the 1st of April yearly, and every the owner or owners of such improvement or cleared land shall give ir natural marks, or artificial marks, and describe their age as near as de to the Town Clark, who is hereby ordered to record the same in a o be kept for that purpose, for which such informer shall receive one g for each such horse or neat cattle, and six pence for each sheep and

Provided always, and be it understood, that such astrays above men- is not one of his near neighbours, which shall be left to the Town to decide ; and the Clark shall send word to the owner or owners if s them by mark or brands, and if unknown to the Town Clark, he ly ordered to advertise them in three different places in the township, nich he shall be entitled to receive from the owner or owners, as eth, viz : For sending word, or writing, or recording, or informing in ay, one shilling and threepence ; if advertised, one shilling and three for each horse or neat cattle, and for each sheep or swine six pence per

And be it further enacted by the authority aforesaid that if any in- nt or householder who shall leave any astrays, as above mentioned, or her cleared lands for eight days from the 20th of November to the April, and neglect to give notice thereof, as by the above Act men- , shall lose the reward for finding or feeding such astrays, and pay the one shilling for each horse or neat cattle, and six pence for each or swine.

And be it further enacted by the same authority that if no owner or s shall appear by the first Monday in April to prove their property, and in that case, the Town Clark shall advertise for sale all such s in three townships, viz : Ameliasburg, Sophiasburg, and Hallowell space of twenty days, describing the marks and brands, color and near as possible, and if no owner or owners shall appear and prove property, then the Town Clark shall proceed to the sale of such s, by appointing the day of sale, to the highest bidder ; and after ing the expenses, to be adjudged by persons hereafter to be ap- d from each parish in this town, the overplus shall be delivered into nds of a treasurer, hereafter to be appointed.

And be it observed : That all well-regulated townships is divided into s. Be it enacted by the majority of votes that this town shall be l into parishes and described as followeth, that is to say : That from . 45, west of Green Point, to lot No. 19, shall be a parish by the f St. John's ; and by the authority aforesaid, that including No. 19 to in the Crown lands, west of Green Point, shall be a parish by the f St. Matthew. And be it further enacted by the authority aforesaid, cluding the tenants on the Crown lands, and including lot No. 28, e a parish by the name of St. Giles, and from Nicholas Wessel's to ell shall be a parish by the name of Mount Pleasant.

Whereas, all the fines and forfeitures that may occur within our limits appropriated to charitable uses ; we the inhabitants of Sophiasburg, own meeting, on the 3rd day of March, 1800, do think it necessary int our treasurer in this town out of the most respectable of its inha- . to be treasurer to this town, to receive all forfeitures or other sums y that is or shall be ordered to be appropriated to charitable uses. treasurer is hereby ordered to serve in that connection during good ur, or until he shall wish a successor.

nd be it enacted by the authority aforesaid, that we do appoint and te Peter Vallen, who is appointed treasurer, who is to keep a book, o is to receive all moneys coming into his hands, and enter from who l, and for what fined ; and when a successor is appointed he shall give he moneys he has belonging to the said town, with the books and , to his successor, and deliver the same on oath if required ; and that rish shall nominate one good and respectable inhabitant, who, to- with the overseer of the poor, shall be inspectors to enquire and see fines and forfeitures of this town is regularly received and delivered reasurer.

nd if any person who comes and proves of any astrays that has been one year and a day, then the treasurer and those parish inspectors, and seers of the poor, shall refund such moneys as was delivered to the r, deducting two shillings on the pound for the fees of said treasurer. nd be it enacted by the authority aforesaid, that when there is any in the hands of the treasurer, and a necessity to lay it out in some ble use, this body corporate shall have the sole management and dis- of, who is to receipt to the treasurer for the same end, have recorded in k, and the use they had applied the same, and the treasurer, over- the poor, and the parish inspector may hold meetings, and adjourn ne when, and as often as they or the major part of them shall to do so, and shall be a body corporate to sue and be sued, anything y appertain to their several offices."

n all of which we conclude that it was not the Thistle question which the grangers of those days, but that "Led Astray" was the great n of the hour with our worthy friends in 1800. But this great effort making seems to have "settled it" with the amateur legislators, as nothing farther "on the authority aforesaid" till 1820. The follow- copy of the proceedings that year :—

n Act passed at the town meeting, Sophiasburg, 3rd January, 1820, relief of the poor of the township of Sophiasburg.

port of the committee on the subject.

e the committee appointed, who have the care of the poor of the nship, have agreed to report that one half-penny on the pound for an's ratable property be paid for the present year, and it is the sense ommittee that when any person is agreed with to keep any of the at they endeavour to get them to take produce in payment for de- said poor.

"Signed on behalf of the committee,
JAMES NOXEN, Chairman.

burg, 3rd January, 1820."

again in 1821 we find the following in reference to the same s— e report of the committee to the care of the poor for the present that one farthing in the pound of each man's rateable property will ient for the present year.

"JAMES NOXEN, Chairman."

he same meeting, which was held January 1st, at Gersline's Inn, wing town officers were appointed :—

Short, Town Clerk ; John R. Way, Moses Thompson, Constables ; ortt, Thomas D. Appleby, Assessors ; Sylvanus Hoxsee, Collector ; V. Fox, Joseph Hazard, Town Wardens.

The same gentlemen continued to fill the above positions for a number of years in succession, being yearly chosen by the people at their regular town meetings.

"We quote from the appendix to the report of the annual town meetings of 1826 :—

"Our laws at present be as they will ;
We have them long and keep them still."

Again, from the proceedings of 1827 :—

"Farmers' town laws as they've heretofore been ;
Hogs not to run at large in Demorest Vill."

Dr. Moore, the present town clerk, was appointed to that position at the first meeting of the first council elected under the Municipal Act, and acted as such a great many years. He was also the first clerk of the County Council of Prince Edward county, and filled the position a number of years in succession.

The town officers for the present year are as follows :—

Samuel N. Smith, Reeve ; Josiah Benson, Deputy Reeve ; Nathaniel J. Boulter, A. B. Foster, John Whiting, Councillors ; Thomas Moore, M.D., Clerk.

AMELIASBURG.

The seventh of the original "Ten Towns" of Upper Canada, and the last of the three original townships of Prince Edward, was named after the seventh daughter of the King. It is the most regularly laid out tract in the county—partly on account of its compact form, and partly on account of the knowledge acquired by the Government surveyors from several years of previous experience of the requirements of the settlers in the way of roads, &c. For instance, in the first surveyed towns there were only "base lines," or concessions established ; and settlers were left to get "across lots" the best way they could, or to locate their own roads in places most convenient to themselves ; whereas, we find on the oldest map of this township, deposited in the Crown Lands' office, a note to this effect :—"The fro. of this township is a continuation of Louis Kotte's survey, "from Green Point (Sophiasburg) to the head of the Bay Quinte ; whose "orders were, in 1785, to have cross-roads between every six lots." This system was afterwards carried out, and enlarged upon in subsequent surveys, till in many of the townships further west side-roads were laid out between every *second* lot.

The surveying of these townships progressed very slowly. The "Fifth Town" was probably laid out before any, or but very few, settlers located within its limits ; but this—the "Seventh Town"—was settled by quite a number, and in various parts, before the surveying was finished, or even commenced. Some members of the Weese family, the first settlers of the county and then living on lot 89, 1 concession, where William F. Weese still resides, were members of the original surveying party, who, under Kotte, laid out the Township.

As originally constituted, Ameliasburg comprised within its limits the present township of Hillier, and a part of Hallowell. Its area is very much less than formerly, though it is still the second township in the county in point of size, wealth, and population, containing, according to the latest revised statistics, an area of 43,092 acres, valued at $1,077,300 and a population of 3,067 souls, among whom are 888 ratepayers, owning personal property assessed at a valuation of $78,250, and possessing a taxable income of $2,600. The township is butted and bounded as follows, that is to say : on the north by Bay Quinte and the township of Murray, in the County of Northumberland ; on the east by the Township of Sophiasburg ; on the south by the Township of Hillier ; and on the west by that portion of Lake Ontario known as Weller's Bay.

The formation, soil, roads, improvements, &c., &c., are so very similar to those of Sophiasburg, already described, that a repetition would simply occupy space, and be subject of monotony to the reader. We need simply mention that this township differs from Sophiasburg, only *eight acres* in extent of area ; and $200 in valuation of real property ; with the assessed personal property being exactly the same ;—but in population there is a greater difference—Ameliasburg leading by 939. There is also another point of resemblance between the two, or rather the same peculiarity in both, each has a beautiful lake near its centre—high up above the Lake Ontario level ; (the difference in altitude is over one hundred feet). The similarity of each to the other is particularly striking ; each was formerly filled with myriads of the very finest specimens and most highly prized varieties of fish ; each is emptied by a beautiful stream flowing into Bay Quinte ; and on each of these streams a mill was built during the time of the very earliest settlements. In the case of Ameliasburg, it was Owen Roblin who erected the mill which formed the nucleus of a village. This afterwards developed into a prosperous and busy little town—called from the mill and its owner, "Roblin's Mills,"—a name it still retains, although the official name of the post-office is Ameliasburg. It is a little singular that this place, Demorestville, and Milford—the most important villages in the three original townships, should have received their names in a precisely similar manner.

ROBLIN'S MILLS at the present time contains several good general stores, a first-class carriage shop, a harness shop, tailor shops, the usual number of blacksmith shops to be found in places of similar size, a very home-like and comfortable temperance hotel, and about three hundred inhabitants.

The Town Hall is here situated. It was built in 1874, by E. Sprague, contractor, at a cost of $4,500. The *material* employed is a beautifully tinted blue limestone, with Kingston gray cut-stone facings, with arched windows and doors—the whole of modern style, superior construction, and considerable claim to architectural beauty. It is designed throughout with due regard not only to appearance—which is of much importance in public buildings, but to comfort and convenience of those whose duty or pleasure it may be to use or visit it. The basement contains a wood-house, store-house and lock-up ; the main floor has high walls and ceiling, is large and very nicely arranged for public business, having a couple of private rooms in the rear end, from one of which opens a large fire-proof vault, built in the latest and most approved style, and with all the modern improvements. Altogether, it is the finest public hall owned by any rural municipality in the country, so far as we are aware.

There is a park, or more strictly speaking a fine level field, attached to the building, and this is provided with all requisite stands, sheds, &c., &c., designed for the use of the Township Agricultural Society, which is in a healthy and prosperous condition. The value of Hall, buildings, ground, &c., is estimated at $5,500.

A Post Office was opened, and first kept by a Mr. Mitchell, for three years, when it was closed, on account of his removal. It was reopened with Owen Roblin as Post Master, who still holds the position. There is a daily mail and stage to and from Belleville, which is ten miles distant ; Picton being eighteen miles, Wellington ten miles, Demorestville fifteen miles, Consecon eight miles, and the Carrying Place twelve miles.

The fourth Division Court of the county of Prince Edward is held here. The officials are Edward Roblin, Clerk, and J. S. Tice, Bailiff.

Captain Peterson's company of the 16th regiment is here, Lieut. Bothwell is second officer.

"Lake" Lodge, A. F. and A.M., No. 215, G. R. C., meet here. The following are the officers for the present year :—W.M., Benj. Bothwell ; S.W., J. G. Johnston ; J.W., H. Grannis ; Secy., A. N. Sprague ; Treas., John Sprague ; Chap., William Anderson ; S.D., E. Sprague ; J.D., John Way ; D.C., William DeLong ; J. G., John Roblin ; Tyler, James Higgins.

The present mills of Owen Roblin were built in 1842. They are fo-stories in height, built of stone, and run by steam, having a capacity of one hundred barrels of flour daily. Considerable merchant milling was done here till the failure of the winter wheat crop, since which time the business has been confined chiefly to custom work. There is a saw-mill, shingle-mill, and cooperage attached; and this being a very favorable locality for apples, of which large quantities are grown, a considerable trade is done at these premises in apple barrels.

The inhabitants are supplied with pure water from Roblin's Lake, on the banks of which the village is built, by means of a windmill supplying a tank, which in turn supplies the various houses through a system of pipage connected therewith.

It appears that Roblin's mills, though now the seat of municipal government was not always the "Capital" of the Township; as will be seen from the following extract from the minutes deposited with the Township Records. But we may here be allowed to digress for a moment, and remark that the earliest preserved Records are not of very ancient date; the oldest in existence being minutes of the first meeting held under the Municipal Act in 1850, at the house of David Coleman, of Roblin's mills. From this we find that the councillors that were to represent the respective Wards were as follows:—Carrying Place Ward, Reuben Young; Centre Ward, William Dempsey; Roblin Lake Ward, Owen Roblin; Rednerville Ward, Samuel S. Wallbridge; Consecon Ward, Thomas G. McGrath; of whom Wm. Dempsey was, on motion, elected Reeve, R. C. Colter was then appointed Clerk, and James Redner and John Russell, Auditors. At the next meeting a motion was carried to hold the next meeting in another place.

Accordingly the next meeting was held February 4th ensuing at the house of John Pultus, of Rednersville, at which the additional town officers were appointed:—Peter Rowe, Peter Dempsey, Sylvanus Sprague, Assessors; Richard J. Clute, Collector; William Brickman, Treasurer; David Bryant, Superintendent of Common Schools. Then come the minutes above referred to. We quote:—

"Mr. Roblin, seconded by Mr. J. G. McGrath, moved *That, in the "opinion of this Council, the session of this Council should be held at a place "as near the centre of the Township as circumstances will permit, and that "the house of David Coleman being the most central is therefore a suitable "place for holding the sessions of this Council*—which was put, and the yeas "and nays were taken on the motion. Yeas, Roblin and McGrath; nays, "Wallbridge, Young, and Dempsey."

"Owen Roblin, seconded by J. G. McGrath, moved *That the moving of "the session of this Council by a bare motion, without the passing of a By-law "for that purpose, from the place where the inhabitants have been accustomed "to meet for the transaction of business, is not only unjust and opposed to the "wishes of a very large majority of the people, but is contrary to the intent "and meaning of the Municipal Act, and therefore illegal.* Yeas, Roblin "and McGrath; nays, Wallbridge, Young, and Dempsey."

In 1852 the township became entitled to a second representative at the County Council Board. From the second year of the operation of the Municipal Act, inclusive (the first having been above given), the following gentlemen served the township as Reeve and Deputy-Reeve:—1851, Wm. Dempsey, Reeve; (no Deputy this year); 1852, Owen Roblin, Clarke Whittier; 1853, Owen Roblin, William F. Weese.

The Reeve and Deputy being then elected from and by the five Councillors, we give an extract from the minutes of the meeting which elected the above gentlemen:—

"Mr. Roblin, seconded by Mr. Redner, moves *That Wm. F. Weese be "appointed Deputy-Reeve.* Mr. Whittier, seconded by Mr. Young, moves "*That James Redner be appointed Deputy-Reeve.* On the vote being taken "there was a tie. Messrs. Roblin and Redner being for Weese, and Messrs. "Young and Whittier for Redner. Mr. Roblin, being the highest on the "assessment roll, gave the casting vote for Weese."

On referring to the assessment roll of that year we find Mr. Roblin assessed for £5,115, or $21,660. As assessments were then made at or below one third the real value, this represented a property worth at least $65,000. The next highest on the roll that year was James Redner, for £2,080 [illegible], or a real value of about $25,000. Though the above mode of deciding municipal matters is quite at variance with present practice, we are not sure but that it was then in accordance with past precedent. At all events, it was in accordance with the letter of the law at that time, and was probably not original with those gentlemen, though it is the only instance of the kind we remember to have seen recorded, and seems quite [illegible] of today.

The list proceeds:—1854, Sylvanus Sprague, Reeve; James Jacques, Deputy-Reeve. Mr. Sprague filled the above position without intermission for a period of twelve years. Mr. Colter was Clerk from 1850 to 1855. Simon Murchison, Charles Dorry, and J. J. Richards followed each other in the clerkship, each retaining the office a number of years. In 1868, the first year in which the Reeves were elected by the popular vote, S. Sprague was chosen to that position, with George Roblin for Deputy, and Charles E. Dorry for Clerk. In 1869 the township was represented in the County Council by George Roblin, Reeve; Henry B. Hunt, Deputy-Reeve; in 1870, by William Anderson and William R. Dempsey; in 1871, by William Anderson and Reuben Young; in 1872, by William G. Stafford and James M. Spencer; this year Mr. Richards succeeded Mr. Dorry as Clerk; in 1873, the same as previous year; in 1874, William G. Stafford and William R. Dempsey; in 1875, William R. Dempsey and Joseph Nightingale, with H. T. Webb for Clerk; and the same in 1876 and also in 1877. For 1878 the Town officers are Joseph Nightingale, Reeve; James A. Johnston, Deputy-Reeve; Peter V. Beech, Robert S. Hunt, Elijah Sprague, Councillors; James Benson, Clerk; Royal W. Brenson, H. T. Webb, Auditors; John B. Cunningham, Assessor; Peter Bonter, Collector; William De Long, Treasurer.

There were collected last year for county purposes, $1,956; for school purposes, $4[illegible]; and for local purposes, $1,301.75. The rate of taxation was three mills in the dollar. The township has a fund of over $15,000 of what is termed the Public School Trust, invested in real estate at eight per cent., payable yearly, which they use for school purposes.

The villages of Ameliasburg, besides Roblin's Mills, already described, are Consecon, Rednersville, Rossmore and Mountain View, of more or less importance in the order named.

Besides these there is a post-office called Albury (though why so called we could not learn) situated on lot 83, 1st concession, four miles from Carrying Place, and ten from Belleville, off which it has a tri-weekly mail. It has been a post-office many years, though there is no village there or thereabouts, the first postmaster being Isaac Dempsey, and the present one William Peck.

Mountain View, so named from its natural elevation, is situated on lot 68, in the 2nd concession, four miles from Roblin's Mills, and six miles from Belleville. There is a cheese-factory here, and a few houses—altogether a very unpretentious "village," though a pleasant hamlet, on the Belleville and Picton road, distant from the latter place about nineteen miles, and from Demorestville about ten.

Rossmore is built upon a rocky point called Point Henessey, across the bay from Belleville, and one mile distant therefrom. Communication is had by means of a steam ferry, which runs every hour. Rossmore contains a store, blacksmith shops and three taverns, and a fluctuating population of [illegible] to 300, according to the "run" of B. F. Baker & Co.'s Mills, near by, which are said by competent judges to be the largest (in sawing capacity) and most complete in the Bay Quinte. These mills employ seventy-five to one hundred men when running full. They were idle last season, on account of the unprecedented depression in the lumber trade, but are now running full again. They keep up the village of Rossmore, which, without them, would be but a half-deserted hamlet of inferior habitations; with the mills to aid it, is but the temporary residence of thos in lumber manufacture during the summer season.

There is a very large amount of travel passing through the pla Belleville and all parts of Prince Edward County, and the ferry a profitable business at very reasonable figures; while some of t notwithstanding the supposed existence of a prohibitory law (Act being in force in Prince Edward), do a rushing trade in t whiskey.

Rednersville is a pleasant little village on the Bay Quinte miles east from Belleville, containing a general store, blacksmit tailor and shoe-shops; hotel, wharf and store-houses; one ch exceptionally fine private houses, and a population of 150 to 200 daily mail off Belleville, and stage to and from that point an Mills. It is surrounded by a delightful agricultural section, large amount of grain is handled here, Mr. James Redner some chasing as high as 100,000 bushels in a season—principally bark

Consecon—derived from the Indian word "con-con," meani erel, from the great abundance of that fish along the shore in the early settlement of the place—is situated at the head of We five miles south-east of the Carrying Place; partly in Amelia partly in Hillier; on both sides the Consecon River, which empt lake of the same name, on which was built the first grist mill i burg, in 1804, by Matthias Marsh, whose son still keeps the The situation of the place is very pleasant, and its location is s undoubtedly make it one day a place of commercial importanc Bay being one of the most safe, commodious, and easily accessib its natural position is concerned) of any harbor upon the lake York and Kingston road runs through the place.

There are a number of very good business houses in the vil ding three general stores, one grocery, one drug-store, and som carriage and blacksmith shops.

There are four fine churches, a graded school, employing tw and two telegraph offices.

It is the seat of the seventh Division Court of the County Edward, of which J. D. Cadman is Clerk, and D. H. Weeks Ba

"Consecon" Lodge, A. F. & A. M., G. R. C., meets here. officers are:—W. M., James Baird; S. W., John Baird; J. McDonald; Secretary, George J. Waddell; Treasurer, S. R. Ja lain, R. J. Osborne; S. D., D. H. Weeks; J. D., J. H. Young; Bedell and S. F. Noxon; Tyler, John Ruttan.

The village has a daily mail, per stage, to and from Picton an It is distant from the former place about twenty-two miles, an latter about fifteen. It contains a population of about 400 original settlement of Ameliasburg forms an interesting chapter tory of the County and of the Province.

Without doubt, the honor of having first settled in the County Edward is due to John Weese, an American Colonist of German from Herkimer County, New York, who espoused the Royal ca two terms in the British army during the Revolutionary War, left the country, and settled, with his family, on lot 89, 1st con the autumn of 1783, and there remained till his death at a ripe

Mr. Weese was the father of a large family, and the an numerous progeny. His grandson, William F. Weese (of wh sketch elsewhere appears), still owns and resides upon the orig stead.

In matters of this kind, when the actual participants in which we seek information have long since passed away, it is difficult to arrive at the exact facts and dates. It is affirmed though denied by some—that Thomas Dempsey, who settled 1st concession, was the second settler in the present limits of A He was, at all events, one of the very earliest; and when th came to have a municipal organisation, he was the first As appointed; and it is related of him that he made yearly visits, years, to the site of the present village of Wellington, *to asses* The distance was about fifteen miles—across the present to Ameliasburg and Hillier, and all through a literally "howling w The one rate-payer, whose taxes probably amounted to less tha of the value of the time and trouble it took to collect the sam tably Daniel Reynolds, mentioned in connexion with the early of Wellington. Mr. Dempsey was born at the manor of the General Schuyler, at Hackensack, near New York, his father b time private secretary to that officer. When quite a lad he w into the Continental army, and served a year against the Britis then discharged, and subsequently joined the Royal troops, fa he was obliged to fly the country and come to Canada, when abandoned New York.

Elijah Wallbridge, a U. E. Loyalist from Bennington, Verm very early settler—the first in the extreme eastern part of the where he drew and purchased large tracts of land, at and near Point. A large quantity of this still remains in the possession scendants.

Further east came William Anderson, the Ways, Morden DeLongs, Robins, Parliament, the Spragues, Bonters, Sagers, an

Among the very first in the township were Elias Alley, Jo and John Belnap. These are said, by some, to have been the fi in the order named, after John Weese.

Mr. Belnap had a son, also named John, who is said to hav first white child born in the township.

The only island of any size within the municipal bounds of A is Huff's Island, immediately to the east of the central part of ship, from which it is separated by a deep morass. It deriv from Solomon Huff, son of Paul Huff, a U. E. Loyalist, from town. Solomon settled on the island in 1825. His nearest n that comparatively late date, were at Demorestville on the on at Wallbridge's, east of Mississauga Point, on the other.

We have heard from that proverbially correct and indisput the "oldest inhabitant," that the courtship and marriage o pioneers was conducted in a manner more sociable than would the present ideas of propriety in our aristocratic church circle Their rather "loose" (as we would now call them) notions of th are sufficiently indicated by the remark of a highly esteemed g the olden time, to the effect that "time was too valuable to "about these matters." Truth to say, there was sufficient ca state of affairs, in the fact that for many years, and as late as only existed in all Upper Canada, *six magistrates* who were au perform the marriage ceremony—besides the clergymen of the England, who were very, very scarce in those days. The con all this was that many so-called marriages were somewhat "irre to remedy this awkward state of affairs, and straighten the "s in this most important business, special Acts of Parliament from time to time.

Very many from Sophiasburg and other parts of the county Ameliasburg, were wont to betake themselves to the Carryi have this ceremony performed. On the way thither, the hospit Weese made his house an open house for all parties going to a always went on foot, and were often obliged by the distance to night on the road.

Mr. Weese's daughter Catherine and David Sager were the married. Elder Winer, a Baptist minister, residing at the Carr

erted hamlet of inferior habitations ; and which,
s but the temporary residence of those employed
uring the summer season.
amount of travel passing through the place between
f Prince Edward County, and the ferry does quite
ery reasonable figures ; while some of the taverns,
posed existence of a prohibitory law (the Dunkin
nce Edward), do a rushing trade in the vilest of

leasant little village on the Bay Quinté shore, five
e, containing a general store, blacksmith, waggon,
hotel, wharf and store-houses ; one church, some
e houses, and a population of 150 to 200. It has a
, and stage to and from that point and Roblin's
l by a delightful agricultural section, and a very
handled here, Mr. James Redner sometimes pur-
0 bushels in a season—principally barley.

from the Indian word "con-con," meaning a pick-
undance of that fish along the shore in the days of
the place—is situated at the head of Weller's Bay,
the Carrying Place ; partly in Ameliasburg, and
th sides the Consecon River, which empties into the
on which was built the first grist mill in Amelias-
ias Marsh, whose son still keeps the Post Office.
ce is very pleasant, and its location is such as will
e day a place of commercial importance. Weller's
st safe, commodious, and easily accessible (so far as
oncerned) of any harbor upon the lake. The old
runs through the place.
of very good business houses in the village, inclu-
, one grocery, one drug-store, and some first-rate
shops.
urches, a graded school, employing two teachers,
s.
seventh Division Court of the County of Prince
Cadman is Clerk, and D. H. Weeks Bailiff.
A. F. & A. M., G. R. C., meets here. Its present
James Baird ; S. W., John Baird ; J. W., Albert
George J. Waddell ; Treasurer, S. R. Jones ; Chap-
D., D. H. Weeks ; J. D., J. H. Young ; S.-S., W.
; Tyler, John Ruttan.
ly mail, per stage, to and from Picton and Trenton.
ner place about twenty-two miles, and from the
contains a population of about 400 souls. The
meliasburg forms an interesting chapter in the his-
of the Province.
nour of having first settled in the County of Prince
Weese, an American Colonist of German extraction,
New York, who espoused the Royal cause, served
army during the Revolutionary War, and finally
tled, with his family, on lot 89, 1st concession, in
there remained till his death at a ripe old age.
father of a large family, and the ancestor of a
s grandson, William F. Weese (of whom a short
s, still owns and resides upon the original home-

kind, when the actual participants in matters of
ion have long since passed away, it is extremely
exact facts and dates. It is affirmed by many—
—that Thomas Dempsey, who settled on lot 93,
econd settler in the present limits of Ameliasburg.
one of the very earliest ; and when the township
ipal organization, he was the first Assessor ever
ated of him that he made yearly visits, for several
present village of Wellington, *to assess his men,*
t fifteen miles—across the present townships of
, and all through a literally "howling wilderness."
se taxes probably amounted to less than a quarter
and trouble it took to collect the same—was pro-
mentioned in connexion with the early settlement
mpsey was born at the manse of the American
Hackensack, near New York, his father being at the
that officer. When quite a lad he was drafted
ny, and served a year against the British. He was
bsequently joined the Royal troops, for which act
e country and come to Canada, when the British

U. E. Loyalist from Bennington, Vermont, was a
ed in the extreme eastern part of the township,
hased large tracts of land, at and near Mississauga
of this still remains in the possession of his de-

William Anderson, the Ways, Mordens, Redners
ment, the Spragues, Rutters, Sagers, and Pecks.
t in the township were Elias Alley, John Blecker,
se are said, by some, to have been the first settlers,
e John Weese.
n, also named John, who is said to have been the
the township.

x size within the municipal bounds of Ameliasburg,
ately to the east of the central part of the town-
eparated by a deep morass. It derives its name
e of Paul Huff, a U. E. Loyalist, from Adolphus-
on the island in 1823. His nearest neighbors, at
date, were at Demorestville on the one hand, and
Mississauga Point, on the other.
that proverbially correct and indisputable source,
," that the courtship and marriage of the early
in a manner more sociable than would accord with
priety in our aristocratic church circles of to-day.
s we would now call them) notions of those matters
d by the remark of a highly esteemed gentleman of
fect that "time was too valuable to make a fuss
Truth to say, there was sufficient cause for this
ct that for many years, and as late as 1814, there
r Canada, *few magistrates* who were authorized to
remony—besides the clergymen of the Church of
, very scarce in those days. The consequence of
o-called marriages were somewhat "irregular," and
d state of affairs, and straighten the "crookedness"
business, special Acts of Parliament were passed

liasburg and other parts of the county, as well as
t to betake themselves to the Carrying Place to
oined. On the way thither, the hospitality of Mr.
n open house for all parties going to and fro, who
d were often obliged by the distance to remain over

r Catharine and David Sager were the first couple
a Baptist minister, residing at the Carrying Place,

performed the ceremony, receiving therefor, one York shilling, which was the usual fee for performing that interesting ceremony.

The longevity of the Pioneers and their immediate descendants has often been a subject of remark. Quite a number of them lived to be over one hundred years of age. Mr. Rush, who departed this life in June of the present year, was the oldest man in the township till his death, at nearly one hundred years. Mrs. De Long, daughter of Daniel Cole, living on lot 92, 2nd concession, is now in her one hundredth year. Mrs. Cole, daughter of Henry Redner, is in the nineties. Both are hearty and vigorous, and in full possession of all their faculties.

The first school in the township was taught by John Smith, in a log building, thatched with marsh-hay, on lot 85, 1st concession. The next was Abijah Benedict.

It is not stated when, or by whom, the first house was built, but it was undoubtedly by John Weese ; as his family were so far ahead of others in point of time, that some members of it never saw the face of a white man—except each other—for more than three years.

The first frame barn raised in the township was on the lot of Mr. Weese, and it stands there still.

For many years before the Napanee mills were built, the Weeses went to Kingston mills to get their gristing done.

Probably the one spot of all others in Ameliasburg—or indeed in the whole Bay Quinte region, in which the greatest historical interest centres, is

THE CARRYING PLACE.—This is a narrow isthmus separating that part of Lake Ontario known as Weller's Bay, on the one side from Bay Quinte on the other ; and connecting Prince Edward County with the main-land. It is called thus from the fact that the Indians in their journeyings between the East and the West, were wont to travel by canoe along the lake shore and up or down Bay Quinte ; and on arriving at the head thereof, or at Weller's Bay—according as they were going East or West—would pick up their canoes, and *carry* them from water to water.

This, as is well known, was the early mode of travel ; not only by the aborigines, but by the first white settlers also ; and not only here but all over the American Continent. To use a modern term, they seemed thoroughly to understand the principle of utilizing the "magnificent water-stretches," in which Canada—of all countries—most abounds. There were consequently a great many "carrying places," in all directions ; in fact there were several in Prince Edward County. But the historical associations connected with *this one*, during the early days of our country's existence as a British Province, made it, of all others, "*The* Carrying Place," and it was always so known by the old settlers ; and yet is, by every one who knows aught of Canada.

Its geographical position seemed to mark the Carrying Place as a very important point, from the earliest date. It was thought that there would arise a great city there some day, and it was in serious contemplation by some of our ruling men to make the capital of the country here, at a day when the present city of Toronto was "a dreary dismal place, not even possessing the characteristics of a village."

Governor Simcoe was one of those who saw through a different light, however, and though he was in favor of making London the capital of Upper Canada—probably for military reasons—a scheme in which he was over-ruled by his superior, Lord Dorchester—still he saw in Toronto a great future ; and was convinced that *it* and *not* the Carrying Place was to become the great commercial city of the Province. This is proven by the advice he gave Robert Young, a personal friend, and ex-captain of the British Navy, who located on the Carrying Place in 1792, there being at that time but one other settler in the neighborhood. The Governor tried to induce him to locate on what is now "Yonge Street," saying that "Little York" would some day be a great place. Young was entitled to 1200 acres of land, and there was plenty then, and for years afterwards unoccupied, where the heart of Toronto now stands. But the captain thought differently, and settled on the lot at the Carrying Place, where Reuben Young now lives ; and drew his 1200 acres there, and on the north shore of Pleasant Bay.

The scheme has been agitated, from time to time, of cutting a canal through the Isthmus. The inducements to do so are the immense benefits which would accrue to the mercantile marine interests, by giving vessels a shorter and perfectly safe passage from the Upper Lake to Kingston ; whereas they now have a more tedious, and—in rough weather—an extremely dangerous course.

It is claimed that the number of wrecks which annually occur along the south of Prince Edward, and other more dangerous points towards the foot of the Lake, would go far towards building a canal ; and it is matter of surprise to the uninitiated, why the scheme has not long ago been carried out.

Also, in case of war, the advantages it would offer, in the way of forming an inland navy, and gaining maritime supremacy on the Lakes, are almost incalculable.

But aside from the many acknowledged advantages of a commercial, and naval or military nature, which its geographical position gives it, it is a place interwoven with the associations and memories of all our country's fathers, east and west.

Even from as far west as the present town of Port Hope (where the first settlement between York and Napanee was made by four families, named Ashford, Harris, Johnson, and Stevens, in 1793)—and as settlements afterwards sprang up along the shore—from many intermediate points came the pioneers in their canoes, with bags of corn to the Napanee mills. They would unload ; carry first their canoes across to Bay Quinte—afterwards their corn, and then re-embark ; repeating the same operation when returning with their meal. This business used to occupy a week at a trip from Port Hope ; this being the farthest point west, from which gristing was brought to these celebrated mills ; which, with the exception of Kingston mills, were the oldest in Upper Canada ; and the inhabitants of that western settlement were as familiar with the outlines and land marks of the Carrying Place, as were settlers themselves.

One of the oldest settlers in this vicinity was Mr. Weller, after whom Weller's Bay was named. W. H. Weller, of Cobourg, now Master in Chancery for the Counties of Northumberland and Durham, Peterboro', and Victoria, and Judge Weller, of Peterboro', are among his descendants. Mr. Weller was a very enterprising and public-spirited citizen. For a great many years he ran a stage line between Kingston and York. The route lay over the old military road known in the east as the York road, and in the west as the Kingston road. It was also called the Danforth road, from an American named Asa Danforth, who contracted with the Government to build it in 179-, and completed it as far west as Ancaster in 1801. It passed through almost the entire length of Prince Edward County, entering it at the east by Dorland's Ferry, between Adolphustown and the "Stone Mills," five miles below Picton. Thus it happened that all who passed through the country, whether by land or water, from east to west, or *vice versa*, were familiar with the Carrying Place.

The exact trail of the old Indian Portage is followed by the present road, laid out when the township was surveyed. It was intended by the surveyor to run a straight line across the Isthmus ; but when he came to that part of the work he was taken ill, and gave his subordinate instructions to lay out a range of 100-acre lots on either side the Carrying Place. These orders were literally obeyed, his deputy following the various deviations of the path, which was then, as it had been from time immemorial, a simple Indian trail ; and it is to this accidental circumstance that we are indebted for the preservation of this ancient historical land-mark in its original shape.

The road itself, though slightly devious, is on the whole pretty direct, the general course being north-east and south-west, and the distance from water to water one and three-quarters miles.

The ranges of lots on both sides were originally a portion of the "Seventh Town;" but at a later date the north-west range was made a part of the township of Murray, in the county of Northumberland, and the "Carrying Place" has been for many years the land boundary between Northumberland and Prince Edward counties.

Robert Wilkins, a captain in the Royal service during the Revolutionary War, came from Shelburne, Nova Scotia, whither he had emigrated after the British abandoned New York, and settled here. He was for many years one of the most enterprising and influential men in the Bay Quinté district.

This gentleman's son, Lieut.-Colonel the Hon. Robert Charles Wilkins, who died there in March, 1866, was for more than half a century one of our leading men in commercial, political, and military affairs, being intimately identified with the history of all matters bearing upon the well-being of the inhabitants of the country at large, and particularly of the Bay Quinté region, throughout which his name has been a familiar household word for the past two generations.

We cannot bid adieu to this romantic spot or to Ameliasburg without offering our endorsement to the sentiments of a well-known and much-admired writer, from whom we quote:—

"The tourist will find abundant food for thought at the Carrying "Place, whether he contemplates the far remote past, ere the Indian was "disturbed in his native abode, or in the days when the French Recollet "missionaries followed in the footsteps of those whom they sought to con"vert; or the time when the pioneer settler and surveyor first trod the "path; or whether he reflect upon the many human beings who have "come and gone on their way of life—now one way, now another; or of "the trader, intent only upon pressing his business into the extreme con"fines of the earliest settlements; or of the soldiers—regulars and militia "—who pressed onward for the conflict to drive off an invading foe; or of "the thousand prisoners carried captive through the Province they came "boastingly to conquer."

HILLIER AND WELLINGTON.

As compared with the other townships of the county, Hillier is fifth in point of age, and fourth in size, wealth and population. Of a somewhat irregular shape, it is surrounded on the north-west, north-east, and south-east, by the townships of Ameliasburg, Sophiasburg, and Hallowell; while Lake Ontario—with its numerous indentations, including Weller's Bay, Pleasure Bay, and Huyck's Bay—laves its shores on its western and southern boundaries. Viewed from a topographical, geological, or agricultural stand point, it is so very similar to the three townships by which it is surrounded, and which have already been described, as to render it unnecessary to dive into minute details. It will suffice to say that it is fully up to the average, and in honest truth a model township in everything which combines to make one locality more worthy of praise than another; or more to be desired as a comfortable home to the settler—welcome retreat to the visitor—or pleasant resort to the traveller and tourist. It was originally settled, as were the townships above referred to, by those noble sons of noble sires, the United Empire Loyalists, whose devoted attachment to their King and Constitution led to their abandonment of riches and power under a Government founded upon political principles to which their loyalty was opposed.

The early history of Hillier is the history of Ameliasburg, of which it formed a part till 1823, when an Act of Parliament was passed, on petition of the residents, to set off that portion as a separate Municipality. This Act came into force on the 1st January, 1824; and the new township received its name from Major Hillier, who was at that time Secretary to Sir P. Maitland, the Governor of the Province.

The oldest preserved official records only date back to 1850, which makes it impracticable to give lists of Municipal officers further back than that date. The first Reeve elected by the popular voice of the whole township (in 1850), was James T. Lane. For the first named year we find that Stephen P. Niles was Reeve; George Jones, Deputy Reeve; William Thorn, John Y. Weeks, and Daniel Y. Williams, Councillors; Allen M. Dorland, and Samuel Pennock, Auditors; William Netherby, Collector; John McFaul, Surveyor; Cornelius Clapp, License Inspector; William Netherby, Peter Valleau, David S. Young, James Jones, Sr., and Samuel Pennock, Commissioners of the Poor; also, thirty-two Pathmasters, twenty Fence Viewers, and eighteen Poundkeepers.

For the present year (1878), the various municipal duties are performed by the following named gentlemen:—

John Young, Reeve; Robert T. Jones, Deputy Reeve; Daniel Howe, Lancelot Nethery, Paul C. Van Horn, Councillors; John Graydon, A. M. Dorland, Auditors; R. A. McFaul, Assessor; Peter C. Ainsworth, Collector; Richard Noxon, Surveyor and Road Commissioner; and thirty-two Path Masters, eleven Fence Viewers, and twenty Poundkeepers. Franklin Jones is Township Clerk, having lately succeeded his father, Samuel Jones, J. P., who held the position for five years. Mr. Jones is an educated gentleman, of Irish descent, a Major of militia, and leading man in local affairs; his father having settled at Pleasant Bay, over a half century ago, where such farms as his son resides upon were considered worth from five to ten dollars an acre. Now the beautiful farms in the neighborhood are worth from $60 to $90 per acre. Both of Mr. Jones' parents lived in Hillier till over ninety years of age.

Among the earliest settlers in this locality were, Ira De Long, Thomas C. Beadle, Richard Van Horn, John Baird, John Tripp, and E. Hawley.

On the north side of Pleasant Bay the earliest settlers were the Youngs, Pietsons, Huffmans, Camerons, Careys, and Fergusons.

In the eastern part of the township the first settlements were made by the Williams's, Valleaus, Hicks's, Dolands, Pearsalls, and Mordens.

In other parts the Clapps, De Longs, Pettengills, Trumpours, Bowermans, Hutchinsons, and Pettits were among the pioneers of the respective localities in which their descendants still reside.

The township contains a number of pleasant and prosperous post villages; also, WELLINGTON, the only incorporated village in the county, is situated within its limits.

CONSECON, the second village in size and importance in the county, is also partially situated in Hillier. The two parts of this village are divided by the Consecon river, which also divides Hillier from Ameliasburg, in which township the major part of the village is situated, and in the sketch of which a short description of Consecon may be found.

PLEASANT VALLEY, situated about two miles east of Pleasant Bay, is really, as its name implies a *pleasant* village, in every sense, not only as regards its situation, and the beauty of the surrounding country, but the air of business, prosperity, improvement, and general comfort to be noticed about the place is more of what is generally supposed to be the special attributes of the growing western towns. It is situated on the old Danforth military road between Consecon and Wellington, five miles distant from the former, and seven miles from the latter. It has a daily mail both ways per stage, to and from Picton and Trenton, Henry Palmer is Postmaster. There are two telegraph offices here, a good country hotel, a couple of stores, carriage and blacksmith shops, churches, school, gristmill, sawmill, &c., besides a number of fine private houses for a village of its size.

The town hall of Hillier is here situated—a large commodious and well arranged building, with grounds and sheds attached. It was erected some ten years since, at a cost of $1,400.

There was formerly a Division Court held here. Under the old system of Administration of Justice, Thomas Flagler, Philip Clapp, and Stephen Niles were the first Commissioners in the township, for holding the Courts of Requests.

The first settlers in the place were Josep
Stapleton, and James Foster. The former bu
day, the first in this part of the country, and
manufactory in the county.

The other villages in the township are ALL
tween 1st and 2nd concessions, north. This
Cyrus Allison, father of W. H. R. Allison, n
leading lawyers of the District. The Rev. Mr.
as well as one of the oldest and most highly
who has ever served the cause of the Church in

MELVILLE, a smart and beautifully locate
con lake, and ROSE HILL, on lots 17 and 18,
villages, each containing a mill, one or more st
the various mechanical enterprises.

From the last Assessment Roll (1878), we
Total number of acres in township, 31,389
23,836; value of real property, $750,550;
$67,305; amount of taxable income, $11,300;
township, 1,242; number of cattle owned in
sheep, 1,945; number of hogs owned in townsh

Hillier is pre-eminently an agricultural cons
vantages already noted as belonging to townsh
here also to be found in a marked degree; an
duties of the writer have seldom brought him in
with a community possessing to a greater ext
moral, intellectual, and material elements of wo
and Christian contentment.

WELLINGTON

The only incorporated village in Prince
mer limits of Hillier, from which it was se
separate municipality, under the general Act
by a By-law of the County Council, passed Octo
of forming an Incorporated village had been agit
opposed somewhat strongly by the majority of t
pally—and in a somewhat less degree by the in
which a small extent of territory was also taken

One condition of incorporation, viz.: that th
tants was doubted by the people of the township
of enumerators was appointed to decide the ques
able to the advocates of incorporation, as we lea

"BY-LAW, No. 8.—*Whereas*, by the cens
"Wellington, in the townships of Hillier and
"appears that there are over seven hundred and
"limits of the said village; and

"*Whereas*, one hundred of the freeholders a
"limits have petitioned this Council to pass
"village; and

"*Whereas*, it is only just and proper to co
"tained in said petition:

"*Therefore the Municipal Corporation of th*
"*enacts*:

"I. That so much of the townships of Hilli
"tained within the following boundaries, shall
"Wellington; and shall, from and after the ti
"effect, be incorporated under the above name,
"meaning of the Statute, 22 Vic., chap. 54, Sec.

"Commencing at the lake shore of Lake Onta
"No. 7, in the first concession of the townshi
"the west boundary line of said lot, to the re
"said lot, and along the rear of Lots Nos. 6, 5,
"rear of those parts of Lots Nos. 2 and 1, as at
"by Daniel Reynolds, J. T. Dorland, and Joseph
"between Hillier and Hallowell; thence along th
"Nos. 1 and 2, in the township of Hallowell, as
"Kessey, Joseph Cummins, and Francis Mand
"of said Lot No. 2, in the township of Hallowell
"boundary line of Lot No. 2 to the waters of W
"water line of said East Lake and Lake Ontari

"II. The first election shall be held in the t
"John T. Dorland shall be Returning Officer.

"III. This By-law shall take effect on and
"December next.

"Passed 30th October, 1862.

"N. BALLARD, *Clerk*.

At the first election held for the purpose of
the following gentlemen were selected to fill
Mahon, Campbell, Cummings, Harris, and Brow
ing was held in accordance with the Statute ma
ing will be found a copy of the public record of t

"Proceedings of the first meeting of the Cou
ton, 19th January, 1863.

"Present, E. D. McMahon, Donald Campbel
Brown, and Joseph Cumming.

"Moved by D. Campbell, sec. by E. D. McN
be appointed secretary of the meeting. Carried.

"Moved by D. Campbell, sec. by E. D. McN
be chairman.

"Moved in amendment by Mr. Cummings, s
Campbell be chairman. Carried.

"Moved by William Harris, sec. by Jonathan
be reeve. Carried.

"Moved and seconded that John T. Dorland

"Moved and seconded that D. S. Hicks be

"Moved and seconded that John H. Ferguso
be assessors. Carried.

"Moved and seconded that Fulton Palen be a

"Moved and seconded that D. S. Hicks and
keepers. Carried.

Moved and seconded that J. B. McGoniga
Carried.

"Moved and seconded that George Herringt
masters. Carried.

"Moved and seconded that William McDon
Cummings be fence-viewers. Carried.

"Moved and seconded that William Tivey a
Commissioners. Carried.

"Moved and seconded that this Council do
School-house, on Wednesday, the 28th inst., at

"19th January, 1863. "Jo

The early history of Wellington is environed
in the romantic incidents related of the first p
pleasures, their sufferings and rejoicings. The f
vicinity was a U. E. Loyalist named Daniel Rey
years among the Indians before his footsteps we
race. The date of his settlement has not been t

One of the early families to settle here was
of whose sons—Amos Hutchinson, about seve
resides in the place.

ORICAL SKETCH OF THE

first settlers in the place were Joseph Dorland, John and William
n, and James Foster. The former built a mill here at a very early
first in this part of the country, and subsequently the first potash
tory in the county.
other villages in the township are ALLISONVILLE, on lots 74, 75, best and 2nd concessions, north. This place was named after Rev.
llison, father of W. H. R. Allison, now of Picton, and one of the
lawyers of the District. The Rev. Mr. Allison was one of the earliest,
s one of the ablest and most highly esteemed Wesleyan Ministers
ever served the cause of the Church in the Bay Quinté District.
LVILLE, a smart and beautifully located village at the head of Conse-
, and ROSE HILL, on lots 17 and 18, near the lake front, are post
each containing a mill, one or more stores, and the usual number of
ous n cel anical enterprises.
n the last Assessment Roll (1878), we take the following figures:—
l number of acres in township, 31,389; number of acres improved,
value of real property, $750,550; value of personal property,
; amount of taxable income, $11,300; number of horses owned in
p, 1,242; number of cattle owned in township, 1,723; number of
,945; number of hogs owned in township, 494.
ier is pre-eminently an agricultural constituency, and the superior ad-
s already noted as belonging to townships previously described, are
o to be found in a marked degree; and—taken in its entirety—the
f the writer have seldom brought him into social or business relations
community possessing to a greater extent all the evidences of the
ntellectual, and material elements of worldly prosperity, rural felicity,
ristian contentment.

WELLINGTON.

only incorporated village in Prince Edward, is within the far-
nits of Hillier, from which it was separated and erected into a
municipality, under the general Act of Parliament in that behalf,
-law of the County Council, passed October 30th, 1862. The project
ng an Incorporated village had been agitated for some time, but was
l somewhat strongly by the majority of the people of Hillier, princi-
and in a somewhat less degree by the inhabitants of Hallowell, from
small extent of territory was also taken to form the new village.
condition of incorporation, viz.: that the place contain 750 inhabi-
as doubted by the people of the townships, and a "mixed commission"
erators was appointed to decide the question. The result was favor-
the advocates of incorporation, as we learn from the following:
Y-LAW, No. 8.—*Whereas*, by the census returns of the village of
ngton, in the townships of Hillier and Hallowell, in this county, it
rs that there are over seven hundred and fifty inhabitants within the
of the said village; and
Whereas, one hundred of the freeholders and householders within such
have petitioned this Council to pass a By-law incorporating said
e; and
Whereas, it is only just and proper to comply with the request con-
d in said petition:
herefore the Municipal Corporation of the County of Prince Edward,
s:
That so much of the townships of Hillier and Hallowell as is con-
l within the following boundaries, shall be designated the village of
ington; and shall, from and after the time when this By-law takes
be incorporated under the above name, according to the intent and
ng of the Statute, 22 Vic., chap. 54, Sec. 10:
mmencing at the lake shore of Lake Ontario, on the west side of Lot
, in the first concession of the township of Hillier; thence along
est boundary line of said lot, to the rear; thence along the rear of
t, and along the rear of Lots Nos. 6, 5, 4, and 3; thence along the
f those parts of Lots Nos. 2 and 1, as at present owned and occupied
niel Reynolds, J. T. Dorland, and Joseph Cummins, to the town line
n Hillier and Hallowell; thence along the rear of those parts of Lots
and 2, in the township of Hallowell, as occupied by Patrick Mac-
y, Joseph Cummins, and Francis Mandeville, to the east boundary
Lot No. 2, in the township of Hallowell; thence along the said east
ary line of Lot No. 2 to the waters of West Lake; thence along the
line of said East Lake and Lake Ontario, to the place of beginning.
The first election shall be held in the town hall in said village, and
T. Dorland shall be Returning Officer.
. This By-law shall take effect on and after the twentieth day of
ber next.
sed 30th October, 1862.
BALLARD, *Clerk.* "G. STRIKER, *Warden.*"
e first election held for the purpose of choosing a village Council,
wing gentlemen were selected to fill the positions: Messrs. Mc-
Campbell, Cummings, Harris, and Brown. The first Council meet-
held in accordance with the Statute made and provided, and follow-
e found a copy of the public record of the same.
ceedings of the first meeting of the Council of the village of Welling-
January, 1863.
sent, E. D. McMahon, Donald Campbell, William Harris, Jonathan
nd Joseph Cumming.
ved by D. Campbell, sec. by E. D. McMahon, that John T. Dorland
ted secretary of the meeting. Carried.
ved by D. Campbell, sec. by E. D. McMahon, that Mr. Cummings,
an.
ved in amendment by Mr. Cummings, sec. by Mr. Harris, that Mr.
be chairman. Carried.
ved by William Harris, sec. by Jonathan Brown, that E. D. McMahon
Carried.
ed and seconded that John T. Dorland be clerk. Carried.
ved and seconded that D. S. Hicks be collector. Carried.
ed and seconded that John H. Ferguson and Jonathan R. Trumpour
rs. Carried.
ed and seconded that Fulton Palen be appointed constable. Carried.
ed and seconded that D. S. Hicks and James McKenna be pound-
Carried.
and seconded that J. B. McGonigal be Inspector of Licenses.
ed and seconded that George Herrington and Wm. Harness be road-
Carried.
ed and seconded that William McDonald, J. D. Clapp, and Joseph
be fence-viewers. Carried.
ed and seconded that William Tirey and Allen Pettangell be Poor
oners. Carried.
ed and seconded that this Council do adjourn, to meet again at the
use, on Wednesday, the 28th inst., at 6.30 p.m. Carried.
January, 1863. "JOHN T. DORLAND, Clerk."
rly history of Wellington is environed with facts which rival fiction
antic incidents related of the first pioneers,—their hardships and
their sufferings and rejoicings. The first white man to settle in the
s a U. E. Loyalist named Daniel Reynolds, who lived a number of
ng the Indians before his footsteps were followed by another of his
date of his settlement has not been to a certainty ascertained.
the early families to settle here was that of Mr. Hutchinson, one
ons—Amos Hutchinson, about seventy-five years of age,—still
he place.

Robert McCartney, Thompson, Cannan, Paul Trumpour, Archibald McFaul, Benjamin Garrett, Robert and William Hubbs, the Ellisses, Dorlands, and Haights, were among the first settlers. Of these, Mr. Trumpour was afterwards a very influential man, and the father of a large family, from whom it is said more of the inhabitants of the district are descended than from any other one man.

The latest assessment statistics of the village give the following result: Total number of acres 1,459⅝; value of real property, $152,985; value of personal property, $7,350; population (very nearly) 500.

This would seem to indicate a considerable retrograde movement in population since incorporation; but when (as a prominent citizen observed to us), it is remembered that the *object* of the enumerators in 1862 was to get the population up to the number necessary to incorporate; and also that the enumeration of assessors has somehow or other come to be looked upon as being *under* estimates, particularly in regard to valuation and population, the *real* difference will be found to be slight. We drop this remark by way of honest explanation, and *not* to cast any reflection upon the original enumerators (who were "all honourable men," no doubt), nor upon that highly useful, disinterested, and patriotic class of our fellow-citizens known as assessors. But the fact remains, and we presume those gentlemen "know how it is themselves," and "don't you forget it." That it *has* gone down somewhat is admitted, though not nearly so much as the figures quoted show, owing to the failure of the Wellington Fisheries, which formerly afforded employment to a large number of people.

Geographically, the village is situated in the south-east corner of the Township of Hillier, and the south-west corner of Hallowell, the metes and bounds thereof being more particularly set forth in the by-law of the County Council assenting to its incorporation, which is above quoted. It is built on a comparatively level area, which slopes gently towards the lake, and in point of location 'tis a most desirable and favourable one for the growth of a prosperous and healthy town, such as Wellington in reality is. It consists for the most part of one main street, over a mile in length, running near to and parallel with the Lake Ontario shore.

For a place of its size it bears unmistakable indications of a thrift and prosperity in direct contrariety to the apparent deductions to be drawn from the admitted decrease in population. It contains three large stores in various special lines, two general stores, cabinet factories, blacksmith and wagon shops, harness shop, tailor and shoe shops, three hotels, two telegraph offices, a number of wharves and store-houses, and a proportionately large number of fine private residences, many of which are of most ample dimensions and extremely handsome proportions. The place has not that bustling, business air observable in some western towns of less size, but it is nevertheless a very pleasant spot as a resort from the cares of business for a brief period; and the general "tone" of everything—the handsome gardens, the handsomer residences, and the dignified and courteous manner of the people themselves, give an outsider an unmistakable idea of the solid comfort and high social standing of the inhabitants, to form the acquaintance of whom it must be a pleasing episode in the experience of the traveller.

There are also a number of churches, some of which are ornaments to the village, from an architectural as well as a moral standpoint.

The Fifth Division Court of the county is held here; William Young being Clerk, and Thomas Jackson bailiff of the same.

Among the leading citizens of the place are Cornelius Clapp, J.P., and Donald Cameron, J.P., the latter of whom is the post-master, doing a heavy mercantile business, and a very large grain dealer. He has been Reeve of the village from the second year of incorporation till the present year, when he declined re-election.

A Town Hall was built by the village about the time of incorporation. It is a commodious two-story frame building, the lower floor being used for council and all public meetings, the upper story being leased to the masonic fraternity, which is represented here by "Star of the East" Lodge, A. F. & A. M., No. 164 G.R.C., of which the following are the principal officers for the present year:—W. M., B. H. Young; S. W., Josiah Murphy; J. W., Robert Stolm; Sec., W. Flagler; Treas., G. J. Chadd.

There is a deservedly popular school here, having two teachers continually employed. A man named John Stewart was the first teacher anywhere in the vicinity, but at so early a day that the exact date cannot at present be ascertained.

We give herewith a list of the municipal officers for the present year:—Samuel Flagler, Reeve; George J. Chadd, Amos Garrett, David J. McG. Cumming, George W. Clarke, Councillors; John H. Osborne, Clerk; John T. Dorland, Treasurer; Andrew De Witta, Collector; Stephen Bowerman, Assessor; Jonathan T. Brown, David E. Clarke, Auditors; Thomas Jackson, Cornelius Clapp, Road Inspectors; Niles S. Herrington, Dennis Donovan, Pound-keepers; Garrett Harris, James Hadden, Poor Commissioners; George W. Herrington, Daniel L. Mastin, Fire Wardens; Amos Hutchinson, Bell-ringer; William Curlett, Constable, and care-taker of Town Hall.

The Treasurer, John T. Dorland, has held that position ever since incorporation. In addition thereto, he also held the office of Clerk from incorporation till 1873, when he was succeeded by John H. Osborne, who still holds the position.

When this village *first* received a "local habitation and a name," that name was "Smokeville," not a very euphonious, but an extremely significant appellation. Those who should know differ slightly as to its origin, some affirming that it was from Mr. Reynolds, the first settler, who was popularly known as "Old Smoke," by the Indians of the neighbourhood, and afterwards by all the white settlers; while others say it was from an Indian chief of the name of "Smoke," with whom Reynolds was, from his first advent, on terms of friendship and intimacy, and from which circumstance the name came to be applied to himself, and afterwards to the village. However this may be, Smokeville retained its suggestive, if not classic, title for many years. The pioneer who named it thus died at a ripe old age, in 1826, leaving several hundred acres of the best land in the county to his descendants, a number of whom still reside in the neighbourhood.

A post-office was established here about 1815, through the influence of Mr. McFaul, who became the first post-master. The high social position which this gentleman occupied, and the deserved influence he exerted in all public affairs—the benefits of which will be felt by succeeding generations,—give him a prominent place in the local history of Wellington, and an affectionate corner in the hearts of its inhabitants which will be fresh and green while memory lasts. An Irishman by birth, of finished education and polished manners, he emigrated to this country when quite young; and although poor, and without friends or influence, he succeeded by his honest energy and industry, in accumulating a very large fortune; while his charity and hospitality were proverbial, and his well-merited reputation for integrity and fair-dealing made him one of the most honored of citizens.

The name "Wellington," which was given to the post-office his influence had been instrumental in establishing, is owing to his love of his native country and his patriotic admiration of that country's noblest son, the "Iron Duke," who was then in the zenith of his fame.

ATHOL.

This township was originally a part of the "Fifth Town," and was afterwards taken therefrom, along with other portions—as well as parts of the sixth and seventh towns—to form the township of Hallowell, from which it was subsequently set off as an independent municipality by the Act, 12 Vic., Cap. 31. The reasons which caused the set-off from Hallowell were precisely the same as those causing the original set-off from the "Fifth Town"; and which the said original set-off was designed to remedy, viz:—to use the words of the original Act setting off Hallowell—that "the in-

"habitants of the said townships experience many difficulties from the uncommon length of such townships."

An agitation was carried on for some time to secure the separation of that part of Hallowell south of East Lake, which—from its geographical position—would have been the proper portion to erect into a separate township. The scheme was opposed by some of the residents on the north side, however; and chiefly by Mr. Worden, who then represented the county in Parliament. This gentleman was an American who came to the country when he and it were yet quite young; and who, by energy and business tact, succeeded in building up an immense fortune—as fortunes in this new country were rated. The foundation of this fortune was laid by industrious and plodding application to his daily toil, in a small tannery, which he erected on the north shore of East Lake, where he at that time resided and carried on the business of tanner and currier. To him, the old saw, that "there's nothing like leather," was doubly exemplified; as he accumulated the means of erecting this tannery by "whipping the wild-cat,"—as it was then called—*i.e.*, going from house to house and making up new brogans, or cobbling old ones, for those who were fortunate enough to possess the reality of the one, or the means of paying for the other. It is related of Mr. Worden that when he came into the country all his earthly possessions consisted of the clothes on his back, and a few of the most necessary tools of a journeyman shoe maker's "kit." But he brought with him a strong physical constitution, an honest nature, and an indomitable will; and he came to a country whose annals from the earliest times were replete with the history of men who by their individual virtues have raised themselves from the lowest rounds of the ladder to be "rulers among men," and of such was Mr. Worden an eminent example.

It would appear, however, that Mr. Worden did not possess a just appreciation of the fact that "the inhabitants of the said township experience many difficulties from the uncommon length of such township," and particularly those south of East Lake. At all events he opposed the division; his ostensible reason being that "there was no one south of the Lake fit for a magistrate;" his *real* reason probably being that he held a grudge against some of the leading men there on account of their political opposition, which adversely affected his popularity in that section; although it would appear that—taken as a whole—he enjoyed the confidence of his constituents in a very marked degree.

Subsequently, a petition was gotten up to include the first Range north of East Lake in the new township; and Mr. Bockus, of Picton, having meantime been elected to represent the county in Parliament, in the room of Mr. Worden, the scheme received his approbation and encouragement, and he had it set apart by the Act intituled as above.

In gratitude to Mr. Bockus, for his interest in a matter so closely affecting their public welfare, the citizens gave him the privilege of naming the new township; and with a patriotism for which the "land o' the heather" has always been proverbial, he called it after the parish where his ancestors first saw light—Athol.

It accordingly took its place in the sisterhood of townships on the 1st January, 1848, and the 3rd of that month the first town meeting was held, with Peter Hare, now of Bloomfield, as chairman, with the following result:

Solomon Spafford, Clerk; William Kinney, Assessor; John H. Young, Collector; William Richards, James Ketcham, Richard Hare, Town Wardens.

The following year, 1849, Solomon Spafford was again elected Clerk, and Samuel Yerwood, David Kelly, and Peter S. Hare, were elected Town Wardens—the last who filled that office—as the Municipal Act came into force in 1850, and the first "Town Council" who met January 21st of that year, under the new system were: William A. Palen, Peter S. Hare, Wilson Bentley, and William F. Blakeley, Wilson Bentley was chosen Reeve, and Dyer Stanton, Town Clerk. Mr. Stanton was succeeded three years later by James Cornwall, who held the position nineteen years, since which the office has been filled by the present incumbent.

The officers are as below, for the present year:

Benjamin A. Hubbs, Reeve; Henry Preston, John Kinney, Eli Ketchum, George Martin, Councillors; William Sayers, Assessor; John Gibson, Collector; Harvey Spafford, Clerk and Treasurer.

The earliest settlers were the Youngs (Col. Henry and family)—Col. James Rogers, and Major David McGregor Rogers, natives of New Hampshire, and officers of the "Queen's Royal," a regiment better known as "Rogers' Rangers"; Augustus Spencer, Samson Striker, Elisha Miller, Caleb Ellsworth, "Squire" Peters, and the Spaffords, all U. E. Loyalists; Jonathan Ferguson, a soldier in the Revolutionary War, who previously served in the British force which captured Canada from the French; Capt. Dyer, an ex-officer of the British Army; James Blakeley, also a retired British officer. The Ketchums, Scotts, and John Ogden were also among the earliest settlers.

James Blakeley had a family of whom three sons afterwards held Commissions of the Peace. One of them still resides on the old homestead, and is a leading and prominent man and License Inspector for the county.

The Rogers, previous to settling at East Lake, abode a short time in Fredericksburg, and after some years residence at the Lake, removed to the Carrying Place, and subsequently farther west. They have a large number of descendants living throughout Northumberland and Peterboro' Counties, many of them men of leading position in the communities in which they respectively reside.

Augustus Spencer was a brother of Col. Spencer, of the celebrated Rogers' Rangers. He was the first militia Captain in the township.

James Spencer, son of the above, and James Ketchum, son of Eli, also above mentioned, are the oldest living men who were born within the limits of the township. They are both very aged, though still in possession of all their faculties.

If anything further were required to prove the salubrity of the climate of Prince Edward, and the adaptability of the surroundings to the longevity of its inhabitants, we might mention the case of Alva Stevens, a native of Otsego County, New York, who came here at the time of the war of 1812-13, when quite young, and settled on lot 5, south of East Lake, where he still resides, a hale, hearty, and exceedingly sociable old gentleman, now past his ninety-fifth birthday. He is the oldest man living in the county, and the early sufferings of the pioneers are fresh in his memory and detailed by him with an unusual degree of minuteness and lucidity.

The settlement of the Youngs at East Lake above referred to, was admittedly the first within the present limits of Athol, while by some it was claimed as the first in Prince Edward County; and as such a brief notice of it might not be here out of place. Col. Henry Young, born at Jamaica, Long Island, in 1737, was the second son of a family of six children of an English gunsmith, from Nottingham, who came to America at an early age, and settled at the above place. Henry joined the British Army when quite young, and served with distinction through a period of six years in the French and Indian wars, under Generals Amherst and Abercrombie. He took part in the battle of Bennington, and afterwards in no less than seventeen battles against the Continentals. He received as a reward for gallant services at the former battle, an Ensigncy in the "King's Royal." His rank of Colonel, by which he was popularly known, was of Provincial appointment. He retired on half-pay at the close of the war, receiving a grant of 3,000 acres of land for himself, besides the usual quantities for the various members of his family. His first residence in Canada was at Cataraqui. With a brother officer, named McCartney, he set out in a canoe in the summer (it is said) of 1790. They came up to what is now Picton Bay, and leaving their canoe, crossed what is now Hallowell and Athol townships; and after a thorough exploration of East and West Lakes, he selected a spot on the north side of the former for his future home. Having returned to [C]
he obtained supplies, and with his sons Daniel and Henry, he s[et]
and located the same year, at the place afterwards and still known
landing—where the trail over the old Indian Carrying Place fro[m]
Bay struck East Lake. Col. Young left his sons here and return[ed]
family at Cataraqui for the winter. The next summer he spent in
them to clear land, erect a log house, &c., then again returned Ea[st]
family, whom he brought up to East Lake in the ensuing autum[n]
His four daughters, Elizabeth, Mary, Catherine, and Sarah, were m[arried]
East Lake settlers and U. E. Loyalists, named respectively Henr[y]
Jonathan Ferguson, William Dyre, and John Miller; and all liv[ed]
over eighty years of age—their father having previously passed awa[y]
Lake in his eighty-fourth year.

The descendants of this branch of the Young family—of wh[ich]
were several branches among the early pioneers—are now quite n[umerous]
and all highly respectable and worthy citizens. The above facts
from one acquainted with their early history, may be considered
way reliable.

The neighborhood of East Lake continued to receive accessio[ns]
population; and we have it on the authority of Rev. G. W. Mille[r]
mitted authority on such matters—that on the 1st January, 1800, t[he settle-]
ment contained thirty-four families—fifteen on the north side, and
on the south side of the Lake as follows:—commencing at wha[t]
Cherry Valley, at the head of the Lake, and proceeding down t[he]
shore, in the following order—Col. John Peters, half-pay office[r]
Rogers, half-pay officer; David Friar and —. Friar, U. E. Loyali[sts]
well Furguson, U. E. Loyalist; Elisha Miller, U. E. Loyalist;
Taylor; Caleb Ellsworth; Lieut. Henry Young, half-pay officer
Young, jr., U. E. L.; Augustus Spencer, half-pay officer; Ge
Benjamin Wait, U. E. Loyalists; William Dyre, U. E. Loyalist; an[d]
Ellsworth.

On the south side were Henry Tufelt, U. E. Loyalist; Jonatha[n Furgu-]
son, sr., U. E. Loyalist; Jonathan Furguson, jr., U. E. Loyalist;
Badgeley, U. E. Loyalist; John Miller, U. E. Loyalist; Farnton F
U. E. Loyalist; William Blakeley, U. E. Loyalist; Samson Strik[er]
Loyalist; Barrett Dyer, U. E. Loyalist; Daniel Baldwin; John O
E. Loyalist; Richard Ogden, U. E. Loyalist; Solomon Spafford
McCartney, half-pay officer; Joseph Lane; William Easley; C
Richards, U. E. Loyalist; James Clapp, U. E. Loyalist; and
Furguson. There were no settlers at that time on either side the [Lake ex-]
cept those dwelling on the shores, above mentioned.

Since that time the East Lake settlement with its thirty-four
and their few patches of clearing reclaimed from the forest, has gro[wn]
splendid township, whose [...] cultivated farms comprise an are[a of]
248 acres, with [...] large [pro]portion of 16,986 improved; which is
at a valuation of $794,[...] supports a population of 1,280 souls
$47,110 of assessable personal property; the whole being geographi[cally de-]
scribed as bounded northerly by the township of Hallowell; eas[t by]
South Marysburg, and on the south and west by Lake Ontario.

There are two post-offices in Athol. The one on lot 17, south
Lake, and the other at C[herr]y V[alley], which is a pleasant little
situated on a stream at the head of East Lake. It is surrounded [by]
agricultural country, with the best of roads in all directions; di[stant]
miles from Picton and three from Milford. It has daily mail be[tween]
to and from Picton and Milford, and contains store, blacksmith, wa[gon and]
shoe shops, and an office of the Montreal Telegraph Company.

The Town Hall is here situated. It was built some six years
cost, including sheds attached, of $1,600.

The soil and formation of Athol is somewhat different from a g[reat part]
of the county. The surface is slightly undulating; and within its
embraced a large area where the soil is much deeper and of bette[r quality]
than in the adjoining township of South Marysburg. This is seen, e[ven with-]
out personal inspection, by the report of the "equalization of asse[ssment"]
Committee of the County Council, which places the whole average
valuation the same as Sophiasburg and Ameliasburg—which is a
acknowledgment of the average superiority of Athol to the average
of the townships named; as the proportionate average of "impro[ved"]
in Athol is the least of the three.

A great number of most promising young orchards, in various
growth, can be seen throughout the township, and large additions
annually made to the acreage devoted to the fruit-raising industry
wisdom of such a course is patent to the most casual observer—both
climate being admirably adapted to the culture of all varieties of fru[it that]
can be grown anywhere in this latitude.

DESCRIPTIONS.

SAINT GREGORY'S (R. C.) CHURCH,

A view of which will be found in another part of this work, was e[rected in]
the year 1832 by the Rev. M. Lalor, and dedicated the same year,
usual ceremonies, by Bishop McDonell and Bishop Gaulin, assi[sted by a]
number of the leading clergymen of the Diocese, among whom
Very Reverend M. Brennan, then of Belleville, Rural Dean;
Angus McDonell, Vicar-General of Kingston Diocese; the Re[v.]
Roche, of Prescott; and the Rev. P. Dollard and Very Rev. Vica[r]
Gordon, of the Diocese of Hamilton.

Previous to the advent of Father Lalor the mission was attend[ed regu-]
larly every fortnight by Father Brennan, of Belleville. The Catho[lic popu-]
lation of Prince Edward was then very small, and could be easi[ly accom-]
modated in the present brick school house, erected by Father Br[ennan in]
1830, and now used as a Separate School.

The present incumbent is the Rev. James Brennan, nephew of
Rev. Father Brennan, whom he assisted at Belleville for eight y[ears pre-]
vious to taking charge of Saint Gregory's church in 1870—the Re[v. Father]
Lalor retiring at that time on account of old age.

Father Lalor ruled the mission with zeal and ability for over
thirty years, during which time he lived deeply in the hearts of h[is people,]
respected and venerated by all. He was ordained by Bishop M[cDonell,]
and was one of the first priests of the Diocese of Kingston, and
Upper Canada. He still lives in strength and vigour, near
Ireland.

Rev. Mr. Brennan is a native of Ireland, having been born
20th, 1830, at Mooncoyn, in Kilkenny. He was educated at St.
College, Kilkenny, and St. Patrick's College, Carlow. He cam[e to this]
country in 1862, and was ordained in December of that year by t[he]
Rev. Dr. Horan, Bishop of Kingston.

He is a highly educated and liberal-minded gentleman, a zealo[us worker]
in his own church, and bears the esteem of all denominations
his instruction and guidance the congregation continues to dev[elop in a]
most encouraging manner.

Among the most pleasing adjuncts to the Church is the Separ[ate School]
in connection therewith. There are now two teachers constantly
in the school, and it boasts an average daily attendance of abo[ut]
pupils.

THE PICTON BREWERY.—Picton is not a manufacturing town
few establishments of that description which it does possess can
a completeness of detail in their respective lines to which—the
being considered—it would be difficult to find an equal.

This is particularly the case with the "Picton Brewery," o[wned and]
operated by W. P. Despard.

his future home. Having returned to Cataraqui
with his sons Daniel and Henry, he came back
at the place afterwards and still known as Indian
over the old Indian Carrying Place from Picton
d, Young left his sons here and returned to his
winter. The next summer he spent in assisting
log house, &c., then again returned East for his
up to East Lake in the ensuing autumn, 1784.
th. Mary, Catherine, and Sarah, were married to
E. Loyalists, named respectively Henry Tufelt,
m Dyre, and John Miller; and all lived to be
heir father having previously passed away at East
ear.
s branch of the Young family—of whom there
ng the early pioneers—are now quite numerous,
and worthy citizens. The above facts, gleaned
their early history, may be considered in every

East Lake continued to receive accessions to its
t on the authority of Rev. G. W. Miller—an ad-
atters—that on the 1st January, 1800, the settle-
families—fifteen on the north side, and nineteen
Lake as follows:—commencing at what is now
d of the Lakes and proceeding down the north
oler—Col. John Peters, half-pay officer; Major
David Friar and —, Friar, U. E. Loyalists; Ros-
yalist; Elisha Miller, U. E. Loyalist; Blaisdell
; Lieut. Henry Young, half-pay officer; Henry
ugustus Spencer, half-pay officer; George and
lists; William Dyre, U. E. Loyalist; and George

Henry Tufelt, U. E. Loyalist; Jonathan Furgu-
onathan Furguson, jr., U. E. Loyalist; Anthony
John Miller, U. E. Loyalist; Farnton Furguson,
Blakeley, U. E. Loyalist; Sauson Striker, U. E.
. E. Loyalist; Daniel Baldwin; John Ogden, U.
len, U. E. Loyalist; Solomon Spafford; Joseph
r; Joseph Lane; William Easley; Col. Owen
; James Clapp, U. E. Loyalist; and Charles
settlers at that time on either side the Lake, ex-
shores, above mentioned.
st Lake settlement with its thirty-four families,
being reclaimed from the forest, has grown into a
cultivated farms comprise an area of 23,-
tion of 16,980 improved; which is assessed
and supports a population of 1,280 souls owning
tal property; the whole being geographically de-
dly by the township of Hallowell; easterly by
the south and west by Lake Ontario.
es in Athol. The one on lot 17, south of East
Va[illegible] which is a pleasant little village,
head of East Lake. It is surrounded by a fine
the best of roads in all directions; distant six
ores from Milford. It has daily mail both ways,
lford, and contains store, blacksmith, wagon, and
the Montreal Telegraph Company.
situated. It was built some six years ago, at a
ed, of $1,000.
of Athol is somewhat different from a great part
e is slightly undulating; and within its limits is
re the soil is much deeper and of better quality
hip of South Marysburg. This is seen, even with-
the report of the "equalization of assessment"
Council, which places the whole average per acre
nasburg and Ameliasburg—which is a practical
erage superiority of Athol to the average of either
as the proportionate average of "improvement"
three.
promising young orchards, in various stages of
ghout the township, and large additions are being
ge devoted to the fruit-raising industry; and the
pparent to the most casual observer—both soil and
dapted to the culture of all varieties of fruit, which
this latitude.

DESCRIPTIONS.

GREGORY'S (R. C.) CHURCH,

ound in another part of this work, was erected in
M. Lalor, and dedicated the same year, with the
op McDonell and Bishop Gaulin, assisted by a
rgymen of the Diocese, among whom were the
nan, then of Belleville, Rural Dean; the Rev.
General of Kingston Diocese; the Rev. E. P.
he Rev. P. Dollard and Very Rev. Vicar-General
Hamilton.
of Father Lalor the mission was attended regu-
ather Brennan, of Belleville. The Catholic popu-
was then very small, and could be easily accom-
ick school-house, erected by Father Brennan in
eparate School.
is the Rev. James Brennan, nephew of the Very
om he assisted at Belleville for eight years, pre-
Saint Gregory's church in 1879—the Rev. Father
e on account of old age.
e mission with zeal and ability for over four and
time he lived deeply in the hearts of his people,
y all. He was ordained by Bishop McDonell,
riests of the Diocese of Kingston, and, indeed,
lives in strength and vigour, near Marlboro',

native of Ireland, having been born January
in Kilkenny. He was educated at St. Kieran's
t. Patrick's College, Carlow. He came to this
ordained in December of that year by the Right
Kingston.
l and liberal-minded gentleman, a zealous worker
bears the esteem of all denominations. Under
nce the congregation continues to develop in a

ng adjuncts to the Church is the Separate School
There are now two teachers constantly employed
ts an average daily attendance of about eighty

—Picton is not a manufacturing town, but the
t description which it does possess can boast of
n their respective lines to which—their extent
l be difficult to find an equal.
s case with the "Picton Brewery," owned and
nl.

The delightful situation and beautiful surroundings of the place need no praise at our hands, as a correct idea can be formed of the location from the view of the establishment, which appears elsewhere; but perhaps a short description of the interior arrangements may not be uninteresting.

Built on a side-hill, from which continually gush forth numerous springs of the purest water, it consists of four stories and basement, or cellar. From the two lower stories direct communication is had with the waggon-road to the rear, which runs along the side of the hill. The springs referred to are of great volume, and of an equal temperature of 48° the year round. Coming from an elevation greater than the highest point of the building, and being connected with every part of the interior by a series of pipes, the whole system of water supply, as well as the various changes of *wateres* from place to place, and from course to course, during the process of brewing, is conducted without the application of mechanical power, simply by the force of gravitation. All the various tanks and vats which go to complete the system are situated, the one a little lower than the other, in the order in which they have to be used during the brewing process; and each is supplied with twined copper worms or coils, through any one or all of which either hot or cold water can be turned at option, by the simple adjustment of the regulating valves; or hot water can be used in one place while cold is being applied in another at pleasure.

The heating power is supplied by a 17-horse-power steam-engine, manufactured by Hyslop & Ronald, of Chatham, Ont. This engine also drives the machinery of the "mill-room," including malt grinder, worts-masher, etc., etc., as well as the elevator for hoisting the barrels for shipment from cellar to shipping-floor, from which a car runs on a tramway to the waggon road in rear of the building.

The fermenting room contains, in addition to the ordinary "cooler" usually found in breweries, a newly-invented patent cooler of the most approved and costly pattern, from the celebrated establishment of Booth & Son, of Toronto.

The "washing floor," where all barrels, casks, etc., etc., are cleansed, is one of the most ingeniously devised and conveniently operated imaginable; and the whole arrangements, from grinding the malt to barrelling and bottling the ale, are of the most complete description, combining cleanliness, convenience, economy, and skilful management throughout the process, which enables the proprietor to compete in all the towns along the north shore of Ontario and the St. Lawrence, with the celebrated Carling and other establishments of similar magnitude.

Mr. Despard, the proprietor, is an Englishman, and originally a Civil Engineer by profession, having been engaged in that capacity on the Grand Trunk Railway during its construction, and afterwards on the Brockville and Ottawa, and other Canadian Public Works.

He afterwards carried on a commission business in Belleville for six years, and subsequently followed the same business in Oswego for three years. He has owned and operated the "Picton Brewery" for the past eight years.

Mr. Despard has a number of relatives in this country occupying prominent positions in society and business circles. A brother is manager of the Bank of Montreal in London; and a cousin of the same name is manager of the Canada Fire and Marine Insurance Co. of Hamilton.

THE "LITTLE GIANT" TURBINE WATER WHEEL WORKS are situated five miles below Picton, on the Marysburgh shore. Our readers will be more familiar with the place when designated by the name so widely known the "Old Stone Mills."

The original building of these celebrated mills by Major Van Alstine, in 1796 is elsewhere referred to. Mr. J. C. Wilson, an enterprising gentleman of Picton, has owned the property for some time past, as also the bulk of the original Van Alstine grant, alongside the Lake of the Mountain, from which the water supply is derived, to drive the machinery of the mills.

Up to within about three years, the "Stone Mills" consisted of large flouring and plaster mills—two separate buildings—which have always since their first operation been the centre of a brisk trade, of a more than local character. About that time Mr. Wilson entered into an arrangement with George H. Jones, then of Auburn, New York, for the manufacture of the celebrated wheel known as the "Little Giant." A large four-story stone building was erected for the purpose, almost immediately alongside the flouring mill; and in this has been placed many thousand dollars' worth of the finest iron-working machinery to be seen in any establishment of similar character in Canada. It is so complete in all its details that one man can execute an amount of work equal to that performed by several mechanics, with such machinery as is generally in use. Eighteen to twenty men are now employed in the machine shop alone; and these, together with the hands employed about the other mills—with their families—make quite a little village.

The works are kept running full, with orders ahead as fast as previous ones can be turned out. We have inspected the Messrs. Wilson & Co's. order book, and find orders from a great number of the leading mill-owners from all the manufacturing centres of Canada—notwithstanding the extraordinary competition in that line of business. These orders are from all points, between Halifax on the one hand, and the great North-West on the other; and we noticed cases where well-known manufacturers had ordered a second—and in a few instances a third—wheel, to replace others—some of which were also "Turbines" of other manufacture—of which there are a great variety in the market.

The "Little Giant" is so essentially *the* Turbine however, being different from all others in construction, as to merit a more lengthy description, than the space at our disposal would justify. Its chief points of peculiarity are that it consists of two wheels keyed to the same shaft, back to back, discharging water in opposite directions, and so arranged by means of planed iron partitions and a planed slide-gate, similar to the slide-valve in a steam engine, that either or both wheels, or any degree of power of either or both can be operated at pleasure, according to the amount of work to be performed. It is so arranged as to render leakage or breakage almost beyond a possibility—and should any stone or other obstruction chance to get into the bucket, there are facilities for removing the same without interfering in any way with the wheel.

An inspection of the "Little Giant" which drives these works will convince the most sceptical that it is the most complete and powerful wheel made. Being rather ignorant of hydraulic principles, as applied to static pressure, the velocity of discharge, &c., &c., under the different degrees of "head," we could not believe it possible till we saw the water turned on and off, and *vice versâ* that the beautiful little piece of mechanism which one might—without the least exaggeration—conveniently place in an ordinary great-coat pocket, could drive with such tremendous force the ponderous machinery with which the works above are supplied. The actual dimensions of the wheel are six inches diameter, the smallest size the company ever made; while they make them up to thirty six, according to the "head," volume of water, and power required.

Though the enterprise is yet in its infancy, it is being rapidly developed, and the "Little Giant" has borne the palm wherever yet tested, having taken—among other honors—the first prize at the last Provincial Fair, in 1877.

———o———

BIOGRAPHICAL SKETCHES.

CHARLES BOOKUS (deceased), was born in Osnabruck, Stormont County, Upper Canada, on the 30th December, 1802.

At an early age he removed to Gananoque, when he engaged in mercantile pursuits, and where in 1828, he married Caroline Mallory, then a member of the family of her step-grandfather, Colonel Stone, a very influential man of the place, and founder of the village of Gananoque. In 1829 Mr. Bockus removed to Picton, where he took rank at once among the most progressive, enterprising, and energetic merchants of that time. His business grew to immense magnitude, and his success was correspondingly great.

But notwithstanding the demands of his private affairs, he still found time to devote to the promotion of all enterprises of a character calculated to benefit the community at large; and was for many years one of the most trusted and prominent men of the District. In 1836 he was returned to Parliament for Prince Edward County. Here he continued to display the same executive abilities which had always distinguished him; and in the second year of his term he was made Chairman of the Finance Committee of Parliament—a position requiring a great amount of sagacity, intelligence, and familiarity with the principles of political economy. He continued to discharge the duties of this position for three sessions—till the end of his Parliamentary term—with much honor and credit to himself, and benefit to the country.

In 1843 the calls of business necessitated his removal to Montreal, and on his departure, he was presented by his fellow-citizens with a valuable service of silver, on which was conveyed to him, in the most affectionate terms, the high respect and esteem in which he was held throughout this, the county where best he was known.

After an absence of nearly thirty years, spent partly in the United States, Mr. Bockus returned to his old home—his youngest daughter having married there. Though now past the prime of life he at once interested himself in the project of a Railway; and after a lengthy and fiercely fought campaign, amid many trials and discouragements, and prosecuted with unwearying labor, untiring vigilance and an extraordinary energy, worthy of the man and the cause, he succeeded in converting hostility into friendliness; and, later, in establishing the long-wished-for Prince Edward Railway on a firm footing.

Before his work was finished, however, the work in which he took so deep an interest, and which he grieved over not being able to complete, he was attacked by a fatal disease, which, after an illness of five months, closed a busy, useful life. Honored by all men, and most deeply mourned by his afflicted family, Charles Bockus passed peacefully to his eternal rest, on the morning of January 10th, 1878, at the residence of his son-in-law, Mr. Walter T. Ross, of Picton.

Dr. Willet Casey Dorland (deceased), the son of Gilbert Dorland, a U. E. Loyalist, who was among the early settlers in Prince Edward, in 1802, was born in that county, on the 30th July, 1805.

The Dorlands were of Dutch descent, their ancestors having been among the first who came from Holland, and settled on Long Island, New York. From this time, the family record is traceable for some eight generations. We find in the time of James II, that among those who took the oath of allegiance to that Sovereign in 1678, were John and Elias Dorland. Captain Thomas Dorland was an officer of the British service at the time of the revolt of the American Colonies, and was taken prisoner by Continental troops; but escaped from their surveillance and fled to Canada about the year 1780; being one of the very first pioneers of the Bay Quinté Region.

We believe we draw a very reasonable conclusion, that the subject of our short memoir was the descendant of some of our early representative families. His great-uncle, Philip Dorland, was the first member returned by Prince Edward Electoral Division (then including Adolphustown, in the County of Lennox), in 1792. His maternal grandfather, Willet Casey, was elected to the fourth parliament in 1807. His uncle, Samuel Casey, was returned to Parliament by the United Counties of Lennox and Addington, in 1820.

Dr. Dorland was elected in the Reform interest for Prince Edward County, by a large majority, to represent that Electoral Division in the Sixth Parliament of Old Canada in 1859; and was one of those who gave his adhesion to the Queen's choice—Ottawa—as the permanent capital. He always took strong and advanced grounds in his political positions in life; was unusually frank and liberal in his opinions, in the intercourse with men and measures of the period—regarding the legal and educational institutions of our country as the groundwork of our future success, and national greatness; and was justly regarded by a very wide circle of friends, both at home and abroad, as a gentleman of private as well as political integrity.

Having received a medical education, he was enabled to successfully apply himself to a wide, ample field of usefulness, and thereby render valuable counsel and assistance to the profession.

He was married to a daughter of the late Stephen More, of Duchess County, New York. He held a commission of Coroner and Justice of the Peace for the County of Prince Edward at the time of his death, which occurred, to the sincere regret of all his acquaintances, and deep affliction of his friends, on the 5th of January, 1874.

James P. Morden (deceased), late of Ameliasburg, came of a family of U. E. Loyalists, on both father's and mother's side, his ancestors having come to Canada in the earliest days of the settlement of the Bay Quinté region. His mother's maiden name was Margaret Parliament, and his father was James Morden, the first settler of Sophiasburg in the vicinity of Northport, and the man who built the first house in that now prosperous village, where the subject of this brief memoir was born, and resided until twenty-one years of age, when he married Miss Catherine Babcock and removed to Ameliasburg, and settled on lot 68, in the first concession, where he raised a family of seven children—one of the daughters marrying Elkhannah Babbitt, who now owns and resides upon the old Morden homestead, a view of which we give elsewhere. Mr. Babbit, though quiet and unostentatious in manner, is among our most intelligent and enterprising farmers, having a high reputation among his neighbours for those qualities which combine to make the popular gentleman and useful citizen.

The Mordens were not only among the pioneers of Prince Edward, but the influence exerted by the family in all affairs of their county and respective townships, was of a character commensurate with their intelligence and high social position. James P. Morden was for nearly half a century one of the leading citizens in the neighbourhood of Rednersville, and always exerted himself in support of all public measures pertaining to the public welfare, though he never sought—nor would he himself accept—any position of public trust.

His good qualities were recognized, however, by his unsolicited appointment to the Commission of the Peace; and in his death, which occurred in March, 1864, at the age of sixty-three years, the community lost one of its best citizens.

Judge Jellett.—In looking over the *Belfast News-Letter* of the date of August 11th, 1797, we find the following obituary:—

"*Died.*—On the 9th inst., Morgan Jellett, of Moira, Esq., in the seventy-fifth year of his age. He was the oldest Magistrate in the county of Down. He was an upright and honourable gentleman, and zealously attached to the King and happy Constitution. He was an advocate of the widow and orphan, truly charitable, without ostentation, and in all his dealings with mankind, just and honorable, and now is gone to enjoy the reward of a well spent life."

The subject of the above notice was grandfather of Morgan Jellett, Esq., who came from Belfast to Canada and settled in Port Hope in 1832, when his son, Robert Patterson Jellett, now Judge of the County Court of Prince Edward county, was a boy only five years of age. Mr. Jellett followed mercantile pursuits for a number of years, and was afterwards for a long time Clerk of the old Court of Commissioners. He subsequently removed to Cobourg, and was appointed Clerk of the County Council of the united counties of Northumberland and Durham, a position which he held till his death.

Robert P. Jellett, his eldest son, and the s
eleven, entered the dry-goods business when thi
employ of Hiram Gillett, then and ever since doi
He followed this occupation for six years, was afte
for two years, was for quite a length of time a s
veying parties laying out new townships, and sp
sailor before the mast on the lakes.

His circumstances did not admit of a college
being commenced and finished at the Port H
entered the law office of the Hon. Sidney Smith,
ary; and while attending to the duties pertaining t
himself in classics and mathematics, and the othe
tion, to enable him to procure admission to the la
the 11th September, 1851.

He finished his studies in the office of Messrs.
was admitted to practise as an attorney, Novemb
the bar 17th November, 1856. He commenced p
& Bell, afterwards doing business by himself, and
with his brother, now Alderman Jellett, who still
Belleville. He was appointed Judge of the Count
County on the 12th July, 1873. In 1854 he m
Quebec, who died in 1869. He was married
daughter of Rev. John Grier, of Belleville, and h
daughter.

From the above account of his youth, it will
had a hard fight with the world. He is indeed a
eminent position now occupied by him neither t
to his own superior natural qualifications, coupled
verance which finally overcame the reverses of fo
foremost position among the jurists of his adopted
Socially, he is very far from that austerity o
wrongly, is the traditional picture of the wearer o
contrary an extremely affable and companionable
of a good joke. He is also a keen sportsman, an
with numerous rare and interesting specimens,
prowess as an expert with the line, the trigger, an

Philip Low, Q.C., of Picton, is a native of
Isles, whence he came to Canada when quite young
of rank, and served with Sir Ralph Abercrombie
name was Villeneuve, a descendant of a very ol
French Huguenots, and cousin to Admiral Villen
French fleet against Lord Nelson at Trafalgar.

After arriving in Canada, Mr. Low chose l
entered upon his studies in the office of the late H
onto. He was afterwards for a time in the office o
Kingston, but returned to Toronto and completed
the Hon. Marshall Bidwell, then Speaker of th
Canada. He was called to the bar in 1836.

During the rebellion of '37-'38, he was a volunt
pany, under Col. FitzGibbon, and commanded
Montgomery's Hill, where the rebels were attacke
the first of his military experience, having held a
Active Militia, and won the Governor's medal an
petitive prizes for marksmanship and other military
force.

Soon after acquiring his profession he remo
chose as his future home, and has practised here
success ever since. He has been Clerk of the Peace
Attorney since 1858, and a Q.C. since July 1st, 18

He was always a man of progressive ideas, an
number of enterprises which have redounded gre
One of these was the laying of a submarine cable
give Picton telegraphic communication with the
accomplished in 1855, while he was president of t
Company,—another enterprise originated and carr
designed, and for a time succeeded, in breaking th
Montreal Telegraph Company, the only other the
vince. He succeeded in getting a paid up stock o
was built and operated from Quebec to Buffalo, wi
Barrie, and many other similar places to which th
not extend. These towns undoubtedly owed thei
solely to the energy and enterprise of Mr. Low. T
six branch offices, with head office at Picton, and
for a number of years. The directors having
Toronto, Mr. Low resigned the presidency, an
Grand Trunk was purchased by and amalgamated w

He was the means of establishing an agency of
this place—of which he is now Solicitor; and he
flax mill at Picton for a number of years; but t
necessarily followed the cheap cotton movement
late American Civil War—coupled with the lack
—rendered the enterprise unprofitable, and it was

He has allowed no amount of professional or
from his services to the "body politic," though, f
"politics," he stands aloof, his services being
political nature. He was chairman of the old Bo
lated town affairs prior to erection of Town Cou
Law in 1850; at the first meeting of which he was
a position which he continued to fill for three co

To his exertions is due in a great measure the
of the Prince Edward County Railway, of which
of the most active promoters.

Mr. Low has a very extensive and lucrative
Prince Edward County, but in other parts of the
highest respect and confidence of his fellow citize
numerous distinctions, he holds the position of M

His beautiful estate, and charming family res
shown in another part of this volume, is called
family name, "Villeneuve."

Lt. Col. Walter Ross, M.P., is a son of W
Fearn, Rosshire, Scotland, where he was born in
emigrated to Canada in 1842. He first settled in
entered the dry goods business in the service of
among the leading wholesale houses of the country
induced his employers to send him to Picton
(during the periodical absence of Mr. McAllister, t
establishment of this place. He remained in this
commenced a business for himself, which has gr
now of an extent seldom seen in Provincial tow
equals even in the large cities, for size, convenie
ness in every detail, the various departments bei
attractive and inviting manner; and every thin
wears an air of "business" only requiring a gla
that the manager is master of his profession.

Mr. Ross exhibits this same trait in all h
common remark, that whatever he undertakes to
attentive and energetic in the prosecution of w
hand to.

That his qualities are appreciated by his fell

)RICAL SKETCH OF THE

rt P. Jellett, his eldest son, and the second child of a family of ntered the dry-goods business when thirteen years of age, in the f Hiram Gillett, then and ever since doing business in Port Hope. wed this occupation for six years, was afterwards employed on a farm rears, was for quite a length of time a chain-bearer in various surarties laying out new townships, and spent one season as a common fore the mast on the lakes.

ircumstances did not admit of a college course, his school education mmenced and finished at the Port Hope Grammar School. He he law office of the Hon. Sidney Smith, of Cobourg, on a small sal while attending to the duties pertaining to that position he instructed n classics and mathematics, and the other higher branches of educaenable him to procure admission to the law society, which he did on September, 1851.

nished his studies in the office of Messrs. Ross & Bell, of Belleville, itted to practise as an attorney, November 23rd, 1852, and called to 7th November, 1856. He commenced practice as a partner of Ross fterwards doing business by himself, and subsequently in partnership brother, now Alderman Jellett, who still practises law in the city of . He was appointed Judge of the County Court of Prince Edward m the 12th July, 1873. In 1854 he married Miss Macnider, of who died in 1869. He was married again, in 1873, to Lilias, of Rev. John Grier, of Belleville, and has a family of one son and a

the above account of his youth, it will be observed that the Judge rd fight with the world. He is indeed a self-made man, owing the position now occupied by him neither to favour nor to fortune, but n superior natural qualifications, coupled with an energy and persewhich finally overcame the reverses of fortune, and placed him in a position among the jurists of his adopted country.

lly, he is very far from that austerity of manner which, rightly or is the traditional picture of the wearer of the ermine; being on the an extremely affable and companionable gentleman, and very fond joke. He is also a keen sportsman, and his residence is adorned nerous rare and interesting specimens, which bear witness to his s an expert with the line, the trigger, and the oar.

P Low, Q.C., of Picton, is a native of Jersey, one of the Channel ence he came to Canada when quite young. His father was an officer and served with Sir Ralph Abercrombie in Egypt. His mother's s Villeneuve, a descendant of a very old and illustrious family of Huguenots, and cousin to Admiral Villeneuve, who commanded the eet against Lord Nelson at Trafalgar.

arriving in Canada, Mr. Low chose law as his profession, and pon his studies in the office of the late Hon. H. J. Boulton, of Tore was afterwards for a time in the office of Mr. George McKenzie, of . but returned to Toronto and completed his course in the office of Marshall Bidwell, then Speaker of the Legislative Assembly of He was called to the bar in 1836.

g the rebellion of '37-'38, he was a volunteer in Captain Cox's conder Col. Fitzgibbon, and commanded a section in the affair at ery's Hill, where the rebels were attacked and routed. This was not f his military experience, having held a commission in the Jersey ilitia, and won the Governor's medal and a number of other comprizes for marksmanship and other military attainments while in that

after acquiring his profession he removed to Picton, which he his future home, and has practised here with an enviable degree of ver since. He has been Clerk of the Peace since 1847, County Crown since 1858, and a Q.C. since July 1st, 1867.

s always a man of progressive ideas, and was the originator of a l enterprises which have redounded greatly to the public welfare. hese was the laying of a submarine cable across the Bay Quinté, to on telegraphic communication with the outside world. This was shed in 1855, while he was president of the Grand Trunk Telegraph ,—another enterprise originated and carried out by him. This was and for a time succeeded, in breaking the grinding monopoly of the Telegraph Company, the only other then in existence in the Prole succeeded in getting a paid up stock of about $100,000, and a line and operated from Quebec to Buffalo, with branches to Peterboro', nd many other similar places to which the Montreal Company would d. These towns undoubtedly owed their early telegraphic facilities the energy and enterprise of Mr. Low. This new company had sixtyh offices, with head office at Picton, and Mr. Low was its president mber of years. The directors having removed the head office to Mr. Low resigned the presidency, and after his retirement the unk was purchased by and amalgamated with the Montreal Company. as the means of establishing an agency of the Bank of Montreal, at e—of which he is now Solicitor; and he also started and operated a at Picton for a number of years; but the decline in linens, which ly followed the cheap cotton movement on the termination of the rican Civil War—coupled with the lack of transportation facilities ed the enterprise unprofitable, and it was abandoned.

s allowed no amount of professional or business duties to detract services to the "body politic," though, from what is strictly termed ," he stands aloof, his services being more of a public, than a nature. He was chairman of the old Board of Police, which regun affairs prior to erection of Town Councils by the change of the 850; at the first meeting of which he was *unanimously* chosen mayor, which he continued to fill for three consecutive years.

s exertions is due in a great measure the present encouraging status ince Edward County Railway, of which he was from the first, one st active promoters.

Low has a very extensive and lucrative law practice—not only in dward County, but in other parts of the country. He enjoys the espect and confidence of his fellow citizens; and in addition to his distinctions, he holds the position of Major of Militia.

eautiful estate, and charming family residence, a view of which is another part of this volume, is called in honor of his mother's me, "Villeneuve."

Col. Walter Ross, M.P., is a son of Walter Ross, of the Parish of osshire, Scotland, where he was born in 1819, and from whence he d to Canada in 1842. He first settled in the city of Hamilton, and he dry goods business in the service of Bryce & McMurrich, then e leading wholesale houses of the country. His aptitude for business his employers to send him to Picton soon after, to take charge he periodical absence of Mr. McAllister, the manager) of their branch ment of this place. He remained in this position till 1845, when he ed a business for himself, which has gradually increased, till it is n extent seldom seen in Provincial towns, while his store has few ven in the large cities, for size, convenience, neatness, and completevery detail, the various departments being all arranged in the most e and inviting manner; and every thing about the establishment air of "business" only requiring a glance to satisfy the observer manager is master of his profession.

Ross exhibits this same trait in all his enterprises; and it is a remark, that whatever he undertakes to do, *he does well*; being very and energetic in the prosecution of whatever task he turns his

his qualities are appreciated by his fellow citizens, is evidenced by the fact that he was elected to the position of Town Councillor for eight successive years, and chosen to fill the Mayor's chair, without opposition, for four consecutive terms.

He was the member for Prince Edward in the last Parliament of Canada, before Confederation; having defeated William Anderson, the previous member, in 1863, by a majority of 312. He was elected to the House of Commons of the Dominion in 1867, over James S. McCuaig by a majority of 337; defeated Mr. McCuaig a second time at the general election of 1872, by a majority of 134, and again defeated the same gentleman in 1872, by a majority of 126. At the Reform Convention recently held, to select a candidate for the next general election, he was the unanimous choice of the party, but declined on account of the pressing demands of his private business upon his time and attention.

In 1863, Mr. Ross was appointed Lt.-Col., of the 16th Regiment of Volunteers, a position which he still holds. If we are to credit the assertions of his subordinate officers and men, he is one of the most popular officers in the Volunteer force. From a knowledge derived from personal acquaintance with the facts, and based upon a comparatively extended military experience, we venture to assert that Col. Ross, commands a regiment which is second to none in the service in any of those qualities which are the admiration of military men. To accomplish such a result, the Col. has of course been assisted by an able and efficient staff, as well as field, line, and subaltern officers who would do credit to any military service.

Mr. Ross is President of the Prince Edward County Railway, in the organization of which he took an active and leading part, and with the success of which he is in a great measure to be credited.

He was married in 1845, to Elizabeth, daughter of Henry Thorpe, Esq., of Fredericksburg, who has since died, leaving a family of two sons and two daughters, three of whom are married; Walter T. Ross, the eldest son being collector of the Port of Picton and of the Prince Edward District. He married again in 1861, a sister of the late Judge Fairfield, and widow of the late Dr. Prynn; and by this marriage there is one daughter.

In addition to his other business Mr. Ross was for years the leading grain merchant in the county, he was also one of the organizers of the Ontario and Quebec Navigation Company, of which he was for many years the largest stockholder, and for a long time president. He is now senior volunteer colonel in this military division.

Lieut.-Colonel Gideon Striker, M.P.P.—This gentleman's grandfather, James Striker, removed from Duchess county, New York, and came into Hallowell when that township was almost a wilderness. The immediate cause of his removal thither was his arrest by the Continental authorities, on the charge of harbouring spies. He was tried by drum-head court-martial, found guilty, and sentenced to be executed for the same; and the carrying out of the sentence was only prevented by the violent interposition of an influential rebel officer of high rank, who was a relative and great personal friend of Mr. Striker. The charge was really a mistaken one, and arose from the fact that two nephews—Royalists—were visiting with him a short time previously, and to avoid possible trouble with the authorities they conducted their movements with just enough secrecy to draw suspicion upon their host. The war was still in progress, and a simple charge of such a nature in the then inflamed state of the public mind, was sufficient to ensure conviction. Mr. Striker had a very narrow escape of his life, and the incident had the effect of hastening his departure to this refuge of his Majesty's loyal subjects—in doing which he was obliged to abandon a splendid and valuable property, which was afterwards confiscated.

Garrett Striker, a son of James, lived in Hallowell for many years, removing thence to Picton, when his son Gideon, the youngest of a family of seven children, was a boy of five years of age, and there remained till his death.

Gideon learned the drug trade and commenced business for himself when quite young. He never married, but being very attentive to business matters he was successful in trade, and retired with a competency some ten years ago.

Mr. Striker has always taken an active and intelligent part in all public affairs—municipal and political. He has served a great number of times as Town Councillor and Reeve, and was Warden for the county for three consecutive years.

He is a Reformer in politics, and has been returned in the interests of that party on three different occasions to the Ontario Legislature against (1) William Anderson, previous member, majority 129; (2) James S. McCuaig, of Picton, majority 19; (3) Robert Clapp, of Milford, majority 63; and is the present sitting member. His activity in party politics has been rewarded by the firm devotion of his political friends, and also by the bitterest opposition of his opponents, who have petitioned against his return no less than three times, but still he "holds the fort," and enjoys the reputation of never yet having been defeated in any municipal or political electoral contest.

He is a Justice of the Peace and a lieut.-colonel of reserve militia, and is justly considered one of the most earnest, energetic, and consistent men of his party, as his return to Parliament three times in succession against very strong and able candidates amply testifies.

Robert Clapp, J.P., of Milford, is the grandson of Joseph Clapp, a U. E. Loyalist, from Poughkeepsie, New York, which place he left in the winter of 1780-1, and drove in a sleigh with his family, then consisting of three persons, *via* Lake Champlain and Montreal to Adolphustown, where he located land and settled on Hay Bay. They were just a month in making the distance from Poughkeepsie, the great part of the route laying through a wilderness without roads, or—where inhabitated at all—by those who were intensely hostile to all of Mr. Clapp's political views.

He raised a family of five sons and three daughters, of whom James, second son, born at Hay Bay in 1800, was father of Robert, the subject of this sketch.

Under the direction of Col. Cartwright, of Kingston, grandfather of the present finance minister of the Dominion, old Mr. Clapp removed with his family in 1808 to the township of Marysburg—a Mr. Garrett being at that time the only other settler in the neighbourhood. He came here for the purpose of engaging in the lumber trade, that township being in a great part covered with dense forests of magnificent pine and oak, which was at that day just beginning to find an outlet to Europe through the Quebec market. For this purpose he had built a mill on Black River previous to the removal of his family from Adolphustown. Being the first, and for many years the only mill in the township, it gave rise to the name of Milford, which the village afterwards springing up at the place, has ever since retained.

Mr. Clapp entered the military service of the Government in 1812, and while in duty at Kingston, under Col. Parker, he contracted a disease of which he died in camp. By the then existing law the property all fell to the eldest son, and Robert's father was one of those left destitute by this untimely death. He was then but thirteen years of age; but by steady application he had succeeded in saving enough money by the time he was twenty-four to buy lot 24, 1st con., N. B. R. He married about this time a Miss Sproule, recently from Ireland; and it may give the reader some idea of the then state of the country to say that when the Sproule family immigrated hither they were obliged to walk on foot all the way from Lachine to Kingston. Mr. Clapp remained on the above lot till his death about three years ago. His energies were unimpaired to the last; and, though over seventy-five years of age, he had never seen a moment's illness till within three hours of his death.

Old Mr. Clapp was the ancestor of a very numerous progeny, many of whom are now settled throughout the old Bay Quinté District, and many more throughout various parts of Ontario. The history of the "Clapp

"Family in America" is an exceedingly interesting and romantic one. It embraces a book of many hundreds of pages, and it took Messrs. D. Clapp & Sons, the celebrated Boston publishing house, from 1816 to 1876 to compile and complete the work.

The original ancestor of this very numerous Clan, as we might call them, was Roger Clapp, one of the Pilgrim Fathers, who was born in Salcombe, Devonshire, England, in 1609, and came to New England in the good ship *Mary and John*, in 1630. The family take a pardonable pride in their ancestry.

Robert Clapp, born in Milford in 1830, still resides in that village. He served his time as a miller's apprentice, and afterwards followed the trade for eighteen years. He now carries on farming extensively, owning and cultivating four hundred acres of land, and has some very fine stock, including some of the best strains of Ayrshire cattle, Leicester sheep, and Berkshire pigs, besides a stud of horses of the draught and roadster classes, which are difficult to surpass.

He also carries on a general agency business in Picton, where he has an office. He was U. S. Consul at that port for over twelve years. He has been a Justice of the Peace for seventeen years; has served in the Township Council for over twenty years, during part of which he represented the township in the County Council, in which deliberative body he occupied the Warden's chair in 1870.

He was a candidate for the Legislature in the Conservative interest in 1875, but was defeated by a small majority.

He is looked upon as one of the leading members of the community in all matters pertaining to the general prosperity of its citizens.

J. Platt Nash, M.D., of Picton, the third of a family of seven children of James and Sarah Nash, was born near the city of Hamilton, Ont., in 1838. His father was an American by birth, having removed to Hamilton from the United States when only six years of age; while on his mother's side he is of U. E. Loyalist descent.

The Doctor was educated in Canada, and, after acquiring his profession, was for some time a member of the Board of Examiners of the Eclectic School of Medicine, to which he belongs, though he is not the slavish adherent of any "ism" or "pathy," being only intent on serving the true interests of his profession, and therefore ready at all times to ingraft in his practice any theory whose virtues highly recommend it, no matter who the discoverer, or to what particular "school" of medicine he may belong.

He has had a great variety of practice, having first located at Wroxeter, Ont., which he was subsequently induced, by the representation of friends, to leave for Marshall, Michigan. In this city he soon acquired an immense practice, but failing health obliged him to abandon it, and he went to Brooklyn, New York, where he practised for a time with flattering success.

The Doctor is a great admirer of the beautiful in Nature; and while on a visit to this section of the country, he became so enamoured of the delightful scenery of the Bay Quinte that he determined to enjoy it more fully than his then distant residence would allow, so he left Brooklyn and settled in Picton some eleven years ago, and has since followed the practice of his profession, with an enviable degree of success attending his labors. We have the authority of disinterested parties for saying that he now does the largest, most successful, and lucrative practice in Prince Edward County.

Though a hard worker—devoting himself assiduously to the calls of his profession—he does not forget his duties as a citizen, but on the contrary, evinces a lively regard for all matters of public interest, having identified himself with the welfare of his adopted home, which he now represents in the Town and County Councils as Deputy Reeve of Picton.

He married, in 1858, Miss Sittzer, an American lady, of Auburn, New York, who still survives, a partner of his joys and sorrows.

In addition to being a Licentiate of the College of Physicians and Surgeons of Ontario, he is a member of the National Medical Eclectic Association of the United States; and the walls of his office are adorned with numerous certificates and diplomas from various Universities throughout Canada and the United States.

W. H. R. Allison, of Picton, Barrister-at-Law, is the grandson of Benjamin Allison, and son of Rev. C. R. Allison. His grandfather came from the United States after the close of the Revolutionary War, and subsequently settled in Marysburg, where the Rev. Mr. Allison was born, who afterwards became one of the most popular and highly respected Methodist ministers ever in the Bay Quinte region. It was he who gave its name to Allisonville, a post-village in the Township of Hillier. He married a daughter of Henry Hoover, an American colonist who joined the British service at the beginning of the Revolutionary War. He was captured in arms by the enemy when but seventeen years of age, and kept in close confinement by them till the close of the war, when he was released. He then came to Adolphustown, where he was one of the first settlers. He raised a family of nine children here, whose descendants continue to be—as he was in the olden time—among the leading citizens of that part of the county.

In following out the principle of itinerancy—which is one of the fundamental doctrines of the Methodist Church—Mr. Allison was called to labor at Brockville, where his eldest son, W. H. R. Allison, was born in 1836.

This gentleman received his education at Victoria College, and studied law in the office of Philip Low, Q. C., of Picton. Thus it will be seen that Mr. Allison's advantages were of a superior character, both as to his university education and the benefits to be derived from a law course in the office of so noted a lawyer as Mr. Low. These advantages, backed by those natural abilities of which the Allison family all possessed more than an ordinary share, enabled Mr. Allison—after being called to the bar in 1864—to at once take a leading position among the lawyers of the country. He commenced practice in Picton, where he has continued since to do business, and where he has succeeded in acquiring a reputation for strict integrity and professional ability which has drawn to his office one of the most lucrative law-practices possessed by any lawyer of the Bay Quinte district.

One son brightens Mr. Allison's home, the fruit of his marriage to the only daughter of John P. Roblin, for many years the representative of Prince Edward in the Parliament of Canada, and one of those political leaders whose courage and patriotism so nobly contributed to the overthrow of the tyrannical domination of the "Family Compact," and gave to Canada what now makes her one of the finest countries on the face of the earth—Responsible Government.

Edwards Merrill, of Picton, Barrister-at-Law, is the grandson of Samuel Merrill, who came from Connecticut during the first years of the present century, and settled in Kingston, where he engaged in mercantile pursuits with considerable success.

His son Samuel (afterwards the father of a family of eleven children, of whom Edwards was the eighth in order of age), having chosen law as his profession, and having studied in the office of Hon. C. A. Hagerman (then of Kingston, but afterwards Chief Justice of Upper Canada), was called to the bar in 1823, and subsequently removing to Picton (in which place he was the first lawyer), he continued to practise his profession there for nearly half a century. For many years—and until failing health had incapacitated him from laborious application to professional duties—he enjoyed an exceedingly large and lucrative practice; and was considered, on all hands, one of the best lawyers—whether as counsel or as advocate—in the whole of the old Midland and Bay Quinte districts. He was first Registrar of the Surrogate Court here, which position he resigned, after having satisfactorily filled it for eighteen years. He was also Master-in-C
a number of years.

When a young man he was of a military, as well as patriotic
volunteered in an artillery company, which did duty at King
the war of 1812-13.

Though himself a strong Conservative, some of the family e
other side of politics quite as strongly. Among these was h
Stephen B. Merrill, for many years proprietor of that veteran R
nal, the *Prescott Telegraph*, and now Inspector of Inland Reven

Edwards Merrill was born in Picton in September, 1841. H
cated here, and studied law in the office of this place. He was
attorney in 1865; called to the bar in 1867; and has practised
since. He married, in 1866, Carrie, daughter of the late Paul V
grand-daughter of the late Dr. J. B. Chamberlain, both widely
highly respected citizens of Napanee; and the present family
three children.

Like his father, he is of a military turn, having been for a
years, and until its disbandment, Captain of No. 2 Company, 16th

As a lawyer, Mr. Merrill is regarded as a sound counsel, a
man in his profession; while, as a gentleman, he commands the
all who have the pleasure of his acquaintance.

Lewis B. Stinson, J. P.—This gentleman is of U. E. Loyal
on both his father's and mother's side. His great-grandfather, J
and father, also named John, came to Hallowell when their ne
bour on the west was at the Carrying Place twenty miles dista
the east at what is now Picton.

Old Mr. Stinson was a major in the Royal army, and, in c
was proscribed by his former neighbours in the vicinity of Dunb
Hampshire, where he lived previous to the Revolutionary Wa
befriended, and secreted in hiding-places in and about the
period of six weeks, by a Rebel officer named Greeley, a devot
friend, at the imminent peril of his own life for the politic
"harbouring a spy," had the major been discovered. A det
Continental troops actually searched the house for him on o
when he was hidden therein, but left without starting their gam

This rebel officer Greeley, above referred to, was an ancestor
Horace Greeley, of immortal memory, and also of a number of G
subsequently came to Canada and settled in Prince Edward coun
whom afterwards attained to great prominence and popularity in
cipal, and provincial politics.

After hiding Mayor Stinson about his place for six weeks,
assisted him to fly to Canada—that land of refuge for unhappy a
loyalists—whence his family followed him in about three years.

After prospecting around the shore of West Lake they ascende
stream now known as Stinson's Creek, and settled on what has
been the Stinson homestead. Lewis B. Stinson's father, Dav
months old the day they landed here. He was the eldest of a f
sons and four daughters, for each of whom land was drawn as U. E
He afterwards married Rachel, daughter of Daniel Young, w
came to East Lake about the same time the Stinsons came to
The fruit of this marriage was five sons and five daughters.

Lewis B. Stinson is the eldest of this family. He was born
homestead, where he has always lived, and carries on farming.
ever, is not his chief occupation, although he has a beautiful, wel
and highly cultivated estate. He carries on a general agency
Picton, which occupies his entire time, and is president and insp
Prince Edward Mutual Fire Insurance Co., whose head office is at

He married in 1842, Sarah Ann, daughter of James P. Spen
descendant of U. E. Loyalists, here from the earliest settlemen
present family consists of four sons and three daughters.

Mr. Stinson has been a Justice of the Peace for many years;
sented his native township as Reeve for over fifteen years, and is
in that capacity. He has presided over the deliberations of
Council as Warden, and in over twenty contests for municipal
has never but once been defeated, though there is an adverse p
jority of over one hundred in his township.

He is Captain of Reserve Militia for more than twenty years,
every rank from private up.

His grandfather was the first representative sent by this co
liament.

Each generation, from the great-grandfather down to himsel
honoured by appointments to the commission of the peace, and
taken an active, leading, and honourable part in local politics an
affairs.

John Prinyer, J. P., of Prinyer—which post office was na
honour—resides on lot 29, Bay front, North Marysburg, and is
leading men of the township in all affairs which relate to the ge
est or public welfare. He is the son of John B. Prinyer, a ge
French descent, who came from Lower Canada at an early age
on the present homestead, afterwards marrying a daughter of Col.
the very first settler in what is now North Marysburg.

John Prinyer was the fourth, and is now the only surviv
family of seven—four sons and three daughters. He has always
influence in matters of public interest which few possess, and t
North Marysburg have him to thank for their success in gettin
ship set off from South Marysburg.

He has served as Reeve of Marysburg and North Marysbur
years, and has been Warden of the county.

He is a farmer by occupation; a Justice of the Peace for
the present Deputy-Reeve of North-Marysburg, and a preventi
Her Majesty's customs.

Wellington Boulter, J. P., of Sophiasburg, is tenth in o
of a family of twenty-one children of George Boulter, of Engl
who was born in Montreal, and removed to Sophiasburg in
twenty-one years of age, afterwards marrying Sarah, daughte
Peck, a U. E. Loyalist, and one of the first settlers in the townsl

He came into the township, then a wilderness, without mone
out friends, and by dint of energy, honesty, perseverance, and
application, he became owner of six of the best farms in Pri
county. He was a Justice of the Peace and Captain of Militia for
He has long since retired from active business, and resides in
Belleville.

His old homestead, lots 37 and 38, Bay front of Sophiasb
owned and occupied by his son Wellington. The place has th
of a delightful situation, opposite Big Island, and near the ston
connecting the Island with the main land. The farm is one o
to be seen, and Mr. Boulter takes a pride in having everythin
place in the best of order, having his farm well stocked with t
chanical appliances and agricultural machinery known to modern
while in his fields may now be seen some very fine imported an
most admired strains of short-horns, of which he owns the fi
the county.

He is President of the Prince Edward County Agricultural
Secretary of the Township Agricultural Society. He is a Ju
Peace, and a leading member of the Orange body—having been
cutive years County Master, and a Grand Lodge officer of that

He is also largely engaged in the insurance and loan busine
Director, and was some time vice-President of the Prince Edw
Mutual Insurance Company; and has been the general agent

;teen years. He was also Master-in-Chancery for

e was of a military, as well as patriotic, turn, and
ry company, which did duty at Kingston during

ong Conservative, some of the family espoused the
site as strongly. Among these was his brother,
many years proprietor of that veteran Reform jour-
sh, and now Inspector of Inland Revenue.
born in Picton in September, 1841. He was edu-
aw in the office of this place. He was admitted as
to the bar in 1867; and has practised here ever
866, Carrie, daughter of the late Paul Wright, and
te Dr. J. B. Chamberlain, both widely known and
of Napanee; and the present family consists of

of a military turn, having been for a number of
ndment, Captain of No. 2 Company, 16th Regiment.
errill is regarded as a sound counsel, and a rising
while, as a gentleman, he commands the respect of
s of his acquaintance.

. P.—This gentleman is of U. E. Loyalist descent
mother's side. His great-grandfather, John Stinson,
John, came to Hallowell when their nearest neigh-
the Carrying Place twenty miles distant, and on
Picton.
a major in the Royal army, and, in consequence,
rmer neighbours in the vicinity of Dunbarton, New
ved previous to the Revolutionary War. He was
al in hiding-places in and about the house for a
a Rebel officer named Greeley, a devoted personal
t peril of his own life for the political crime of
ad the major been discovered. A detachment of
ally searched the house for him on one occasion
rein, but left without starting their game.
eley, above referred to, was an ancestor of the late
rtal memory, and also of a number of Greeleys who
nada and settled in Prince Edward county—some of
d to great prominence and popularity in local, muni-
ities.
Stinson about his place for six weeks, Mr. Greeley
nada—that land of refuge for unhappy and defeated
nily followed him in about three years.
und the shore of West Lake they ascended the little
inson's Creek, and settled on what has ever since
stead. Lewis B. Stinson's father, David, was six
landed here. He was the eldest of a family of six
for each of whom land was drawn as U. E. Loyalists.
Rachel, daughter of Daniel Young, whose people
t the same time the Stinsons came to West Lake.
ge was five sons and five daughters.
he eldest of this family. He was born on the old
always lived, and carries on farming. This, how-
upation, although he has a beautiful, well-improved,
state. He carries on a general agency business in
is entire time, and is president and inspector of the
Fire Insurance Co., whose head office is at Picton.
Sarah Ann, daughter of James P. Spencer, another
yalists, here from the earliest settlements, and the
f four sons and three daughters.
a Justice of the Peace for many years; has repre-
ip as Reeve for over fifteen years, and is now acting
has presided over the deliberations of the County
in over twenty contests for municipal honours he
defeated, though there is an adverse political ma-
d in his township.
ive Militia for more than twenty years, having held
ip.
the first representative sent by this county to Par-

n the great-grandfather down to himself, has been
ts to the commission of the peace, and has always
and honourable part in local politics and municipal

P., of Prinyer—which post office was named in his
9, Bay front, North Marysburg, and is one of the
ship in all affairs which relate to the general inter-
He is the son of John B. Prinyer, a gentleman of
me from Lower Canada at an early age and settled
, afterwards marrying a daughter of Col. McDonald.
what is now North Marysburg.
e fourth, and is now the only surviving son of a
s and three daughters. He has always exerted an
public interest which few possess, and the people of
im to thank for their success in getting the town-
Marysburg.
eve of Marysburg and North Marysburg for twelve
den of the county.
cupation; a Justice of the Peace for many years;
ve of North-Marysburg, and a preventive officer in

n, J. P., of Sophiasburg, is tenth in order of age
ne children of George Boulter, of English descent,
real, and removed to Sophiasburg in 1820, when
, afterwards marrying Sarah, daughter of James
and one of the first settlers in the township.
nship, then a wilderness, without money and with-
of energy, honesty, perseverance, and uncommon
owner of six of the best farms in Prince Edward
e of the Peace and Captain of Militia for many years.
from active business, and resides in the city of

lots 37 and 38, Bay front of Sophiasburg, is now
his son Wellington. The place has the advantage
opposite Big Island, and near the stone causeway
ith the main land. The farm is one of the finest
lter takes a pride in having everything about his
r, having his farm well stocked with the best me-
gricultural machinery known to modern husbandry,
ow be seen some very fine imported animals of the
short-horns, of which he owns the finest herd in

e Prince Edward County Agricultural Society, and
hip Agricultural Society. He is a Justice of the
mber of the Orange body—having been four conse-
ster, and a Grand Lodge officer of that society.
gaged in the insurance and loan business. He is a
time vice-President of the Prince Edward County
any; and has been the general agent for Central
Ontario East, of the Mutual Life Association of Canada since its first organization. He is also valuator and agent for two of the wealthiest loan companies in the country. He is a very active and energetic business man, and the above named companies have been able to secure in him an agent who has added greatly to their business in the section of country over which he operates.

Of the family of twenty-one children, Wellington and his brother James, of Big Island, are the ones now left in Prince Edward. The latter is a man of wealth and influence, and he served many terms in the Township and County Councils. Dr. Boulter, of Stirling, the present member for North Hastings in the Ontario Legislature, is another brother.

Mr. Boulter is an active and earnest worker in the Church, and in that connection does not shrink from those material duties our Great High Priest calls upon his disciples to perform. He has been for many years a deacon of the Presbyterian Church, and secretary of the society.

SAMUEL N. SMITH, J. P., who resides near Demorestville, on the Marsh front of Sophiasburg, is descended from U. E. L. stock, his grandfather being a loyal refugee, and one of the pioneers of this section of country. His father erected "Smith's Mills"—popularly so called—among the earliest industrial enterprises of the township, which are still successfully operated by one of the family.

Mr. Smith has always taken a foremost part in the municipal affairs of his native township, and displayed an honest zeal—unbiassed by party prejudices—in furthering what he considered the material interests of his constituents. It is principally to him that the inhabitants are indebted for their beautiful town hall, the finest, with a single exception, in the county, and one of the best in the Province.

Mr. Smith has been for some years a Justice of the Peace; is the present Reeve of Sophiasburg; and the Warden of Prince Edward county. He has been twice married—first to a daughter of Isaac Hamilton, Esq., having one daughter by this marriage; afterwards to Miss Olive Weeks, a very highly esteemed and much respected lady. In all matters pertaining to the general welfare of the community, Mr. Smith's energies have always been exerted.

JOHN YOUNG, REEVE OF HILLIER.—The Youngs appear to have come of Irish extraction. At least this is the case with the paternal grandmother of the subject of this sketch, for we have it on undoubted authority, that *her two* grandfathers were among the brave and loyal few who held out to the last at the siege of Derry. Mr. Young's grandfather, Robert Young, was born in Boston, "Massachusetts Bay," then a British colony. The Youngs seem to have been a family of considerable influence there, as we find Robert Young, at the time of the outbreak of the Revolutionary War, in command of a British man-of-war, with the rank of captain in the navy. He had rendered important service long before this, having been present in command of his ship at the capture of Louisburg, and was also present, though not actually engaged, at the capture of Quebec from the French. He took part in many naval engagements during the Revolution, resigning at its close, and removing to the Annapolis Valley, Nova Scotia, whither he came west with his family of seven sons and one daughter, and settled at the Carrying Place in 1792, there being only one other settler there when he arrived.

He located first on lot 1, 1st concession, Ameliasburg, where Reuben Young now lives, and drew nearly 2,000 acres of land for himself and family, 1,200 of which was along the shore of Pleasant Bay. The farm now owned and occupied by John Young is part of the above.

Thomas Young, John's father, was born in Annapolis, N. S., and was about twelve years of age when he came to the Carrying Place. He married in 1810, Nancy, daughter of Hugh Robinson, of Ameliasburg (now Hillier). He had a family of four sons and six daughters, of whom John, the ninth in order of age, married in 1862 Miss Letitia Jane Whittier, a cousin of Right Hon. Sir John A. Macdonald, the result of the marriage being two sons.

Mr. Young, though scarcely yet of middle age, has already come to the front in public affairs, having been for a long time Township Councillor, and for two years Reeve, a position he now holds. He is also a Captain of Militia, is a very agreeable as well as a popular gentleman of liberal and progressive ideas, and is looked upon by the citizens of the township in which he resides as one of the most promising and rising young men of the community.

JOSEPH PIERSON, J. P., who owns and occupies lot 31, 3rd concession of Hillier, is a cousin of John Young above-mentioned, being a grandson of Captain Young, of Revolutionary fame. His father emigrated from New Jersey in 1802 and settled on the lot where Mr. Pierson now lives. Here he married the only daughter of Captain Young, and here Joseph was born—the second child and eldest son of a family of two sons and two daughters.

Old Mr. Pierson was in the military service of the Government during the war of 1812-13, and again during the rebellion of 1837-38, when he held command, with the rank of Lieut.-Colonel, of a force whose duty it was to convey Government supply boats between the Carrying Place and Burlington (now Hamilton). In this capacity he rendered very important service. He retained his military rank till his death, at Christmas, 1877, at the age of ninety-five years and twenty-five days. In point of physique and intellect, he was so much above the average as to merit the title of an extraordinary man. He lived highly respected, and died deeply lamented, retaining his remarkable faculties to the last.

His son Joseph was married in 1851 to Caroline, daughter of Conrad Huffman, and has a family of three sons and one daughter.

For many years he has identified himself with all enterprises of general interest or a progressive nature, always espousing the side of the public weal, independent of personal considerations.

His present business is farming and dealing in grain. He handles a very large amount of probably the finest barley grown in America. It was barley grown and exhibited by himself which took highest award at the Centennial.

He was at one time engaged in ship-building, and built the first vessel (The Lille—brigantine) that ever crossed the Atlantic from the great lakes.

Mr. Pierson is a Justice of the Peace of many years standing, and a Captain of Militia. He was Reeve of his township (with the intermission of a single term) for twenty-one years; has been Warden of the county, and is looked upon by the inhabitants of the entire district as one of the most enterprising, liberal, intelligent, and influential men in his native county of Prince Edward.

LIEUT.-COL. REUBEN YOUNG, of the Carrying Place, is a grandson of Captain Young of the British navy, above-mentioned, and the son of James Young, who was born in Annapolis, Nova Scotia, and came to the Carrying Place, with his father and family, in 1792. He married Catherine Weller, the daughter of one of the oldest settlers in this part of Prince Edward county, who gave it the name of Weller's Bay. Reuben was born at the Carrying Place in 1805, and his father and mother both died there in 1830.

During the rebellion of 1837-38, Col. Young was a subordinate officer under Col. Wilkins, who commanded a detachment of troops at the Carrying Place. He was subsequently promoted step by step till he received the Lieut.-Colonelcy of the 5th Regiment of Prince Edward Militia—a rank which he still retains. The whole Young family seem to have been possessed of that military spirit which the troublous times of '76 had stamped upon their ancestors. The Colonel's father was an officer of volunteers during the war of 1812-13, and was stationed at Kingston, having charge of American prisoners-of-war.

Lieut.-Colonel Young is a farmer by occupation. He cultivates about

330 acres of choice land, and is what is termed an advanced agriculturist in every sense of the word. He has always taken a deep interest in all affairs of public concern, and a perusal of the township and county records show his name in connection with the offices of Councillor, Deputy-Reeve, and Reeve, for a period extending over a great many years.

He was married to Miss Nancy Bryant, who still survives, past seventy years of age, while the Colonel is about three years her senior. They have one son, who resides at Trenton.

He was appointed Issuer of Marriage Licenses by Sir John Coleman, and is an officer of the Department of Marine and Fisheries—having charge of the Carrying Place light, which is situated on his estate.

James Redner, of Rednersville, Lot 77, 1st concession of Ameliasburg, is grandson of Henry Redner, who, with his family of three sons and three daughters, came to Canada immediately after the close of the revolutionary war, and settled at the above place. He was in the royal military service during that war, and was dangerously wounded in battle and captured by the enemy. While still under surveillance, and when yet but partially recovered, he escaped from his captors, and, eluding his pursuers, succeeded in reaching the British lines. He was a man of considerable wealth and influence previous to the revolt of the colonies; and the success of the revolutionists obliged him to flee to Canada for his life, abandoning a beautiful estate in Duchess County, New York, which was subsequently confiscated. He and his children (except one), have long since passed away. Mrs. Cole, of Ameliasburg, one of his daughters, still survives, in the enjoyment of good health and possession of all her faculties, over 93 years of age, probably the oldest woman, with one exception, in the county.

James Redner is the son of Henry Redner, jr., one of the above, who raised a family of two sons and three daughters, of whom James is the only one now surviving. He was born at Rednersville, which was named after his family, who were the first settlers in the place, and subsequently—as now—among the most influential families in the community. He has been closely identified, ever since his arrival at mature years, with the material interests of his native township, and the particular locality where he first saw light. He carried on mercantile business, and purchased grain for over a quarter of a century in Rednersville, where he has always lived, and where he still owns a beautiful farm, a wharf, storehouses, etc., and carries on a very extensive business in grain.

Mr. Redner retains the military and patriotic spirit of his ancestors, having served when quite a youth in '37, '38, and '39 in Lt.-Col. (then Captain), Wilkins' troop of Prince Edward Cavalry, in which he was orderly-sergeant. He was afterwards an officer of militia, and was a number of times appointed to the Commission of the Peace, but refused to qualify.

He has frequently been an incumbent of the various offices within the power of his fellow-citizens to bestow, which he has never failed to fill with satisfaction to his constituents and as honor and credit to himself, for which his natural and acquired abilities eminently fit him.

William F. Weese, now living on Lot 89, 1st concession of Ameliasburg, is a grandson—and one of a very large number of descendants—of John W. Weese, who came into the country and settled, first in Adolphustown, with a family of one daughter and three sons, viz., John, Francis, and Henry, the latter of whom was the father of the subject of this sketch.

Old Mr. Weese was of German extraction, his ancestors being among the Knickerbockers who first gave Manhattan a "local habitation and a name." He appears to have been a man past middle age at the time of his settlement in Canada, as his eldest son John had already served two terms in the Royal American contingent of the British army during the revolutionary war; and he himself was a captain in the secret service of the king's army. Captain Hogle, a cousin of his, and an officer of the British regulars, was killed at Bunker Hill, and buried by the Hon. Col. Wilkins' father, a personal friend, then an officer of the same regiment. Some of the Hogles afterwards fled to Canada to avoid persecution at the hands of the victorious Continentals, and the name is now a familiar one along the mainland shore of Bay Quinte.

It will be surmised from the above record that Herkimer County, New York, was too hot a place for the Weeses after the Americans got the upper hand. They saved their lives only by abandoning valuable possessions, and taking a hurried departure to Canada. The old man and family came to Adolphustown a little before the final close of the war; and on the 26th of November of the same year (1783), crossed the Bay, and located and settled upon the lot where his grandson, William F. Weese, now lives, being, without doubt, *the first actual settler* in what is now Prince Edward County.

His son John went at the same time—or very shortly thereafter, directly across the Bay, and settled on the main land, in the township of Sidney; and it is claimed by those acquainted with the facts, that he was the *first actual settler* in the present county of Hastings. He removed some years subsequently to the site of the present village of Stirling, in Hastings County, and became the first settler in the limits of the township of Rawdon.

Old Mr. Weese, his son Henry and grandson William F. have lived from the first settlement, to the present time, or till the date of their several deaths, on the same farm located in 1783.

William F. Weese has been engaged extensively, and with varying success, in commercial pursuits, and in the lumber trade, but farming is now his chief occupation. He owns and works 550 acres of land, a large portion of which is of a very superior description.

The military and patriotic spirit proverbial in the descendants of the original Loyalists, was not wanting in Mr. Weese. When the Rebellion broke out, he enlisted in Captain Flagler's company of the "Queen's Own," and stood guard at York when Lount and Matthews were hung. When discharged by expiration of time, he again enlisted in Lt.-Col. (then Captain) Wilkins' Horse, and was engaged in despatch bearing, and similar service, till the disbandment of the forces, on the establishment of peace.

Though considerably beyond three score years, Mr. Weese is still a strong, hearty, active man, and of a long-lived race—nearly all of the name having lived to an exceptionally great age.

He has served in various public capacities, having been Councillor or Deputy Reeve of his native township for eleven consecutive years. He has also held a number of official appointments in the militia, and been thrice appointed to the Commission of the Peace, a position for which he has thus far declined to qualify.

William DeLong, J.P., of Ameliasburg, is descended from U. E. Loyalist ancestors, on both father's and mother's side. The DeLong family supplied many good men and true to the Royal cause in time of Britain's sorest need; and they were obliged in consequence to abandon all worldly possessions, and fly for their lives—some to Nova Scotia—others to Canada. They formerly lived in Duchess County, New York, in a pleasant and prosperous community, and in affluence; they came here in poverty, and settled in an uninviting wilderness, which, by the exertions of themselves, and men of similar stamp, has been converted into one of the finest countries of the present time.

Henry DeLong was one of our country's pioneers. He settled in Ameliasburg at a very early day of the township's history. He had a family of eleven children, of whom Simon, the fourth in order of age, married a daughter of William Dempsey, Esq., whose father was the second settler in Ameliasburg, and whose family have always exercised a telling influence in all local affairs since the earliest days.

The eldest child of this marriage was Wil
burg in 1821, and now one of the leading ag
County—as well as one of the most influenti
carries on a farm of nearly 800 acres in exte
everything of the most approved pattern in
line, and a large lot of superior stock, in
coach and draught horses.

In politics he has always been strongly id
party; while in matters of general public in
efforts to aid and assist all useful enterprises
and firmest supporters of the Prince Edward
location—taken in connexion with that of
vent his receiving so much personal benefi
work, as many others who opposed the sche
ground of its cost, or gave it but an indiffer

For the greater part of his life he has bee
Methodist Church, having been Circuit Ste
quarter of a century.

He was married in 1844, to Letitia, daugh
iasburg—for very many years one of the lea
County in educational, religious, munic
matters; in all of which he distinguished hi
honorable positions. The above marriage ha
three sons and four daughters.

During a period extending over nearly ha
filled many positions of honour and public tr
tice of the Peace and Treasurer of the towns

James Peck, residing on lot 93, 1st co
seventy-five years of age, and is the son of Ja
the son of Jacobus Peck, one of the oldest
first settled in Sophiasburg, and subsequent
and located on the above lot, which has si
family. His maternal grandfather was also s
of Jacobus. Thus his father and mother w
name.

Samuel and Jacobus Peck emigrated from
the close of the Revolutionary War, where
subsequently removing thence to Sophiasburg
the ancestors of a very large number of des
being one of the most numerous of any to be
District.

Quite a number of the family served the R
Revolution. Jacobus and Samuel, above nam
secret service. James Peck, Sr., was an offi
regiment at the evacuation of New York, and a
body servant to Major André when that office
as a spy by the Americans. As such the Ame
Major's private personal effects. The larger
quently turned over to André's relatives; bu
as mementoes. Among these is a solid silve
which James Peck still retains as a relic of
quite a curiosity in its way, and extravagant
ous times been offered for it by curiosity
grandfathers were own-cousins to that gall
officer.

Mr. Peck was second son, and is now the
family of ten children. Though now over sev
hale and hearty. We found him at work in t
and could scarce believe that a man of s
preserved frame had passed the allotted three-
and has ever since lived, on the farm where he
relatives of the same name have lived to
toil by.

He holds a commission as Major of Mil
Captains appointed at the inspection of that n

Daniel Y. Williams, J.P., is one of the
His maternal grandfather was Daniel Young,
ancestry, who had settled near Troy, New Yo
the American Colonies, and was a Major in t
war, afterwards removing to Canada, as a res
about 1783 or 1784 at East Lake.

The Williams family are of Welsh extracti
father, Samuel Williams, was living near N
the Revolution, and was obliged by the result
an adherent to the Royal cause, to go into exi
sentiments, abandoning property and possessi
to Canada, settled at West Lake, near the sa
was literally a "howling wilderness" for mil
of him that his only earthly possessions when
were an axe, a jack-knife, and one tin dish. he
went about clearing land, subsisting on fish a
ful. He suffered untold hardships, howeve
pioneers; but his clearing grew larger, and
time. His first patch of wheat was covered
rake improvised with his axe, and he went f
Mills to get his wheat ground into flour.

Caleb Williams, father of Daniel, was bor
died there in 1870.

Daniel, the eldest of a family of three sons
six are still living in the District, was born
1821, and settled on his present farm in 1842
even at that late day a vast wilderness, with
clearing to relieve the monotony of the forest
He married, 3rd May, 1841, Sarah, dau
Hillier, a U. E. Loyalist. Mrs. Williams d
leaving a family of three sons and four daught

Mr. Williams owns a beautiful piece of two
is his present occupation, though he was for
profitably engaged in patent and proprietary m
widely in connexion with that business.

He has been for many years a Justice of the
ous positions of public trust in the gift of the

Daniel Pettit, J. P., of Hillier, is of
grandfather being a Loyalist from Long Island
the Royal army. On the establishment of the
nies he removed to Canada, being still quite y
Lake, in Hallowell. He married a daughter o
Loyalist, and had a family of five sons and five
the second son, had six sons and six daughters.

Daniel, the second son of this family, was
When nine years old his father removed to Lo
where Daniel remained till in his twenty-fourt
Lot 2, same range, where he still lives.

He was married in 1842 to Mary Anne, da
Mrs. Pettit died in 1866, leaving four of a
quently married Catherine, daughter of Sin
daughter by the last marriage.

TORICAL SKETCH OF THE

e eldest child of this marriage was William DeLong, born in Amelias- 1821, and now one of the leading agriculturists of Prince Edward —as well as one of the most influential, and highly esteemed. He on a farm of nearly 800 acres in extent, upon which is to be found bing of the most approved pattern in the machine and implement nd a large lot of superior stock, including short-horn cattle, and and draught horses.

politics he has always been strongly identified with the Conservative ; while in matters of general public interest he yields to none in his to aid and assist all useful enterprises. He was one of the foremost most supporters of the Prince Edward County Railway ; though his n—taken in connexion with that of the proposed route—will pre- is receiving so much personal benefit from the completion of the as many others who opposed the scheme entirely, on the niggardly l of its cost, or gave it but an indifferent support.

r the greater part of his life he has been actively connected with the dist Church, having been Circuit Steward of the same for over a r of a century.

was married in 1841, to Letitia, daughter of Col. Peterson, of Amel g—for very many years one of the leading men of Prince Edward y in educational, religious, municipal, military and judiciary rs ; in all of which he distinguished himself by taking prominent and ble positions. The above marriage has been blessed with a family of sons and four daughters.

ring a period extending over nearly half a century, Mr. De Long has many positions of honour and public trust. He is at present a Jus- the Peace and Treasurer of the township.

MES PECK, residing on lot 93, 1st concession, Ameliasburg, is now ty-five years of age, and is the son of James Peck, and he was in turn n of Jacobus Peck, one of the oldest settlers in the county, having ettled in Sophiasburg, and subsequently removed to Ameliasburg, cated on the above lot, which has since been in possession of the . His maternal grandfather was also a Peck—Samuel—and brother obus. Thus his father and mother were cousins, and of the same

muel and Jacobus Peck emigrated from New Jersey to Nova Scotia on ose of the Revolutionary War, where they remained a short time, quently removing thence to Sophiasburg, as above noted. They were ncestors of a very large number of descendants—the name of Peck one of the most numerous of any to be found in the Prince Edward ct.

ite a number of the family served the Royal cause in arms during the ution. Jacobus and Samuel, above named, were both Captains in the service. James Peck, Sr., was an officer in Colonel Van Buskirk's nt at the evacuation of New York, and an uncle named John Peck was servant to Major André when that officer was captured and executed py by the Americans. As such the Americans turned over to him the 's private personal effects. The larger portion of these were subse- ly turned over to André's relatives ; but a few trifles were retained mentoes. Among these is a solid silver "scale" sketching pencil, James Peck still retains as a relic of Revolutionary times. It is a curiosity in its way, and extravagant sums of money have at vari- imes been offered for it by curiosity hunters. Both Mr. Peck's lfathers were own-cousins to that gallant but unfortunate British r.

r. Peck was second son, and is now the only surviving member of a y of ten children. Though now over seventy-five years of age, he is nd hearty. We found him at work in the field, on going to visit him, could scarce believe that a man of so fresh appearance and well ved frame had passed the allotted three-score and ten. He was born, as ever since lived, on the farm where he now resides. Several of his ves of the same name have lived to see almost a century of time y.

holds a commission as Major of Militia, and was one of the first ins appointed at the inspection of that military organization.

ANIEL Y. WILLIAMS, J.P., is one of the leading citizens of Hillier. aternal grandfather was Daniel Young, a U. E. Loyalist, of Dutch ry, who had settled near Troy, New York, previous to the revolt of merican Colonies, and was a Major in the King's service during the fterwards removing to Canada, as a result of the war, and settling 1783 or 1784 at East Lake.

e Williams family are of Welsh extraction. Daniel's paternal grand- , Samuel Williams, was living near New York on the outbreak of evolution, and was obliged by the results of the war, in which he was herent to the Royal cause, to go into exile, a martyr to his political ents, abandoning property and possessions by so doing ; and coming ada, settled at West Lake, near the sand banks, when all around terally a "howling wilderness" for miles on every hand. It is told that his only earthly possessions when he arrived at West Lake n axe, a jack-knife, and one tin dish, besides his gun. He at once bout clearing land, subsisting on fish and game, which were plenti- He suffered untold hardships, however, as did all our country's rs ; but his clearing grew larger, and his prospects brighter with His first patch of wheat was covered with his hands, and a rude mprovised with his axe, and he went for many years to Kingston o get his wheat ground into flour.

eb Williams, father of Daniel, was born in Hallowell in 1793, and here in 1870.

niel, the oldest of a family of three sons and five daughters, of whom still living in the District, was born in Bloomfield, Hallowell, in and settled on his present farm in 1842, that part of Hillier being that late day a vast wilderness, with only here and there a small g to relieve the monotony of the forest gloom.

married, 3rd May, 1841, Sarah, daughter of Paul Trompour, of , a U. E. Loyalist. Mrs. Williams died about three years ago, a family of three sons and four daughters.

Williams owns a beautiful place of two hundred acres, and farming present occupation, though he was for some years extensively and bly engaged in patent and proprietary medicines, and travelled very in connexion with that business.

has been for many years a Justice of the Peace, and has filled numer- sitions of public trust in the gift of the people.

NIEL PETTIT, J. P., of Hillier, is of U. E. descent, his paternal ather being a Loyalist from Long Island, New York, who served in yal army. On the establishment of the Independence of the Colo- removed to Canada, being still quite young, and settled on West n Hallowell. He married a daughter of John Platt, another U. E. t, and had a family of five sons and five daughters, of whom James, ond son, had six sons and six daughters.

iel, the second son of this family, was born in Hallowell in 1820. nine years old his father removed to Lot 5, 3rd concession, Hillier, Daniel remained till in his twenty-fourth year, when he removed to ame range, where he still lives.

was married in 1842 to Mary Anne, daughter of Jacob Osterhoat. ettit died in 1866, leaving four of a family. Mr. Pettit subse- married Catherine, daughter of Simon DeLong, and has one r by the last marriage.

His grandfather drew two hundred acres of land from Government, for himself and each of his children, as U. E. Loyalists.

Farming is Mr. Pettit's occupation. His place contains two hundred acres of superior and highly cultivated land, with fine house, barns, and buildings of every description, suitable to carrying on a first-class farm in a first-class manner. He spares no pains nor expense in improving his stock, and now owns some very fine short-horn cattle, pure Leicester sheep, and a superior class of roadsters and draught-horses.

He is the oldest Justice of the Peace in the township, and has held various municipal offices and positions of public trust for a period extending over the whole of his life since arriving at mature years.

Nor is the military spirit of his ancestors degenerating in him, as we find that before he was eighteen years of age he volunteered in Captain Wilkins' Prince Edward Cavalry, and served till the close of the rebellion.

Mr. Pettit has been what is popularly known as a successful man in his own business ; and what is better still, he is highly respected and univer- sally esteemed by the entire community.

WILLIAM OWENS, J. P., of Picton, is an Irishman by birth, and came to Canada in 1837, when sixteen years of age. His first act after coming here was to enlist in the service of the Government of his newly adopted country, against the rebellion which then threatened its political existence. He volunteered in a company of Prince Edward infantry, which marched to York, and was put into Colonel Taylor's regiment. The regiment then marched to Burlington, and subsequently, *via* London and Sandwich, to Fort Malden (Amherstburg), arriving there just too late to participate in an affair between the 32nd Regulars and a body of Rebels, on an island in Detroit River a short distance below the fort. Mr. Owens served with Colonel Taylor's regiment till the expiration of his term of service ; then enlisted in the Prince Edward Cavalry, and served with them till their dis- bandment, on the reestablishment of tranquillity.

Mr. Owens is a carpenter by trade, but has followed steam-boating, and shipbuilding, and is now engaged in farming and the manufacture of potash.

He holds commissions as Justice of the Peace, and Captain of Militia, and is a good, honest specimen of a self made man, having been unusually successful in trade, and succeeded by strict business habits and fair dealing, in acquiring the confidence of his fellow citizens, to such an extent that they have made him a member of the Town Council for twenty-three con- secutive years.

He has also filled the chair of Mayor of Picton for eight years, without interruption.

He takes a lively interest in educational affairs, and has been for many years an efficient member of the Board of Public School Trustees.

STEPHEN D. CRANDALL, J.P., of Athol, is the son of James, and grand- son of Palmer Crandall, a U. E. Loyalist, from Duchess County, New York, who came to Canada soon after the close of the Revolutionary war, and settled in the County of Northumberland, where the village of Colborne now stands ; being among the first settlers in that part of the country. He had a numerous family, among whom was James, Stephen's father, who was born at Colborne in 1801, marrying in due time, Fanny, daughter of Cornelius White, another U. E. Loyalist from Duchess County, New York. The issue of this marriage was a family of two sons and two daughters, of whom Stephen, the youngest, was borne in 1838, in the Township of Hillier, to which place his father had previously removed from Colborne.

Stephen D. Crandall now owns and occupies 114 acres bordering the south shore of East Lake, being a part of Lot No. 14. His farm is an ex- ceptionally fine one ; and everything about it—from the large and beautiful house (of which we give a view elsewhere), to the smallest out-building—is in keeping with the character of the farm. The place is also supplied with an abundance of superior stock, and a full complement of the newest and most approved patterns of labor-saving machinery.

A great interest is manifested in fruit-culture, and there are three large and unusually fine apple orchards on the place, in various stages of growth ; and a proportionate quantity of all descriptions of small fruits.

Mr. Crandall is married to Fanny, daughter of J. P. Spencer, of East Lake, a U. E. Loyalist, and veteran of the war of 1812-13 ; and three bright little children bless their home.

Though quite a young man, Mr. Crandall is a Justice of the Peace, and an active and energetic participator in public and political affairs, being a thorough organizer, and zealous worker in the cause of Reform ; and he is acknowledged on all hands as one of the most progressive and intelligent men of his political party, or of the community in which he resides.

CHARLES A. MCDONELL, of Picton, is the son of the late Alexander McDonell, who emigrated to Marysburg, from London, England, in 1792, and settled on Black River, where he remained till his death, in 1859. Mr. McDonell was a native of Ireland, but lived in London from the time he was nine, till he was seventeen years of age, at which time he came to this country. He married a Miss Pierce, who came out in the same ship with him. He was an official in the Quarter-Master's Department, and stationed at Kingston during the war of 1812-13 ; and was one of the old commission- ers under whom Township affairs were managed in those days.

His son, Charles A. McDonell, carried on business as a farmer and speculator, in Marysburg, till about twelve years ago, when he removed to Picton. He is now engaged in the loan, insurance, grain and produce busi- ness, in addition to which he carries on a large general store.

He has been Town Councillor a number of years, and is one of the lead- ing men of the place in financial and business affairs.

JOHN A. SPRAGUE is a resident of Big Island, lot No. 4, south front, and is married to a daughter of R. Badgley, also of Big Island.

The Sprague family were originally from Long Island, New York. A branch of the family also resided in Rhode Island, and the name of at least one of them is familiar even to Canadians, and a household word among Americans, viz., Senator Sprague, of Rhode Island, who as Governor of that State, was one of the most influential and powerful men in the whole Union during the late civil war.

The Sprague family in Canada comprise among their number many of the most influential citizens in Prince Edward County, where they were among the earliest settlers, and where a large number of them still reside.

Mr. Sprague's mother is a sister of the Rev. Cyrus Allison, so well and favourably known throughout the country as one of the oldest and ablest Methodist Ministers in the District. It was from him that Allisonville, a post village in the township, received its name.

Mr. Sprague, though still a young man, has already occupied prominent positions in public and municipal affairs—first as Councillor, afterwards as Deputy-Reeve and Reeve. He is at present engaged in farming, and owns and cultivates 150 acres of excellent land. He is an active, energetic, and rising young man, and a leading spirit in the Reform Party.

MATTHEW BENSON, of Sophiasburg, is of U. E. Loyalist descent, his grandfather, Richard Benson, and grand-uncles, John and Jacob Benson, having left the United States and come to Canada at an early day in its history. These men all raised large families.

Mr. Benson himself is one of a family of twelve. He is what we have many brilliant examples of in Canada—a self-made man, having commenced poor and at an early age, to work for himself. He labored for a long time in farm service, by the month ; but by uncommon application, untiring in- dustry, and prudent management, he has succeeded in placing himself in a position of comparative affluence.

He owns and cultivates 150 acres, comprising a beautiful and valuable farm, situated on the main telegraph road, midway between Picton and Demorestville. He has lately built a splendid brick house on his farm, and is embellishing the place with tastefully laid-out grounds, etc. It is one of the finest farm residences in the township, or perhaps in the county.

Mr. Benson is engaged in hop-raising to some extent, having over ten acres of that crop under cultivation. He married in 1859, Lydia, third daughter of David Barker, of Sophiasburg, a leading person amongst the Society of Friends, performing the duty of minister in that religious body for a number of years.

The Bensons of Sophiasburg are a very numerous family. They are all descendants of the three first above-named, and among them are to be found many of the leading men of the county.

CAPTAIN W. H. MORDEN, of Peterson's Ferry, is, like many of the leading men of Prince Edward, a descendant of U.E. Loyalists, on both father's and mother's side. His father's ancestors were originally of Welsh extraction, and among the pioneers of the New World, where they resided for many generations as British Colonial subjects previous to the revolt of the thirteen American colonies. This intestine war, which found the nearest relatives in many cases arrayed in mortal combat against each other, brought at least four branches of the Morden family to the front, in the defence of the established authority, and what they considered the rightful rulers. Of these four Mordens one was James, elsewhere noticed, who was the first settler in the village of Northport; another of the four, Richard, also settled in Sophiasburg, where he raised a large family, one of whom, James R., subsequently married Miss Mary Betsky. Of the fruits of this marriage, which comprised a numerous family, Capt. Morden is the youngest.

The Mordens are now a very numerous family, and one of the most influential in the Prince Edward District; many of them having at various times held high and honorable positions of public trust in the respective localities in which they reside.

Captain Morden has been twice married, first to Miss Henrietta Savage, who afterwards died, then to Miss Sarah Anne Peterson. The Petersons were one of the most prominent representative families in the county, some of them having held distinguished positions in both the civil and military branches of the public service. Col. Peterson, who died but a few years since, was an uncle of Mrs. Morden, and distinguished for his military services and abilities; while two of her father's uncles of the same name, were among a mere handful of British and Loyalists who successfully defended an old block-house, on the present site of the upper part of New York city, against a vastly superior force of Continental troops. In the siege, the Petersons especially distinguished themselves, and were mentio quently in general orders for their gallant conduct.

Capt. Morden has been for many years a Justice of the Peace, a a sailor by profession, and possessing no innate love of the "sol retains nevertheless the patriotic spirit which animated the Loy century ago; and in him, and such as him, the descendants of th of Upper Canada show no sign of degeneration. He raised a c volunteers on two different occasions, when volunteering meant more than periodical parades in gold lace and "panoply of war," the time of the "Trent" affair, and afterward at the time of t invasion.

His father was also a veteran of 1812, having served as lie Col. (then Captain) Solmes' company. It is said that those two first officers to tender their services to the Government at that tim Edward County.

Captain Morden is a large owner of steamboat property, havi several successful navigation companies. He is captain of the steamer *Picton*, running on the Toronto and Port Dalhousie r bears the reputation of an experienced, able, and careful officer, un prudent management the enterprise which he commands is me deserved success.

JAMES BENSON, of Roblin's Mills, is the son of Henry Benson fries, Scotland, who migrated to Montreal, Canada, in the y where he remained till 1833, when he removed to Ontario and the county of Middlesex. Here his son James was born in 1836, about twenty years of age he came to Prince Edward county and Ameliasburg.

Mr. Benson has devoted the greater part of his life to the im of the youth of the county, having been a public school teach county for over twenty years, during eight of which he has been of the "Union," the largest union school in the county. In the complying with the "School Law Improvement Act," he was o from this county who succeeded in obtaining Provincial certific received his education chiefly from his father, who was a finish and a man of superior intellectual attainments. He has filled va tions of trust in the county, and is at the present time Township Ameliasburg and Registrar of vital statistics.

Though teaching is his profession, Mr. Benson is also a far farm is beautifully situated, in close proximity to the pleasant Roblin's Mills, and overlooking that delightful sheet of water Lake and the view from his residence (which is shown in anoth this work) is a most commanding and enchanting one.

nguished themselves, and were mentioned subse- for their gallant conduct.

n for many years a Justice of the Peace, and though l possessing no innate love of the "soldier's art," atriotic spirit which animated the Loyalists of a and such as him, the descendants of their fathers o sign of degeneration. He raised a company of it occasions, when volunteering meant something ules in gold lace and "panoply of war,"—once at affair, and afterward at the time of the Fenian

veteran of 1812, having served as lieutenant in s' company. It is said that these two were the ir services to the Government at that time in Prince

arge owner of steamboat property, having stock in tion companies. He is captain of the passenger on the Toronto and Port Dalhousie route; and ...rienced, able, and careful officer, under whose enterprise which he commands is meeting with

blin's Mills, is the son of Henry Benson, of Dum- grated to Montreal, Canada, in the year 1828, 833, when he removed to Ontario and settled in Here his son James was born in 1836, and when he came to Prince Edward county and settled in

ed the greater part of his life to the improvement ity, having been a public school teacher in this ears, during eight of which he has been Principal est union school in the county. In the year 1871, hool Law Improvement Act," he was one of three cceeded in obtaining Provincial certificates. He hiefly from his father, who was a finished scholar ellectual attainments. He has filled various posi- ty, and is at the present time Township Clerk of ar of vital statistics.

is profession, Mr. Benson is also a farmer. His d, in close proximity to the pleasant village of ooking that delightful sheet of water—Roblin's his residence (which is shown in another part of manding and enchanting one.

In 1859 he married Angeline, daughter of Henry Parliament, of Ameliasburg. They have a family of two sons living.

Mr. Benson is justly deemed by all his acquaintances as a superior man in his profession, while socially, he is a truly hospitable and highly companionable gentleman; and in all matters relating to business he is commendably exact, and extremely courteous and obliging.

SAMUEL B. BROOKS, of Northport, is one of the leading farmers of Sophiasburg. He emigrated from Bennington, Vermont, just previous to the breaking out of the rebellion of 1837, and settled on lot 27, Marsh front of Sophiasburg, which he still owns. At that day it was a comparatively worthless forest—now it is a magnificent stretch of beautiful fields of waving grain and prolific and healthy fruit orchards, dotted with the finest buildings. Mr. Brooks has retired from the farm, and now resides in the village of Northport.

He was married in 1849, being then twenty years of age, to Rebecca D., daughter of Richard Solmes, of Sophiasburg. The Solmes family were U. E. Loyalists; among the earliest settlers in the township, and there and ever since among the most influential and highly respected citizens in the community.

The fruit of this marriage was two sons, both of whom are married and now occupy high positions in the esteem of their acquaintances and in the social relations of the neighbourhood in which they reside.

DAVID ROWE, of Ameliasburg, is the grandson of William and the son of John Rowe. The former was a U. E. Loyalist, a former resident of New Jersey, from which place he emigrated to Canada about the year 1800.

Mr. Rowe's father was a volunteer during the war of 1812-13. His detachment was stationed at Kingston in charge of American prisoners-of-war.

Mr. Rowe is a carpenter and joiner by trade, but follows farming as his chief occupation. He is one of the most advanced agriculturists in this section of country, and his beautiful and highly cultivated farm bears testimony to good taste and careful management. He lives upon the farm first taken by his grandfather when he came to the country—lots Nos. 107 and 108, 2nd concession, south of Carrying Place.

Of a large family, five brothers and a sister still survive. They all reside in Prince Edward county; and among the descendants of the original Jersey Loyalist are to be found a number of men of wealth, refinement, and influence; second to none in the glorious old Bay Quinté District in the respect and esteem in which they are justly held by their fellow-citizens.

COURT HOUSE & GAOL, PICTON

COURT HOUSE & GAOL, BELLEVIL

...GE & GAOL, PICTON.

... & GAOL, BELLEVILLE.

L A K E
CONSECON
Con. 4
Con. 3
Con. 2

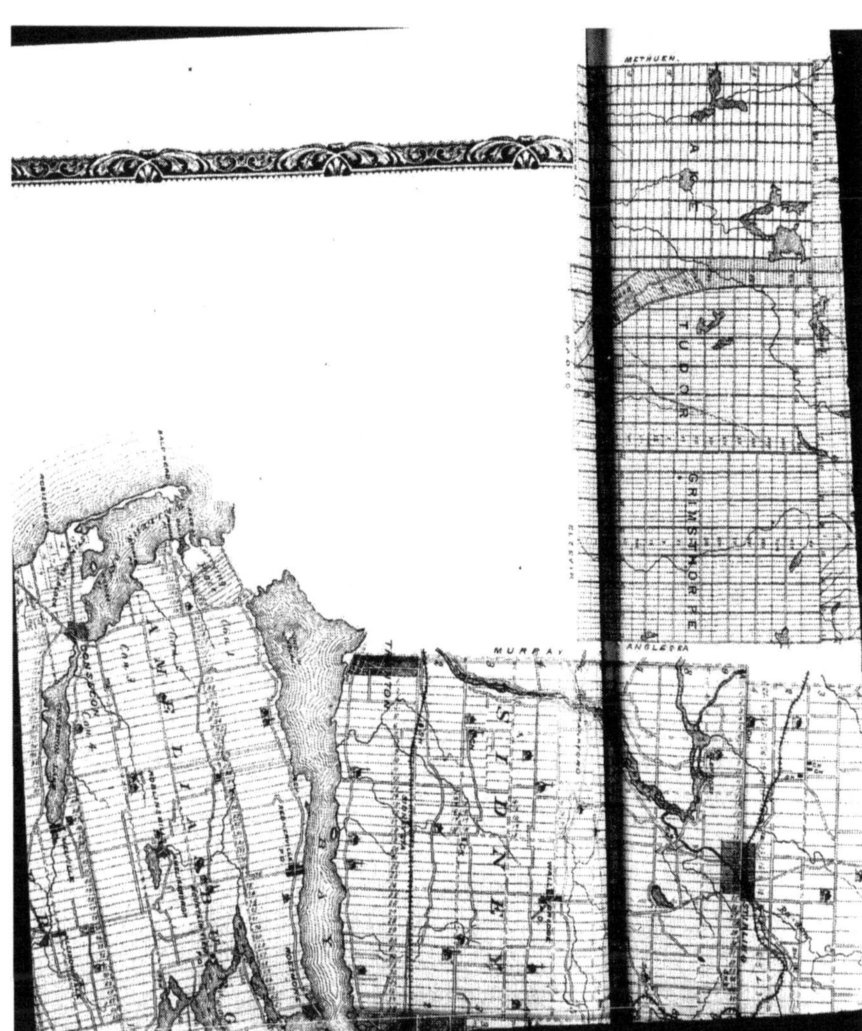

METHUEN
TUDOR
GRIMSTHORPE
MURRAY
ANGLESEA
SIDNEY

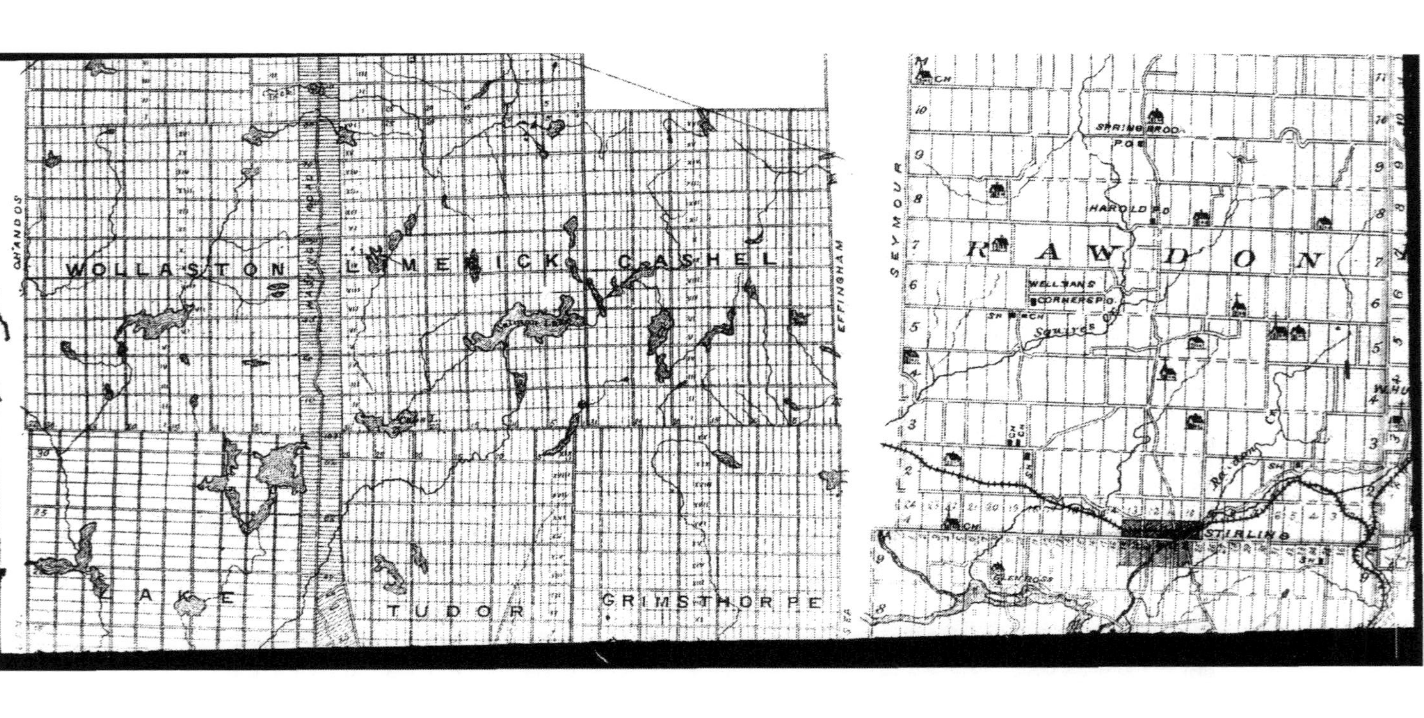
CHANDOS
WOLLASTON
LIMERICK
CASHEL
LAKE
TUDOR
GRIMSTHORPE
HASTINGS ROAD
EFFINGHAM
SEYMOUR
RAWDON
SPRING BROOK P.O.
HAROLD P.O.
WELLMANS CORNERS P.O.
STIRLING
GLEN ROSS

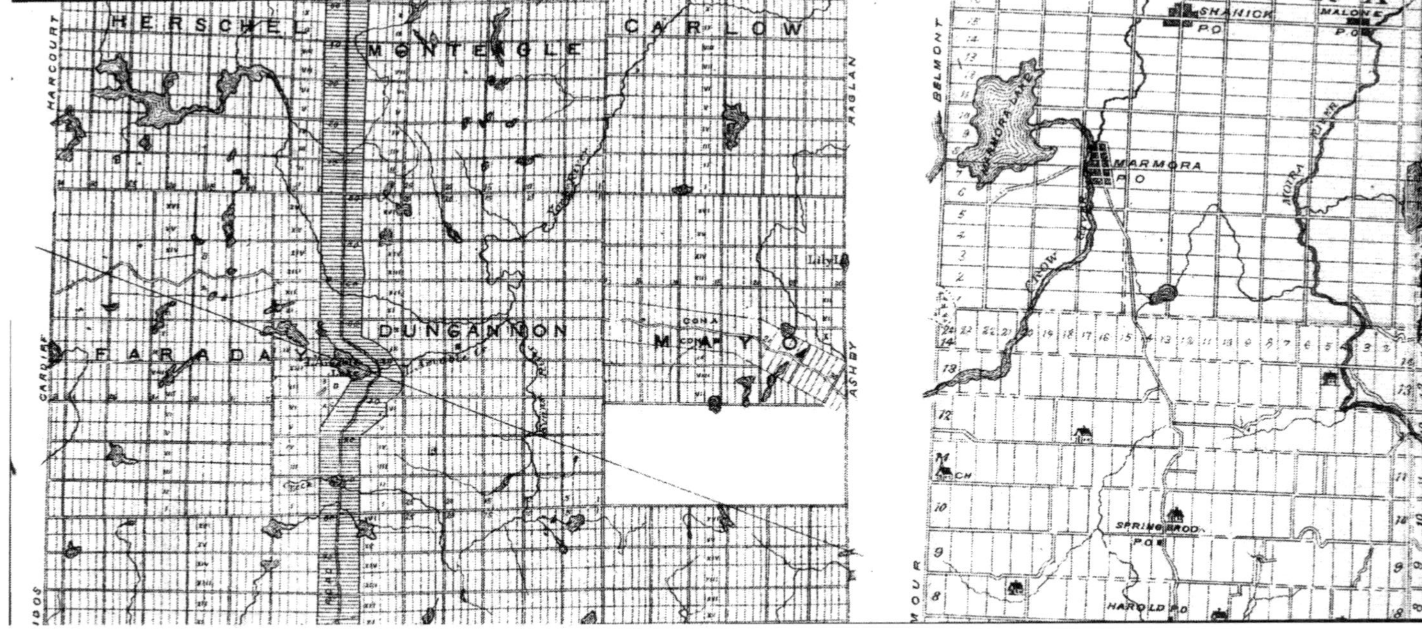
HERSCHEL
MONTEAGLE
CARLOW
FARADAY
DUNGANNON
MAYO
HARCOURT
CARDIFF
RAGLAN
ASHBY
BELMONT
MARMORA LAKE
MARMORA P.O.
CROW RIVER
MOIRA RIVER
SHANICK P.O.
MALONE P.O.
SPRINGBROOK P.O.
HAROLD P.O.

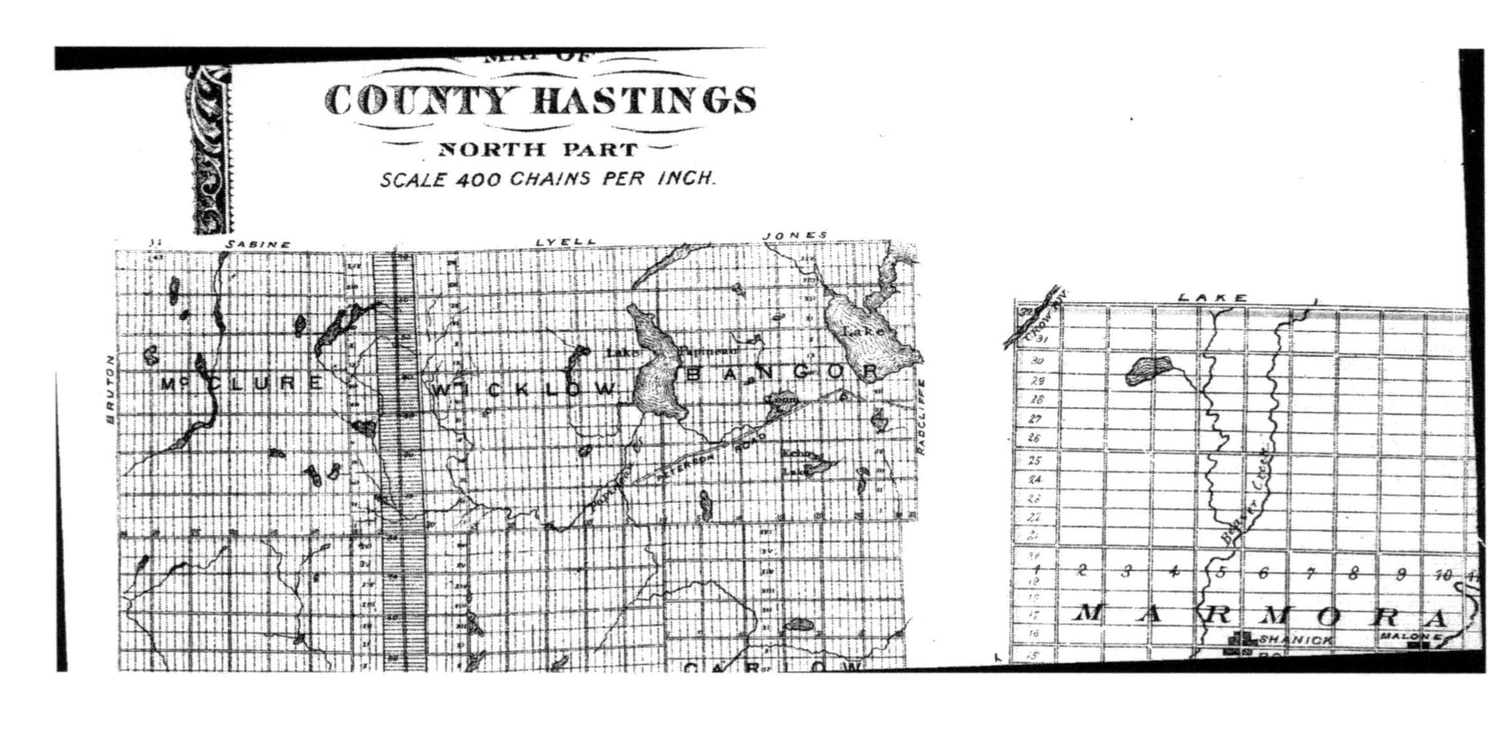

MAP OF
COUNTY HASTINGS
NORTH PART
SCALE 400 CHAINS PER INCH.
SABINE
LYELL
JONES
BRUTON
RADCLIFFE
Mc CLURE
WICKLOW
BANGOR
CARLOW
PETERSON ROAD
LAKE
CROW RIV.
Beaver Creek
MARMORA
SHANICK
MALONE

MAP OF

COUNTY HASTINGS

NORTH PART

SCALE 400 CHAINS PER INCH.

ONTARIO
BAY OF QUINTE
EAST LAKE
ATHOL
MARYSBURGH
CHERRY VALLEY
MILFORD
Long Point

QUINTE
HIG ISLAND
DEMORESTVILLE
SOPHIASBURGH
3rd Con.
LONG REACH
ADOLPHUSTOWN
HAY BAY
PICTON
BAY OF QUINTE
NORTH
MARYSBURGH
MILL POINT
COUNCIL

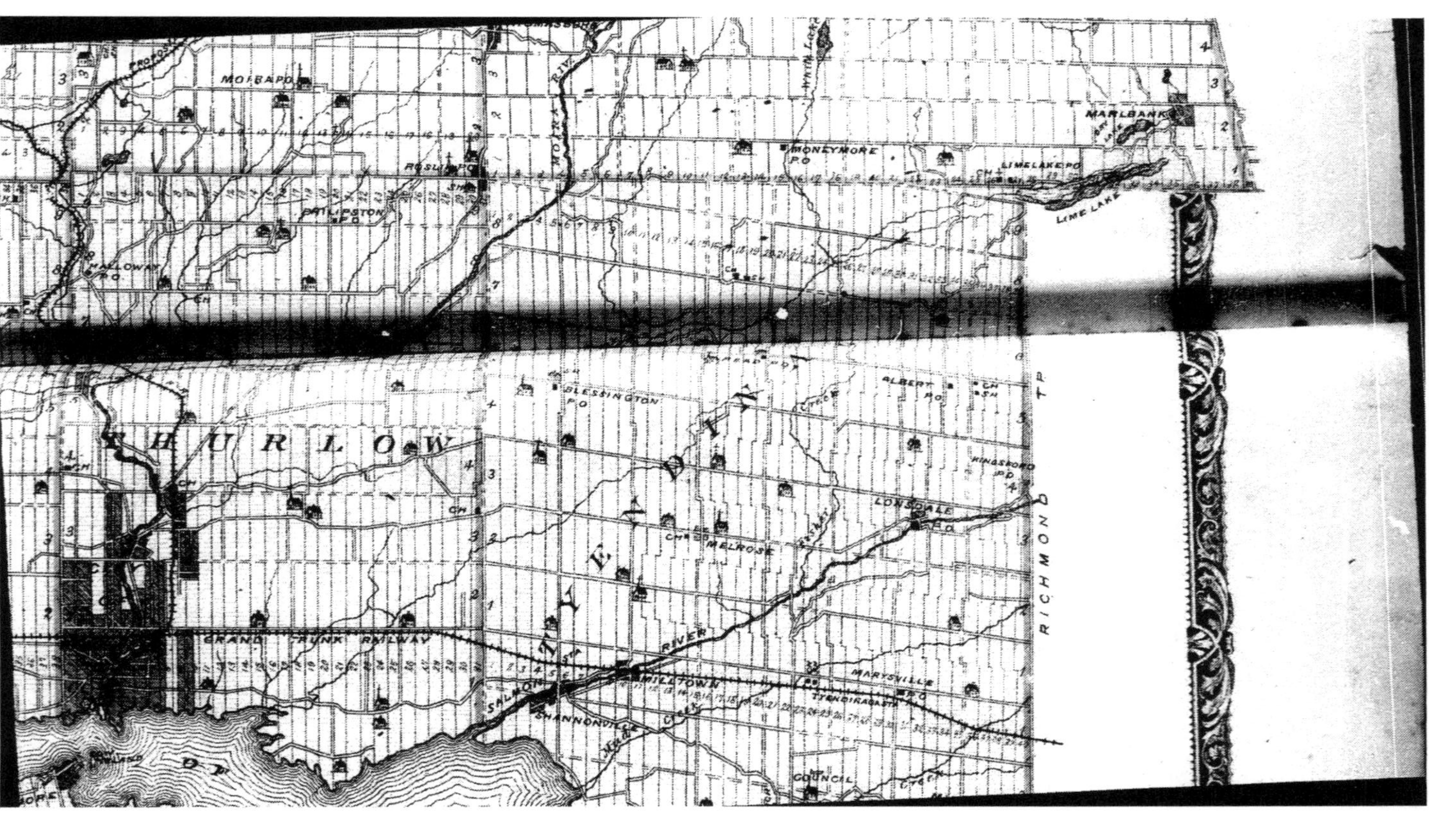

THURLOW
MOIRA PO
MONEYMORE PO
MARLBANK
LIMELAKE PO
LIME LAKE
ROSLIN PO
PHILIPSTON PO
HALLOWAY PO
BLESSINGTON PO
ALBERT PO
KINGSFORD PO
LONSDALE PO
MELROSE
GRAND TRUNK RAILWAY
SALMON RIVER
MILLTOWN
SHANNONVILLE
MARYSVILLE
COUNCIL
RICHMOND TP

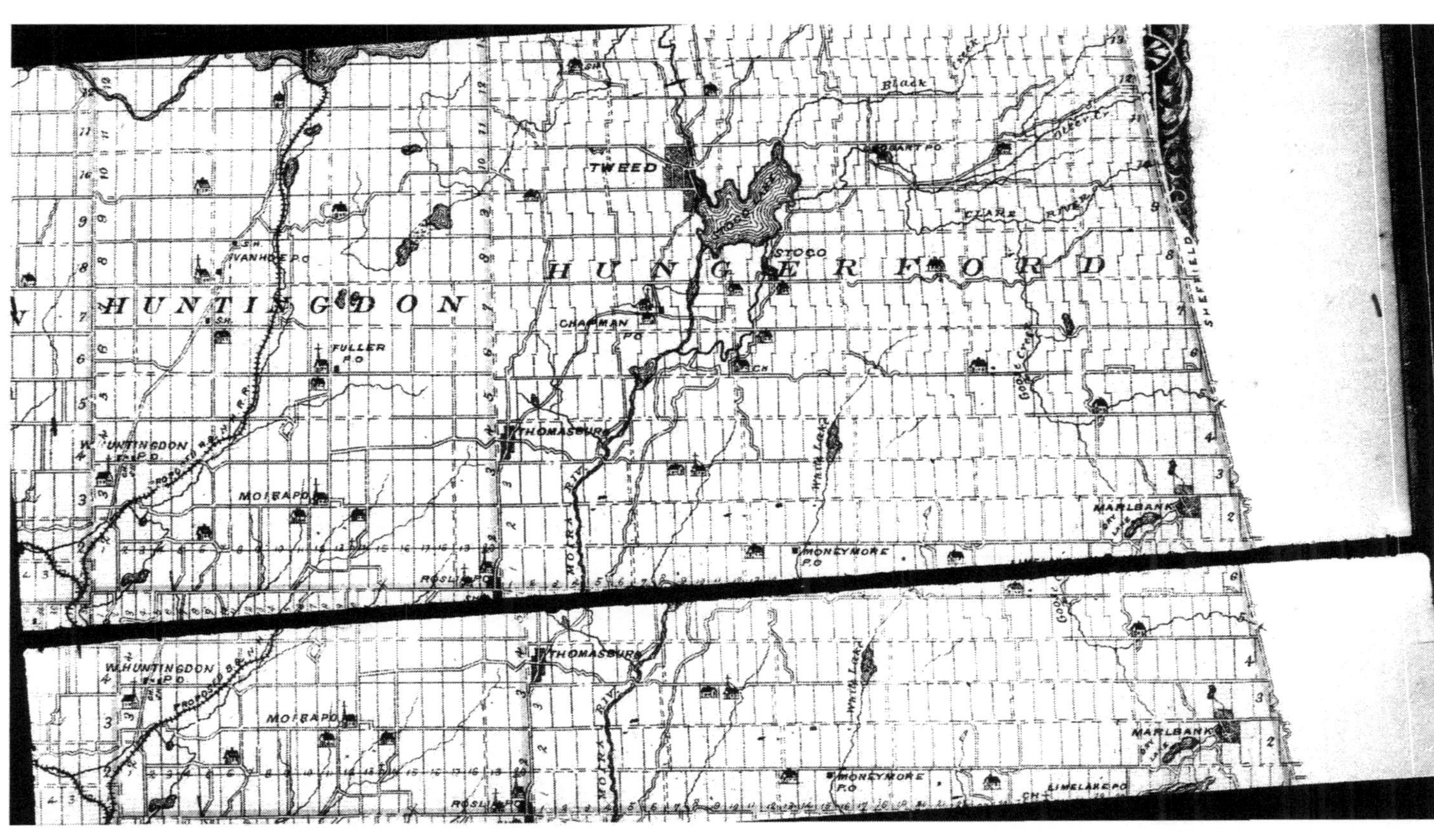
HUNTINGDON
HUNGERFORD
TWEED
STOCO
IVANHOE P.O
FULLER P.O
CHAPMAN P.O
THOMASBURG
W. HUNTINGDON P.O
MOIRA P.O
MOIRA RIV
ROSLIN P.O
MONEYMORE P.O
White Lake
Goose Creek
MARLBANK
LIMELAKE P.O
Black Creek
CLARE RIVER
SHEFFIELD

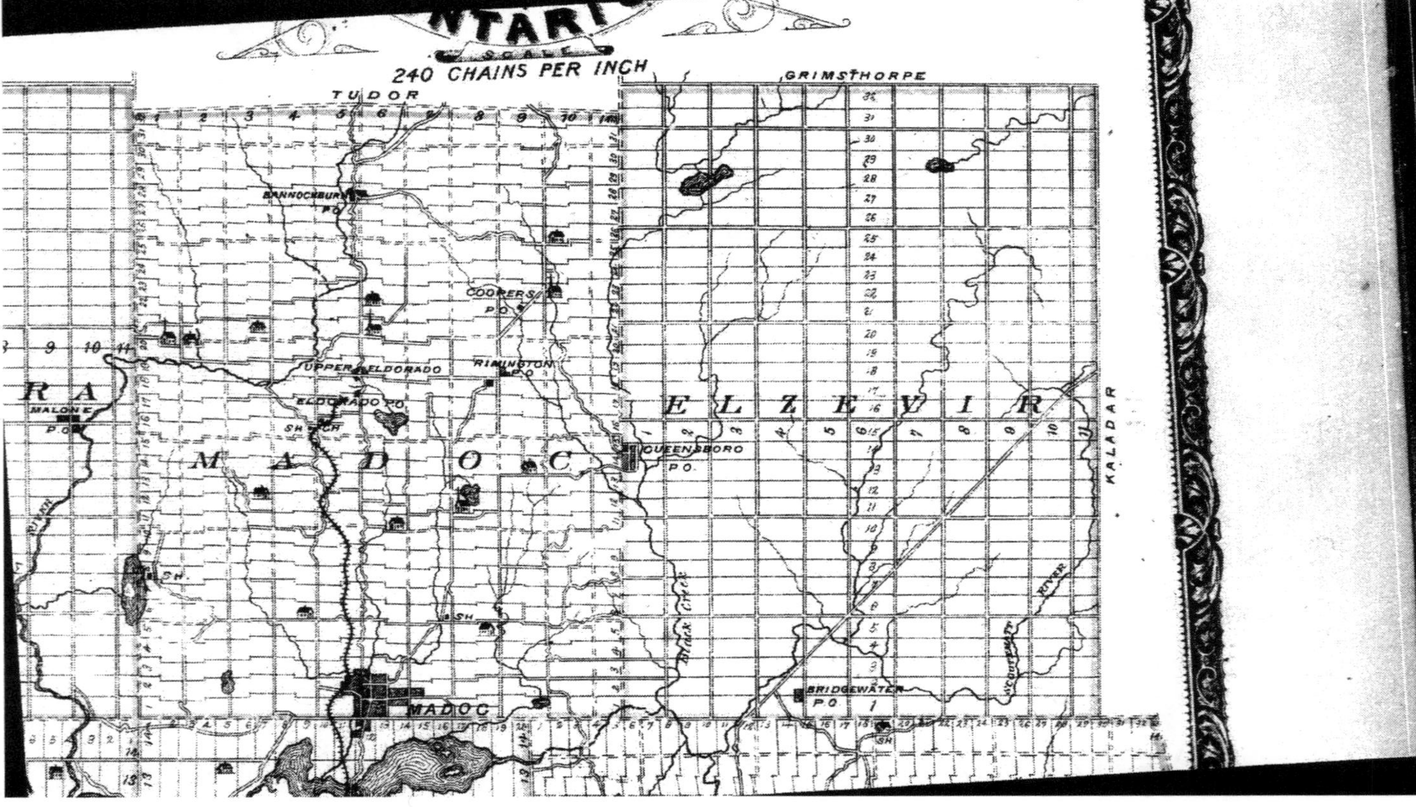
NTARI
SCALE
240 CHAINS PER INCH
TUDOR
GRIMSTHORPE
KALADAR
BANNOCKBURN P.O.
COOPERS P.O.
UPPER ELDORADO
RIMINGTON P.O.
ELDORADO P.O.
MALONE P.O.
R A
M A D O C
E L Z E V I R
QUEENSBORO P.O.
BRIDGEWATER P.O.
MADOC
Black Creek
SKOOTAMATT RIVER
SH.

MAP OF THE COUNTIES
OF
HASTINGS
AND
PRINCE EDWARD
ONTARIO.
SCALE
240 CHAINS PER INCH
TUDOR
GRIMSTHORPE

Walter Ross.
Picton M.P. 1878.

R. Low Q.
Picton.

Dr. W. C. Dorland, Ex. M. P. P.
(Deceased) Hallowell Tp.

G. Striker,
M.P.P. Picton.

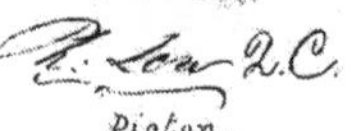

R. Low Q.C.
Picton.

Chas. Bockus.
Deceased, Picton.

"ADJIDAUMO," THE HOMESTEAD OF THE LATE HON. EDMUND MURNEY, NOW THE RES. OF HIS SON-IN-LAW, N. B. FALKNER, BARRISTER AT LAW, BELLEVILLE, ONT.

5

IMAGE EVALUATION
TEST TARGET (MT-3)

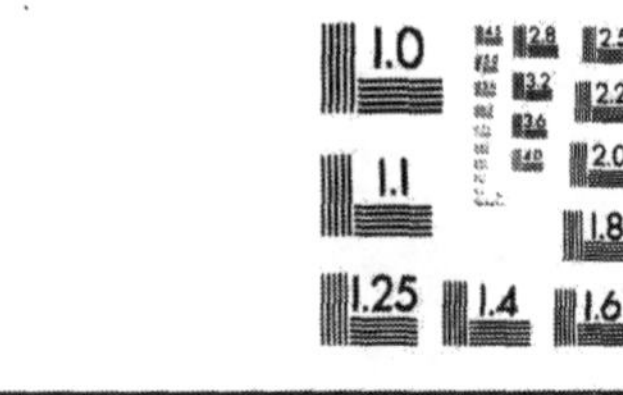

6"

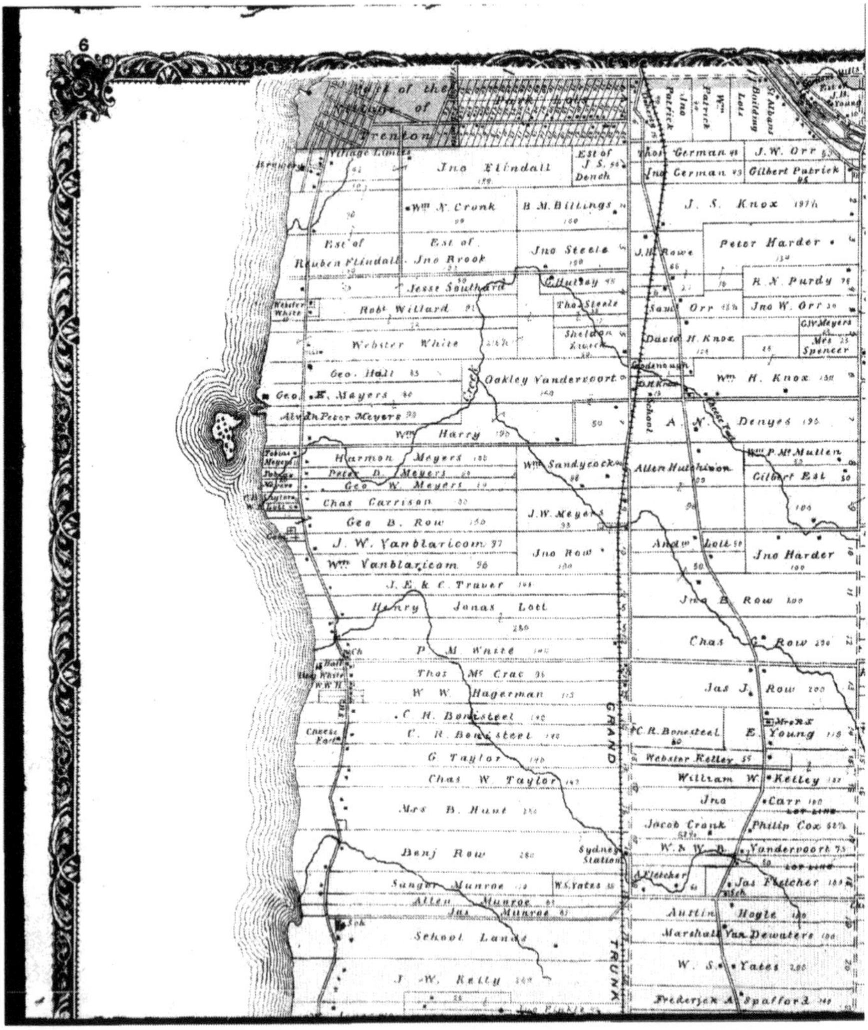

Part of the Village of Trenton

Village Limits

Brewery

Jno Flindall 180

Est of J. S. Dench 50

Thos German 41

J. W. Orr 53

Jno German 43

Gilbert Patrick 65

Jno Patrick

Patrick

Wm

Lots

Building

St Alban's

J. H. Young

Wm N. Cronk 90

B. M. Billings 100

J. S. Knox 197½

Est of Reuben Flindall

Est of Jno Brook

Jno Steele 190

J. H. Rowe

Peter Harder 134

Jesse Southard 50

C. Hulsey 48

R. N. Purdy 76

Robt Willard 92

Thos Steele

Saml Orr 114

Jno W. Orr 30

Webster White

G. W. Meyers

Webster White 226½

Sheldon Knox

David H. Knox

Mrs Spencer

Geo. Hall 85

Oakley Vandervoort 140

Wm H. Knox 150

Creek

Geo. E. Meyers 80

Alvah Peter Meyers 90

A. N. Denyes 195

School

Wm Harry 190

Tobias Meyers

Harmon Meyers 100

Wm Sandycock 98

Allen Hutchison

Wm P. McMullen

Peter D. Meyers

Gilbert Est 50

Geo W. Meyers 19

Chas Carrison 100

J. W. Meyers 93

Geo B. Row 150

J. W. Vanblaricom 97

Andw Lott 50

Wm Vanblaricom 96

Jno Row 100

Jno Harder 100

J. E. & C. Traver 145

Jno B. Row 100

Henry Jonas Lott 280

Chas C. Row 200

P. M. White 100

Thos Mc Crae 96

Jas J. Row 200

W. W. Hagerman 113

C. H. Bonisteel 140

C. R. Bonisteel 140

C. R. Bonesteel 80

Mrs E. Young 118

Cheese Factory

G. Taylor 140

Webster Kelley 55

Chas W. Taylor 142

William W. Kelley 152

Jno Carr 100

Mrs B. Hunt 250

Jacob Cronk 52½

Philip Cox 52½

Benj Row 280

Sydney Station

W. & W. B. Vandervoort 75

Sanger Munroe

W. S. Yates 55

A. Fletcher

Jas Fletcher 189

Allen Munroe

Jas Munroe

Austin Hoyle 100

School Lands

Marshall Van Dewaters 100

W. S. Yates 200

J. W. Kelly 240

Frederick A. Spafford 140

GRAND TRUNK

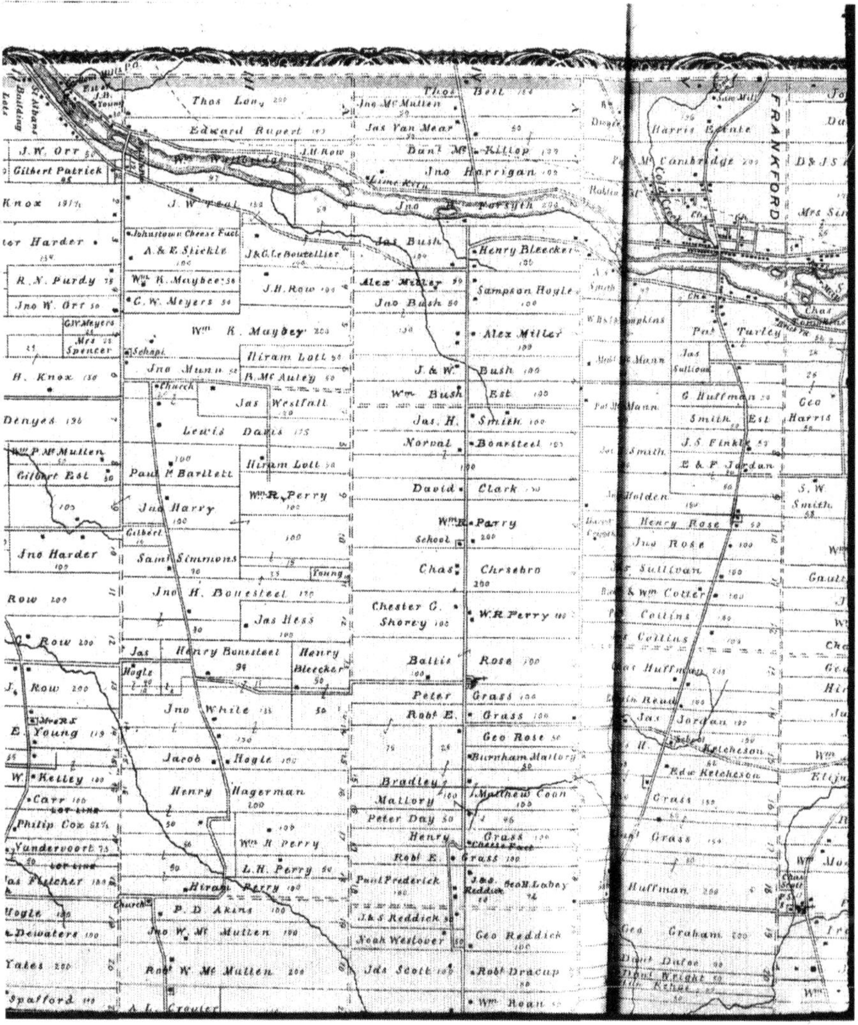

Thos Long 200
Jno McMullen
Thos Bell 184
Edward Rupert
Jas Van Mear
Harris Estate
FRANKFORD
J. W. Orr 50
J.H. Row
Danl McKillop
McCambridge
Gilbert Patrick
Wm Weatherdge
Jno Harrigan 100
Knox
Lime Kiln
J. W. Neal 150
Jno Forsyth 200
Mrs Sin
Peter Harder
Johnstown Cheese Fact
A. & E. Stickle 100
J.&G. Le Boutellier
Jas Bush 100
Henry Bleecker
R. N. Purdy 75
Wm K. Maybee 50
Alex Miller 50
Sampson Hogle 100
J.H. Row 100
Jno Bush 50
Jno W. Orr 50
C. W. Meyers 50
G.W. Meyers
Mrs Spencer
Wm K. Maybey 200
Alex Miller 100
Pat Turley
Hiram Lott 50
School
Jno Munn 50
Jas Sullivan
H. Knox 150
B. McAuley 50
J. & Wm Bush 100
Church
Jas Westfall
Wm Bush Est 100
G. Huffman 50
Geo Harris
Denyes 196
Jas. H. Smith 100
Smith Est
Lewis Davis 175
Wm P. McMullen
Norval Boarsteel
J. S. Finkle
Gilbert Est
Hiram Lott 50
E & F Jordan
Paul F. Bartlett
Wm R. Perry
David Clark
Jas Harry 100
S. W. Smith
Holden
Wm R. Perry
School
Henry Rose
Jno Harder 100
Gilbert
Sam Simmons
Jno Rose
Young
Chas Chesebro 200
Sullivan
Row 200
Jno H. Bonesteel 170
Wm Cotter
Chester C. Shorey 100
Jas Hess 100
W. R. Perry 100
Collins
Row 200
Jas Hogle
Henry Bonesteel
Henry Bleecker
Collins
Battis Rose 100
Huffman
Row 200
Peter Grass 100
Jno White
Robt E. Grass 100
Jas Jordan
E. Young 119
Geo Rose 50
Ketcheson
Jacob Hogle 100
Burnham Mallory
W. Kelley 100
Bradley
Edw Ketcheson
Carr 100
Henry Hagerman
Mallory
Matthew Coon 100
Grass
Philip Cox
Peter Day 50
Grass
Vandervoort 75
Henry Grass
Wm H Perry
Cheese Fact
Robt E. Grass 100
Jas Fletcher 100
L. H. Perry 50
Paul Frederick
Jas Reddick
Geo H. Lahey
Hiram Perry 100
Huffman 200
Church
Hogle
P. D. Akins 100
J.&S. Reddick
Dewaters 100
Jno W. McMullen 100
Noah Westover
Geo Reddick 100
Geo Graham 200
Yates 200
Robt W. McMullen 200
Jas Scott 100
Robt Dracup
Danl Dafoe
Danl Wright
Spafford
A. L. Crouter
Wm Roan 50

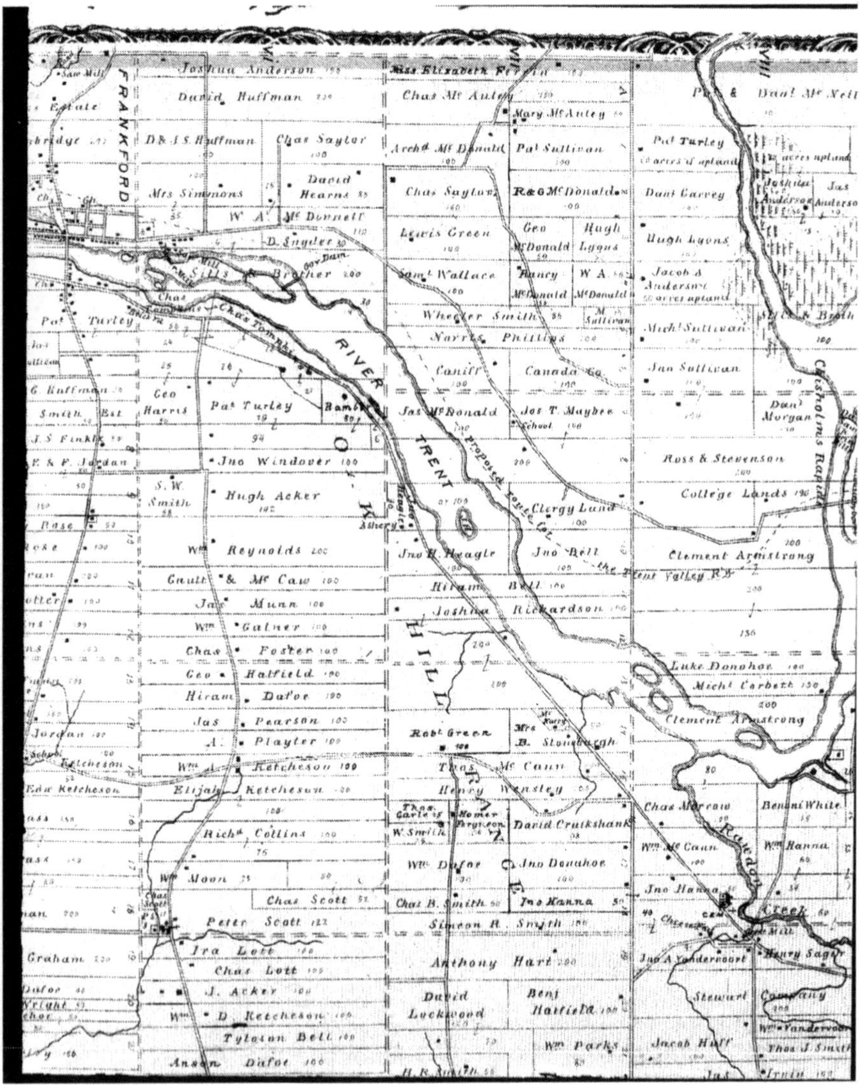

Saw Mill
Estate
FRANKFORD
Joshua Anderson
David Huffman
D & J. S. Huffman
Chas Saylor
Mrs Simmons
David Hearns
W. A. Mc Donnell
D. Snyder
Sills & Brother
Gov Dam
Pat Turley
Chas Tompkins
G. Huffman
Smith Est
J. S. Finkle
E. & F. Jordan
Geo Harris
Pat Turley
Bamber
Jno Windover
S. W. Smith
Hugh Acker
Wm Reynolds
Gault & Mc Caw
Jas Munn
Wm Galner
Chas Foster
Geo Hatfield
Hiram Dafoe
Jas Pearson
A. Playter
Wm A. Ketcheson
Elijah Ketcheson
Ketcheson
Edw Ketcheson
Jordan
School
Richd Collins
Wm Moon
Chas Scott
Peter Scott
Graham
Ira Lott
Chas Lott
J. Acker
Wm D. Ketcheson
Tyleson Bell
Anson Dafoe
Dafoe
Wright
RIVER
TRENT
OAK
HILL
RANGE
Miss Elizabeth Ferris
Chas Mc Auley
Mary Mc Auley
Archd Mc Donald
Pat Sullivan
Chas Saylor
R & G Mc Donald
Lewis Green
Geo McDonald
Hugh Lyons
Saml Wallace
Nancy McDonald
W. A. McDonald
Wheeler Smith
M. Sullivan
Norris Phillips
Caniff
Canada Co.
Jas Mc Donald
Jos T. Maybee
School
Proposed route of the Trent Valley R.R.
Clergy Land
Ashery
Jno H. Haagle
Jno Bell
Hiram Bell
Joshua Richardson
Robt Green
Mrs B. Stoneburgh
Thos Mc Cann
Henry Wensley
Thos Carlisle
Homer Ferguson
W. Smith
David Cruikshank
Wm Dafoe
Jno Donohoe
Chas B. Smith
Jno Hanna
Simeon R. Smith
Anthony Hart
David Lockwood
Benj Hatfield
Wm Parks
H. R. Smith
Pat & Danl Mc Neil
Pat Turley
acres of upland
Joshua Anderson
Jas Anderson
Danl Garvey
Hugh Lyons
Jacob & Anderson
acres upland
Mich Sullivan
Sills & Broth
Jas Sullivan
Chisholms Rapids
Danl Morgan
Ross & Stevenson
College Lands
Clement Armstrong
Luke Donohoe
Michl Carbett
Clement Armstrong
Chas Morrow
Benjn White
Wm Mc Cann
Wm Hanna
Jno Hanna
Rawdon Creek
Cheese
Saw Mill
Jno A. Vandervoort
Henry Sager
Stewart Company
Wm Vandervoort
Jacob Huff
Thos J. Smith
Irwin

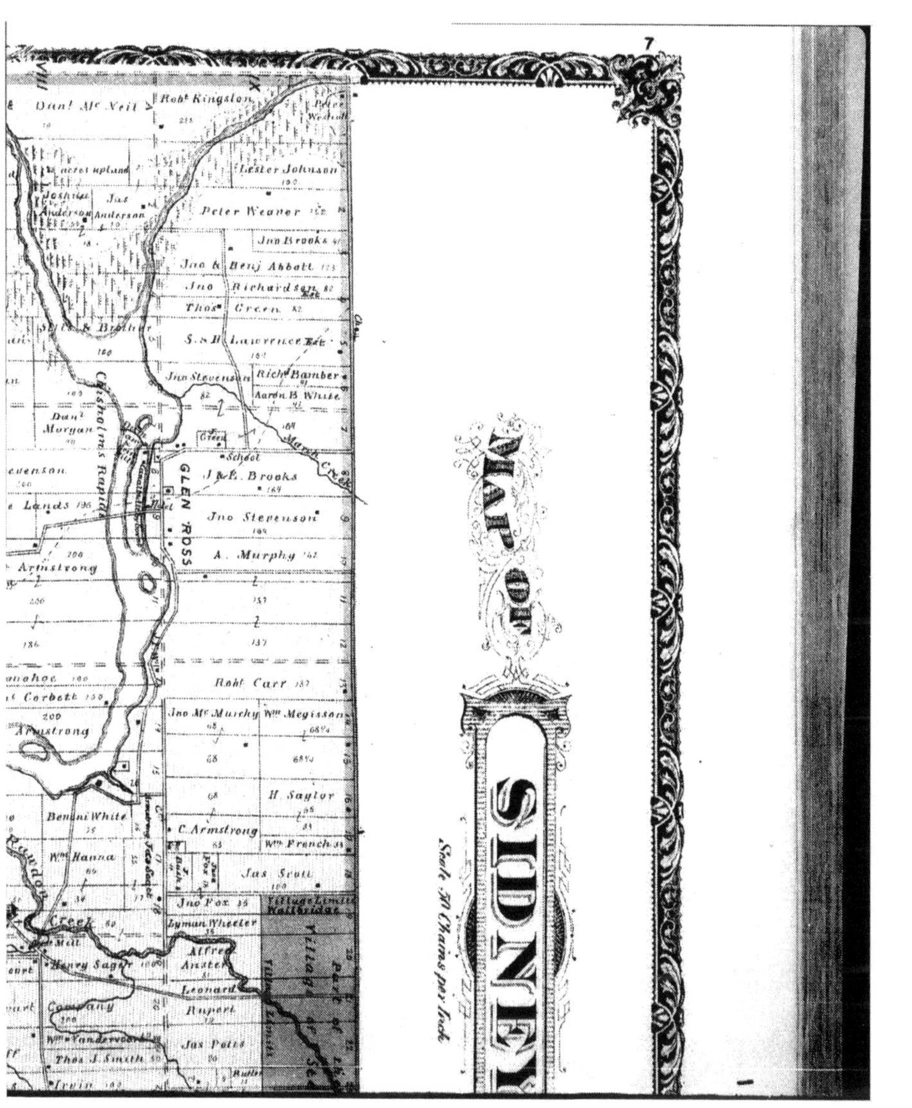
MAP OF
SIDNEY
Scale 50 Chains per Inch
Danl Mc Neil
Robt Kingston
Lester Johnson
Joshua Anderson
Jas Anderson
Peter Weaver
Jno Brooks
Jno & Benj Abbott
Jno Richardson
Thos Green
S. & B. Lawrence
Jno Stevenson
Richd Bamber
Aaron B White
Danl Morgan
Chisholms Rapids
GLEN ROSS
School
J & E. Brooks
Jno Stevenson
A. Murphy
Armstrong
Robt Carr
Corbett
Armstrong
Jno Mc Murchy
Wm Megisson
H. Saylor
C. Armstrong
Wm French
Benjni White
Wm Hanna
Jas Scott
Jno Fox
Lyman Wheeler
Village Limits Wallbridge
Alfred Anster
Henry Sager
Leonard
Rupert
Company
Wm Vandervoort
Jas Potts
Thos J. Smith
Irvin
Marsh Creek
Creek
Mill

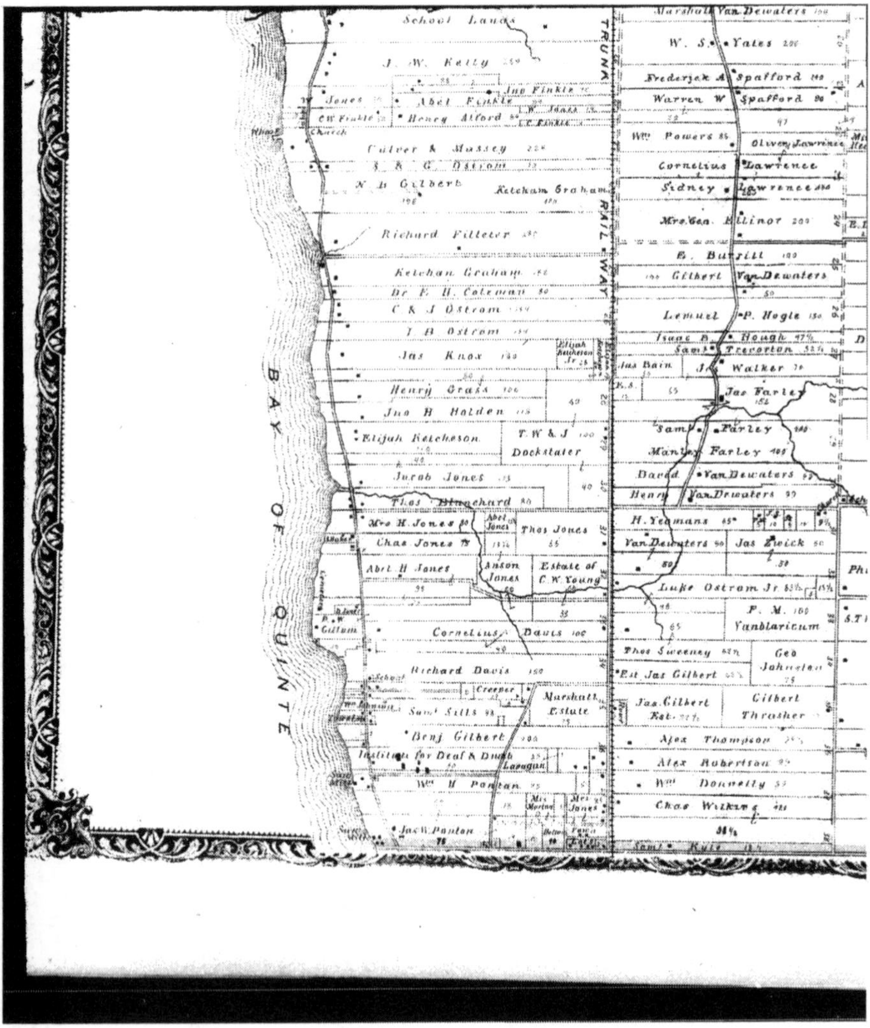

BAY OF QUINTE

TRUNK RAILWAY

School Lands

J. W. Kelly 250

Jno Finkle

W. Jones

Abel Finkle

C.W. Finkle

Henry Alford

Church

Culver & Massey 228

S. & G. Ostrom

N. B. Gilbert 198

Ketcham Graham

Richard Filleter

Ketchan Graham

Dr F. H. Coleman 50

C. & J. Ostrom

J. B. Ostrom

Elijah Ketcheson Jr

Jas Knox 140

Henry Grass 100

Jno B. Holden

Elijah Ketcheson

T. W. & J. Dockstater

Jacob Jones

Thos Blanchard 80

Mrs H. Jones

Abel Jones

Thos Jones

Chas Jones

Abel H. Jones

Anson Jones

Estate of C. W. Young

P. W. Gillum

Cornelius Davis 100

Richard Davis 150

School

Creeper

Marshall Estate

Saml Sills

Benj Gilbert

Institute for Deaf & Dumb

Lanagan

Wm H. Panton

Mrs Morton

Mrs Jones

Jas W. Panton 75

Marshall Van Dewaters 100

W. S. Yates 200

Frederick A. Spafford

Warren W. Spafford 90

Wm Powers

Oliver Lawrence

Cornelius Lawrence

Sidney Lawrence

Mrs Geo. Ellinor 200

E. Burrill 100

Gilbert Van Dewaters

Lemuel P. Hogle 150

Isaac B. Hough 47½

Saml Treverton

Jas Bain

Jas Walker

Jas Farley

Saml Farley

Manley Farley

David Van Dewaters

Henry Van Dewaters

H. Yeomans

Van Dewaters

Jas Zwick

Luke Ostrom Jr.

P. M. Vanblaricum

Thos Sweeney

Geo Johnston

Est Jas Gilbert

Jas Gilbert Est.

Gilbert Thrasher

Alex Thompson

Alex Robertson

Wm Donnelly

Chas Wilkins

Saml Kyle

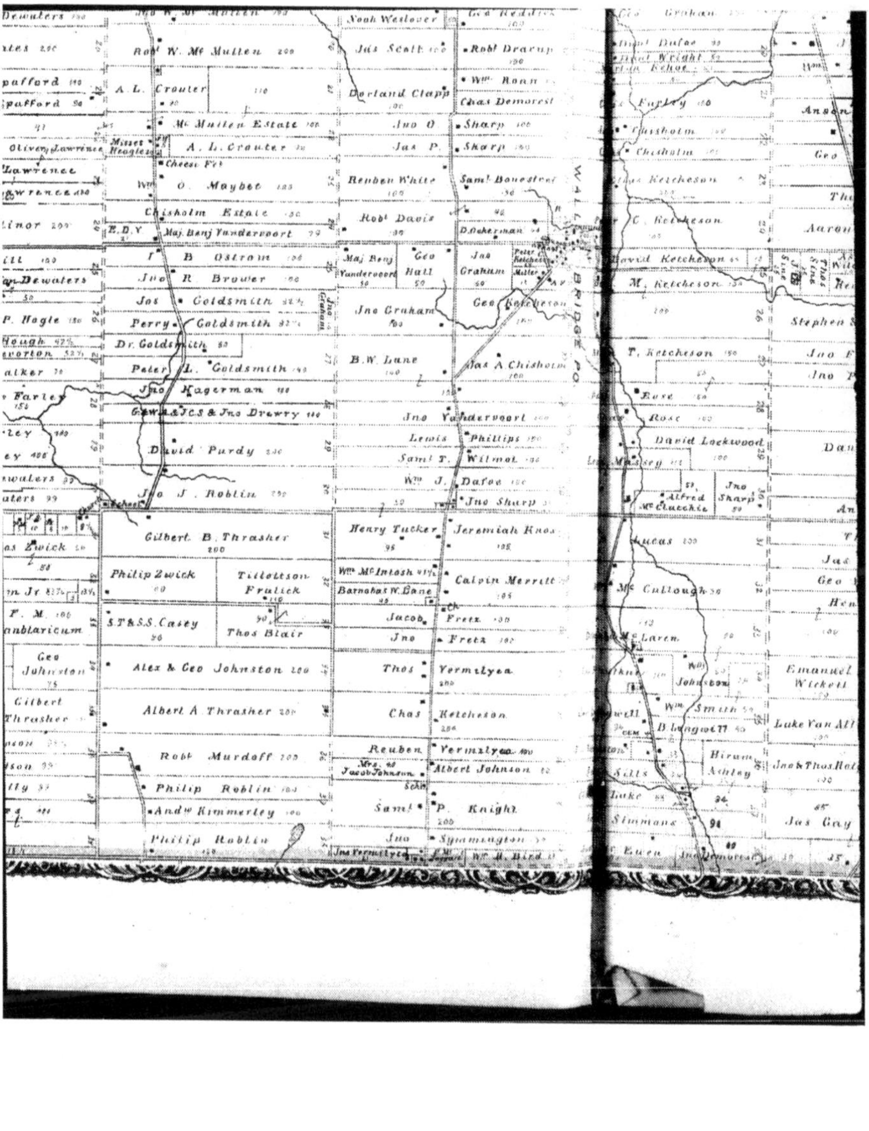

Dewaters 150
Spafford 110
Spafford 90
Oliver Lawrence
Lawrence
Lawrence 150
Minor 200
Dewaters
P. Hagle 150
Hough 92½
Overton 52½
Walker
Farley
Dewaters 99
Zwick
F. M.
Canbloricum
Geo Johnston
Gilbert Thrasher

Robt W. Mc Mullen 200
A. L. Crouter
Mc Mullen Estate
Misset Heagles
A. L. Crouter
Cheese Fct
Wm O. Maybee
Chisholm Estate
E. D. Y.
Maj. Benj Vandervoort
I. B. Ostrom 100
Jno R. Brower 100
Jos Goldsmith 82½
Perry Goldsmith 82½
Dr. Goldsmith
Peter L. Goldsmith
Jno Hagerman
G. W. & J. C. S. & Jno Drewry
David Purdy
Jno J. Roblin
School
Gilbert B. Thrasher 200
Philip Zwick
Tillotson Frulick
S. T. & S. S. Casey
Thos Blair
Alex & Geo Johnston 200
Albert A. Thrasher 200
Robt Murdoff 200
Philip Roblin
Andw Kimmerley
Philip Roblin

Noah Westover
Jas Scott 100
Dorland Clapp
Jno O. Sharp
Jas P. Sharp
Reuben White
Robt Davis
Maj Benj Vandervoort
Geo Hall
Jno Graham
Jno Graham
B. W. Lane
Jno Vandervoort
Lewis Phillips
Saml T. Wilmot
Wm J. Dafoe
Henry Tucker
Wm Mc Intosh
Barnabas W. Bane
Jacob Fretz
Jno Fretz
Thos Vermilyea
Chas Ketcheson
Reuben Vermilyea
Mrs. Jacob Johnson
Saml P. Knight
Jno Symmington
Jno Vermilyea

Geo Reddick
Robt Drarup
Wm Roan
Chas Demorest
Sharp
Sharp
Saml Bonestreet
D. Ackerman
Peter Ketcheson
Miller
Jno Graham
Geo Ketcheson
Jas A. Chisholm
Jno Sharp
Jeremiah Knox
Calvin Merritt
Albert Johnson
Wm H. Bird

WALLBRIDGE P.O.

Geo Graham
Danl Dafoe
Danl Wright
Kehoe
Farley
Chisholm
Chisholm
Ketcheson
C. Ketcheson
David Ketcheson
M. Ketcheson
T. Ketcheson
Rose
Rose
David Lockwood
Massey
Alfred Mc Cutcheon
Jno Sharp
Lucas 200
Mc Cullough
Mc Laren
Johnston
Wm Smith
Blangwell
Hiram Ashley
Sills
Lake
Simmons
Euen
Jno Demorest

Anson
Geo
Aaron
Thos Sine
Stephen
Jno
Jno
Dan
Jas
Geo
Emanuel Wickett
Luke Van Al
Jno & Thos Holt
Jas Gay

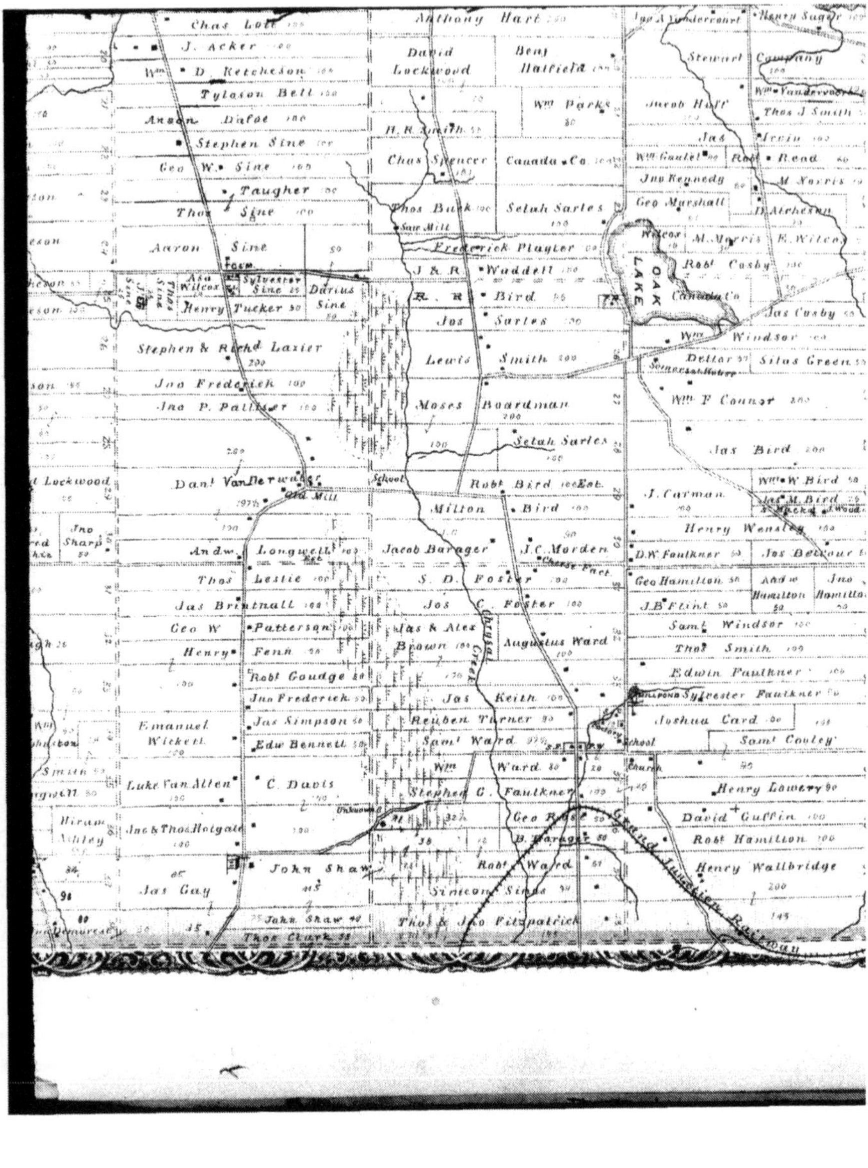

Chas Lott 100
J. Acker 100
Wm D Ketcheson 100
Tyloson Bell 100
Anson Dafoe 100
Stephen Sine 100
Geo W. Sine 100
Thos Taugher Sine 100
Aaron Sine
Asa Wilcox
Sylvester Sine
Darius Sine
Henry Tucker 50
Stephen & Richd Lazier 200
Jno Frederick 100
Jno P. Pallister 100
Dan Van Derwater
Old Mill
Lockwood
Jno Sharp 50
Andw Longwell
Thos Leslie 100
Jas Brintnall 100
Geo W Patterson
Henry Fenn 50
Robt Goudge
Jno Frederick
Jas Simpson
Emanuel Wickett
Edw Bennett
Luke Van Allen
C. Davis
Hiram Ashley
Jno & Thos Holgate
John Shaw
Jas Gay
John Shaw
Thos Clark
Demorest
Smith
Johnston
Anthony Hart 200
David Lockwood
Benj Hatfield
Wm Parks
H. R. Smith
Chas Spencer
Canada Co
Thos Buck
Saw Mill
Selah Sarles 100
Frederick Playter
J & R Waddell
R. R. Bird
Jos Sarles
Lewis Smith 200
Moses Boardman 200
Selah Sarles
School
Robt Bird Est.
Milton Bird
Jacob Barager
J. C. Morden
Cheese Fact.
S. D. Foster
Jos C. Foster
Jas & Alex Brown
Augustus Ward
Chrysal Creek
Jas Keith
Reuben Turner
Saml Ward
Wm Ward
Stephen C. Faulkner
Geo Rose
B. Barager
Robt Ward
Simeon Sills
Thos & Jno Fitzpatrick
Jno A. Vandercourt
Henry Sager
Stewart Company
Wm Vandervoort
Jacob Huff
Thos J Smith
Jas Irvin
Wm Gaulet
Robt Read
Jno Kennedy
M Norris
Geo Marshall
D Aitcheson
Wilcox
M. Morris
E. Wilcox
OAK LAKE
Robt Casby
Canada Co
Jas Casby
Wm Windsor
Dettor
Silas Green
Wm F Connor
Jas Bird 200
J. Carman
Wm W Bird
Jas M Bird
Henry Wensley
D. W. Faulkner
Jas Belfour
Geo Hamilton
Andw Hamilton
Jno Hamilton
J. B. Flint
Saml Windsor
Thos Smith
Edwin Faulkner
Sylvester Faulkner
Joshua Card
School
Saml Cooley
Church
Henry Lowery
David Guffin
Robt Hamilton
Henry Wallbridge 200
Grand Junction Railway

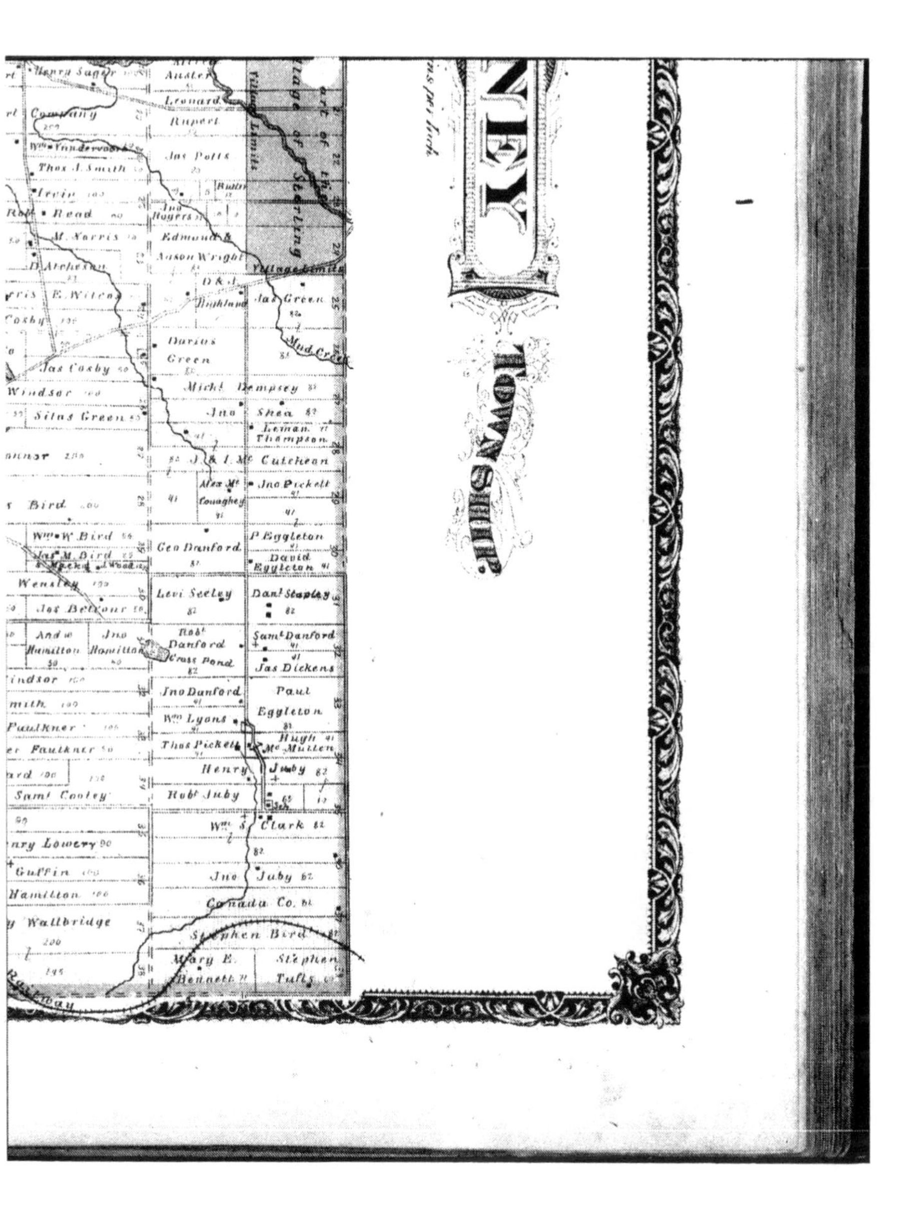

Henry Sager
Auster
Leonard
Rupert
Jas Potts
Thos J. Smith
Irvin
Robt Read
M. Norris
Edmond &
Anson Wright
D Atcheson
E. Wilcox
Cosby
D & J.
Highland
Jas Green
Mud Creek
Darius
Green
Jas Cosby
Windsor
Micht Dempsey
Jno Shea
Silas Green
Leman Thompson
J & I. Mc Cutcheon
Jno Pickett
Alex Mc
Coughey
Bird
Wm W Bird
Jas M. Bird
Geo Danforth
P Eggleton
David
Eggleton
Wensley
Levi Seeley
Danl Stapley
Jas Belfour
Andw
Hamilton
Jno
Hamilton
Robt
Danford
Cross Pond
Saml Danford
Jas Dickens
Windsor
Jno Danford
Paul
Eggleton
Smith
Wm Lyons
Faulkner
Hugh
Mc Mullen
Thos Pickett
Henry Juby
Saml Cooley
Robt Juby
Wm S. Clark
Henry Lowery
Guffin
Jno Juby
Hamilton
Canada Co.
Wallbridge
Stephen Bird
Mary E.
Bennett
Stephen
Tufts
Railway
Village Limits
Part of the Village of Sterling
TOWNSHIP
NEY

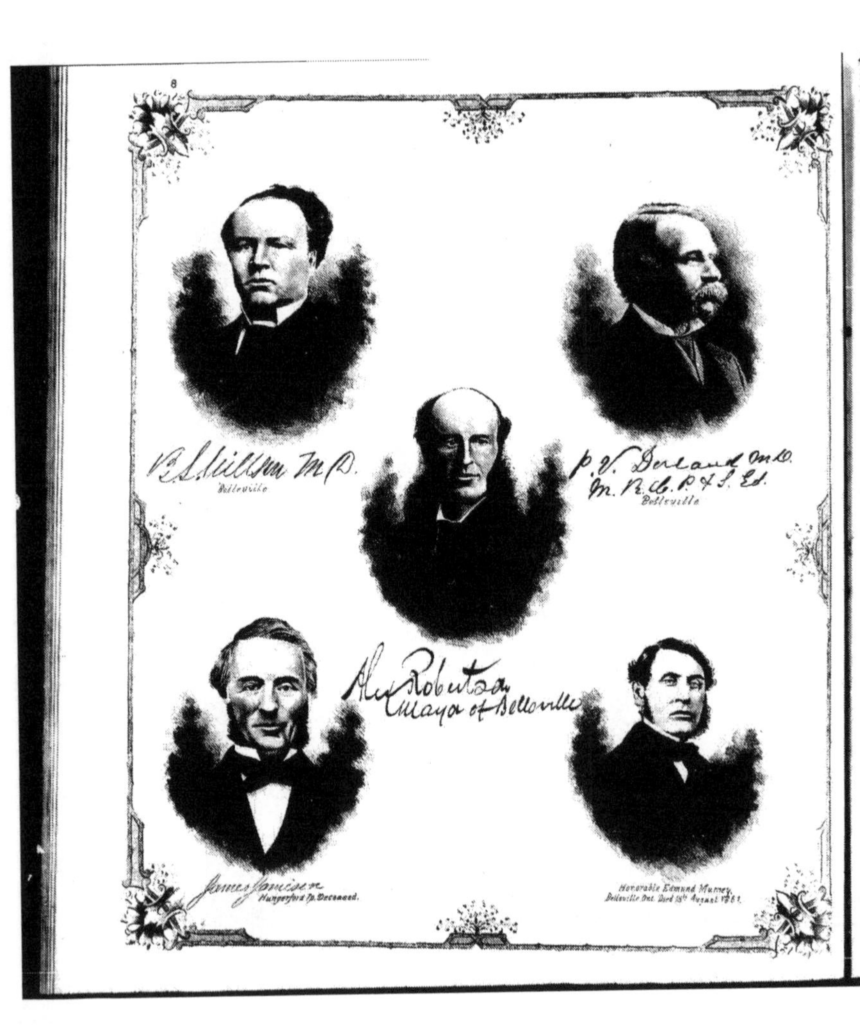
Belleville
P. V. Dorland M.D.
M.R.C.P.&S. Ed.
Belleville
Alex Robertson
Mayor of Belleville
James Jamieson
Hungerford Tp. Deceased.
Honorable Edmund Murney,
Belleville Ont. Died 15th August 1861.

"SUMMER HILL", RES. OF J. PLATT NASH, M.D. PICTON, ONT.

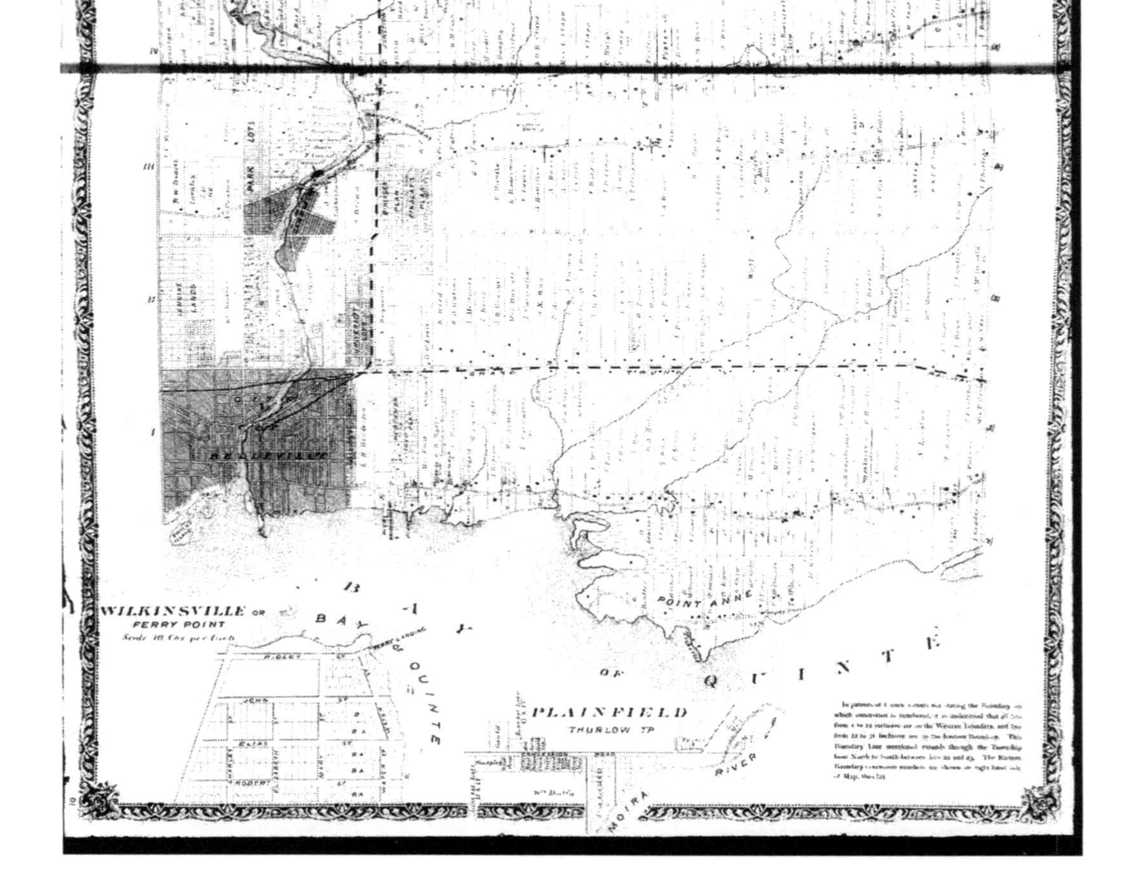
WILKINSVILLE OR FERRY POINT
BAY OF QUINTE
PLAINFIELD
THURLOW TP
POINT ANNE
MOIRA RIVER
PARK LOTS
UNIVERSITY LOTS
JOHN
WATER ST
ROBERT

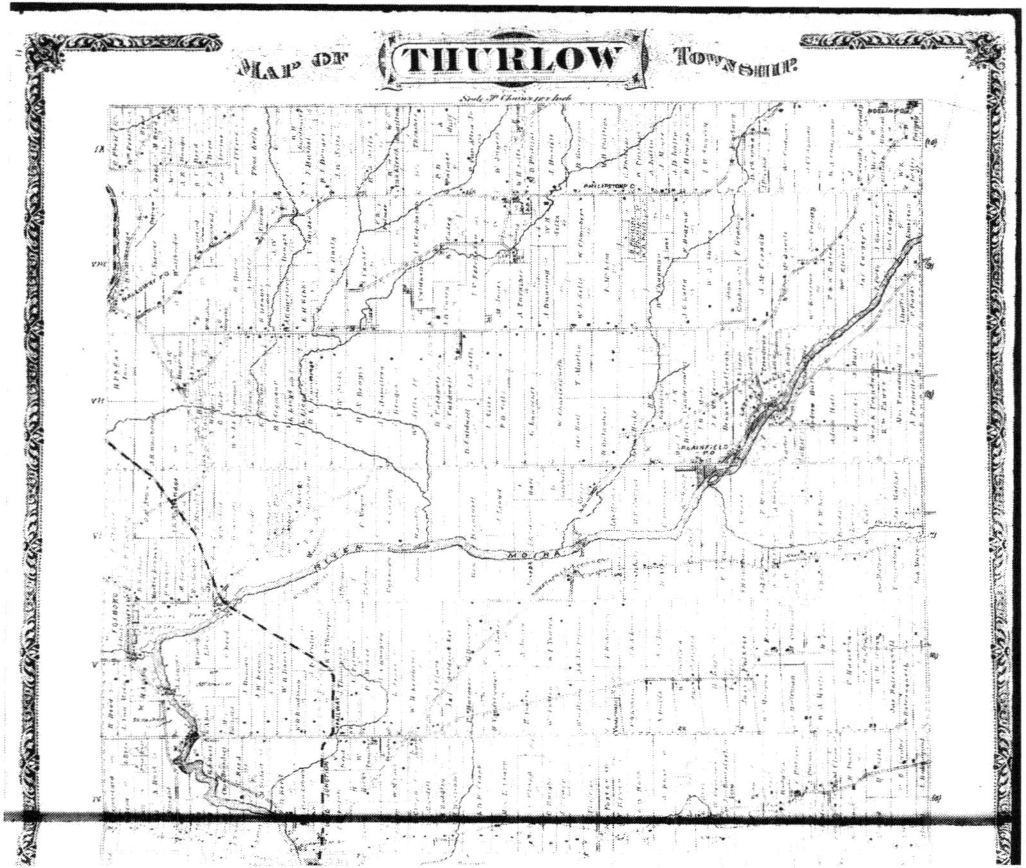

MAP OF THURLOW TOWNSHIP

Scale 80 Chains to 1 Inch

"VILLENEUVE HOUSE", RES. OF PH. LOW, ESQ. Q.C. PICTON, ONTARIO CANADA.

"THE CLAPP PLACE", RES. OF ROBT CLAPP, CON. 1. LOT 26, N.B.R. SOUTH MARYSBURG TP. ONT.

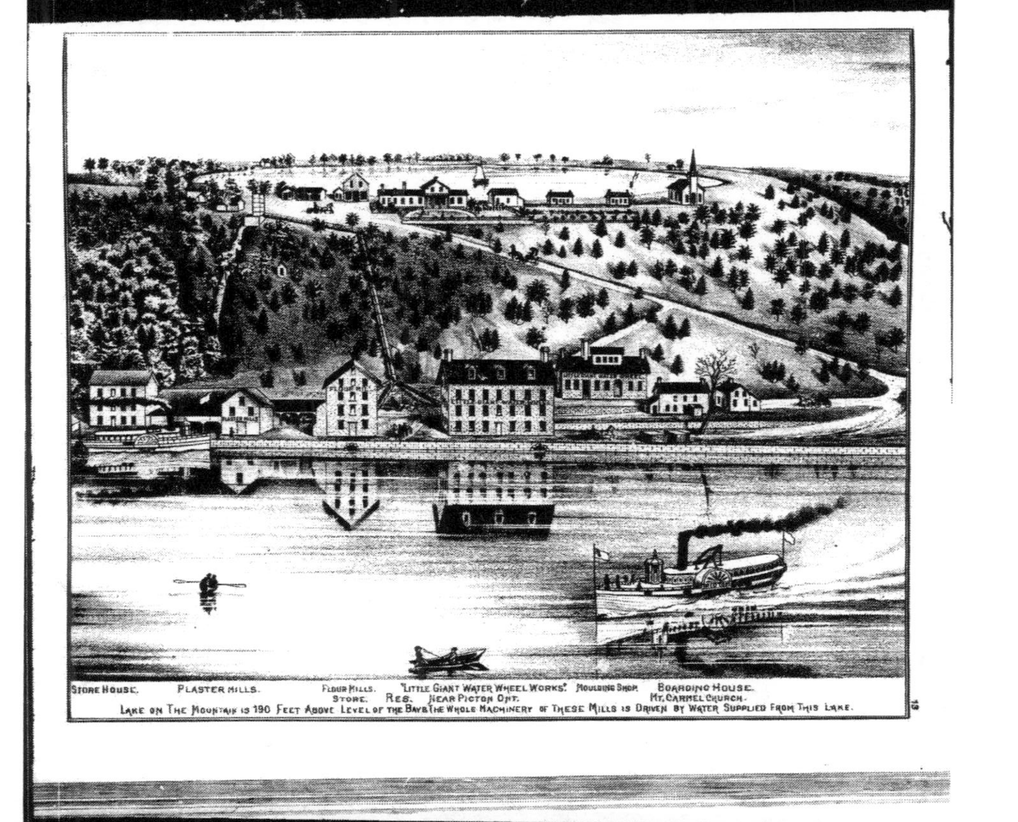
STORE HOUSE.
PLASTER MILLS.
FLOUR MILLS.
STORE.
"LITTLE GIANT WATER WHEEL WORKS."
RES. NEAR PICTON ONT.
MOULDING SHOP.
BOARDING HOUSE.
MT. CARMEL CHURCH.
LAKE ON THE MOUNTAIN IS 190 FEET ABOVE LEVEL OF THE BAY & THE WHOLE MACHINERY OF THESE MILLS IS DRIVEN BY WATER SUPPLIED FROM THIS LAKE.

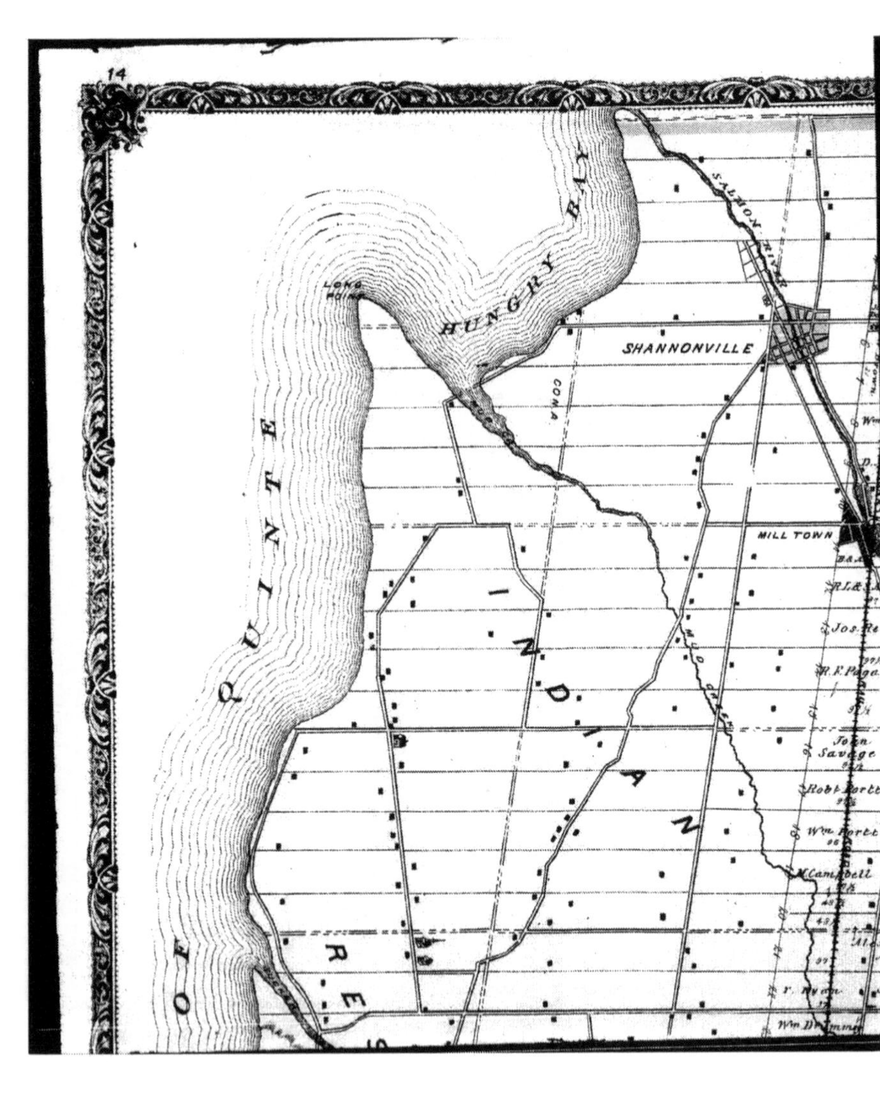
14
HUNGRY BAY
LONG POINT
SHANNONVILLE
SALMON RIVER
COMA
MILL TOWN
QUINTE
OF
INDIAN
RES
MUD CREEK
Jos. Re
R. F. Paga
John Savage
Robt. Portt
Wm. Portt
M. Campbell
Wm. Drummond

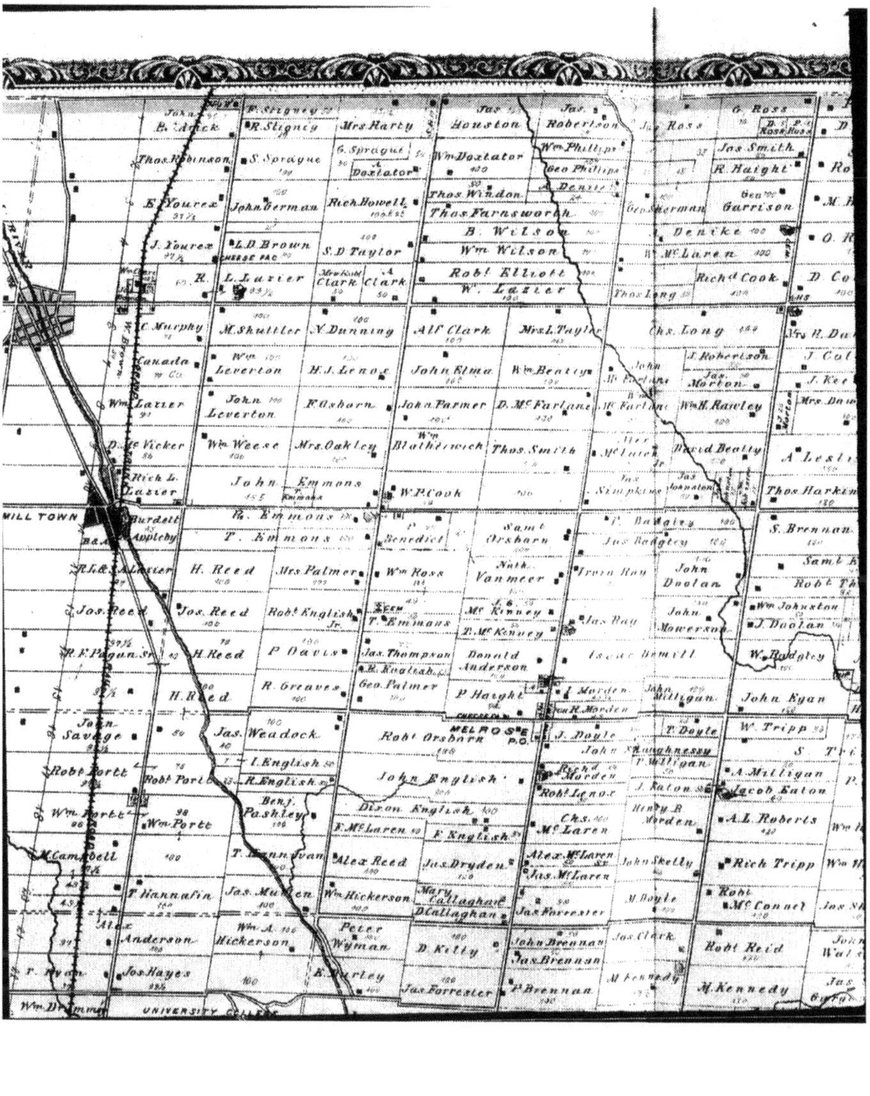

John B. Arick
F. Sligney
R. Sligney
Mrs Harty
Jas. Houston
Jas. Robertson
Jas. Ross
G. Ross
D. S. Ross
P. S. Ross
Thos. Robinson
S. Sprague
G. Sprague
Doxtator
Wm Doxtator
Wm Phillips
Geo Phillips
A. Denike
Jas Smith
R. Haight
E. Yourex
John German
Rich Howell
Thos. Windon
Thos Farnsworth
Geo. Sherman
Geo. W. Garrison
J. Yourex
L. D. Brown
Cheese Fac.
S. D. Taylor
B. Wilson
Wm Wilson
Denike
Wm McLaren
R. L. Lazier
Clark
A. Clark
Robt Elliott
W. Lazier
Richd Cook
Thos Long
C. Murphy
M. Shuttler
N. Dunning
Alf Clark
Mrs L. Taylor
Chs. Long
Canada Co.
Wm Leverton
H. J. Lenox
John Elma
Wm Beatty
John McFarlane
J. Robertson
Jas. Morton
Wm Lazier
John Leverton
F. Osborn
John Farmer
D. McFarlane
McFarlane
Wm H. Rawley
D. McVicker
Wm Weese
Mrs Oakley
Wm Blathdwich
Thos. Smith
Alex McLutch
David Beatty
A. Leslie
Rich L. Lazier
John Emmons
T. Emmons
W. P. Cook
Jas Simpkins
Jas Johnston
Thos. Harkin
MILL TOWN
Burdett & Appleby
R. Emmons
T. Emmons
P. Benedict
Saml Orsborn
P. Badgley
Jas Badgley
S. Brennan
R. L. & S. A. Lazier
H. Reed
Mrs Palmer
Wm Ross
Nath. Vanmeer
Irvin Ray
John Doolan
Saml
Robt
Jas. Reed
Jos. Reed
Robt English Jr.
T. Emmons
J. G. McKinney
Jas Ray
John Mowerson
Wm Johnston
J. Doolan
R. F. Pagan Sr.
H. Reed
P. Davis
Jas. Thompson
R. English
Donald Anderson
Isaac Bemrill
W. Badgley
R. Greaves
Geo. Palmer
H. Reed
P. Haight
J. Morden
R. Morden
John Milligan
John Egan
John Savage
Jas. Weadock
Robt Orsborn
MELROSE P.O.
Cheese Fac.
J. Doyle
P. Doyle
W. Tripp
I. English
John Shaughnessy
T. Milligan
Robt Portt
Robt Portt
R. English
John English
Richd Morden
Robt Lenox
J. Eaton
A. Milligan
Jacob Eaton
Wm Portt
Wm Portt
Benj. Pashley
Dixon English
Henry R. Morden
A. L. Roberts
M. Campbell
F. McLaren
F. English
Chs. McLaren
T. Hanniva
Alex Reed
Jas. Dryden
Alex McLaren
Jas McLaren
John Skelly
Rich Tripp
T. Hannafin
Jas. Mullen
Wm Hickerson
Mary Callaghan
D. Callaghan
Jas Forrester
M. Doyle
Robt McConnel
Alex Anderson
Wm A. Hickerson
Peter Wyman
D. Kelly
John Brennan
Jas Brennan
Jas Clark
Robt Reid
Jos Hayes
E. Hurley
Jas Forrester
P. Brennan
M. Kennedy
M. Kennedy
T. Ryan
Wm Drummer
UNIVERSITY COLLEGE

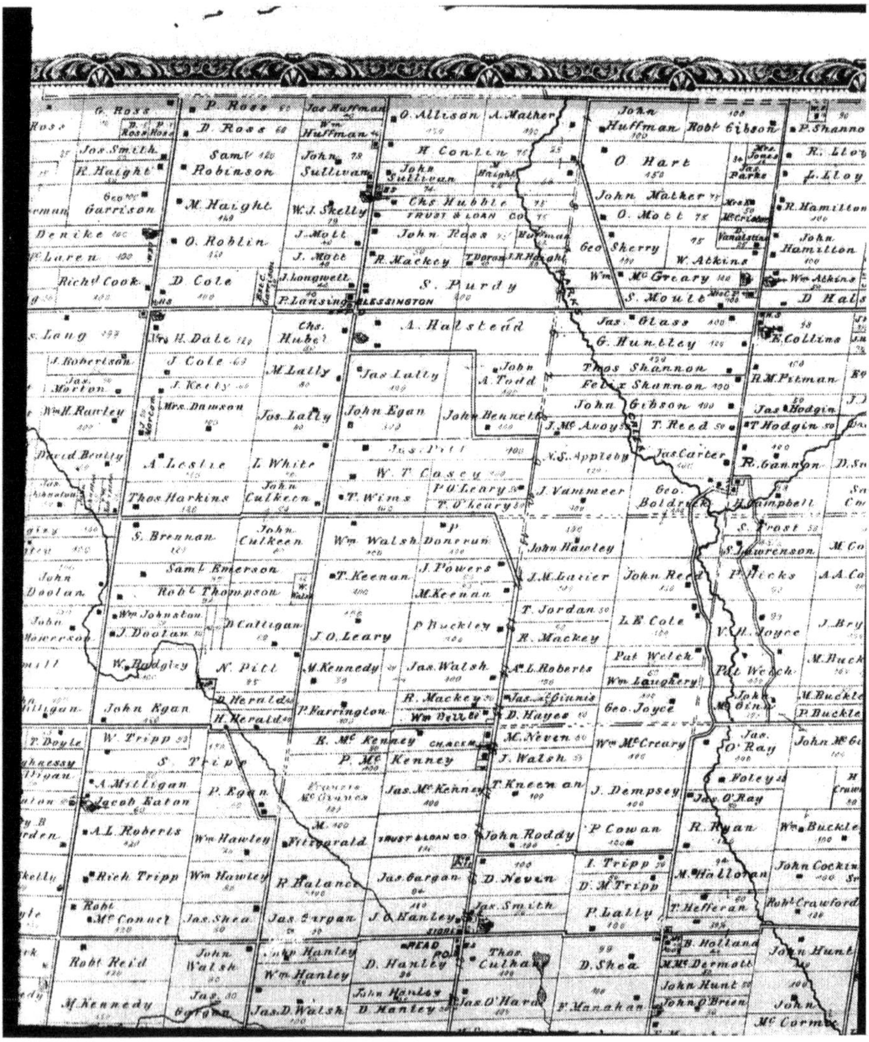

G. Ross
P. Ross
Jas. Huffman
Wm Huffman
O. Allison
A. Mather
John Huffman
Robt Gibson
P. Shannon
D. Ross
D. Ross
Jas. Smith
R. Haight
Geo Garrison
Saml Robinson
M. Haight
John Sullivan
W. J. Skelly
John Sullivan
H. Conlin
Haight
Chs. Hubble
TRUST & LOAN CO.
John Ross
O. Hart
Mrs. Jones
Jas. Parks
John Mather
O. Mott
R. Lloyd
L. Lloyd
R. Hamilton
John Hamilton
Denike
McLaren
Richd Cook
O. Roblin
D. Cole
J. Mott
J. Mott
J. Longwell
P. Lansing
BLESSINGTON
R. Mackey
T. Doran
J. R. Haight
S. Purdy
Geo Sherry
W. Atkins
Wm McGreary
S. Moutt
Wm Atkins
D. Hals
Lang
Mrs H. Dale
Chs. Huber
A. Halstead
Jas. Glass
G. Huntley
E. Collins
J. Robertson
J. Cole
Jas. Morton
J. Keely
M. Lally
Jas. Lally
John A. Todd
Thos Shannon
Felix Shannon
R. M. Pitman
Wm H. Rawley
Mrs. Dawson
Jos. Lally
John Egan
John Bennett
John Gibson
Jas Hodgin
J. McAvoy
T. Reed
T. Hodgin
David Beatty
A. Leslie
I. White
Jas. Pitt
N. S. Appleby
Jas. Carter
R. Gannon
W. T. Casey
P. O'Leary
T. O'Leary
Thos Harkins
John Culkeen
T. Wims
J. Vammeer
Geo. Boldrick
H. Campbell
S. Frost
S. Brennan
John Culkeen
Wm Walsh
P. Donerun
John Hawley
S. Lawrenson
John Doolan
Saml Emerson
Robt Thompson
T. Keenan
J. Powers
M. Keenan
J. M. Latier
John Reed
P. Hicks
John Howerson
Wm Johnston
J. Doolan
D. Calligan
J. O. Leary
P. Buckley
T. Jordan
R. Mackey
L. E. Cole
V. H. Joyce
J. Bry
W. Hadgley
N. Pitt
M. Kennedy
Jas. Walsh
A. L. Roberts
Pat Welch
Wm Laughery
Pat Welch
M. Buck
John Egan
Herald
H. Herald
P. Farrington
R. Mackey
Wm Bell
Jas. McGinnis
D. Hayes
Geo. Joyce
John McGinn
M. Buckley
P. Buckley
T. Doyle
W. Tripp
S. Tripp
R. McKenney
P. McKenney
M. Neven
J. Walsh
Wm McCreary
Jas. O'Ray
John McG
A. Milligan
Jacob Eaton
P. Egan
Francis McGuiness
Jas. McKenney
T. Kneenan
J. Dempsey
Foley
Jas. O'Ray
A. L. Roberts
Wm Hawley
M. Fitzgerald
TRUST & LOAN CO.
John Roddy
P. Cowan
R. Ryan
Wm Buckley
Rich Tripp
Wm Hawley
R. Halance
Jas. Gargan
D. Neven
I. Tripp
D. M. Tripp
M. Halloran
John Cockin
Robt McConnel
Jas. Shea
Jas. Gargan
J. O. Hanley
Jas. Smith
P. Lally
T. Hefferan
Robt Crawford
Robt Reid
John Walsh
John Hanley
D. Hanley
Thos Culhane
D. Shea
B. Holland
M. McDermott
John Hunt
John Hunt
Wm Hanley
M. Kennedy
Jas. Gargan
Jas. D. Walsh
John Hanley
D. Hanley
Jas. O'Hara
F. Manahan
John O'Brien
John McCormick

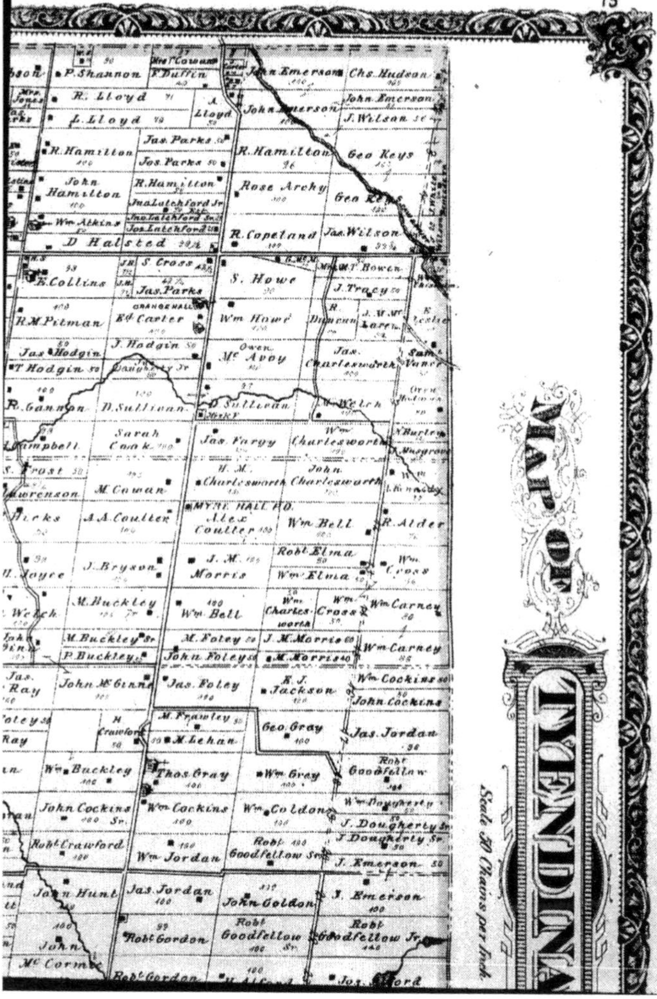

P. Shannon 50 · Mrs. F. Cowan · F. Duffin 40 · John Emerson 100 · Chs. Hudson 100
Mrs. Jones · R. Lloyd 71 · A. Lloyd 50 · John Emerson · John Emerson · J. Wilson 50
L. Lloyd 70
R. Hamilton 100 · Jas. Parks 50 · Jos. Parks 50 · R. Hamilton 96 · Geo Keys 100
John Hamilton 100 · R. Hamilton · Jno. Latchford Jr. · Jno. Latchford Sr. · Jos. Latchford · Rose Archy 100 · Geo Keys 100
Wm Atkins · D Halsted 99½ · R. Copeland 100 · Jas. Wilson 99½
E. Collins 98 · S. Cross 42½ · Jas. Parks 42½ · S. Howe 90 · M. T. Bowen · J. Tracey 50
R. M. Pitman 100 · ORANGE HALL · Ed Carter 100 · Wm Howe 100 · R. Duncan · J. M. McLaren · E. Leslie
Jas Hodgin 50 · J. Hodgin 50 · Owen Mc Avoy 100 · Jas. Charlesworth 100 · Saml Vance
T Hodgin 50 · Daugherty Jr.
R. Gannon 100 · D. Sullivan 100 · D. Sullivan 97 · Welch · Owen Hudson
J. Campbell · Sarah Cook 100 · Jas. Fargy · Wm Charlesworth 100 · N. Burley · Musgrove
S. Frost 50 · M. Cowan 100 · H. M. Charlesworth 100 · John Charlesworth 100 · Kennedy
Lawrenson · A. A. Coulter 100 · MYRT HALL P.O. · Alex Coulter 100 · Wm Bell 100 · R. Alder 75
Hicks · J. Bryson · J. M. Morris · Robt Elma 50 · Wm Elma 50 · Wm Cross 56
H. Joyce
Welch · M. Buckley Jr. · Wm Bell 100 · Wm Charlesworth · Wm Cross · Wm Carney 50
M. Buckley Sr. · P. Buckley · M. Foley 50 · John Foley 50 · J. M. Morris 50 · M. Morris 50 · Wm Carney 85
Jas. Ray 100 · John Mc Ginn 100 · Jas. Foley 100 · E. J. Jackson 100 · Wm Cockins 50 · John Cockins 50
Foley 50 · H Crawford 50 · M. Frawley 50 · M. Lehan 50 · Geo Gray 100 · Jas. Jordan 50
Ray
Wm Buckley 100 · Thos Gray 100 · Wm Gray 100 · Robt Goodfellow
John Cockins 100 Sr. · Wm Cockins 100 · Wm Coldon 100 · Wm Dougherty 50 · J. Dougherty Jr. 50
Robt Crawford 100 · Wm Jordan 100 · Robt Goodfellow Sr. 100 · J. Dougherty Sr. 50 · J. Emerson 50
John Hunt · Jas. Jordan 100 · John Goldon 100 · J. Emerson 100
John 100 · Robt Gordon 99 · Robt Goodfellow Sr. 100 · Robt Goodfellow Jr. 100
Mc Cormick · Robt Gordon · Jos. Alford

MAP OF TYENDINAGA

Scale 40 Chains per Inch

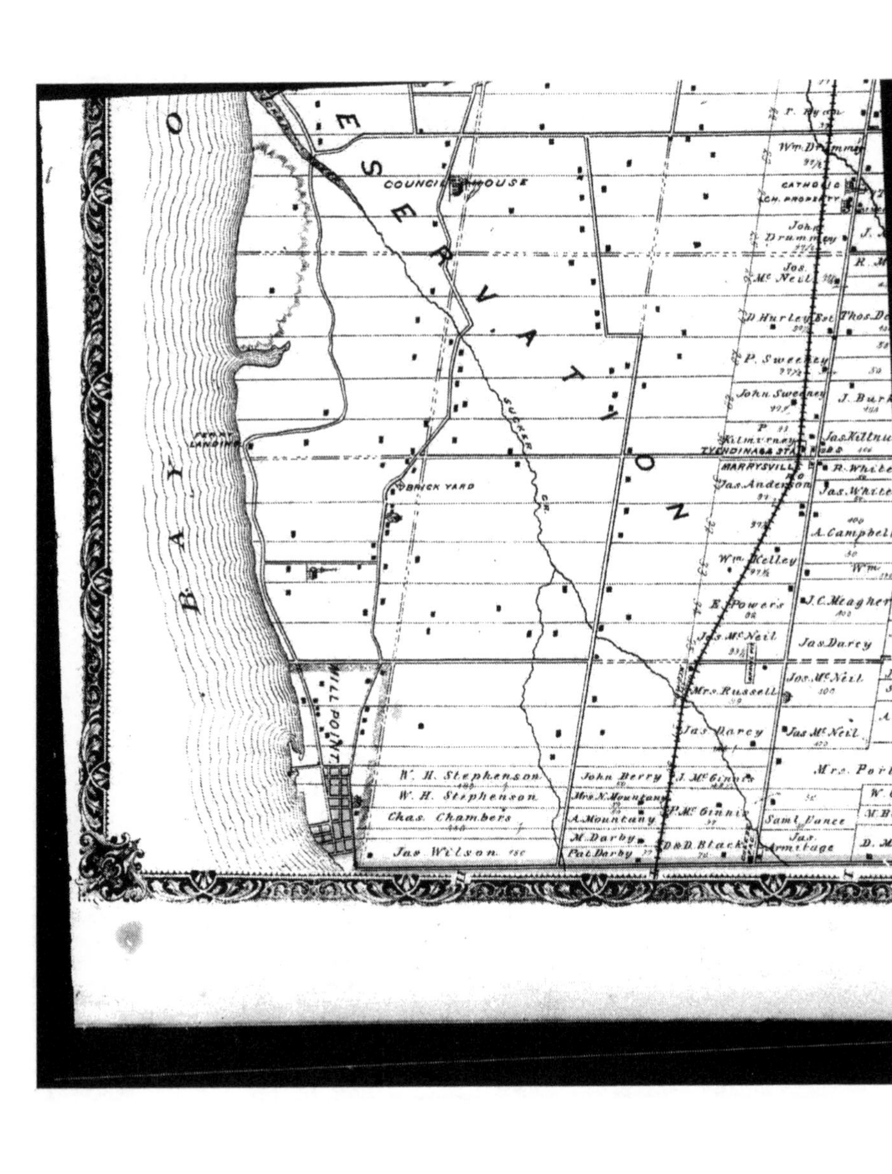

B A Y O
E S E R V A T I O N
COUNCIL HOUSE
CATHOLIC CH. PROPERTY
FERRY LANDING
BRICK YARD
SUCKER CR.
MILL POINT
MARRYSVILLE
P. Ryan
John Drummey
Jos. McNeil
D Hurley Est
P. Sweeney
John Sweeney
Jas. Anderson
Jas. White
R. White
A. Campbell
Wm Kelley
E. Powers
J.C. Meagher
Jas. McNeil
Jas. Darcy
Jos. McNeil
Mrs Russell
Jas. Darcy
Jas McNeil
W. H. Stephenson
W. H. Stephenson
Chas. Chambers
Jas Wilson
John Berry
Mrs N. Mountany
A. Mountany
M. Darby
Pat Darby
D&D Black
J. McGinnis
P. McGinnis
Jas. Armitage

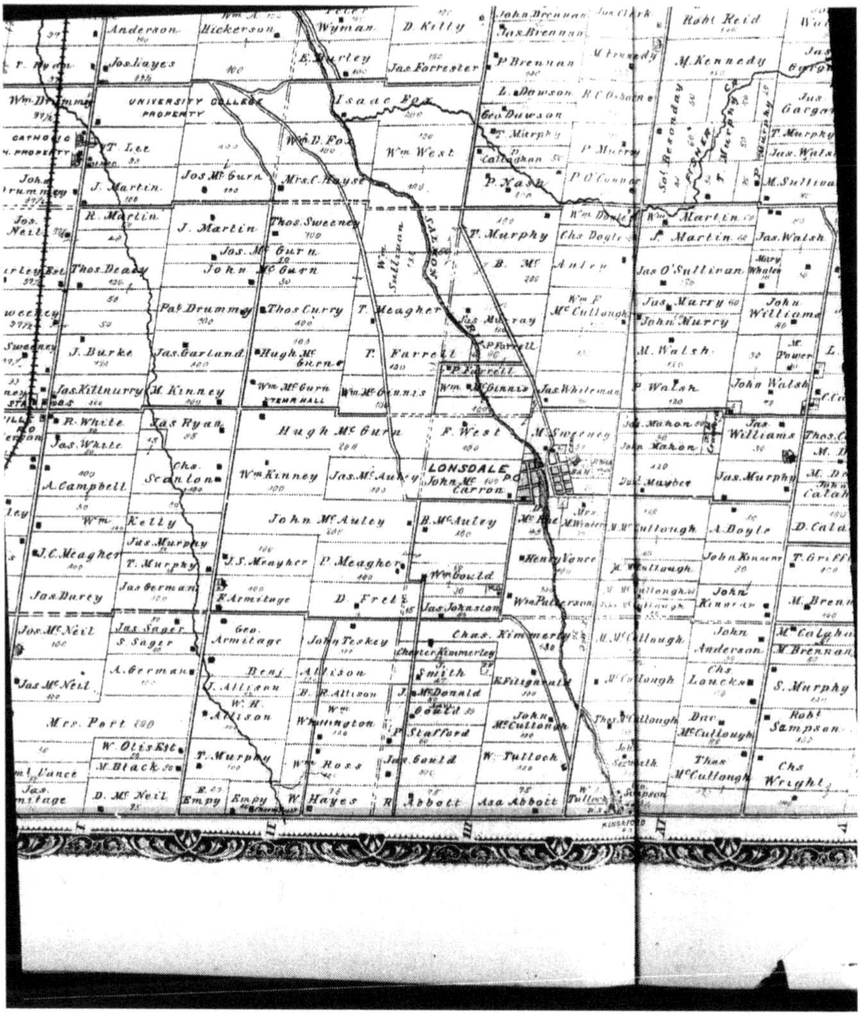

Anderson
Wm. H. Hickerson
Wyman
D. Kelly
John Brennan
Jos. Clark
Robt. Reid
Jos. Hayes
E. Burley
Jas. Forrester
Jas. Brennan
P. Brennan
M. Kennedy
M. Kennedy
Wm. Drummy
UNIVERSITY COLLEGE PROPERTY
Isaac Fox
L. Dawson
R. C. Osborn
Geo. Dawson
CATHOLIC PROPERTY
T. Lee
Wm. D. Fox
Wm. West
T. Murphy
P. Callaghan
P. Murray
Sol. Besonday
Fisher
T. Murphy
Jas. Gargan
T. Murphy
Jas. Walsh
John Drummey
J. Martin
Jos. Mc Gurn
Mrs. C. Hayse
P. Nash
P. O'Connor
M. Sullivan
Jos. Neil
R. Martin
J. Martin
Thos. Sweeney
Wm. Sullivan
SALMON R.
T. Murphy
Chs. Doyle
Wm. Doyle
Wm. Martin
J. Martin
Jas. Walsh
Jas. Mc Gurn
Thos. Deacy
John Mc Gurn
B. Mc Auley
Jas. O'Sullivan
Mary Whalen
Pat. Drummy
Thos. Curry
T. Meagher
Jas. Murray
Wm. F. Mc Cullough
Jas. Murry
John Murry
John Williams
J. Burke
Jas. Garland
Hugh Mc Gurn
T. Farrell
P. Farrell
M. Walsh
M. Power
P. Farrell
Wm. Mc Ginnis
Jas. Killnurry
M. Kinney
Wm. Mc Gurn
TEMP. HALL
Wm. Mc Ginnis
Jas. Whiteman
P. Walsh
John Walsh
R. White
Jas. Ryan
Hugh Mc Gurn
F. West
M. Sweeney
Jas. Mahon
Jas. Williams
Jas. White
John Mahon
A. Campbell
Chs. Scanlon
Wm. Kinney
Jas. Mc Auley
LONSDALE P.O.
John Mc Carron
Dan. Mayber
Jas. Murphy
Wm. Kelly
John Mc Auley
B. Mc Auley
Mrs. M. Winter
M. Mc Cullough
A. Doyle
D. Calahan
J. C. Meagher
Jas. Murphy
T. Murphy
J. S. Meagher
P. Meagher
Henry Vance
M. Mc Cullough
John Kinnear
T. Griffin
Jas. Darcy
Jas. German
F. Armitage
D. Frets
Wm. Gould
Jas. Johnston
Wm. Patterson
John Kinnear
M. Brennan
Jos. Mc Neil
Jas. Sager
S. Sager
Geo. Armitage
John Teskey
Chas. Kimmerley
Chester Kimmerley
M. Mc Cullough
John Anderson
M. Calahan
M. Brennan
A. German
Benj. Allison
J. Allison
Smith
J. Mc Donald
K. Fitzgerald
Chs. Loucks
Jas. Mc Neil
B. R. Allison
W. H. Allison
Wm. Whittington
J. Gould
John Mc Cullough
Thos. Mc Cullough
Dan. Mc Cullough
S. Murphy
Robt. Sampson
Mrs. Port
W. Otis Est.
P. Stafford
M. Black
T. Murphy
Wm. Ross
Jas. Gould
W. Tulloch
John Seymour
Thos. Mc Cullough
Chs. Wright
Wm. Vance
Jas. Armitage
D. Mc Neil
Empy
Empy
W. Hayes
R. Abbott
Asa Abbott
W. Tulloch
Sampson
KINGSFORD
II
III
IV

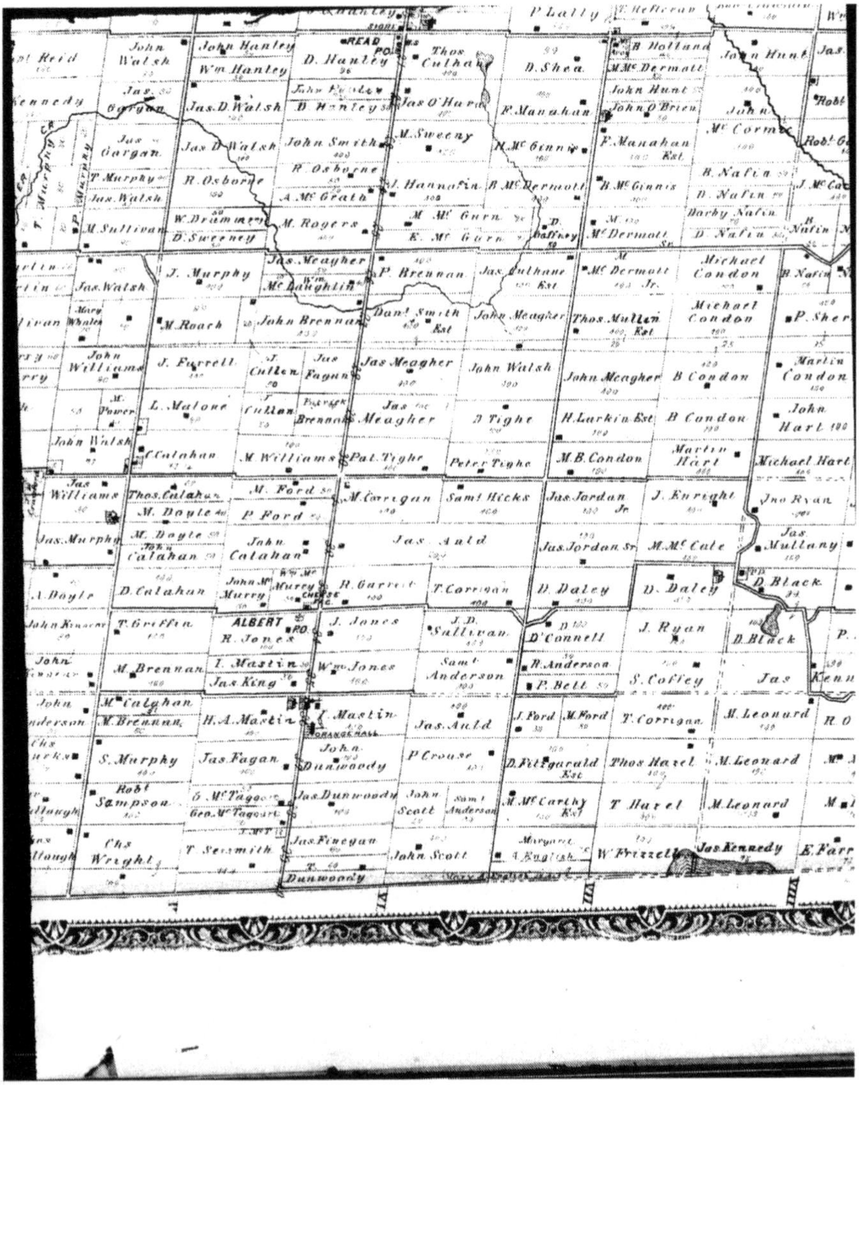

P. Lally
T. Hefferan
John Walsh
John Hanley
READ P.O.
Thos Culhane
D. Shea
B. Holland
M. McDermott
John Hunt
Wm Hanley
D. Hanley
Jas. Gargan
Jas. D. Walsh
D. Hanley
Jas O'Hara
F. Manahan
John Hunt
John O'Brien
John McCormick
Jas Gargan
Jas. D. Walsh
John Smith
M. Sweeny
H. McGinnis
F. Manahan Est.
T. Murphy
R. Osborne
R. Osborne
J. Hannafin
R. McDermott
B. McGinnis
B. Nafin
Jas. Walsh
A. McGrath
M. McGurn
D. Nafin
M. Sullivan
W. Drummond
M. Rogers
E. McGurn
D. Gaffney
M. McDermott Sr.
Darby Nafin
D. Nafin
D. Sweeney
Jas. Walsh
J. Murphy
Jas. Meagher
Wm McLaughlin
P. Brennan
Jas. Culhane Est
M. McDermott Jr.
Michael Condon
Mary Whalen
M. Roach
John Brennan
Danl Smith Est
John Meagher
Thos. Mullen Est
Michael Condon
P. Sher
John Williams
J. Farrell
J. Cullen
Jas Fagan
Jas Meagher
John Walsh
John Meagher
B. Condon
Martin Condon
M. Power
L. Malone
J. Cullen
Patrick Brennan
Jas Meagher
D. Tighe
H. Larkin Est
B. Condon
John Hart
John Walsh
C. Calahan
M. Williams
Pat. Tighe
Peter Tighe
M. B. Condon
Martin Hart
Michael Hart
Jas Williams
Thos. Calahan
M. Ford
M. Corrigan
Saml Hicks
Jas. Jordan Jr.
J. Enright
Jno Ryan
M. Doyle
P. Ford
Jas. Murphy
M. Doyle
John Calahan
John Calahan
Jas. Auld
Jas. Jordan Sr.
M. McCabe
Jas. Mullany
A. Doyle
D. Calahan
John McMurry
Wm McMurry
CHEESE FAC.
R. Garrett
T. Corrigan
D. Daley
D. Daley
D. Black
John Kinnear
T. Griffin
ALBERT P.O.
R. Jones
J. Jones
J. D. Sullivan
D. O'Connell
J. Ryan
D. Black
John Kinnear
M. Brennan
I. Mastin
Jas. King
Wm Jones
Saml Anderson
R. Anderson
P. Bell
S. Coffey
Jas Kennedy
John Anderson
M. Calahan
M. Brennan
H. A. Mastin
I. Mastin
ORANGE HALL
Jas. Auld
J. Ford
M. Ford
T. Corrigan
M. Leonard
S. Murphy
Jas. Fagan
John Dunwoody
P. Crouse
D. Fitzgerald Est
Thos Hazel
M. Leonard
Robt Sampson
G. McTaggart
Geo. McTaggart
Jas. Dunwoody
John Scott
Saml Anderson
M. McCarthy Est
T. Hazel
M. Leonard
Chs Wright
T. Sexsmith
Jas. Finegan
John Scott
Margaret A. English
W. Frizzell
Jas. Kennedy
E. Farr
Dunwoody
VI
VII
VIII

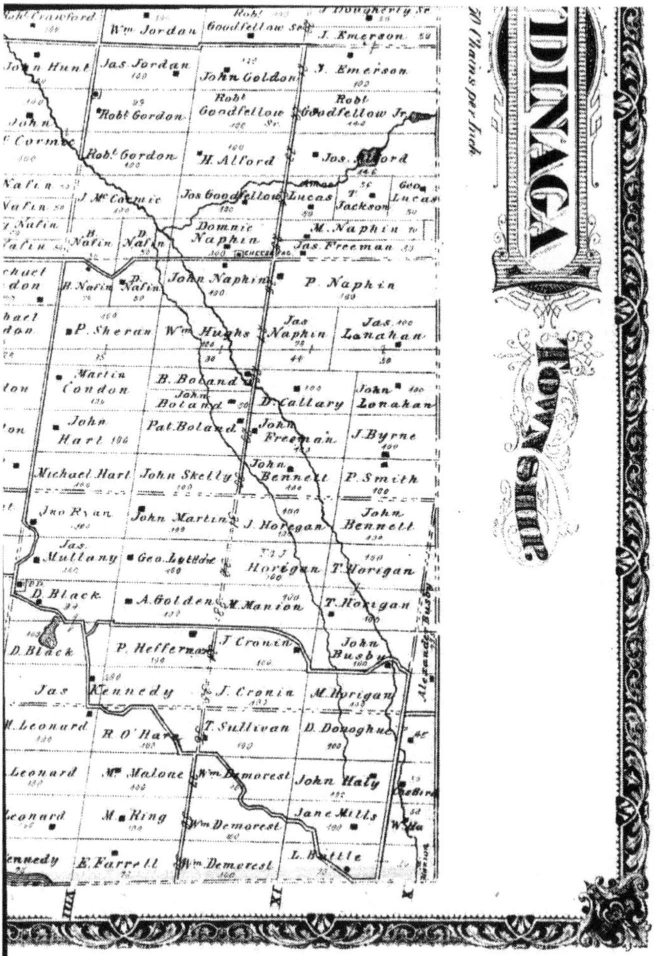

DINAGA

TOWNSHIP

10 Chains per Inch

Robt Crawford 100 | Wm Jordan | Robt Goodfellow Sr | J. Dougherty Sr | J. Emerson 50

John Hunt | Jas. Jordan 100 | John Golden | J. Emerson 100

John McCormic 100 | Robt Gordon | Robt Goodfellow Sr. | Robt Goodfellow Jr.

Robt Gordon 100 | H. Alford 100 | Jos. Alford 146

Nafin 50 | Nafin 50 | J. McCormic 100 | Jos Goodfellow 100 | Amos Lucas 50 | T. Jackson 35 | Geo Lucas 50

Nafin | Nafin | B. Nafin 50 | D. Nafin 50 | Dominic Naphin 100 | M. Naphin 70 | Jas. Freeman 30

Michael Condon | B. Nafin | P. Nafin 50 | John Naphin 100 | P. Naphin 100

Michael Condon | P. Sheran 100 | Wm Hughs 100 | Jas Naphin 44 | Jas. Lanahan 100

Martin Condon 136 | B. Boland | John Boland 50 | D. Callary 100 | John Lanahan 100

John Hart 100 | Pat. Boland | John Freeman | J. Byrne 100

Michael Hart 100 | John Skelly 100 | John Bennett 100 | P. Smith 100

Jno Ryan 100 | John Martin 100 | J. Horegan 100 | John Bennett 130

Jas. Mullany 100 | Geo. Lutdee 100 | T. & J. Horigan 100 | T. Horigan 100

D. Black 94 | A. Golden 100 | M. Manion 100 | T. Horigan 100

D. Black 100 | P. Heffernan 100 | J. Cronin 100 | John Busby 100

Jas Kennedy 100 | J. Cronin 100 | M. Horigan 100 | Alexander Busby

H. Leonard 100 | R. O'Har 100 | T. Sullivan 100 | D. Donoghue 100

Leonard 100 | Mrs Malone 100 | Wm Demorest | John Haly 100

Leonard | M. King 100 | Wm Demorest 100 | Jane Mills 100

Kennedy 75 | E. Farrell 75 | Wm Demorest 100 | L. Battle 75

VIII | IX | X

RES OF JOHN H. YOUNG. "LAKE VIEW FARM," LOT 17, STINSON BLOCK, HILLIER TP. ONT. RES OF JAS. YOUNG.

"RICKARTON," RES. OF A. W. HEPBURN, PICTON, ONT.

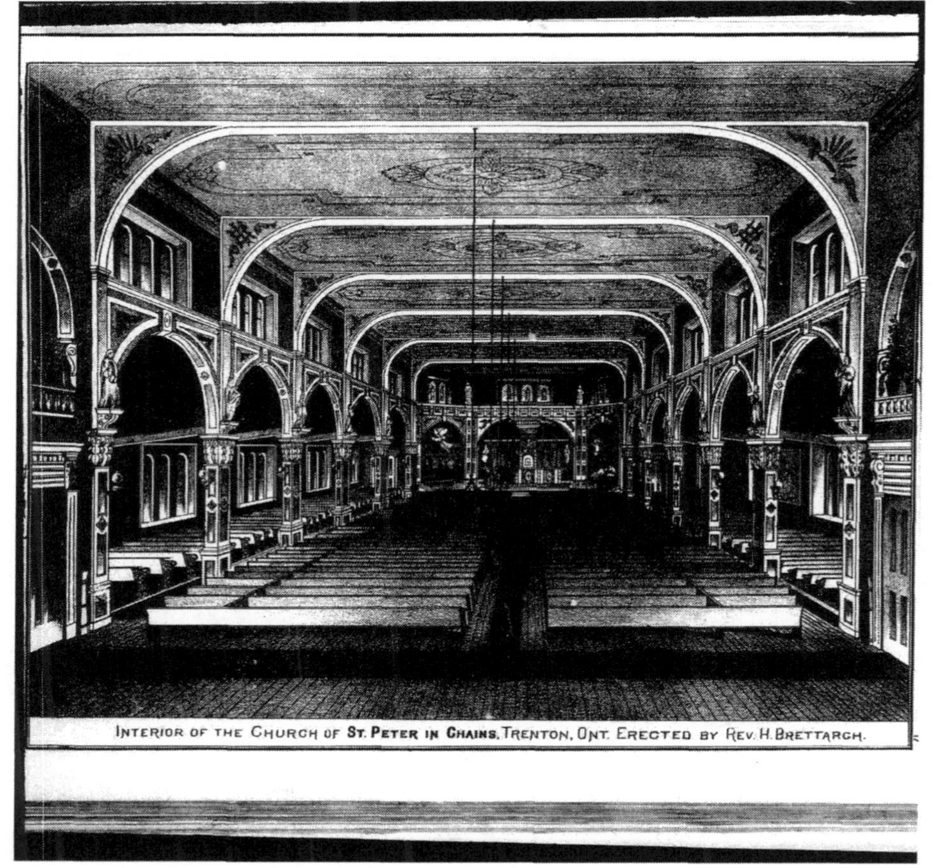

INTERIOR OF THE CHURCH OF **ST. PETER IN CHAINS**, TRENTON, ONT. ERECTED BY REV. H. BRETTARGH.

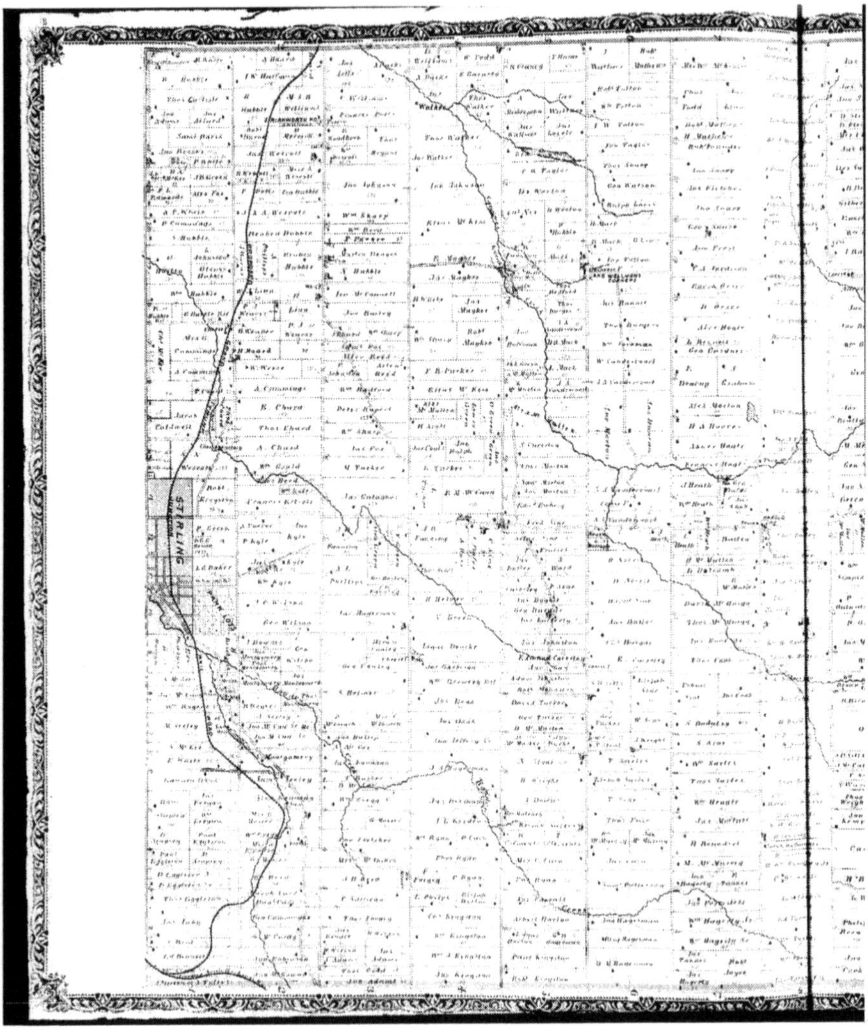

STIRLING

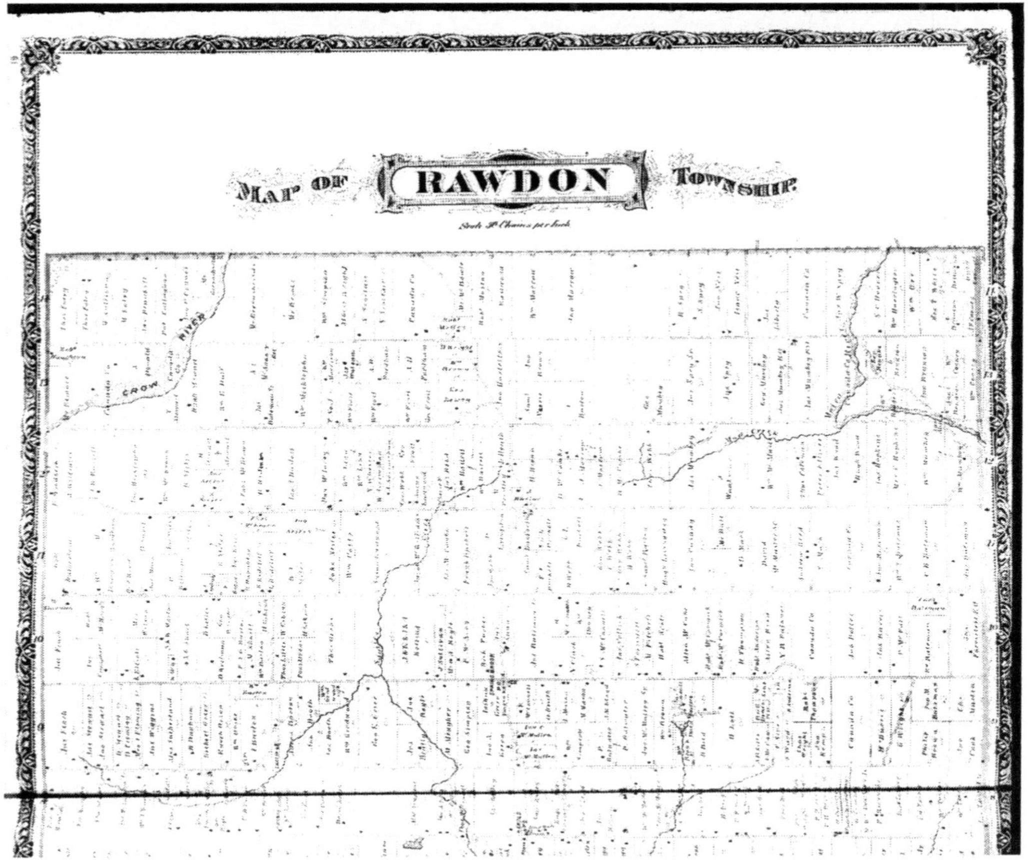

MAP OF RAWDON TOWNSHIP

Scale 80 Chains per Inch

CROW RIVER

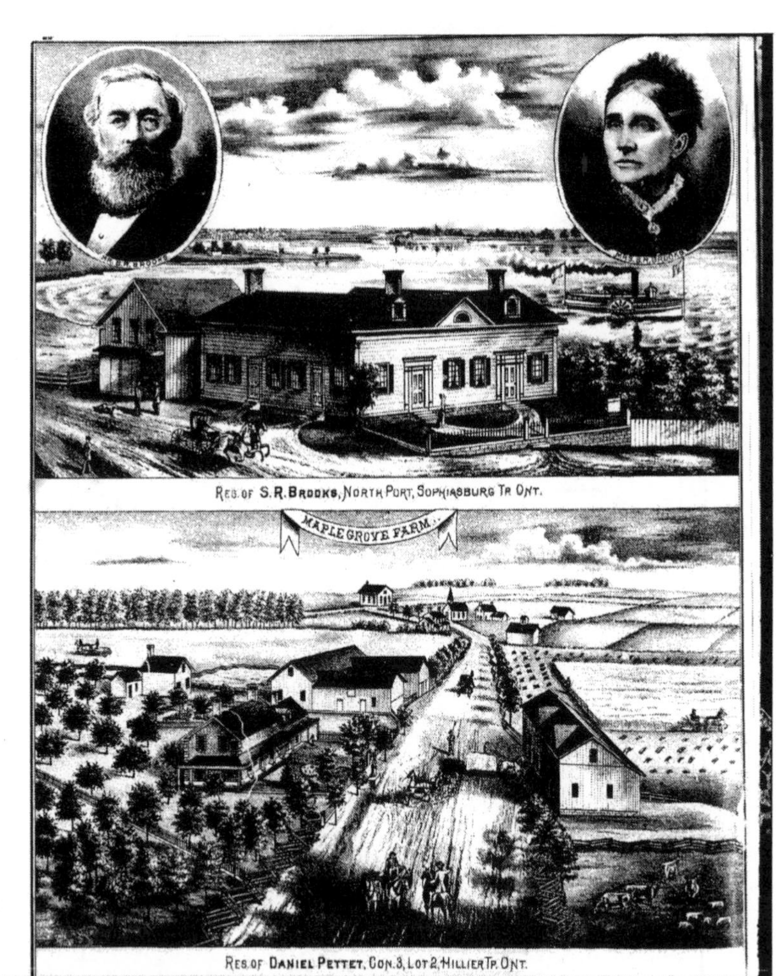

RES. OF S. R. BROOKS, NORTH PORT, SOPHIASBURG TP. ONT.

RES. OF DANIEL PETTET, CON. 3, LOT 2, HILLIER TP. ONT.

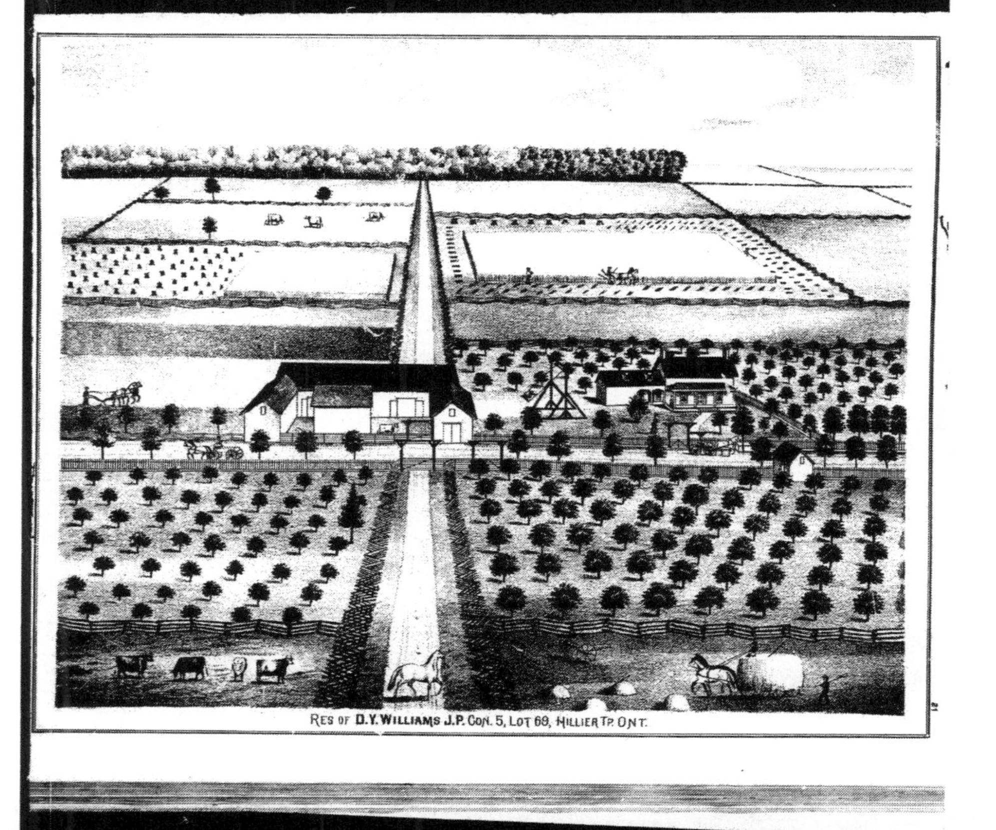

RES OF D.Y. WILLIAMS J.P. CON. 5, LOT 69, HILLIER TP. ONT.

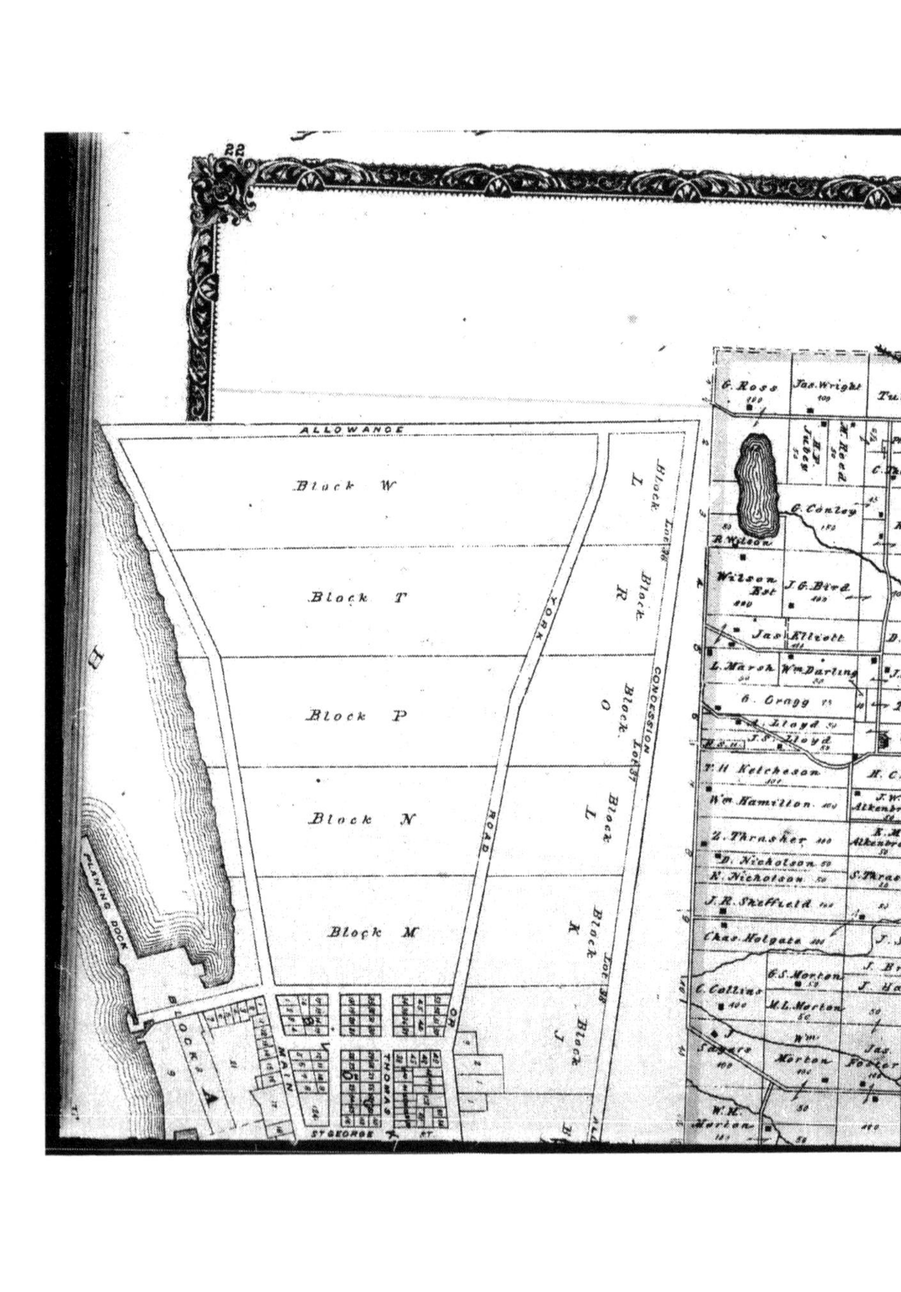
22
ALLOWANCE
Block W
Block T
Block P
Block N
Block M
Block L
Block R
Block O
Block L
Block K
Block J
YORK ROAD
CONCESSION
PLANING DOCK
MAIN
THOMAS
ST GEORGE ST.
G. Ross
Jas. Wright
G. Conley
R. Wilson
Wilson Est
J. G. Bird
Jas. Elliott
L. Marsh
Wm Darling
G. Cragg
J. Lloyd
T. H. Ketcheson
Wm Hamilton
Z. Thrasher
D. Nicholson
E. Nicholson
J. R. Sheffield
Chas. Holgate
G. S. Morton
M. L. Morton
C. Collins
Morton
Jas. Foster
W. H. Morton

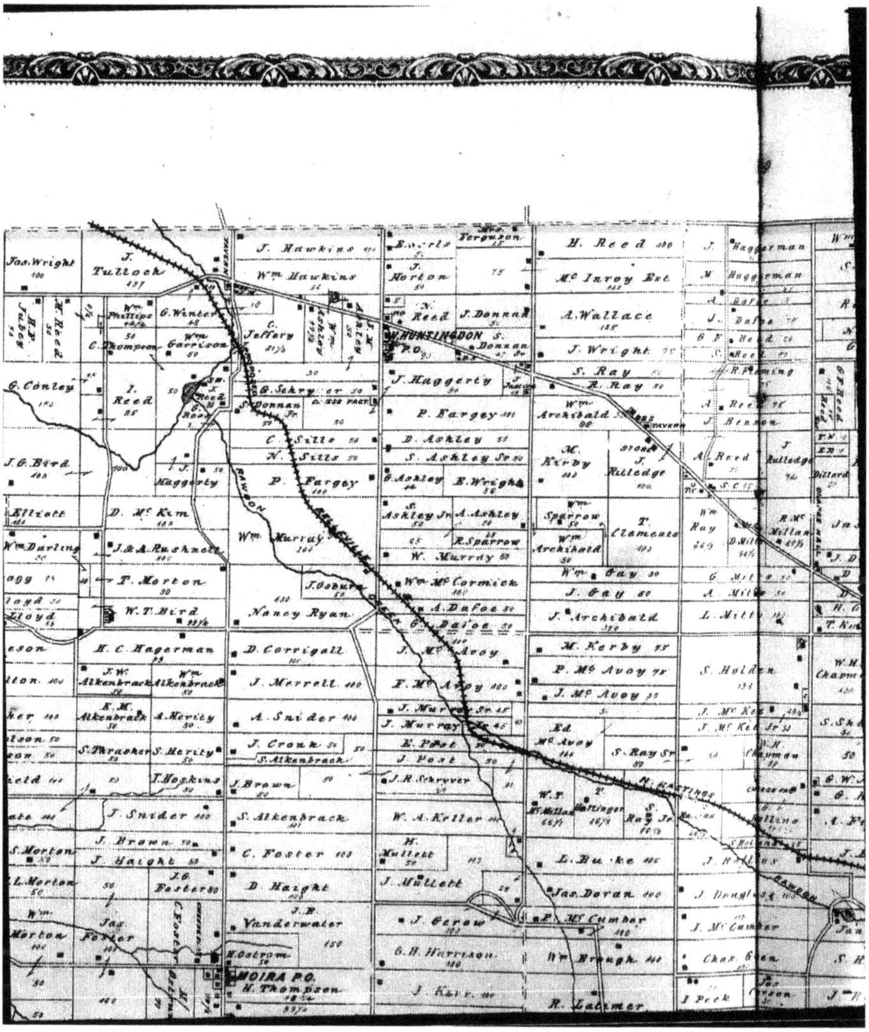

J. Hawkins
Ferguson
H. Reed 100
Jas. Wright 100
J. Tulloch 137
Wm. Hawkins
J. Horton 50
Mc Inroy Est.
J. Haggerman
Wm. Phillips
G. Winter
C. Jeffery
N. Reed
J. Donnan
A. Wallace
C. Thompson
Wm. Garrison
HUNTINGDON P.O.
S. Donnan
J. Wright 75
S. Ray
R. Ray
G. Conley
J. Reed 85
G. Schryver
J. Haggerty
Wm. Archibald
A. Reed
J. Benson
P. Fargey 100
C. Sills 50
D. Ashley 50
N. Sills 50
S. Ashley Sr. 50
J.G. Bird 100
M. Kerby 100
J. Rutledge 100
A. Reed
J. Rutledge
J. Haggerty
P. Fargey
E. Wright 50
Elliott
D. Mc Kim 100
S. Ashley Jr.
A. Ashley
Wm. Sparrow 50
T. Clements 100
Wm. Ray
Wm. Darling
J. & A. Rushnell
Wm. Murray
R. Sparrow
W. Murray 50
Wm. Archibald 50
T. Morton 90
J. Osburn
Wm. Mc Cormick
Wm. Gay 50
J. Gay 50
A. Dafoe 50
Lloyd
W.T. Bird
Nancy Ryan
J. Archibald 150
H.C. Hagerman
D. Corrigall
J. Mc Avoy
M. Kerby
P. Mc Avoy
S. Holden
J.W. Alkenbrack
Wm. Alkenbrack
J. Merrell 100
F. Mc Avoy
J. Mc Avoy
E.M. Alkenbrack
A. Herity
A. Snider 100
J. Murray
Ed. Mc Avoy
S. Ray Sr.
J. Mc Kee
J. Mc Kee Jr.
W.H. Chapman
S. Thrasher
S. Herity
J. Cronk 50
S. Alkenbrack
E. Post
J. Post
T. Hoskins
J. Brown
J.R. Schryver
W. Hastings
J. Snider 100
S. Alkenbrack
W.A. Keller
S. Ray Jr.
S. Morton
J. Brown
J. Haight
C. Foster 100
H. Mullett
L. Burke 100
L. Morton
J.G. Foster
D. Haight
J. Mullett
Jas. Doran
J. Douglass
J.B. Vanderwater
J. Gerow
P. Mc Cumber
J. Mc Cumber
Wm. Morton
Jas. Foster
H. Ostrom
G.H. Harrison
Wm. Brough
Chas. Green
MOIRA P.O.
H. Thompson
J. Kerr
R. Latimer
J. Peek

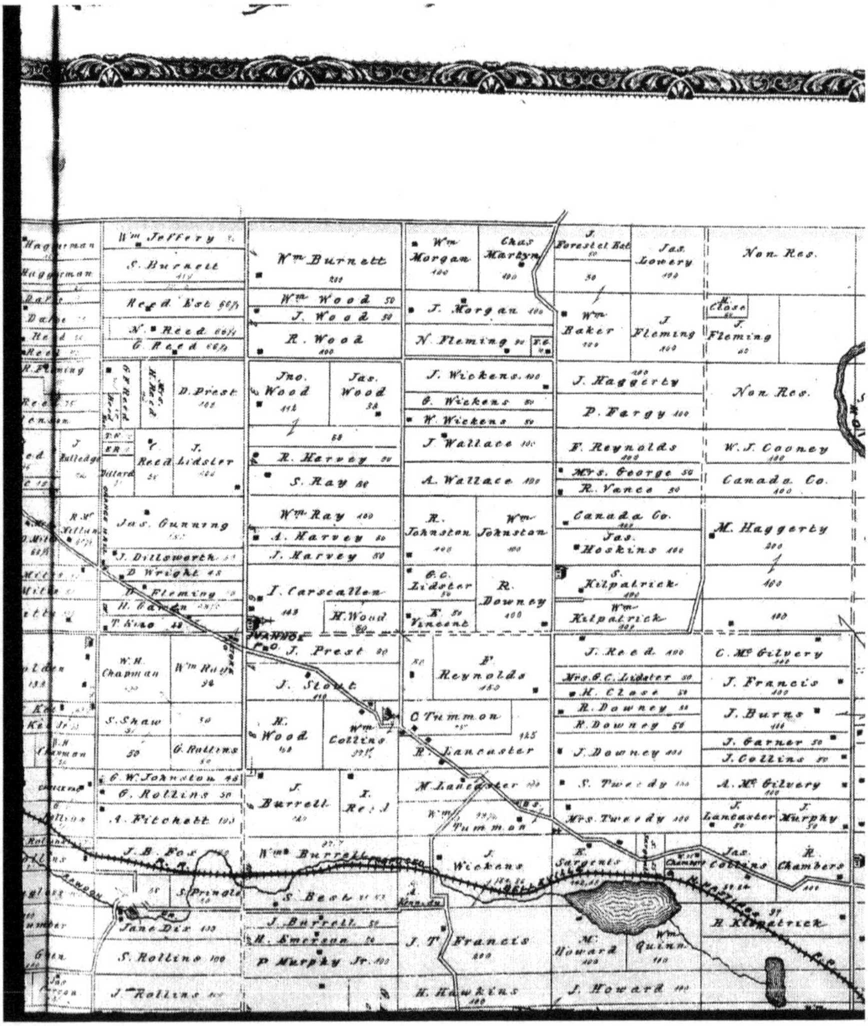

Wm Jeffery
S. Burnett
Wm Burnett
Wm Morgan 100
Chas Martyn 100
J. Forestel Est.
Jas. Lowery 100
Non Res.
Reed Est. 66⅔
Wm Wood 50
J. Wood 50
J. Morgan 100
Wm Baker 100
J. Fleming 100
H. Close
J. Fleming
N. Reed 66⅔
G. Reed 66⅔
R. Wood 100
N. Fleming
R. Fleming
D. Prest 100
Jno. Wood 112
Jas. Wood
J. Wickens 100
G. Wickens 50
W. Wickens 50
J. Haggerty
P. Fargy 100
Non Res.
J. Rutledge
Reed
J. Lidster
R. Harvey 50
S. Ray
J. Wallace 100
A. Wallace 100
F. Reynolds 100
Mrs. George 50
R. Vance 50
W. J. Cooney 100
Canada Co. 100
Jas. Gunning
Wm Ray 100
A. Harvey 50
J. Harvey 50
R. Johnston 100
Wm Johnston 100
Canada Co.
Jas. Hoskins 100
M. Haggerty 200
J. Dillsworth
D. Wright 48
N. Fleming
H. Garden
T. King
J. Carscallen 148
H. Wood 50
G. C. Lidster 50
R. Downey 100
E. Vincent
S. Kilpatrick 100
Wm Kilpatrick 200
IVANHOE P.O.
J. Prest
W. H. Chapman
Wm Ray 94
J. Stout 110
F. Reynolds 150
J. Reed 100
C. McGilvery
Mrs. G. C. Lidster 50
H. Close 50
J. Francis 100
S. Shaw
H. Wood
Wm Collins
C. Tummon
R. Downey
R. Downey 50
J. Burns
J. Garner 50
J. Collins 50
G. Rollins
R. Lancaster
J. Downey 100
G. W. Johnston 46
G. Rollins 50
J. Burrell
J. Reed
M. Lancaster 100
S. Tweedy
A. McGilvery 100
A. Fitchett 103
Wm Tummon
Mrs. Tweedy 100
J. Lancaster 50
J. Murphy 50
J. B. Fox
Wm Burrell
J. Wickens
E. Sargents
Jas. Collins
R. Chambers
S. Pringle
S. Best
R. Kilpatrick
Jane Dix 100
J. Burrell 50
H. Emerson
J. T. Francis 200
M. Howard 100
Wm Quinn 110
S. Rollins 100
P. Murphy Jr. 100
Jno Rollins
H. Hawkins 100
J. Howard 100

MAP OF HUNTINGD

Non-Res. Non-Res. O'Connell Ellen O'Connell 100 M. O'Connell Mary Declare
Wm Harrington 200 T. Nevell 50 R. Goggins L. Doyle 200
T. Connell 100
Non Res. B.F. Sanders 100 J. Foley 100 R. Connell 125 J. Connell 100 J. Taylor 100
J. Connell 75
J. Cooney 100 Jas. O'Connell 200 Jas. Foley 100 S. Reynolds 100 J. O'Reilly
Canada Co. 100
Moira River
Haggerty 200 Non-Res. A. Cooney 200 Jas Lahey 100 J Rylott Wm Rylott
C. Gifford 100 J. Ash R. Collins Wm Orr 200
E. Phillips 100 Mrs. Cottom Wm Harris
McGilvery Mrs. E. Downey 200 Capt Norman A. Vanorman 100
Francis 100 F. Harris
Burns 100 J. Rodgers
Garner 50 A. Cutle 50
Collins S. McGuire 75 J. Baley
McGilvery 100 J. Hough 87 L. Keenew
Murphy 50 Wm Blakely 99 P. Keene 40
J. Kane
Canada Co.
R. Chambers H. Blakely 200 J. Baley C.& P. Coleman 200
Chas. McGuire 150 M.& T. Perry M. Waters P. Waters T. Baley
J. Archer 75 D. O'Donnell J. Baley

Scale 80 Chains per In.

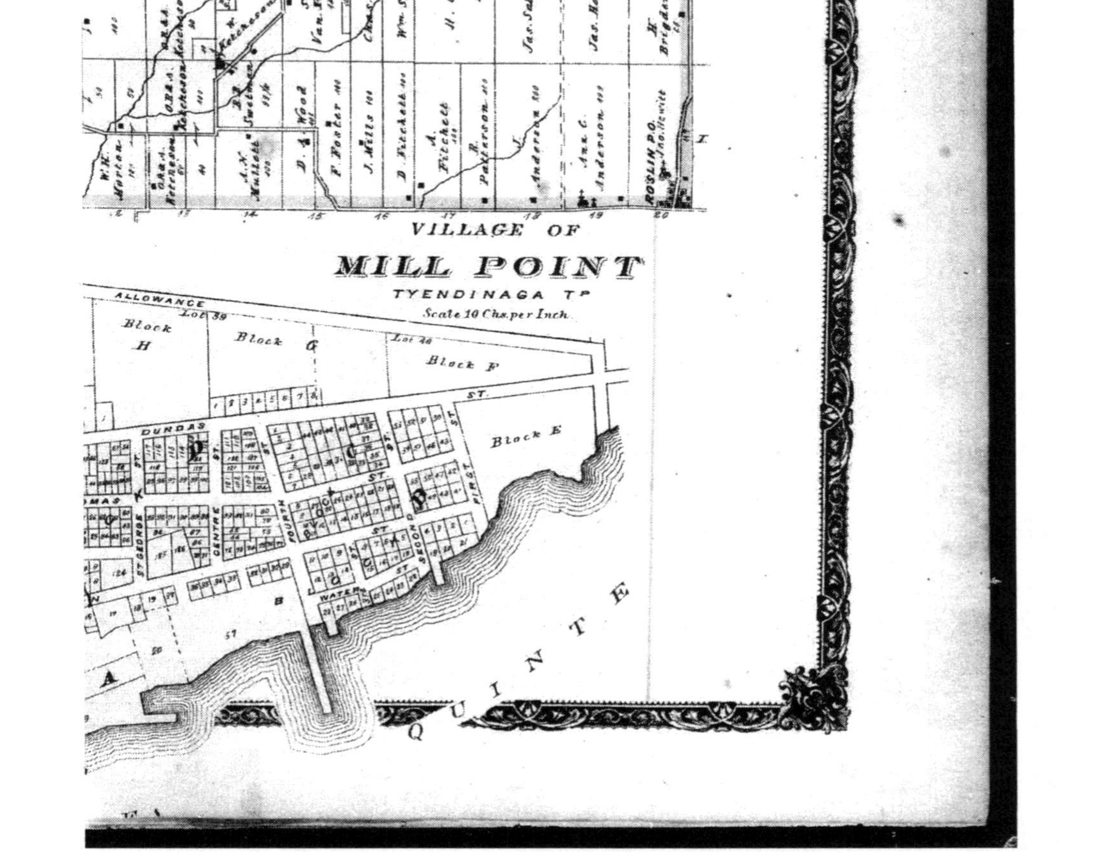

VILLAGE OF
MILL POINT
TYENDINAGA Tp
Scale 10 Chs. per Inch.
ALLOWANCE
Lot 39
Lot 40
Block H
Block G
Block F
Block E
DUNDAS ST.
ST GEORGE ST.
CENTRE ST.
FOURTH ST.
SECOND ST.
FIRST ST.
WATER ST.
QUINTE
W. H. Morton
Chas. A. Ketcheson
O. R. & A. Ketcheson
A. N. Mullett
R. B. Swetman
D. A. Wood
F. Foster
J. Mills
D. Fitchett
A. Fitchett
R. Patterson
J. Anderson
Ann C. Anderson
ROSLIN P.O.
Jno. Hewitt
H. Brigdan
Jas. Salt

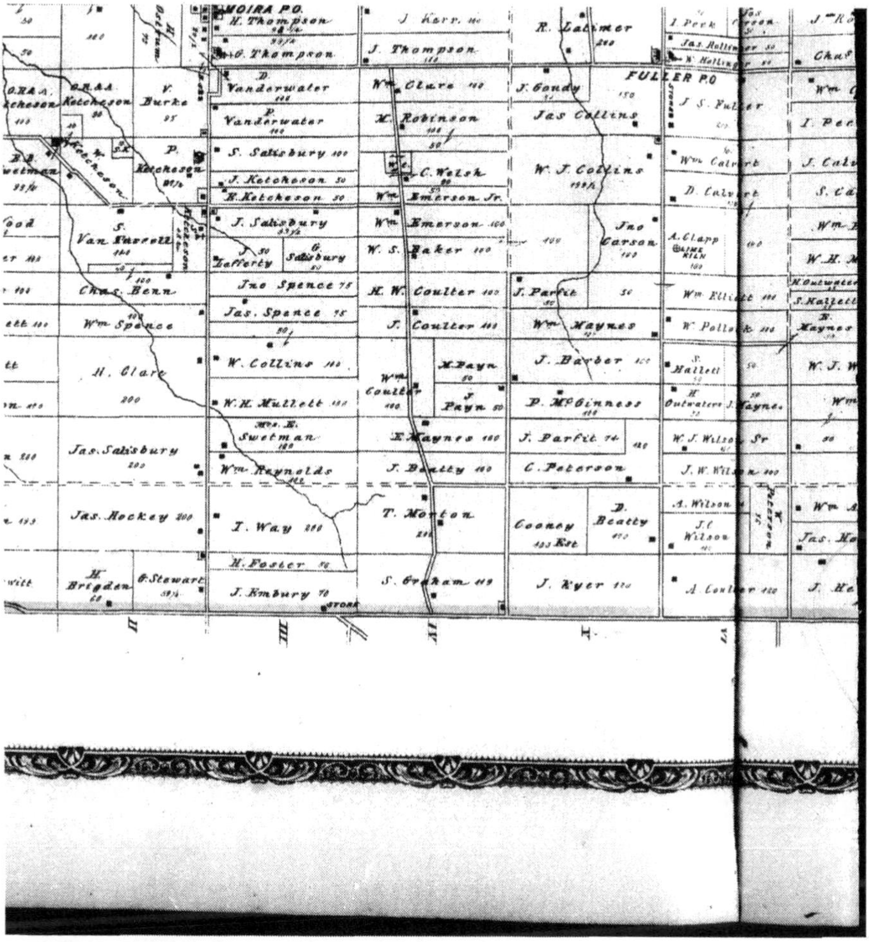

MOIRA P.O.

H. Thompson 98¾

G. Thompson

J. Kerr. 100

J. Thompson 100

R. Latimer 200

I. Peck

Jas. Rollins

W. Hollinger

FULLER P.O.

O.H. & A. Ketcheson

V. Burke 95

D. Vanderwater 100

P. Vanderwater 100

Wm. Clare 100

M. Robinson 100

J. Gondy

Jas. Collins 150

J. S. Fuller

R.B. Swetman 99½

P. Ketcheson 99½

S. Salisbury 100

J. Ketcheson 50

R. Ketcheson 50

C. Welsh

Wm. Emerson Jr.

W. J. Collins 199½

Wm. Calvert

D. Calvert

J. Calvert

S. Van Varcott 100

J. Salisbury 83½

J. Lafferty 50

G. Salisbury

Wm. Emerson 100

W. S. Baker 100

Jno. Carson 100

A. Clapp

Lime Kiln 100

W. H.

Chas. Benn

Jno. Spence 75

Jas. Spence 75

H. W. Coulter 100

J. Parfit 50

Wm. Elliott 100

H. Outwater

S. Hazzell

Wm. Spence 100

J. Coulter 100

Wm. Maynes

W. Pollock 100

R. Maynes

H. Clare 200

W. Collins 100

M. Payn 50

J. Barber 100

S. Hallett 50

W. J. W

W. H. Mullett 100

Wm. Coulter 100

J. Payn 50

D. McGinness 100

H. Outwater

J. Maynes

Jas. Salisbury 200

Mrs. E. Swetman 100

E. Maynes 100

J. Parfit 74

W. J. Wilson Sr.

Wm. Reynolds

J. Beatty 100

C. Peterson

J. W. Wilson 100

Jas. Hockey 200

I. Way 200

T. Morton 200

Cooney Est. 193

D. Beatty 100

A. Wilson

J. C. Wilson

Peterson

Wm. A

Jas. Ho

H. Foster 50

H. Brigden 60

G. Stewart 59½

J. Embury 70

Store

S. Graham 119

J. Ryer 120

A. Coulter 120

J. He

II

III

IV

V

VI

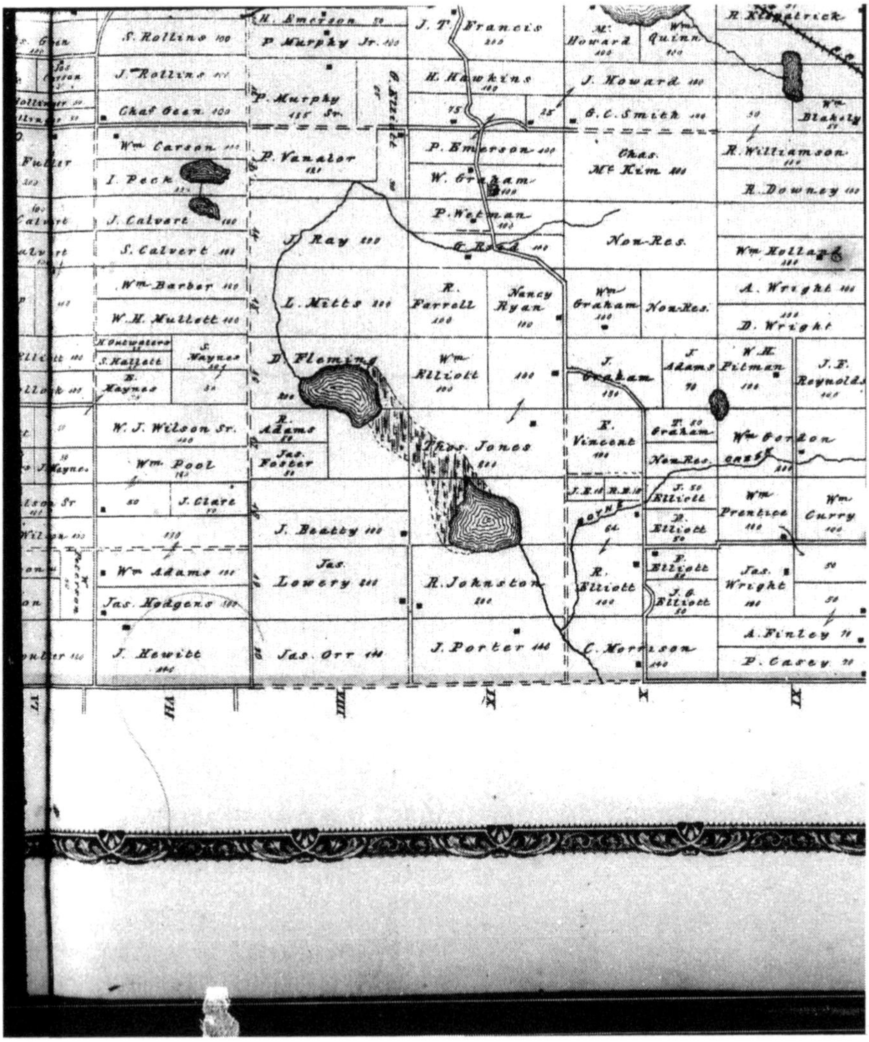

S. Rollins 100
J. Wm Rollins 100
Chas Geen 100
Wm Carson 100
I. Peck
J. Calvert 100
S. Calvert 100
Wm Barber 100
W. H. Mullett 100
S. Hallett
S. Maynes
B. Maynes
W. J. Wilson Sr. 100
Wm Pool
J. Clare
Wm Adams 100
Jas. Hodgens 100
J. Hewitt
H. Emerson 50
P. Murphy Jr. 100
P. Murphy 155 Sr.
P. Vanator
J. Ray 200
L. Mitts 200
D. Fleming
R. Adams
Jas. Foster
J. Beatty 100
Jas. Lowery 200
Jas. Orr 100
J. T. Francis 200
H. Hawkins 100
P. Emerson 100
W. Graham 100
P. Wetman 100
G. Reed 100
R. Farrell 100
Nancy Ryan 100
Wm Elliott
Thos. Jones 200
R. Johnston 200
J. Porter 100
Mc Howard 100
Wm Quinn 100
J. Howard 100
G. C. Smith 100
Chas. Mc Kim 100
Non-Res.
Wm Graham 100
Non-Res.
J. Graham
J. Adams
F. Vincent 100
T. Graham
Non-Res.
J. Elliott
D. Elliott
R. Elliott 100
P. Elliott
J. G. Elliott
C. Morrison 100
R. Kilpatrick
Wm Blakely
R. Williamson 100
R. Downey 100
Wm Holland 100
A. Wright 100
D. Wright
W. H. Pitman 100
J. F. Reynolds
Wm Gordon
Wm Prentice 100
Wm Curry 100
Jas. Wright 100
A. Finley
P. Casey
VII
VIII
IX
X
XI

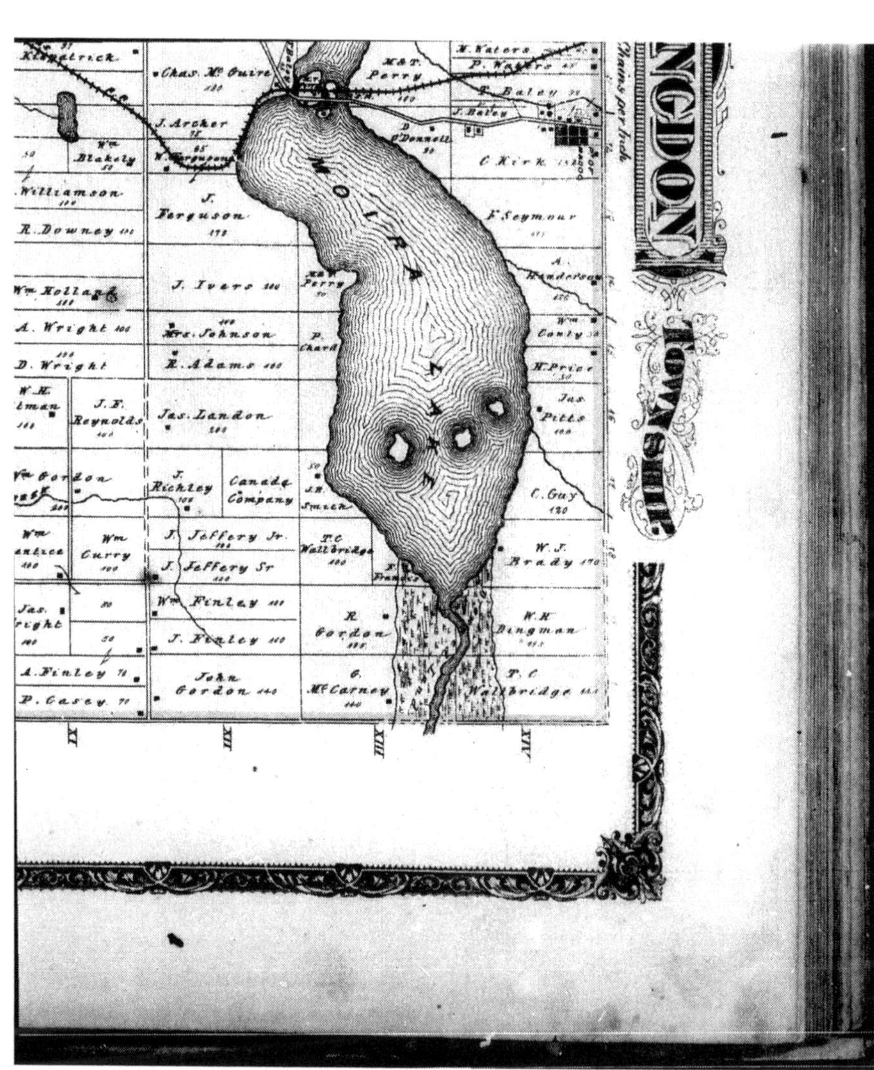

MOIRA LAKE
NGDON
TOWNSHIP
Chains per Inch
Chas. Mc Guire
J. Archer
Wm Blakely
J. Ferguson
Williamson
R. Downey
Wm Holland
J. Ivers
A. Wright
Mrs. Johnson
D. Wright
R. Adams
J. F. Reynolds
Jas. Landon
Wm Gordon
J. Richley
Canada Company
J. Jeffery Jr.
J. Jeffery Sr
Wm Curry
Wm Finley
J. Finley
A. Finley
P. Casey
John Gordon
R. Gordon
G. Mc Carney
T. C. Wallbridge
M. Waters
P. Waters
Perry
J. Baley
D. O'Donnell
C. Kirk
F. Seymour
A. Henderson
Wm Conly
H. Price
Jas. Pitts
C. Guy
W. J. Brady
W. H. Bingman
XI
XII
XIII
XIV

J. Matt Hack M.D.
Picton.
W.H.R Allison, Picton.
Wm Owens
Ex Mayor, Picton.
Barrister-at-Law, Picton, Ont.
A. W. Hepburn
Picton.

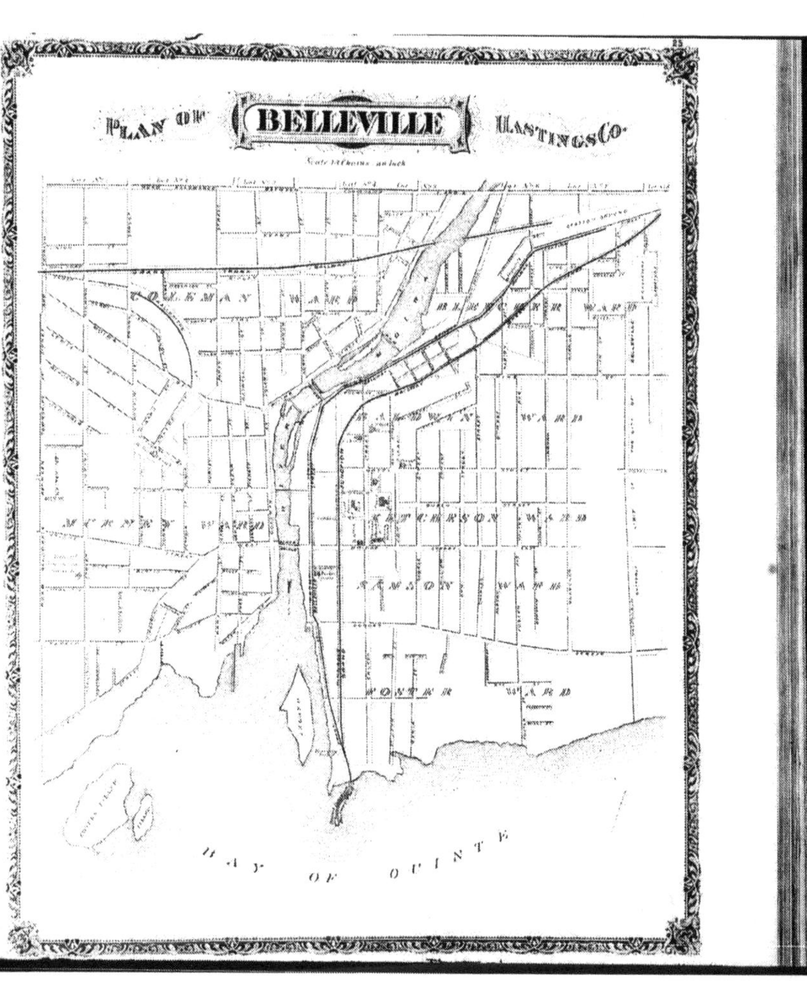
PLAN OF BELLEVILLE HASTINGS CO.
COLEMAN WARD
BLEECKER WARD
BALDWIN WARD
KETCHESON WARD
SAMSON WARD
FOSTER WARD
BAY OF QUINTE

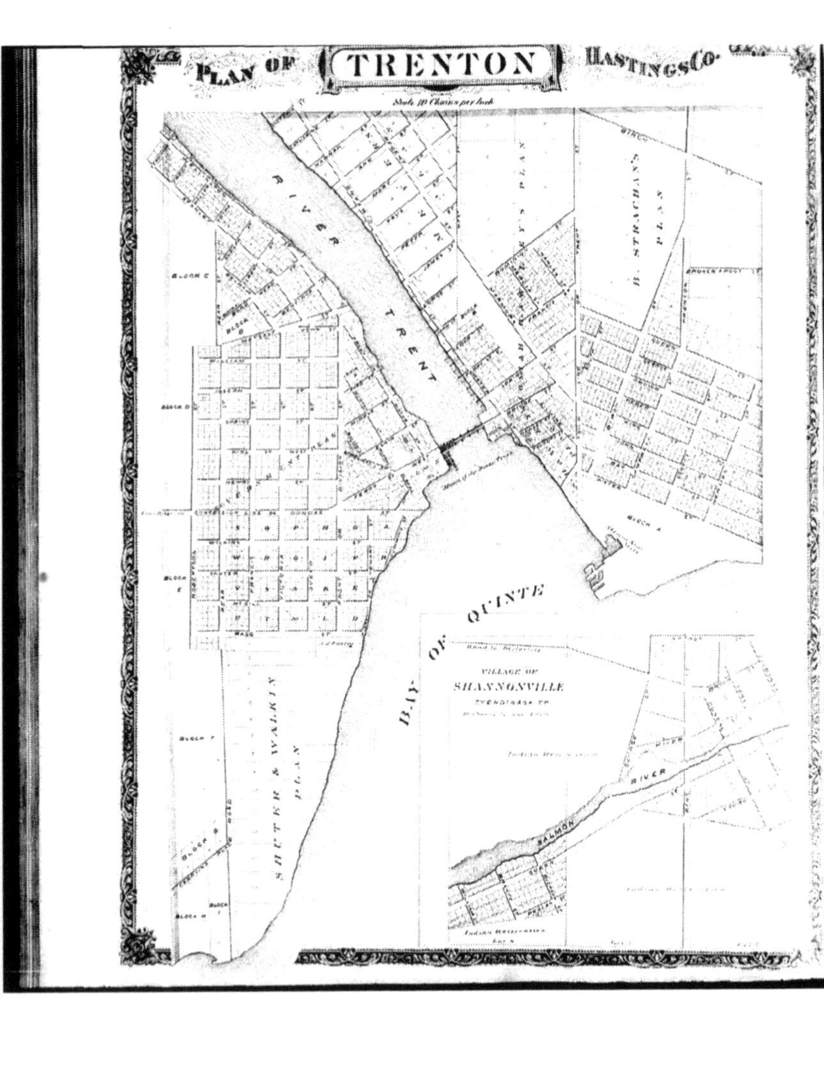
PLAN OF TRENTON HASTINGS CO.
Scale 10 Chains per Inch
RIVER TRENT
B. STRACHAN'S PLAN
BLOCK C
BLOCK D
BLOCK E
BLOCK F
BLOCK G
BLOCK H
BLOCK A
SHUTER & WALKIN PLAN
BAY OF QUINTE
VILLAGE OF
SHANNONVILLE
TYENDINAGA TP.
SALMON RIVER

"THE WILLOWS", RES. OF MOSES BOARDMAN ESQ. CON. 7, LOT 27 & 28, SIDNEY TP. HASTINGS CO. ONT.

BREWERY OF W. P. DESPARD, PICTON, ONT.

IMAGE EVALUATION TEST TARGET (MT-3)

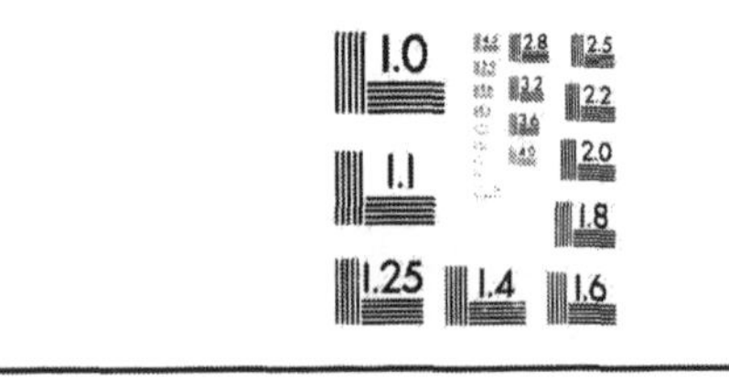

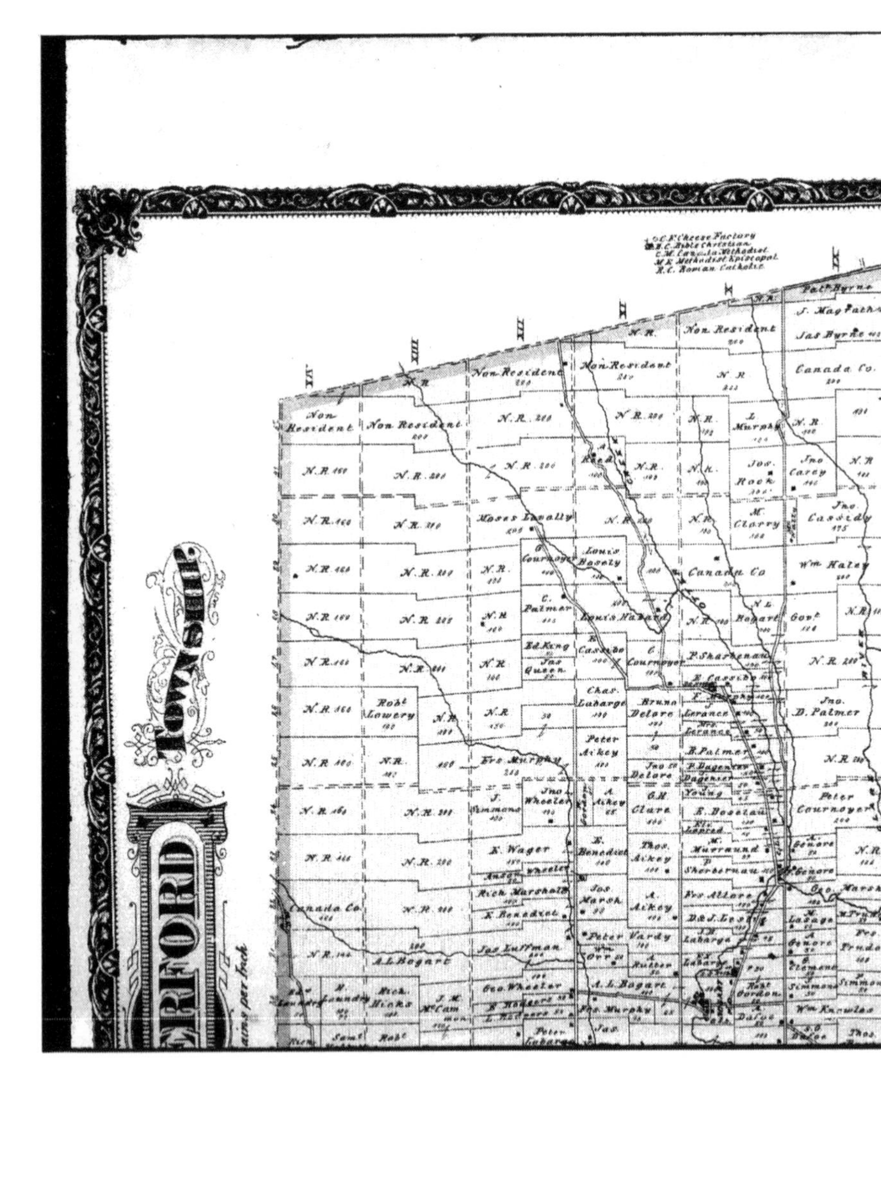

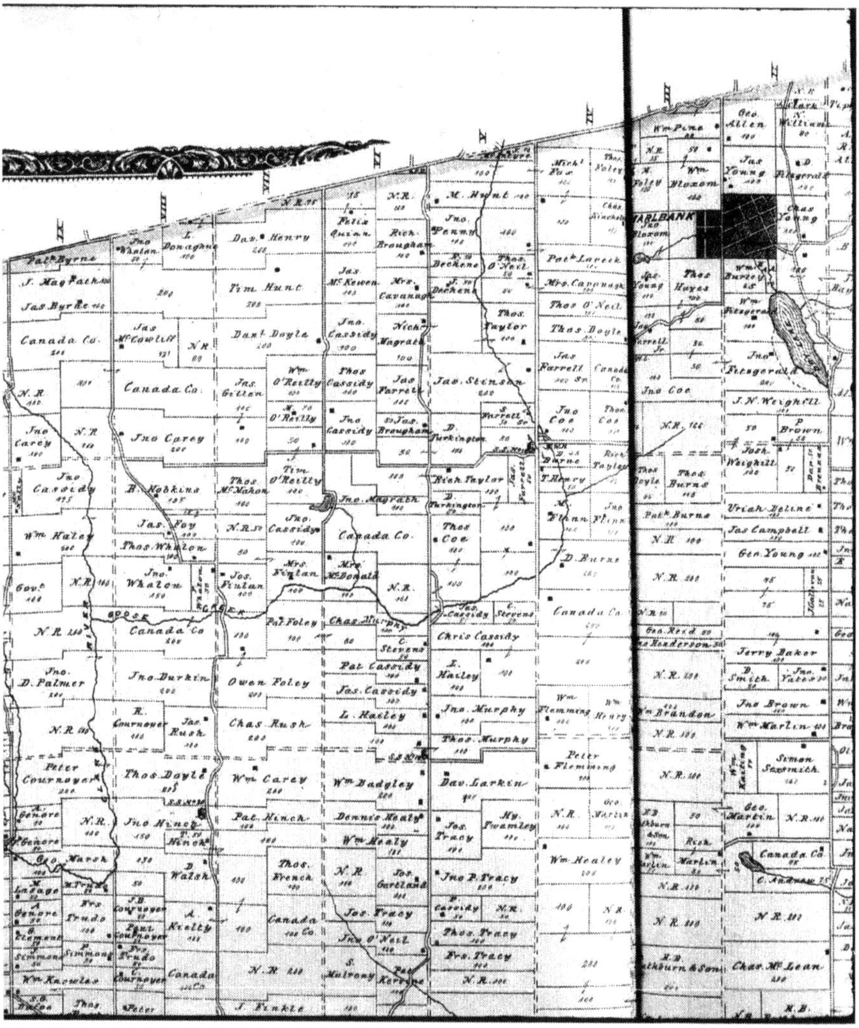

Pat. Byrne
J. Magrath
Jas. Byrne
Canada Co.
N. R.
Jno. Carey
Jno. Cassidy
Wm. Haley
Govt.
N. R.
Jno. D. Palmer
N. R.
Peter Cournoyer
Genore
Geo. Marsh
Lasage
Simmons
Wm. Knowles
Jno. Whalen
L. Donaghue
Jas. McCowliff
Canada Co.
Jno. Carey
B. Hobkins
Jas. Foy
Thos. Whalen
Jno. Whalen
Canada Co.
Jno. Durkin
R. Cournoyer
Jas. Rush
Thos. Doyle
Jno. Hinch
D. Walsh
Fro. Trudo
A. Rielly
Canada Co.
Dan. Henry
Tim. Hunt
Danl. Doyle
Jas. Gillen
Thos. McMahon
N. R.
Jos. Finlan
Owen Foley
Chas. Rush
Wm. Carey
Pat. Hinch
Thos. French
N. R.
J. Finkle
Felix Quinn
Jas. McKewen
Jno. Cassidy
Wm. O'Reilly
O'Reilly
Tim O'Reilly
Jno. Cassidy
Mrs. Finlan
Pat. Foley
Chas. Murphy
Jno. Magrath
Canada Co.
Mrs. McDonald
C. Stevens
Pat Cassidy
Jas. Cassidy
L. Hailey
Wm. Badgley
Dennis Healy
Wm. Healy
Canada Co.
Jos. Tracy
Jno. O'Neil
S. Mulroney
Pat. Kerrens
Rich. Brougham
Mrs. Cavanagh
Nich. Magrath
Thos. Cassidy
Jas. Farrell
Jas. Brougham
Jos. Gartland
M. Hunt
Jno. Penny
R. Deckens
J. Deckens
Thos. O'Neil
Thos. Taylor
Jas. Stinson
D. Turkington
S. Farrell
Rich. Taylor
D. Turkington
Thos. Coe
Jas. Cassidy
C. Stevens
Chris Cassidy
L. Hailey
Jno. Murphy
Thos. Murphy
Dav. Larkin
Jos. Tracy
Hy. Twamley
Jno. P. Tracy
P. Cassidy
Thos. Tracy
Fro. Tracy
N. R.
Michl. Fox
Thos. Foley
Chas. Kinchela
Pat. Lareck
Mrs. Cavanagh
Thos. O'Neil
Thos. Doyle
Jas. Farrell
Canada Co.
Jno. Coe
Thos. Coe
D. Burne
Rich. Taylor
T. Henry
M. Flinn
Jno. Flinn
D. Burne
Canada Co.
Wm. Flemming
Wm. Henry
Peter Flemming
N. R.
Geo. Martin
Wm. Healey
Wm. Pine
N. Foley
Wm. Bloxom
MARLBANK
Jno. Bloxom
Jas. Young
Thos. Hayes
Farrell Jr.
Jno. Coe
N. R.
Thos. Doyle
Thos. Burns
Patk. Burns
N. R.
N. R.
Geo. Reed
Jno. Henderson
N. R.
Wm. Brandon
N. R.
N. R.
Rich. Martin
N. R.
N. R.
Rathburn & Son
Geo. Allen
N. Williams
Jas. Young
D. Fitzgerald
Chas. Young
Wm. Burley
Wm. Fitzgerald
Jno. Fitzgerald
J. N. Weighill
P. Brown
Josh. Weighill
Uriah Helime
Jas Campbell
Geo. Young
Jerry Baker
D. Smith
Jno. Yates
Jno. Brown
Wm. Martin
Wm. Keating
Simon Sexsmith
Geo. Martin
N. R.
Canada Co.
C. Andrews
N. R.
Chas. McLean

VILLAGE OF
MARLBANK
HUNGERFORD TP.
Scale 10 Chs-an Inch
CON VII
JAMES ST.
CONCESSION ROAD
ANN ST.
WILLIAM
MOIRA
CON VII
LOT 16
LOT 35
LIME LAKE P.O.
B. Morton
Alx Summerville
Wm Brandon
Thos Brandon
Thos. Germain
Jno. Maxwell
Nathan Dunn Sr
Geo Henderson
Jno Henderson
Wm Henderson
Oliver Wallace
Jas Henderson
Jas. McFarland
Nathan Dunn Jr
Jno Johnson
Jas. Martin
Jas. Alexander
Dan'l Larkin

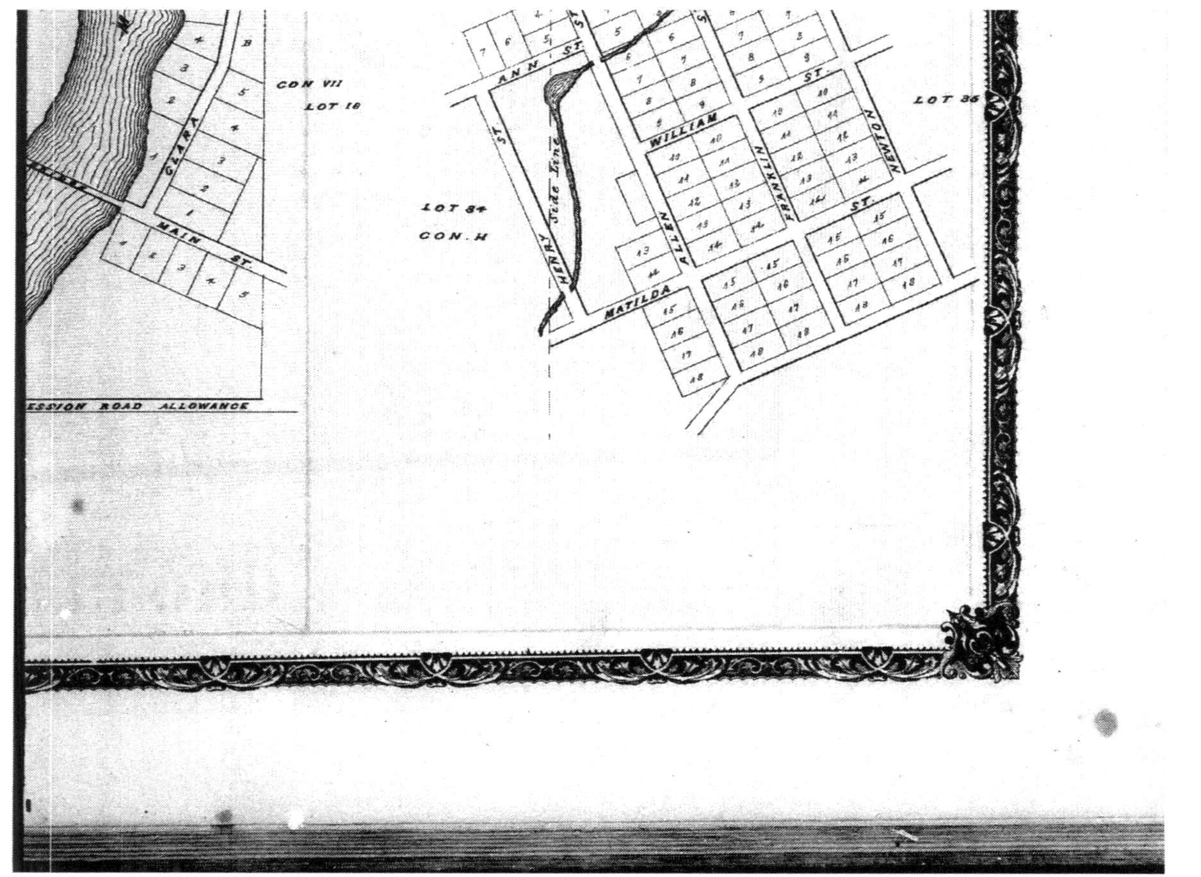
CON VII
LOT 18
LOT 34
CON. II
LOT 35
ANN ST.
HENRY ST.
Side Line
WILLIAM
FRANKLIN
NEWTON
ALLEN
MATILDA
ST.
CLARA
MAIN ST.
ESSION ROAD ALLOWANCE

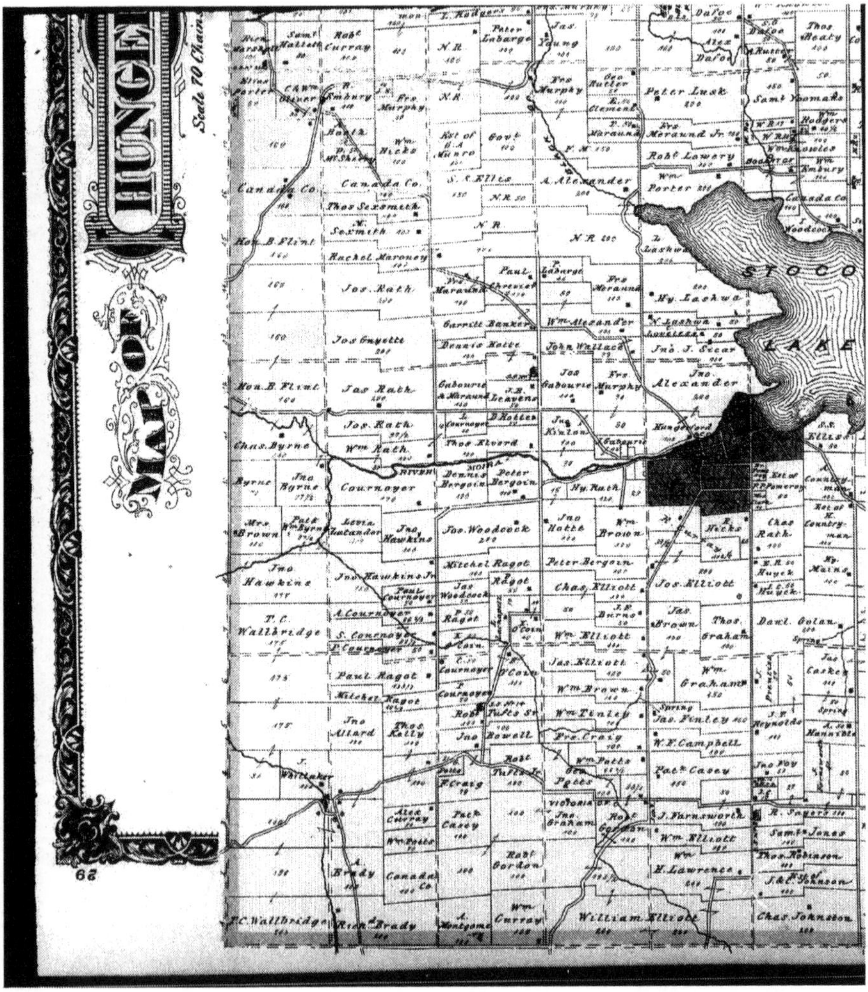

MAP OF HUNGE

Scale 70 Chains

STOCO LAKE

MOIRA RIVER

VICTORIA

Saml Hallett
Robt Curray
Peter Labarge
Jas. Young
Dafoe
Thos Beaty
Ch^s Wm Oliver
R. Embury
Fra. Murphy
N.R.
Fra Murphy
Geo Rutter
Peter Lusk
Saml Yeomans
Wm Hicks
Est of G.A. Munro
Gov^t
Fra Meraund Jr
Robt Lawery
Canada Co.
Thos Sexsmith
S. S. Ellis
A. Alexander
Wm Porter
Canada Co
N. Sexsmith
Hon. B. Flint
Rachel Maroney
Jos. Rath
Paul Chrevier
P. Labarge
Fra Meraund
Hy. Lashwa
Jos Guyette
Garrit Banker
Wm Alexander
N. Lashwa
Dennis Hatte
John Wallace
Jno. J. Secar
Hon. B. Flint
Jas Rath
Gabourie
J.B. Leavens
Jos Gabourie
Fra. Murphy
Jno. Alexander
Jos. Rath
Chas. Byrne
Wm Rath
Thos Elverd
Hungerford
S.S. Ellis
Jno Byrne
Cournoyer
Dennis Bergoin
Peter Bergoin
Hy. Rath
Est of E. Pomeroy
Mrs. Brown
Levia Latandon
Jno Hawkins
Jos. Woodcock
Jas Hatte
Wm Brown
Hicks
Chas Rath
Jno Hawkins
Jno Hawkins Jr
Mitchel Ragot
Peter Bergoin
Chas Elliott
Jos. Elliott
T.C. Wallbridge
A. Cournoyer
S. Cournoyer
P. Cournoyer
Jas Woodcock
J.B. Burns
Jas. Brown
Thos. Graham
Danl. Golan
Wm Elliott
Jas. Elliott
Wm Graham
Jas Caskey
Paul Ragot
Mitchel Ragot
Cournoyer
Wm Brown
Wm Tinley
Jas. Finley
Jno Allard
Thos Kelly
Jno Bowell
Fra. Craig
Wm F. Campbell
J. Wallaker
F. Craig
Robt Tufts Jr
Geo Potts
Wm Potts
Pat^k Casey
Jno Foy
Alex Curray
Pat^k Casey
Jno. Graham
J. Farnsworth
Saml Jones
Wm Potts
Robt Gordon
Wm Elliott
Thos Robinson
A. Brady
Canada Co.
Wm H. Lawrence
F.C. Wallbridge
Richd Brady
A. Montgomery
Wm Curray
William Elliott
Chas Johnston

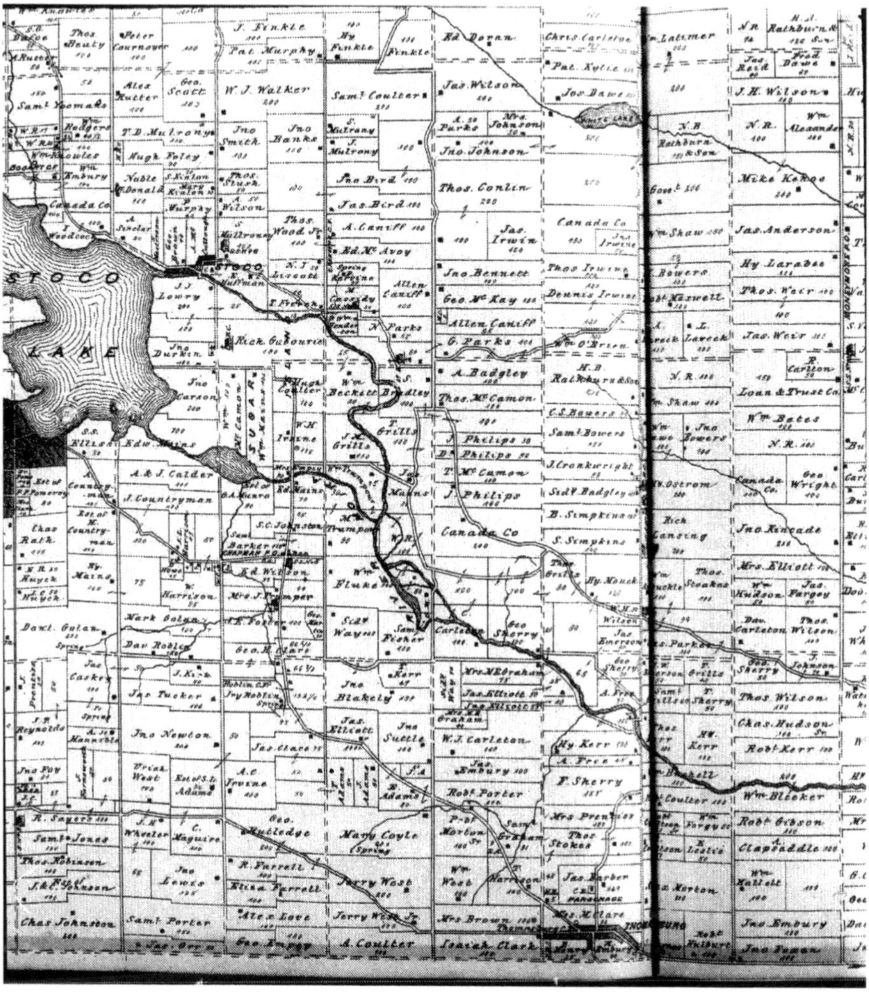

STOCO LAKE

J. Finkle
Pat. Murphy
Hy Finkle
Finkle
Ed. Doran
Chris. Carleton
Latimer
N.R.
Rathburn & Son
Thos. Beaty
Peter Cournoyer
Alex Rutter
Geo. Scott
W. J. Walker
Samˡ. Coulter
Jas. Wilson
Pat. Kylie
Jos. Dawe
Jas. Reid
Fred Dawe
J. H. Wilson
Samˡ. Yeomans
T. D. Mulrony
Jno Smith
Jno Banks
S. Mulrony
J. Mulrony
A. Parks
Mrs. Johnson
Jno. Johnson
N.R. Rathburn & Son
Wᵐ Alexander
Hugh Foley
Noble McDonald
Jno Bird
Jas. Bird
Thos. Conlin
Mike Kehoe
Canada Co
A. Sinclair
Thos. Wood Jr.
A. Caniff
Ed. McAvoy
Jas. Irwin
Canada Co
Jas. Irwine
Wᵐ Shaw
Jas. Anderson
STOCO
J. J. Lowry
Jno Bennett
Thos Irwine
Hy Larabee
Allen Caniff
Geo. McKay
Dennis Irwine
Robᵗ Maxwell
Thos. Weir
Jno Durkin
Rich. Gabourie
Allen Caniff
N. Parks
G. Parks
Wᵐ O'Brien
Jas. Weir
L. Laveck
A. Badgley
H. B. Rathburn & Son
Jno Carson
Wᵐ Beckett
Bradley
Thos. McCamon
Wᵐ Shaw
Loan & Trust Co.
R. Carlton
SUGAR
C. S. Bowers
Wᵐ Bates
S.S. Ellison
Edw. Mains
W. H. Irvine
J. M. Grills
T. Grills
J. Philips
Samˡ. Bowers
Jno Bowers
Dʳ Philips
J. Cronkwright
Wᵐ Ostrom
Canada Co.
Geo Wright
A. & J. Calder
Ed. Mains
Jas. Mains
T. McCamon
J. Countryman
J. Philips
Sidʸ Badgley
Chas Rath.
Est of M. Countryman
S. C. Johnston
Mʳ Trumpour
B. Simpkins
Rich Lansing
Jno Kincade
Canada Co
S. Simpkins
Samˡ Barker
Hy. Maine
Ed. Wilson
Wᵐ Fluke
Thos. Grills
Hy. Mouck
Thos. Stoakes
Mrs. Elliott
W. Harrison
Mrs. J. Trumper
Wᵐ Hudson
Jas. Fargey
Danˡ. Golan
Mark Golan
A. E. Foster
Sidʸ Wayne
Samˡ Fisher
C. Carleton
Geo Sherry
Wilson
Jas. Emerson
Dav. Carleton
Thos. Wilson
Dav Robles
Geo. H. Clare
S. Parker
Geo Sherry
Johnson
Jno Caskey
J. Kerr
Jno Blakely
Kerr
Mrs. McGraham
Jas. Elliott
Geo Sherry
Grills
Jas Tucker
T. Sherry
Thos. Wilson
J. P. Reynolds
Jno Newton
Jas. Elliott
Jno Suttle
Graham
W. J. Carleton
Hy. Kerr
Chas. Hudson
Robᵗ Kerr
Jas. Clare
Jas. Embury
A. Free
F. Sherry
Bushell
Jno Foy
Urick West
Est of S. L. Adams
A. C. Irvine
Adams
Robᵗ Porter
Coulter
Wᵐ Bleeker
R. Sayers
J. H. Wheeler
C. Maguire
Geo. Rutledge
Mary Coyle
Robᵗ Morton
Samˡ Graham
Mrs Prentice
Robᵗ Gibson
Samˡ Jones
Thos Stokes
A. Clapsaddle
Thos. Robinson
Jno Lewis
R. Farrell
Wᵐ West
Harrison
Jas. Barber
Wᵐ Hallett
Eliza Farrell
Jerry West
A. Morton
Chas Johnston
Samˡ. Porter
Alex Love
Jerry West Jr.
Mrs Brown
Mrs McClare
Jno Embury
Geo Embury
A. Coulter
Isaiah Clark
THOMASBURG
Robᵗ Hulburt
Jno Fagan

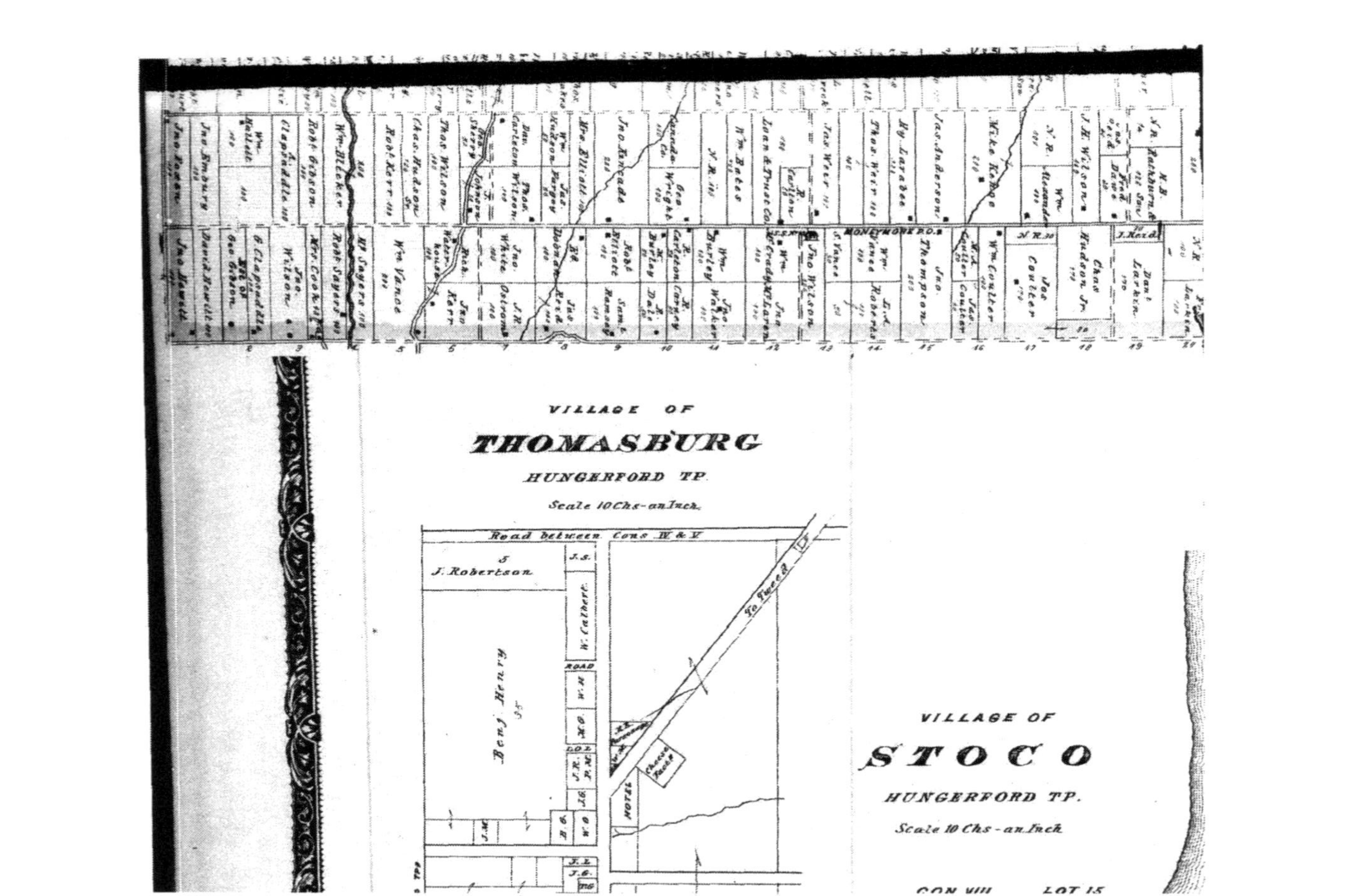

VILLAGE OF
THOMASBURG
HUNGERFORD TP.
Scale 10 Chs - an Inch
Road between Cons. IV & V
J. Robertson
Benj. Henry
W. Calbert
ROAD
To Tweed
Cheese Factory
HOTEL
VILLAGE OF
STOCO
HUNGERFORD TP.
Scale 10 Chs - an Inch
CON. VIII LOT 15
N.B. Rathburn & Son
Fred Dawe
J.H. Wilson
Wm Alexander
Mike Kehoe
Jas. Anderson
Hy. Larabee
Thos. Weir
Jas. Weir
R. Carlton
Loan & Trust Co.
Wm Bates
Canada Co.
Geo. Wright
Jno. Kincade
Mrs. Elliott
Wm Hudson
Jas. Fargey
Dav. Carleton
Thos. Wilson
Geo. Sherry
J. Johnson
Chas. Hudson Sr.
Robt. Kerr
Wm Blecker
Robt. Gibson
A. Clapsaddle
Wm Hallett
Jno. Embury
Jno. Foden
Dan. Larkin
Chas Hudson Jr.
Jos Coulter
Wm Coulter
M.A. Coulter
Jas. Coulter
Jno. Thompson
Wm Vance
L.A. Roberts
S. Vance
Jno. Wilson
Wm McCrady
Jno. McLaren
Wm Burley
Jno. Walker
R. Carleton
R. Carney
W. Burley
R. Dale
Robt. Elliott
Saml. Ramsey
Ed. Dobnan
Jas. Reed
Jno. White
J.R. Ostrom
Rich. Waterhouse
Jno. Kerr
Wm Vance
Mr. Sayers
Robt. Sayers
Mrs. Cook
Jno. Wilson
G. Clapsaddle
Est. of Geo. Gibson
David Hewitt
Jno. Hewett
MONEYMORE RD.

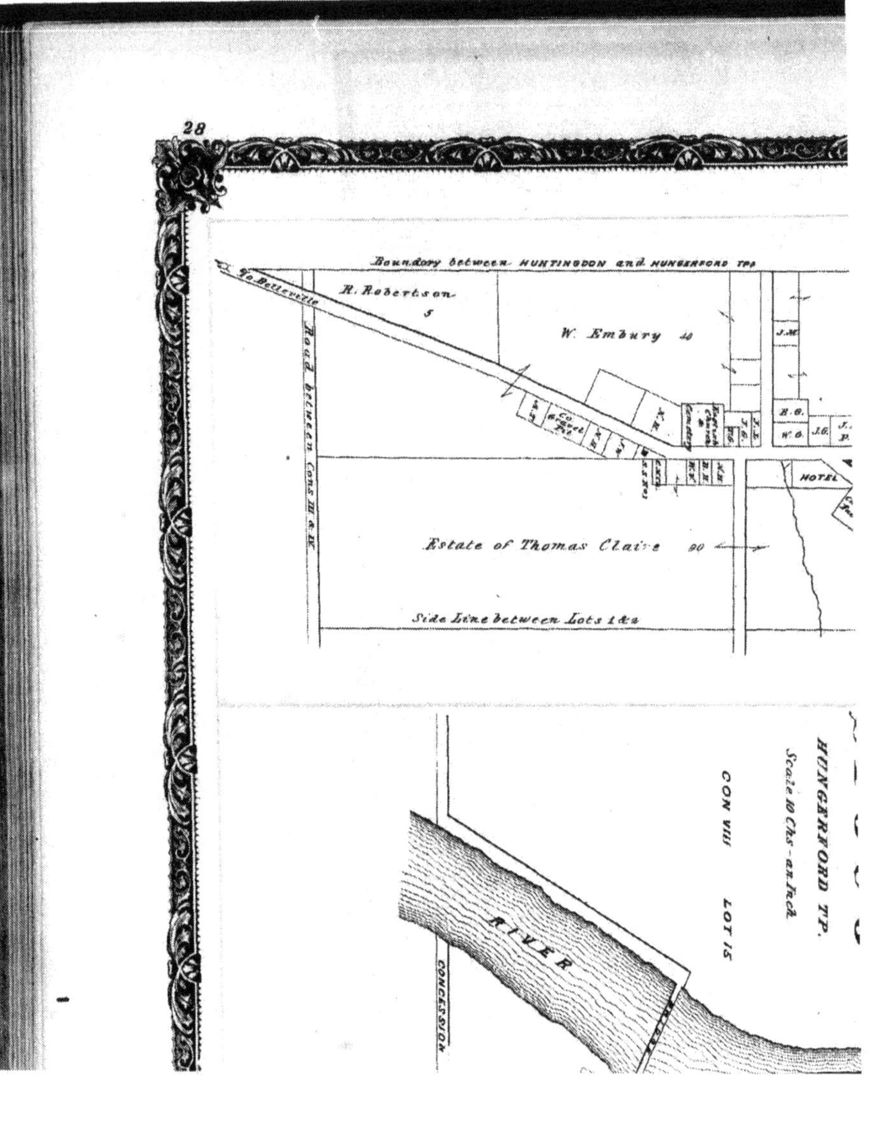
Boundary between HUNTINGDON and HUNGERFORD TPS
To Belleville
R. Robertson
5
W. Embury 40
Road between Cons III & IV
Estate of Thomas Claire 90
Side Line between Lots 1 & 2
HOTEL
English Church & Cemetery
HUNGERFORD TP.
Scale 10 Chs - an Inch
CON VIII
LOT 15
RIVER
CONCESSION

"PROSPECT HOUSE", THE RES. OF C.F. FRANCES ESQ. BARRISTER & ATTORNEY AT LAW, TRENTON, ONT.

PRESBYTERY. ST. GREGORY'S R.C. CHURCH, PICTON, ONT. SCHOOL.

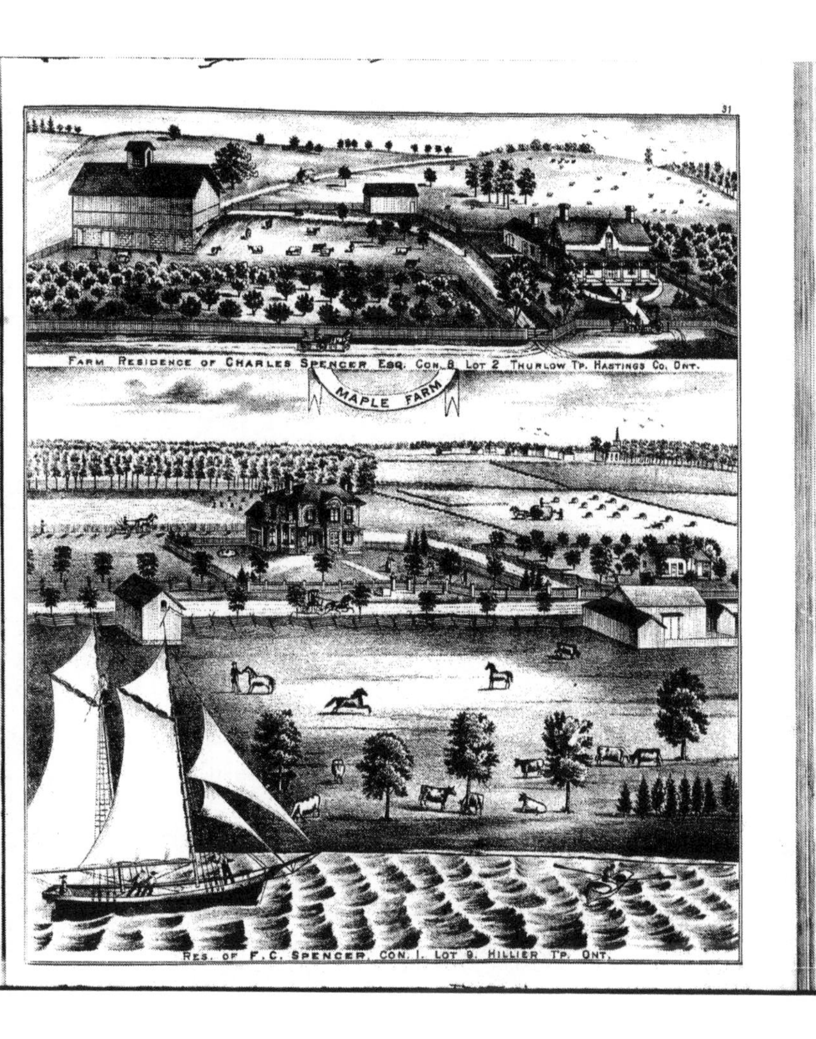

Farm Residence of Charles Spencer Esq. Con. 8 Lot 2 Thurlow Tp. Hastings Co. Ont.

Res. of F. C. Spencer. Con. I. Lot 9. Hillier Tp. Ont.

Mrs M. Plane 100
Jno Artis 100
Jno Bateman 100
Jas Jarvis 100
S.S. No 9
Non Resident 200
A. G. Northup 100
Canada Co. 200
Crown Dept 150
L. Wallbridge 100
Crown Dept 122
Non Resident 100
L. Wallbridge 100
Wm McBeath 100
W. F. Powell 100
W. P. Powell 100
J. Vestervelt 122
Jno Cooney 150
Canada Co. 200
Robt Sheldrake 100
Jno Ivers 100
Henry O'Hara 100
Non Resident 100
Crown Dept 200
Canada Co. 200
Non Resident 200
Jno Cooney 100
Jas Devlin 50
Robt Lyle 50
Alex Palmer 100
Jas O'Hara Jr. 50
Jas O'Hara Sr. 50
Dougall & Clure 100
S. Rollins 50
Hugh McKenzie 50
Jas Fox 100
Mrs M. McCoy 100
Neil McLure 100
Jas O'Hara 150
Canada Co. 250
Crown Dept 250
Chas Lewis 100
Jas Fitzgerald 100
Leonard Kane 100
Thos Rodgers 100
Saml Kyle 200
Alex McLure 100
Chas O'Hara 50
J. Lingham 25
Alex Trotter 75
Jno Rodgers 100
Philip Clapp 200
Canada Co. 200
Margt McLure 20
Wm West 150
Jno McLure 100
Wm West 100
Alex Cross 100
Jas Bailey 50
Est of C. Bailey 50
S.S. No 2
Wm McBeath 100
Ann McBeath 100
Jno Bateman 100
Donald McKenzie 100
Canada Co. 100
Jno Bateman 100
Isaac Canniff 50
Chas Lewis 50
Mrs C. Conner 100
Jno Haight
Clapp 99
Jno Elliss
Jno Adams 50
Jno McKinnon 197
Lime Kiln
Malcolm McIntosh 100
Geo Miller 97
Henry O'Hara 50
Geo Haggerty 100
Jas O'Hara Sr. 100
Hugh McKenzie 50
Rodk McKenzie 52
Saml Rollins 175
Brick Yard
Jas Elliss 50
Geo Miller 97
Mrs Vanderwater
Jno Stewart
Archd Harvey
Mrs M. Vallier 97
Patk Gillan 50
Seymour Wallbridge 100
Hugh Cameron 100
Jno Embury 100
Mrs McKay 200
Patk Feeny 100
Jno Ross 100
Charles F. Aylesworth
Jno St Charles
Wallbridge 100
Canada Co.
L. Wallbridge 100
Jno Welsh 100
Chas St Charles 100
Mrs McBeath
H. Seymour

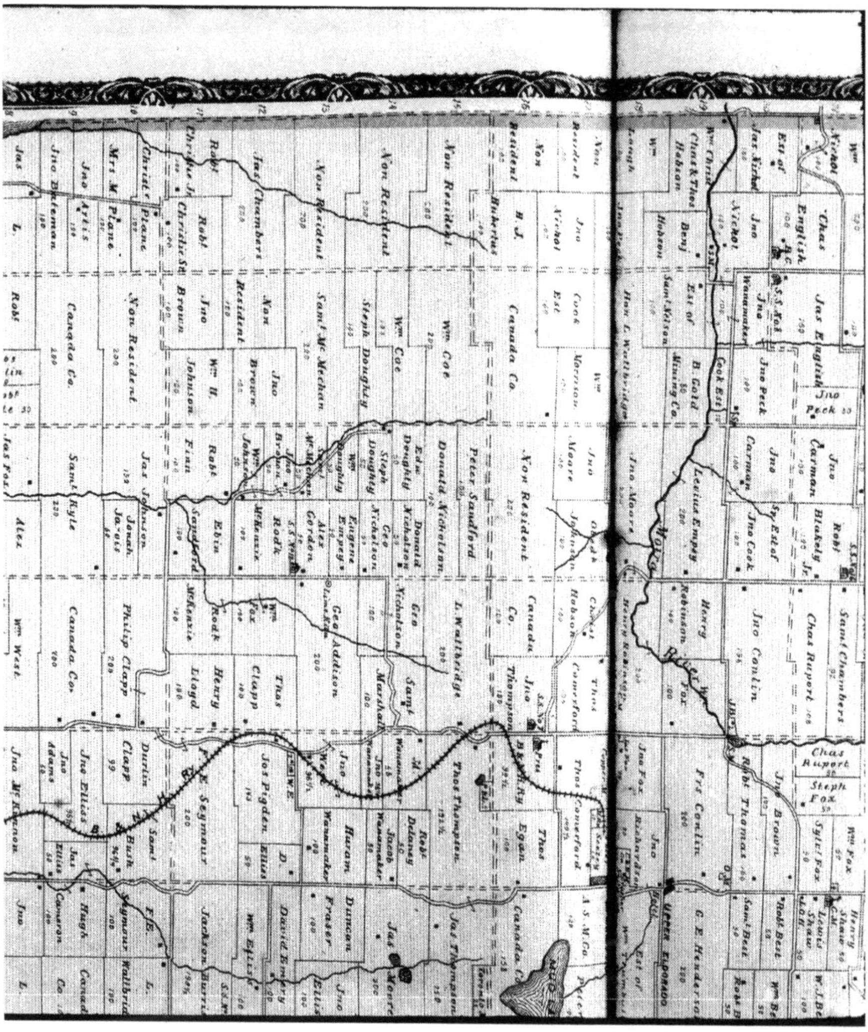

Wm Nichol
Chas English 100
Jas English 160
Jno Peck 50
Jno Carman 150
Robt Blakely Jr. 195
S.S. No. 6
Saml Chambers 92
Chas Rupert 108
Chas Rupert 50
Steph Fox 50
Wm Fox 50
Sylvr Fox 50
Henry Shaw 50
Lewis Shaw 50
W. J. Be
Est of Jas Nichol 150
Jno Nichol 150
S.S. No. 8
Jno Wanamaker
Jno Peck 100
Jno Carman 100
Sr. Est of Jno Cook 100
Jno Conlin 196
Jno Brown 100
Robt Thomas 100
Robt Best 50
Saml Best 50
Cook Est
Wm Christ
Chas & Thos Hobson
Benj Hobson
Est of Saml Nelson 100
B. Gold Mining Co. 50
Levius Empey 200
Henry Robinson 100
Moira River
Fox 100
Frs Conlin 200
G. E. Henderson 200
UPPER ELDORADO
Wm Lough
Jno Peck
Hon L. Wallbridge
Jno Moore 200
Henry Robinson
Jno Fox
Jno Richardson
Est of
Non Resident 100
Jno Nichol 100
Cook Est 100
Wm Morrison 100
Jno Moore 100
Obedh Johnson 100
Christ Hobson 100
Thos Comerford 105
Thos Comerford 149½
A. S. M. Co. 120
Mud Lake
Non Resident 100
H. J. Hubertus 100
Canada Co.
Non Resident 200
Canada Co. 100
Jno Thompson 100
S.S. No. 7
Pres
B. & N. H. Ry
Thos Egan 92½ 100
Canada Co. 158
Non Resident 200
Wm Coe 200
Peter Sandford 100
Donald Nicholson 100
L. Wallbridge 200
Thos Thompson 192½
Jas Thompson 150
Non Resident 200
Wm Coe 193
Edw Doughty 50
Donald Nicholson 50
Geo Nicholson 100
Saml Marshall 100
Jd Wanamaker 50
Robt Delaney 50
Jacob Wanamaker 50
Jno Wm Wanamaker
Jas Moore 200
Steph Doughty 145
Steph Doughty 50
Geo Nicholson 50
Wm Doughty
Eugene Empey 50
Non Resident 200
Saml Mc Mechan 200
Saml Mc Mechan 50
Alex Gordon 50
Lime Kiln
Geo Addison 200
Jno West 96½
Huram Wanamaker 100
Duncan Fraser 100
Jno Ellis
S.S. No. 5
Jno Brown 50
Rodk McKenzie 100
Wm Fox 100
Thos Clapp 100
W. E.
Jos Pigden 143
D. Ellis 50
David Emery 100
Jas Chambers 200
Non Resident 100
Jno Brown 100
Wm Johnson 50
Wm Ellis 100
Robt Christie Jr 100
Robt Christie Sr 100
Jno Brown 100
Wm H. Johnson 100
Robt Finn 100
Ebin Sandford 100
Rodk McKenzie 100
Henry Lloyd 100
F. E. Seymour 200
Jackson Burris 190½
S.S. No.
Christr Plane
Non Resident 200
Jas Johnson 150
Jonah Jarvis 50
Philip Clapp 200
Durlin Clapp 99
Saml Bush 94½
F. E. Seymour 100
L. Wallbridge 100
Mrs M. Plane 100
Jno Artis 150
Canada Co. 200
Saml Kyle 200
Canada Co. 200
Jno Ellis
Jas Ellis 50
Hugh Cameron 100
Canada Co.
Jno Bateman 100
Jno Adams 50
Jas
L.
Robt
Jas Fox
Alex
Wm West
Jno Mc Kinnon
Jno
L.

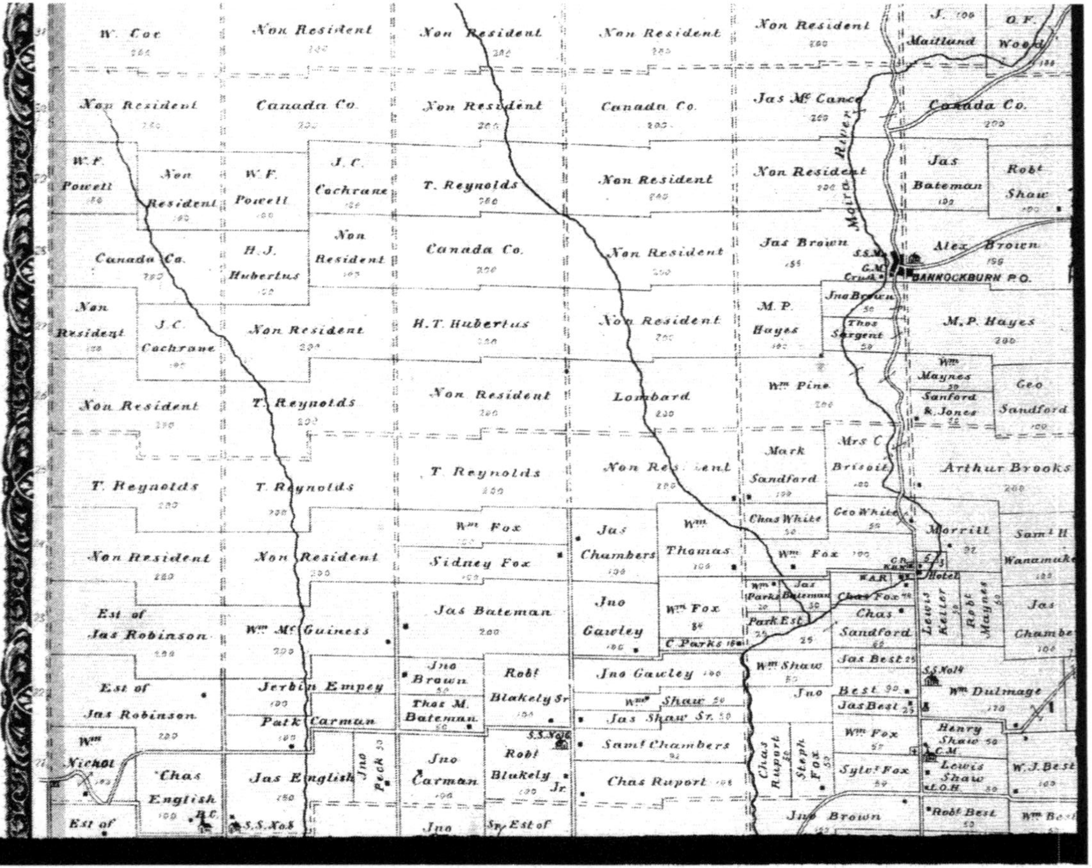

W. Coe 200
Non Resident 200
Non Resident 200
Non Resident 200
Non Resident 200
J. 100 Maitland
O. F. Wood

Non Resident 200
Canada Co. 200
Non Resident 200
Canada Co. 200
Jas McCance 200
Moira River
Canada Co. 200

W. F. Powell 100
Non Resident 100
W. F. Powell 100
J. C. Cochrane 100
T. Reynolds 200
Non Resident 200
Non Resident 200
Jas Bateman 100
Robt Shaw 100

Canada Co. 200
H. J. Hubertus 100
Non Resident 100
Canada Co. 200
Non Resident 200
Jas Brown 155
S.S.No.
C.M.
Cr'y.
BANNOCKBURN P.O.
Alex Brown 156

Non Resident 100
J. C. Cochrane 100
Non Resident 200
H. T. Hubertus 200
Non Resident 200
M. P. Hayes 100
Jno Brown 50
Thos Sargent 50
M. P. Hayes 200

Non Resident 200
T. Reynolds 200
Non Resident 200
Lombard 200
Wm Pine 200
Wm Maynes 50
Sanford & Jones 50
Geo Sandford 100

T. Reynolds 200
T. Reynolds 200
T. Reynolds 200
Non Resident 200
Mark Sandford 100
Mrs C. Brisoit 100
Arthur Brooks 200

Non Resident 200
Non Resident 200
Wm Fox 100
Sidney Fox 100
Jas Chambers 100
Wm Thomas 100
Chas White 50
Geo White 50
Wm Fox 100
Morrill 92
Saml H. Wanamaker 100

Est of Jas Robinson 200
Wm McGuiness 200
Jas Bateman 200
Jno Gawley 100
Wm Fox 84
C. Parks 16
Wm Parks 20
Jas Bateman 30
Park Est 25
25
Chas Fox
W.A.R.
Hotel
Chas Sandford 50
Lewis Keller 50
Robt Maynes 50
Jas Chambers 100

Est of Jas Robinson 200
Jerbin Empey 100
Patk Carman 100
Jno Brown 50
Thos M. Bateman 50
Robt Blakely Sr. 100
Jno Gawley 100
Wm Shaw 50
Jas Shaw Sr. 50
Saml Chambers 92
Wm Shaw 50
Jno Best 90
Jas Best 25
Jas Best 25
S.S.No.14
Wm Dulmage 178

Wm Nichol 100
Chas English 100
R.C.
Jas English 150
Jno Peck 50
Jno Carman 100
Robt Blakely Jr. 100
S.S.No.6
Chas Ruport 100
Chas Ruport 50
Steph Fox 50
Wm Fox 50
Sylvr Fox 50
Henry Shaw 50
C.M.
Lewis Shaw 50
L.O.H.
W. J. Best 100

Est of
S.S.No.8
Jno
Sr. Est of
Jno Brown 100
Robt Best 50
Wm Best

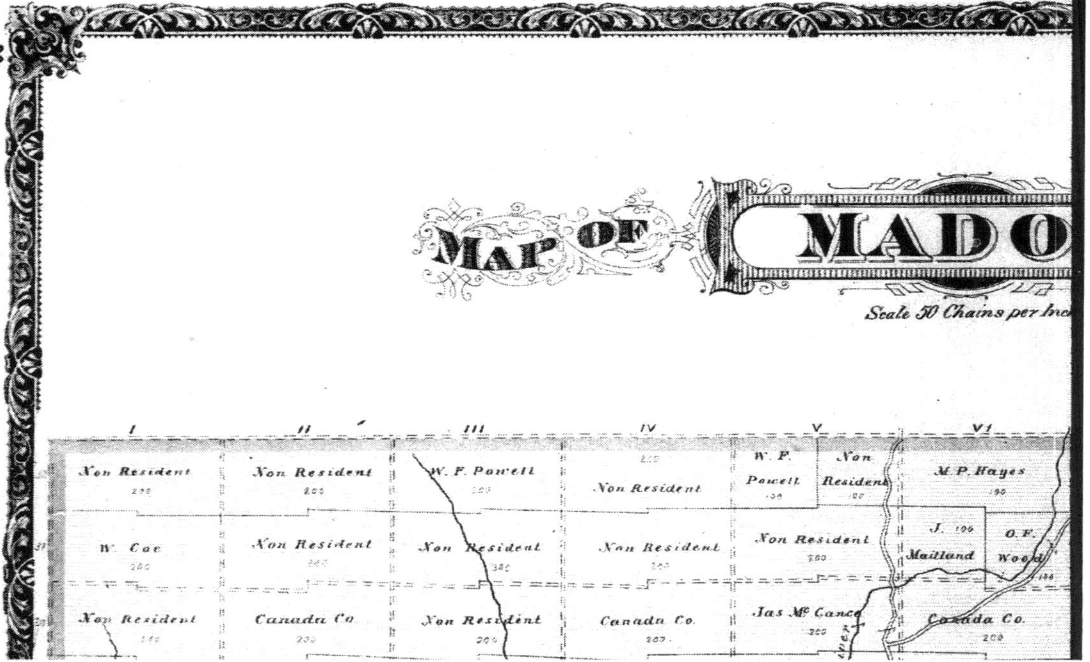
MAP OF
MADO
Scale 50 Chains per Inc
I
II
III
IV
V
VI
Non Resident
200
Non Resident
200
W. F. Powell
Non Resident
W. F.
Powell
Non
Resident
M. P. Hayes
W. Coe
200
Non Resident
Non Resident
Non Resident
Non Resident
J.
Maitland
O. F.
Wood
Non Resident
Canada Co.
Non Resident
Canada Co.
Jas Mc Cance
Canada Co.

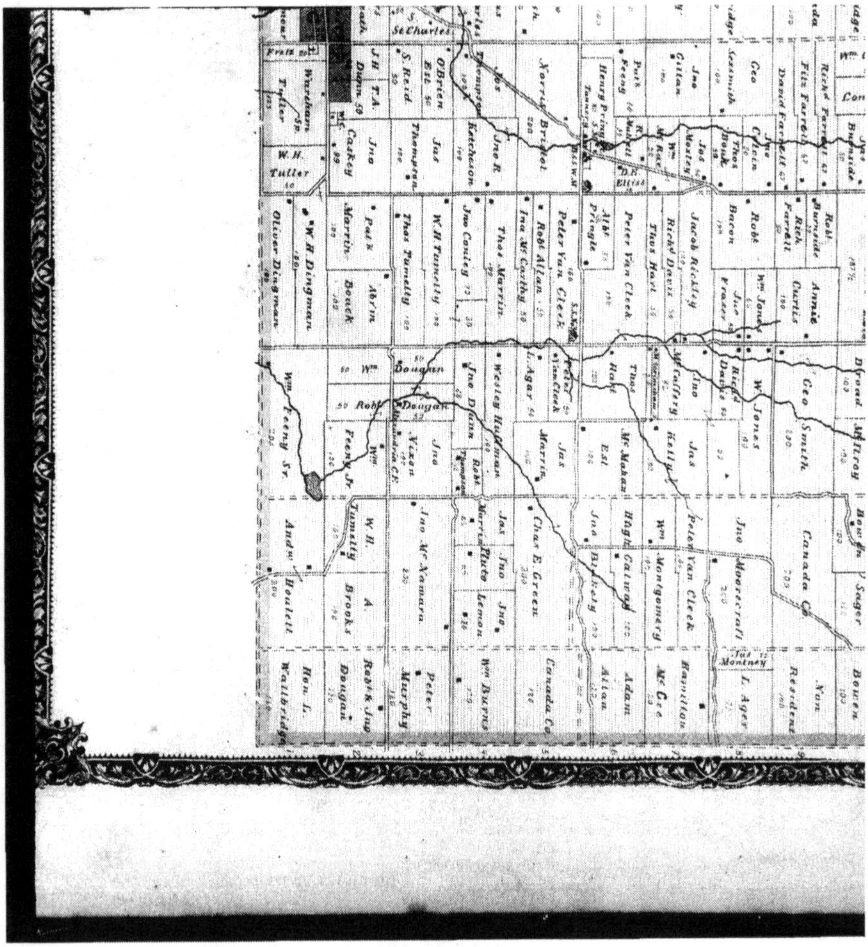

Jas Burnside
Richd Farrell
Fitz Farrell
David Farrell
Geo Sexsmith
Jno Colvin
Thos Bouk
Jno Gillan
Jas Moxley
Wm McRae
Patk Feeny
R. Mullett
D.R. Elliss
Henry Pringle
Tannery
Robt Burnside
Rich Farrell
Robt Bacon
Annie Curtis
Wm Jones
Jno Fraser
Jacob Rickley
Richd Davis
Thos Hart
Peter Van Cleek
Albt Pringle
B. Broad
McIlroy
Geo Smith
W. Jones
Richd Davis
Jno McCaffery
Jas Kelly
Thos Hart
McMahan Est.
Bowen
Sager
Canada Co.
Jno Moorecraft
Jas Montney
Peter Van Cleek
Wm Montgomery
Hugh Catway
Jno Blakely
Bowen
Non Resident
L. Ager
Hamilton
McCre
Adam Allan
Norris Bristol
Jos Thompson
St. Charles
O'Brien Est.
S. Reid
J.H. Dunn
T.A. Dunn
Jno R. Ketcheson
Jas Thompson
Jno Caskey
Wareham Sr.
Tuller
W.H. Tuller
Peter Van Cleek
Robt Allan
Ina McCarthy
Thos Marrin
Jno Conley
W.H. Tumelty
Thos Tumelty
Patk Marrin
Abr'm Bouck
W.H. Dingman
Oliver Dingman
Peter Van Cleek
L. Agar
Jas Marrin
Wesley Huffman
Jno Dunn
Robt Thompson
Wm Dougan
Robt Dougan
Jno Nixon
Alexandria C.E.
Wm Feeny Jr.
Wm Feeny Sr.
Chas E. Green
Jas Marrin
Jno Pluto
Jno Lemon
Jno McNamara
W.H. Tumelty
A. Brooks
Andw Houlett
Canada Co.
Wm Burns
Peter Murphy
Robt & Jno Dougan
Hon. L. Wallbridge

Jno Vye
Jno Hunter
Canada Co 100
Henry Woods
Jno Hazzard
Hugh Cooper
Geo Rollins
Jno Blair 200
Canada Co
Capt R Narman
A. B. Merrick
Thos Rimington
Danl Philips
Jno Broad
Geo Rollins
Saml D. Caskey
Non Resident
Robt Moore
Jno N. Moore
Jno McCoy Jr.
Wm McCoy
Thos Elliott
Est of Jos Kennedy
Hugh McKenzie
Jno Rhodes
Non Resident
RIMINGTON P.O.
Thos Preston
Patk Feeny
Wm Thompson
Benj Remington
Hiram Squires
Frs Elliott
Dan O'Connor
Kanegham Est
Non Resident
Est of Thos Bacon
Thos Bacon Sr
Chas Bacon
Chas Bacon
Canada Co.
Chas Bacon
E. H. Holmes
Steph Caldwell
Elias Holmes
Jas Berry
Jas Kincaid
Frs Martin
Wm Irwin
Jno McCoy
Chas Bacon
Jas McCaw
Jno Love
R.T. Ferguson
Alex Wright
Gilbert McCoy
Jas Berry
Abrm Diamond
Danl C. Ketcheson
J. Hayton
Jos Hayton
Thos & Rich Keene
Mrs Love
R.T. Ferguson
Robt Reid
Frank Trotter
Irwin Love
Jas Roy
Alex Wilson
Saml Kintons
Saml Colay
Thos Keene Jr.
Thos Keene Sr.
Jas Blakely
Mrs Fretz
Wm Kincaid
Mrs Love
Benj Swales
Jno Reid
Jno Eagleson
Thos Sympson
Jas Sympson
Wm Keene
Jno McCoy
MUD LAKE
C. Harris
Foot
Jno Bateman
Chas Bacon
LOH
Donald McKinnon
Fred McKinnon
Alex Glover
L. McKinnon
Wm Elliott
Jas Dixon
J. Lloyd
Alex McKinnon
Jos Jonin
Moses Lloyd
Jno McClure Jr.
Thos Badgley
Jos Hazzard
P. Young
Josh Broad
M. Kirk
L. Kirk
Mrs Wm Keene
Alex Kincaid
Wm Sager
Jos Montney
Canada Co.
Wm Gunn
Conners
Richd Farrell
Jas Burnside
Geo W. Rose
Jos Broad
Patk McIlroy
Mrs M Bowen
Jno Sager
Victor Bowen
Richd Farrell
Fitz Farrell
David Farrell
Robt Burnside
Rich Farrell
Annie Curtis
Geo Smith
Canada Co
Non Resident
Jno Colvin
Geo
Robt
Wm Jones
W. Jones
L. Auer

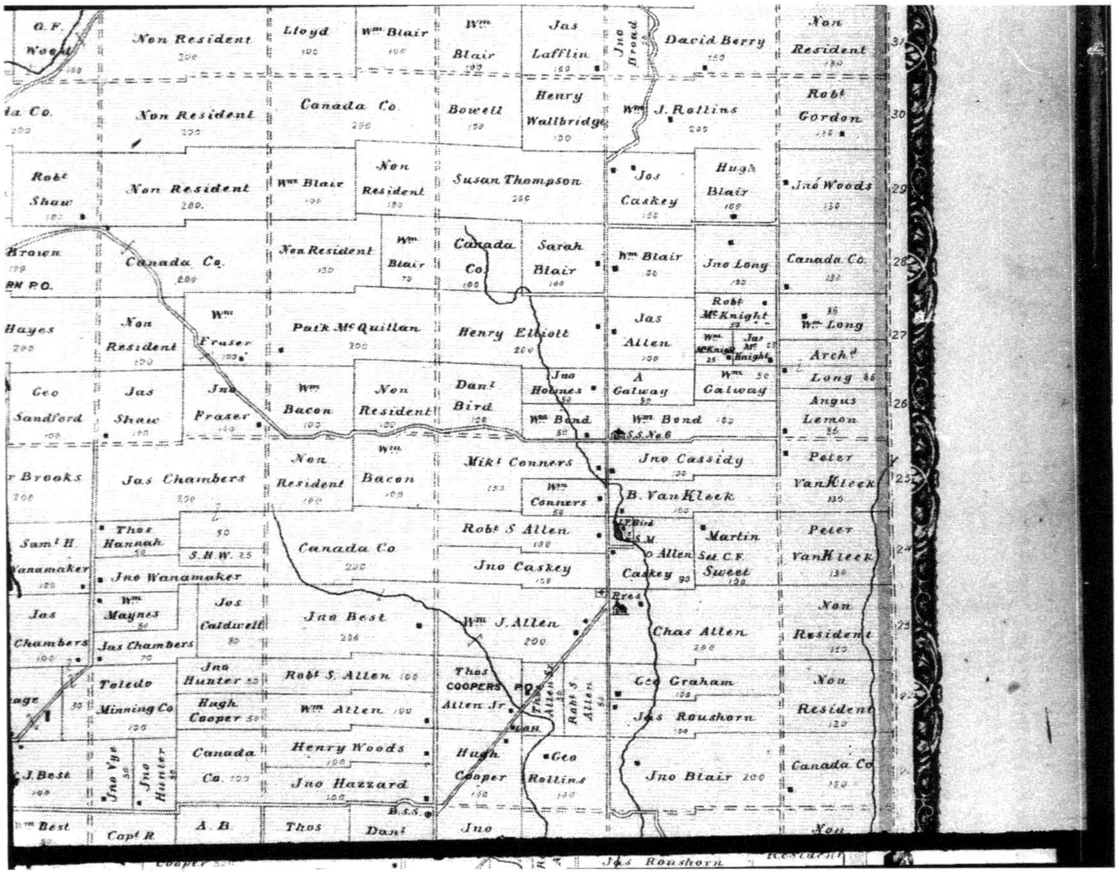

O. F. Wood 100
Non Resident 200
Lloyd 100
Wm Blair 100
Wm Blair 100
Jas Lafflin 100
Jno Broad 50
David Berry 150
Non Resident 150
31

Canada Co. 200
Non Resident 200
Canada Co. 200
Bowell 100
Henry Wallbridge 100
Wm J. Rollins 250
Robt Gordon 150
30

Robt Shaw 100
Non Resident 200.
Wm Blair 100
Non Resident 100
Susan Thompson 200
Jos Caskey 100
Hugh Blair 100
Jno Woods 150
29

Brown 100
Canada Co. 200
Non Resident 150
Wm Blair 70
Canada Co. 100
Sarah Blair 100
Wm Blair 100
Jno Long 100
Canada Co. 100
28

Hayes 200
Non Resident 100
Wm Fraser 100
Patk McQuillan 200
Henry Elliott 200
Jas Allen 100
Robt McKnight 50
Wm McKnight 25
Jas McKnight 25
Wm Long 50
Archd Long 50
27

Geo Sandford 100
Jas Shaw 100
Jno Fraser 100
Wm Bacon 100
Non Resident 100
Danl Bird 100
Jno Holmes 50
A Galway 50
Wm Galway 50
Wm Bond 50
Wm Bond 100
S.S. No. 6
Angus Lemon 50
26

Brooks 200
Jas Chambers 200
Non Resident 100
Wm Bacon 100
Mikl Conners 150
Jno Cassidy 100
Wm Conners 50
B. VanKleek 100
Peter VanKleek 150
25

Saml H. Wanamaker 100
Thos Hannah 50
S.H.W. 25
Canada Co 200
Robt S Allen 100
Jno Caskey 100
S.M.
Allen
Caskey 90
Martin Sweet 120
Peter VanKleek 150
24

Jas Chambers 100
Jno Wanamaker
Wm Maynes 50
Jas Chambers 70
Jos Caldwell 80
Jno Best 200
Wm J. Allen 200
Chas Allen 200
Non Resident 150
23

Toledo Mining Co 100
Jno Hunter 50
Hugh Cooper 50
Robt S. Allen 100
Wm Allen 100
Thos Allen Jr.
COOPERS P.O.
Thos Allen Sr. 50
Robt S Allen 50
Geo Graham 100
Jas Roushorn 100
Non Resident 150
22

J. Best 100
Jno Vye 50
Jno Hunter 50
Canada Co. 100
Henry Woods 100
Jno Hazzard 100
Hugh Cooper 100
Geo Rollins 100
Jno Blair 200
Canada Co 150

Wm Best 50
Capt R.
A. B.
Thos
Danl
B.S.S.
Jno
Non

Cooper
Jas Roushorn
Resident

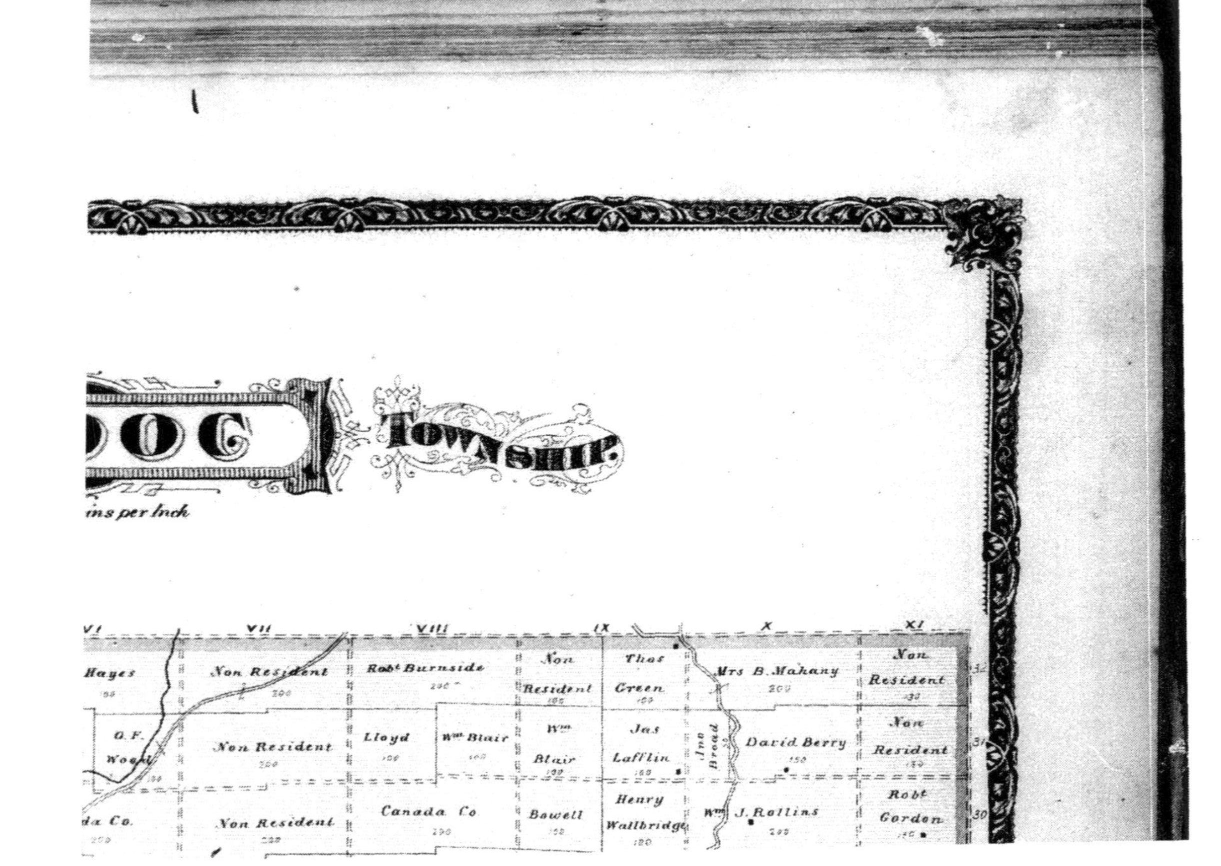

DOG
TOWNSHIP
ins per Inch
VI
VII
VIII
IX
X
XI
Hayes
Non Resident
Robt Burnside
Non Resident
Thos Green
Mrs B. Mahany
Non Resident
O. F. Wood
Non Resident
Lloyd
Wm Blair
Wm Blair
Jas Lafflin
Jno Broad
David Berry
Non Resident
da Co.
Non Resident
Canada Co
Bowell
Henry Wallbridge
Wm J. Rollins
Robt Gordon
32
31
30

RES. OF MATHEW BENSON, CON. 2. LOT 14. SOPHIASBURGH TP. ONT.

"COTE DU BAIE." RES. OF CAPT W. H. MORDEN, GREEN POINT, SOPHIASBURGH TP. ONT.

RES. OF S. D. CRANDALL CON. 1, LOT 14, ATHOL TP. ONT.

FARM RES. OF JOHN C. VERMILYEA ESQ. CON. 3, LOT 2. THURLOW TP. HASTINGS CO.

RES. OF REUBEN YOUNG ESQ. CARRYING PLACE, LOT 1. PRINCE EDWARDS CO. ONT.

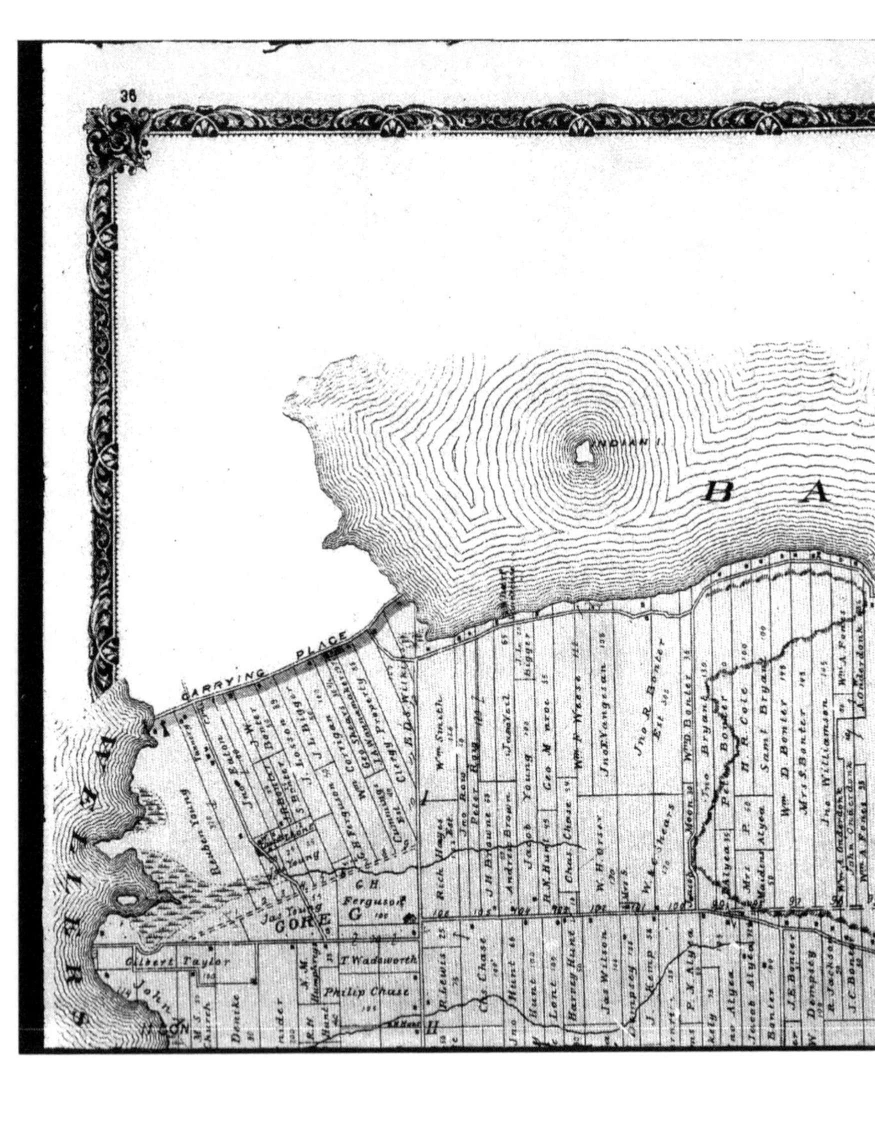

INDIAN I.
B A
CARRYING PLACE
GORE
Reuben Young
Jno Eaton
Wm Smith
Jno Row
Peter Row
J H Brount
Andrew Brown
Jacob Young
Geo Munroe
Chas Chase
Wm P Weese
Jno R Bonter
Sam'l Bryant
Wm D Bonter
Mrs S Bonter
Jno Williamson
Wm A Fones
Rich Hayes
G H Ferguson
Jas Young
Gilbert Taylor
T Wadsworth
Philip Chase
R Lewis
Chs Chase
Jno Hunt
Lont
Harvey Hunt
Jas Wilson
Dempsey
J Kemp
P. H. Alyea
Jacob Alyea
Bonter
J. E. Bonter
Dempsey
R. Jackson
J. C. Bonter

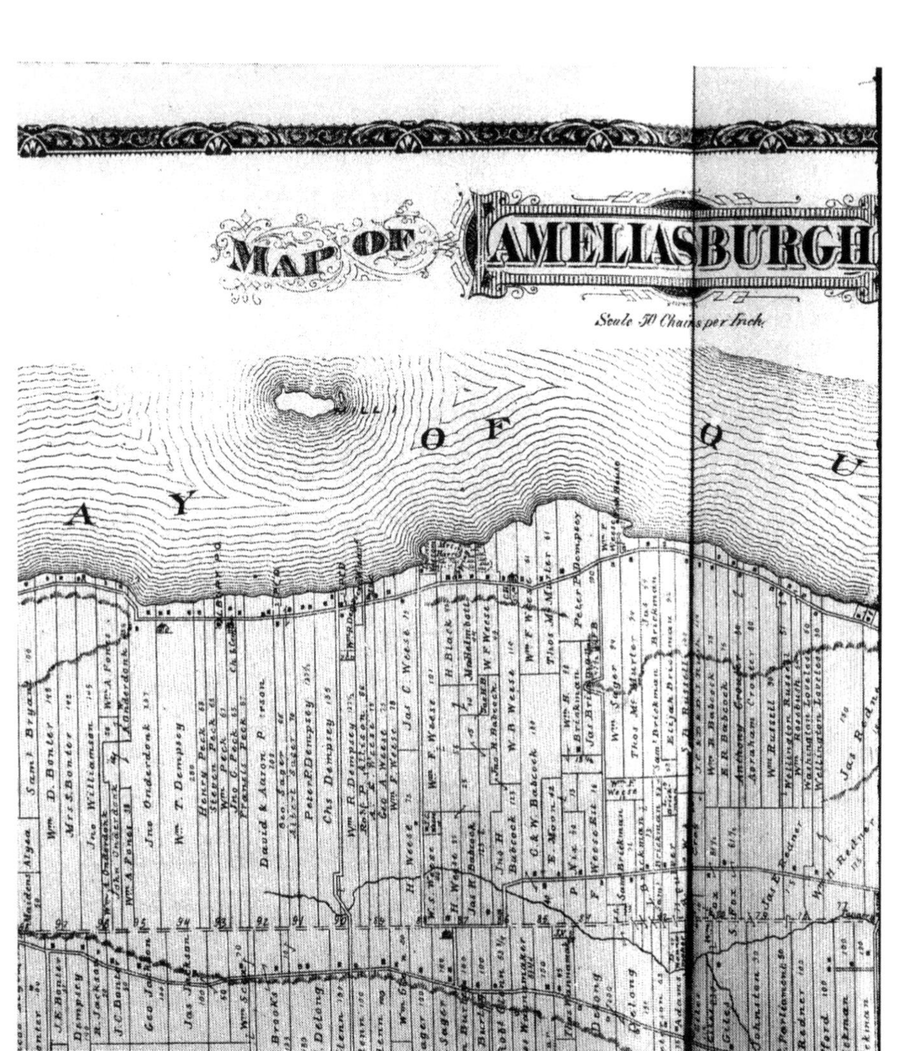
MAP OF AMELIASBURGH
Scale 50 Chains per Inch.
A Y O F Q U
Wm T. Dempsey
Chs Dempsey
Wm R. Dempsey
W. B. Weese
C & W Babcock
Thos Mc Murter
Jas Redner

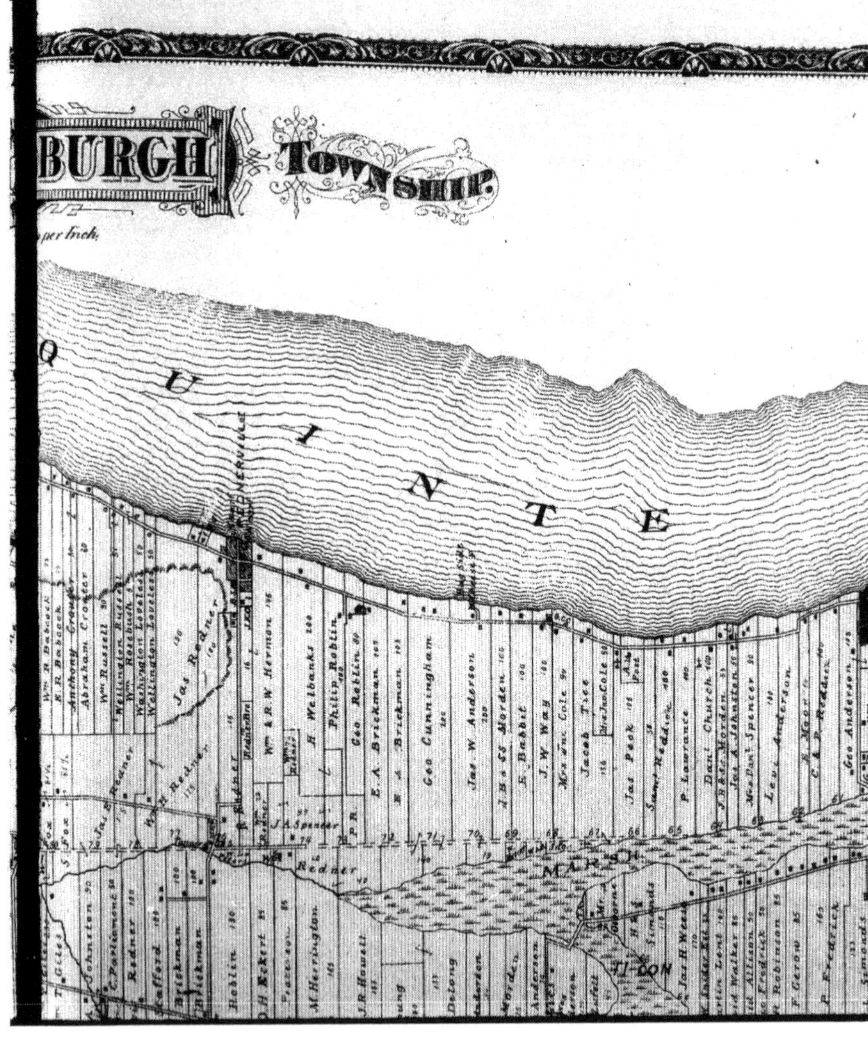
BURGH TOWNSHIP.
per Inch.
Q U I N T E
MARSH

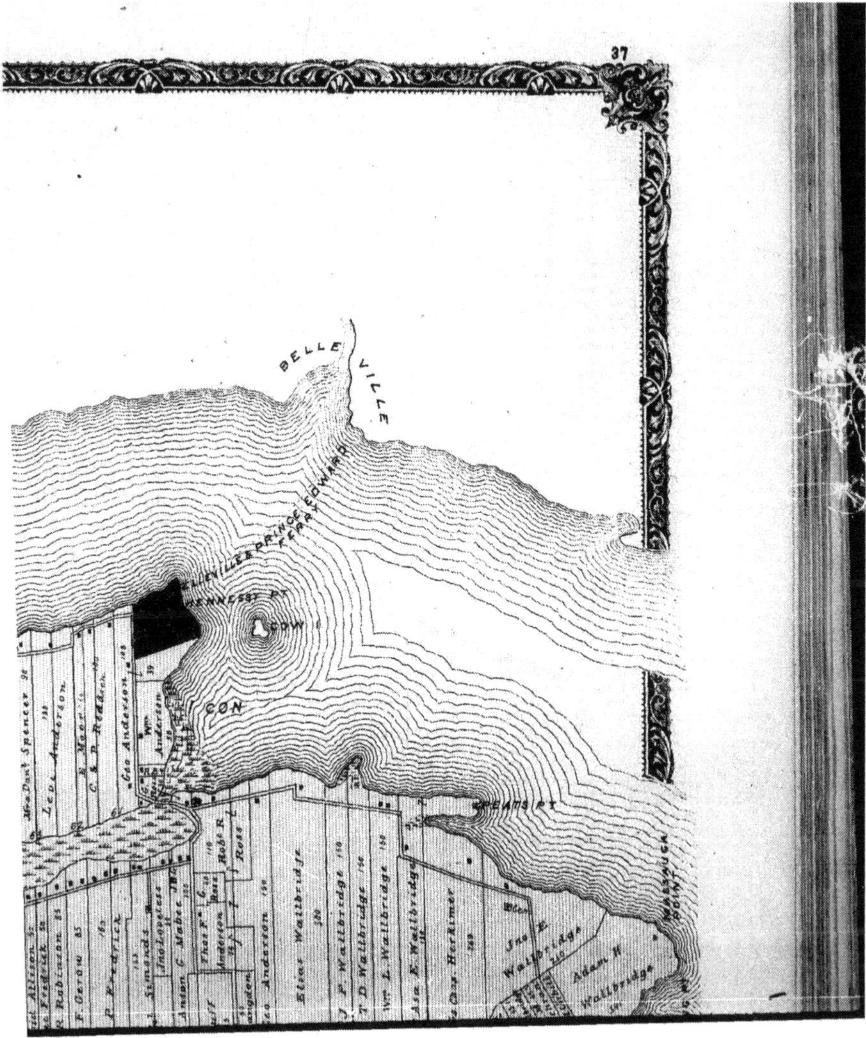
BELLE
VILLE

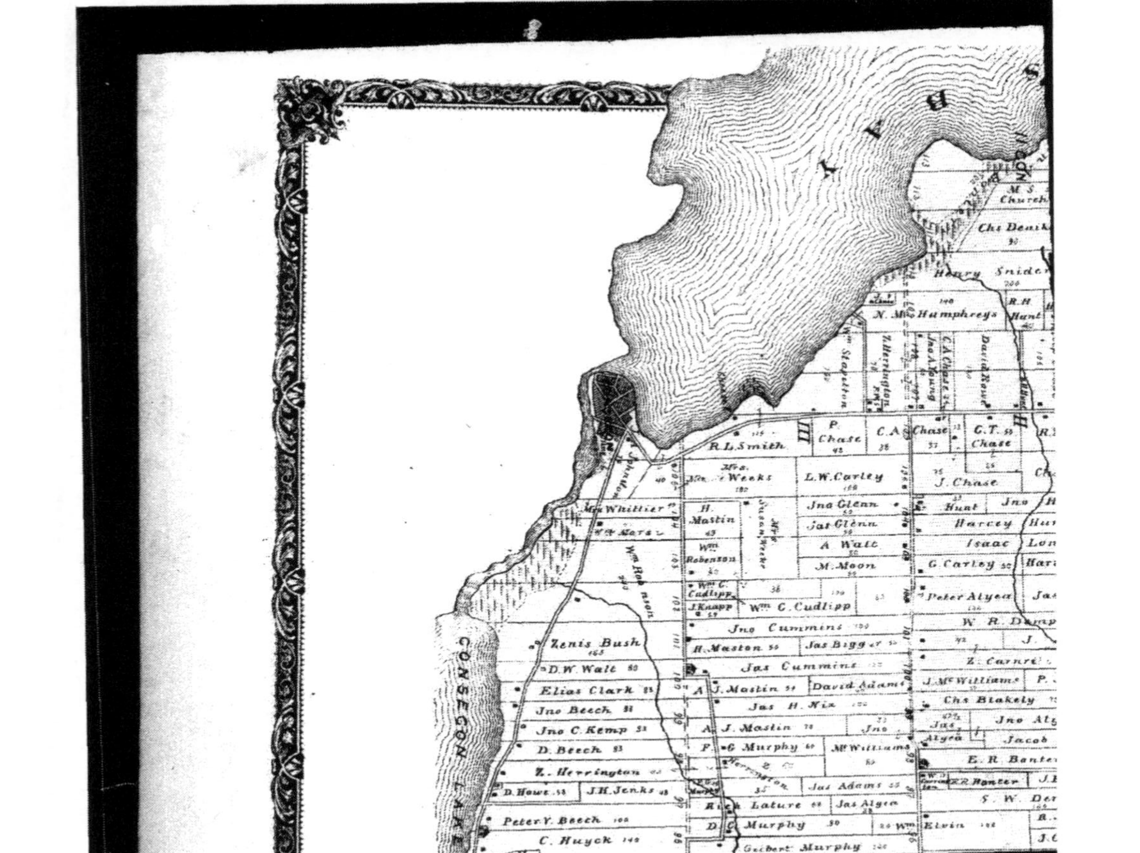

II CON
BAY
CONSECON LAKE
M.S. Church
Chs Denike
Henry Snider
N. McHumphreys
R.H. Hunt
Wm Stapelton
Z. Herrington
C.A. Chase
Jno A. Young
David Rowe
R.L. Smith
P. Chase
C.T. Chase
J. Chase
L.W. Carley
Jno Glenn
Jas Glenn
A. Walt
M. Moon
G. Carley
Peter Alyea
W.R. Demp
H. Mastin
Wm Robenson
Mrs Susan Weeks
Wm Robinson
Wm C. Cudlipp
J. Knapp
Jno Cummins
H. Maston
Jas Bigger
Jas Cummins
Zenis Bush
D.W. Walt
Elias Clark
Jno Beech
Jno C. Kemp
D. Beech
Z. Herrington
D. Howe
J.H. Jenks
Peter V. Beech
C. Huyck
J. Mastin
David Adams
Jas H. Nix
G. Murphy
McWilliams
Jas Adams
Lature
Jas Algea
Murphy
Gilbert Murphy
J. McWilliams
Chs Blakely
Z. Carnrite
Jacob
E.R. Banter
S.W. Der
Elvin
Harcey
Isaac

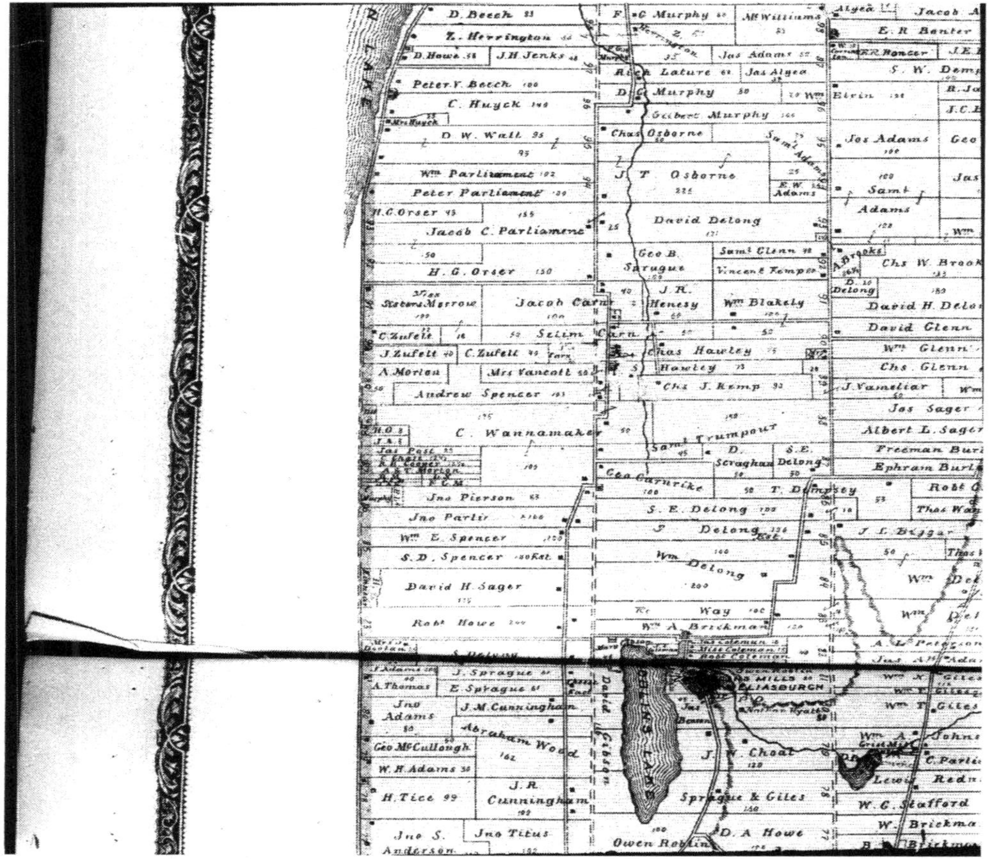

D. Beech
Z. Herrington
D. Howe
J.H. Jenks
Peter Y. Beech
C. Huyck
D.W. Wall
Wm Parliament
Peter Parliament
H.C. Orser
Jacob C. Parliament
H.G. Orser
Misses Sisters Morrow
Jacob Carn
C. Zufelt
Selim Carn
J. Zufelt
C. Zufelt
A. Morton
Mrs Vancott
Andrew Spencer
C. Wannamaker
Jas Post
Jno Pierson
Jno Parlir
Wm E. Spencer
S.D. Spencer
David H. Sager
Robt Howe
J. Sprague
A. Thomas
E. Sprague
Jno Adams
J.M. Cunningham
Abraham Wood
Geo McCullough
W.H. Adams
H. Tice
J.R. Cunningham
Jno S. Anderson
Jno Titus
N LAKE
F. G. Murphy
Mr Williams
Jas Adams
Rich Lature
Jas Algea
D. G. Murphy
Gilbert Murphy
Chas Osborne
Saml Adams
J. T. Osborne
E.W. Adams
David Delong
Geo B. Sprague
Saml Glenn
Vincent Kemper
J.R. Henesy
Wm Blakely
Chas Hawley
Hawley
Chs J. Kemp
Saml Trumpour
D. Scraghan
S.E. Delong
Geo Carnrike
T. Dempsey
S.E. Delong
Delong
Wm Delong
Way
Wm A. Brickman
Miss Coleman
Robt Coleman
ELIASBURGH
David Gibson
ROBLINS LAKE
J. W. Choat
Sprague & Giles
D.A. Howe
Owen Roblin
Algea
Jacob A
E.R. Banter
E.R. Roncer
S.W. Demp
Elvin
Jos Adams
Saml Adams
A. Brooks
Chs W. Brook
D. Delong
David H. Delo
David Glenn
Wm Glenn
Chs. Glenn
J. Vanseliar
Jas Sager
Albert L. Sager
Freeman Burl
Ephram Burl
Robt C
Thos Wa
J.L. Biggar
A. L. Peterson
Wm A.
C. Parlia
Lewis Redn
W.G. Stafford
W. Brickma

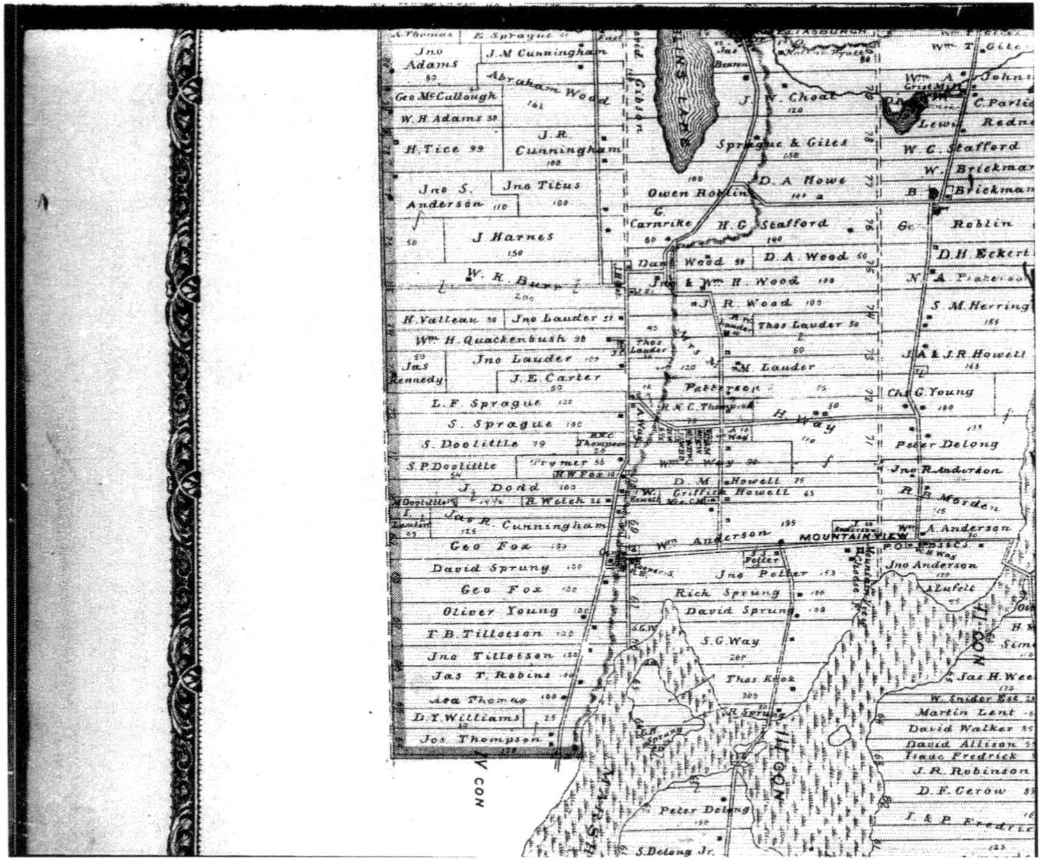

A. Thomas
E. Sprague
Jno Adams
J. M Cunningham
Abraham Wood
Geo McCullough
W. H. Adams
J. R. Cunningham
H. Tice
Jno S. Anderson
Jno Titus
J. Harnes
W. R. Burr
H. Valleau
Jno Lauder
Wm H. Quackenbush
Jas Kennedy
Jno Lauder
J. E. Carter
L. F. Sprague
S. Sprague
S. Doolittle
S. P. Doolittle
H. W. Fox
J. Dodd
R. Welch
Jas R. Cunningham
Geo Fox
David Sprung
Geo Fox
Oliver Young
T. B. Tillotson
Jno Tillotson
Jas T. Robins
Asa Thomas
D. Y. Williams
Jos. Thompson
IV CON
David Gibson
J. N. Choat
Sprague & Giles
Owen Roblin
D. A Howe
G. Carnrike
H. G. Stafford
D. A. Wood
Jno & Wm H. Wood
J. R. Wood
Thos Lauder
M. Lauder
Patterson
H. N. C. Thompson
H. Way
A. Way
Wm C. Way
D. M. Howell
W. Griffith Howell
Wm Anderson
MOUNTAIN VIEW
Jno Potter
Rick Sprung
David Sprung
S. G. Way
Thos Kook
R. Sprung
Peter Delong
S. Delong Jr.
MARSH
III CON
Wm T. Giles
Wm A. Johnson
C. Parliament
Lewis Redner
W. C. Stafford
W. Brickman
B. Brickman
Roblin
D. H. Eckert
N. A. Peterson
S. M. Herring
J. A & J. R. Howell
Chs G. Young
Peter Delong
Jno R. Anderson
R. B. Morden
Wm A. Anderson
P.O.
Jno Anderson
A. Zufelt
II CON
Jas H. Weese
W. Snider Est.
Martin Lent
David Walker
David Allison
Isaac Fredrick
J. R. Robinson
D. F. Cerow
I. & P. Fredrick

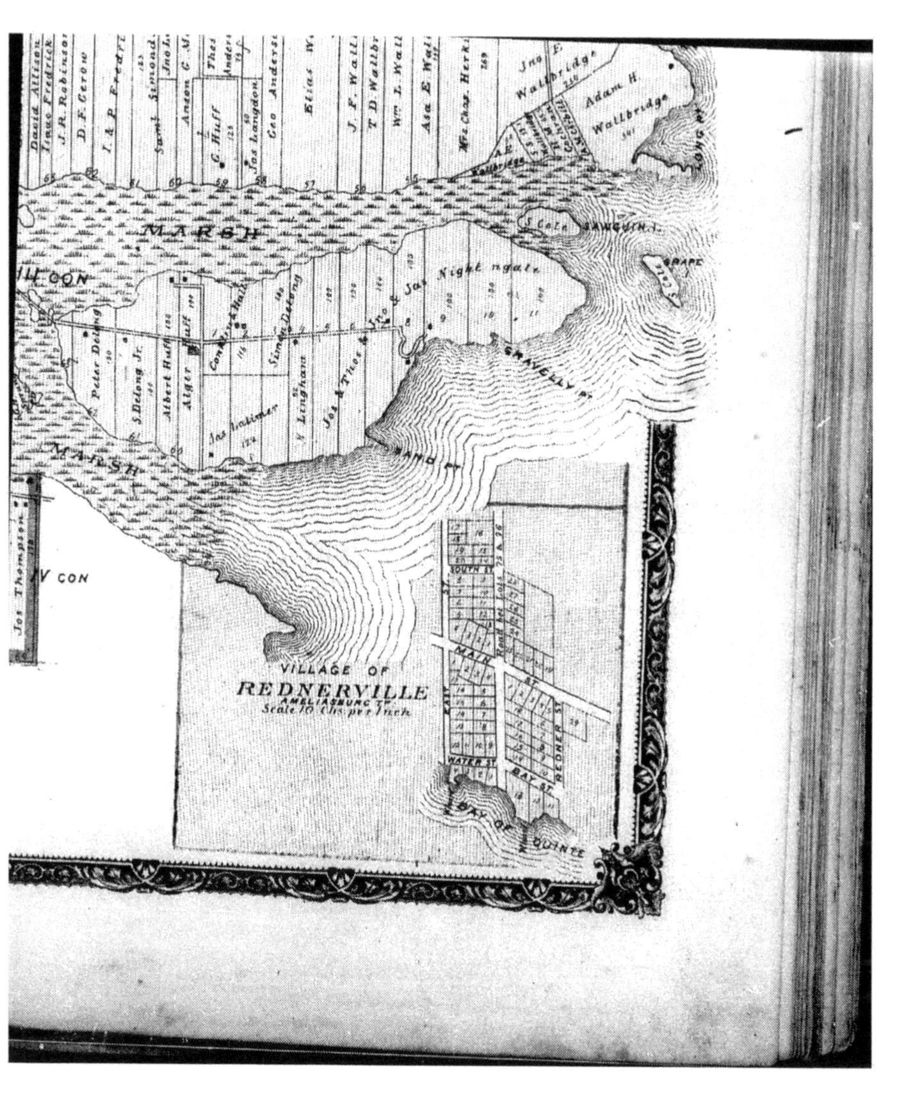

David Allison
Isaac Fredrick
J.R. Robinson
D.F. Gerow
Jas Langdon
Adam H. Wallbridge
MARSH
Peter Delong
S.Delong Jr.
Albert Huff
Simon Delong
Jas Latimer
Jos Thompson
IV CON
GRAVELLY PT
VILLAGE OF
REDNERVILLE
AMELIASBURG TP.
Scale 10 Chs per Inch
SOUTH ST
MAIN ST
WATER ST
BAY ST
REDNER ST
BAY OF QUINTE

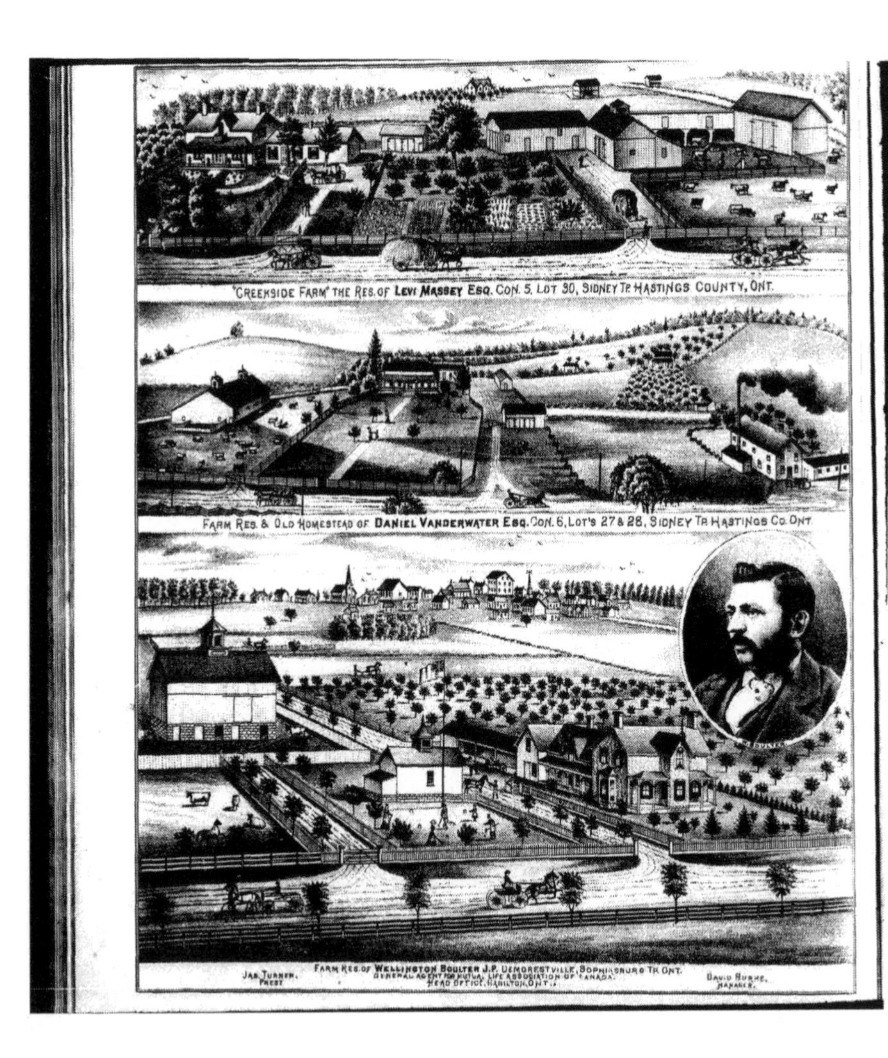

"CREEKSIDE FARM" THE RES. OF LEVI MASSEY ESQ. CON. 5, LOT 30, SIDNEY TP. HASTINGS COUNTY, ONT.

FARM RES. & OLD HOMESTEAD OF DANIEL VANDERWATER ESQ. CON. 6, LOT'S 27 & 28, SIDNEY TP. HASTINGS CO. ONT.

FARM RES. OF WELLINGTON BOULTER J.P. DEMORESTVILLE, SOPHIASBURG TP. ONT.
GENERAL AGENT FOR MUTUAL LIFE ASSOCIATION OF CANADA.
HEAD OFFICE, HAMILTON, ONT.

JAS. TURNER, PREST.

DAVID BURKE, MANAGER.

RES. OF WM. CLINTON, ERVINDOHE, LOT 72, HALLOWELL TP. ONT.

"WILBETH COTTAGE" RES. OF MRS. E. W. BEDELL, CON. 1, LOT 1, HALLOWELL TP. ONT.

RES. OF LEWIS LAZIER, CON. 2, LOT 16, HALLOWELL TP. ONT.

RES. OF S. M. CONGER, PICTON, ONT.

"BAY LODGE", THE RES. OF JAMES S. RYAN ESQ. & ELEVATOR, TRENTON, ONT.

L. T. VORCE PROPRIETOR OF THE ONLY FIRST CLASS LIVERY IN PICTON.
PARTICULAR ATTENTION GIVEN TO COMMERCIAL TRAVELLERS & EXCURSIONS TO SAND BANKS.

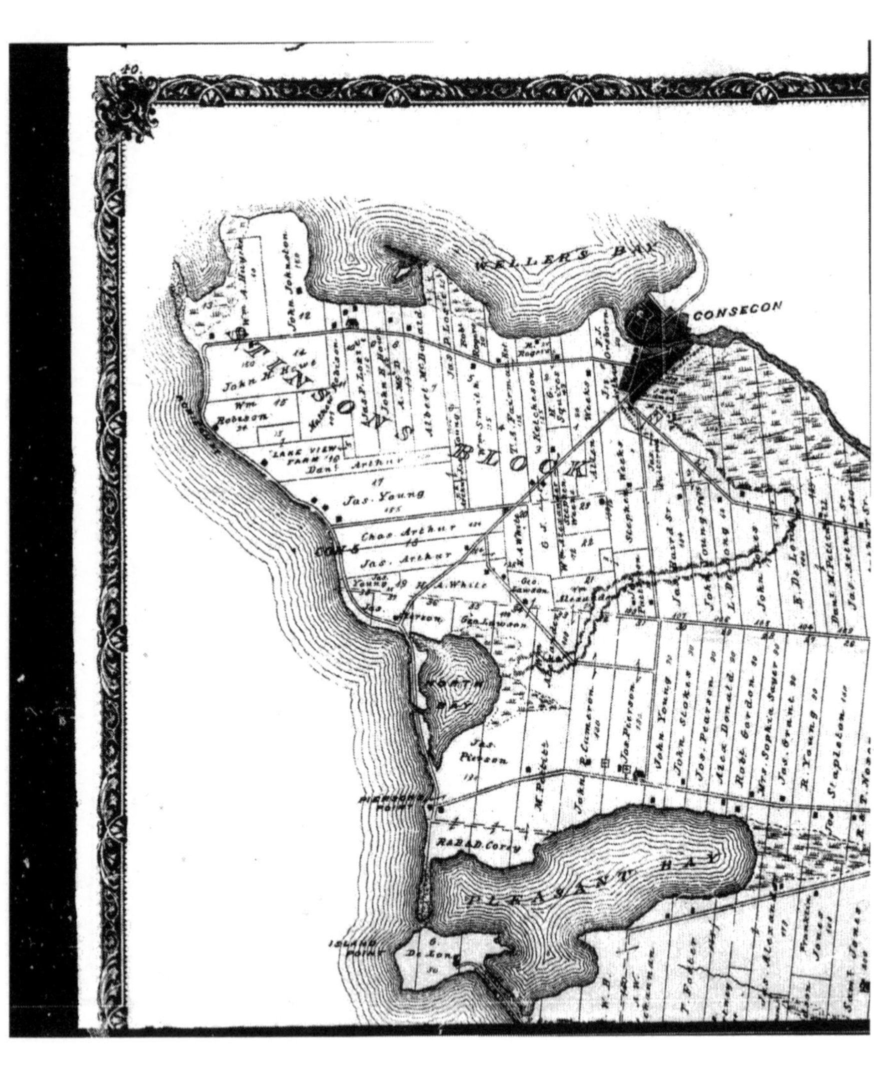
WELLER'S BAY
CONSECON
John Johnston
John H. Howe
Wm Robson
LAKE VIEW FARM
Dan'l Arthur
Jas. Young
Chas. Arthur
Jas. Arthur
H. A. White
Geo. Lawson
Albert McDonald
BLOCK
Stephen Weeks
NORTH BAY
Jas. Pierson
John R. Cameron
Joe Pierson
John Young
John Stokes
Jos. Pearson
Alex Donald
Robt Gordon
Mrs. Sophia Seyer
Jas. Grant
R. Young
Stapleton
R.&D.D. Corey
PLEASANT BAY
ISLAND POINT
De Long
T. Foster

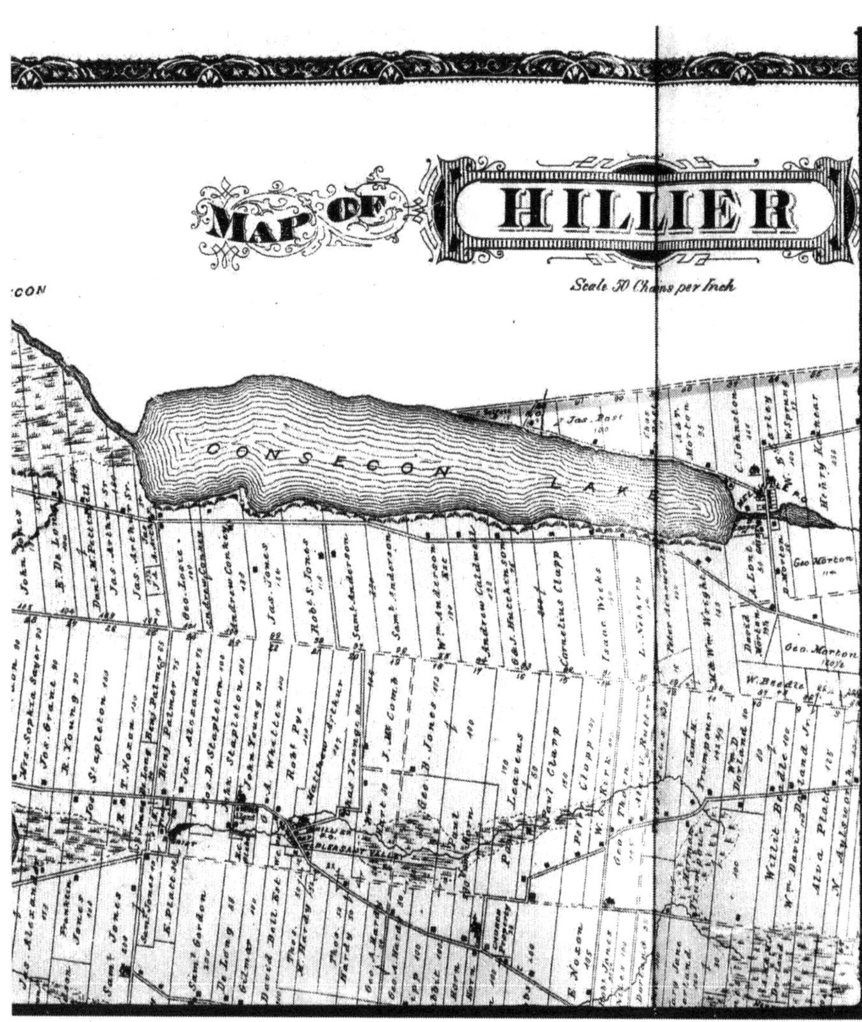
MAP OF
HILLIER
Scale 50 Chains per Inch
CONSECON LAKE

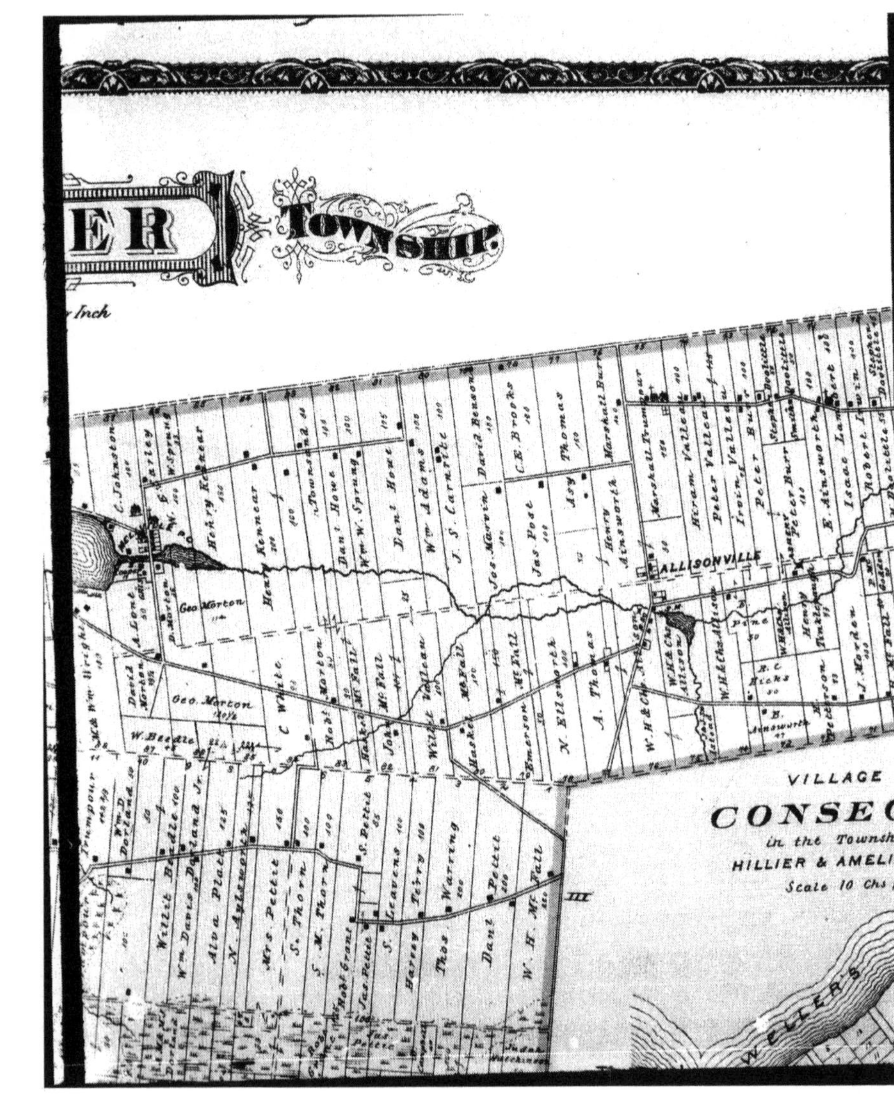

TOWNSHIP
ALLISONVILLE
VILLAGE
HILLIER & AMELI
Geo. Morton
C. Johnston
Henry Kennear
Dan Howe
Wm W. Sprung
Wm Adams
J. S. Carnrite
Jos. Marvin
David Benson
C. E. Brooks
Jas. Post
Thomas
Henry Ainsworth
Marshall Trumpour
Hiram Vallean
Peter Vallean
Peter Burr
E. Ainsworth
Isaac Lambert
Robert Irwin
C. White
N. Ellsworth
A. Thomas
J. Harden
W. Beadle
Willit Beadle
Alva Platt
Mrs. Pettit
S. Thorn
S. M. Thorn
Harvey
Thos
Dan Pettit
W. H. McFall
Jas. Pettit
WELLERS

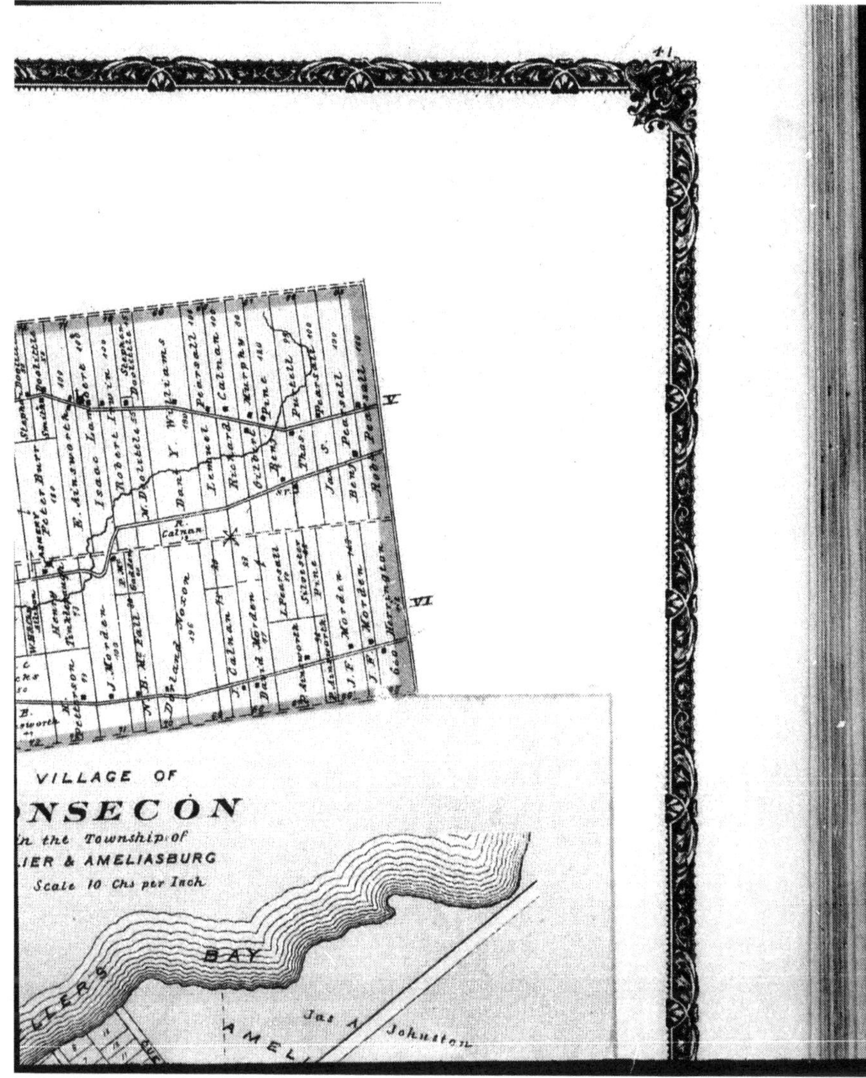
VILLAGE OF
NSECON
in the Township of
LIER & AMELIASBURG
Scale 10 Chs per Inch
BAY
Jas A. Johnston
V.
VI.

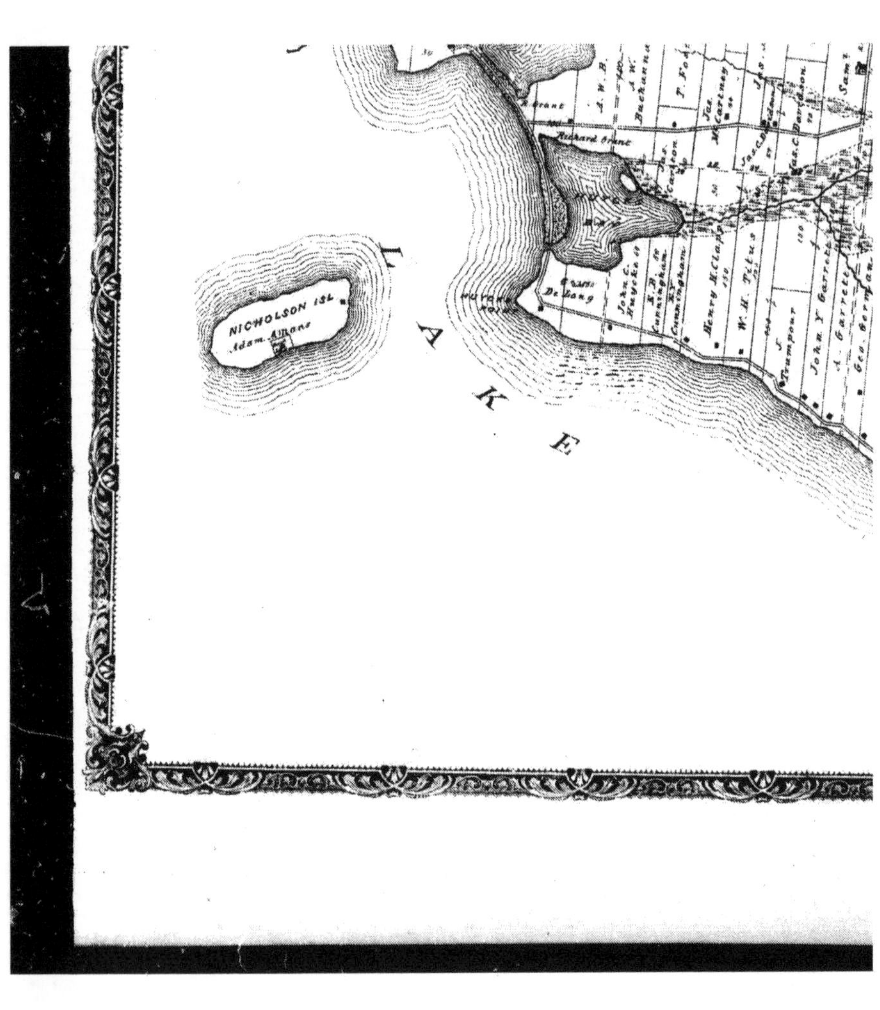
NICHOLSON ISL.
L A K E

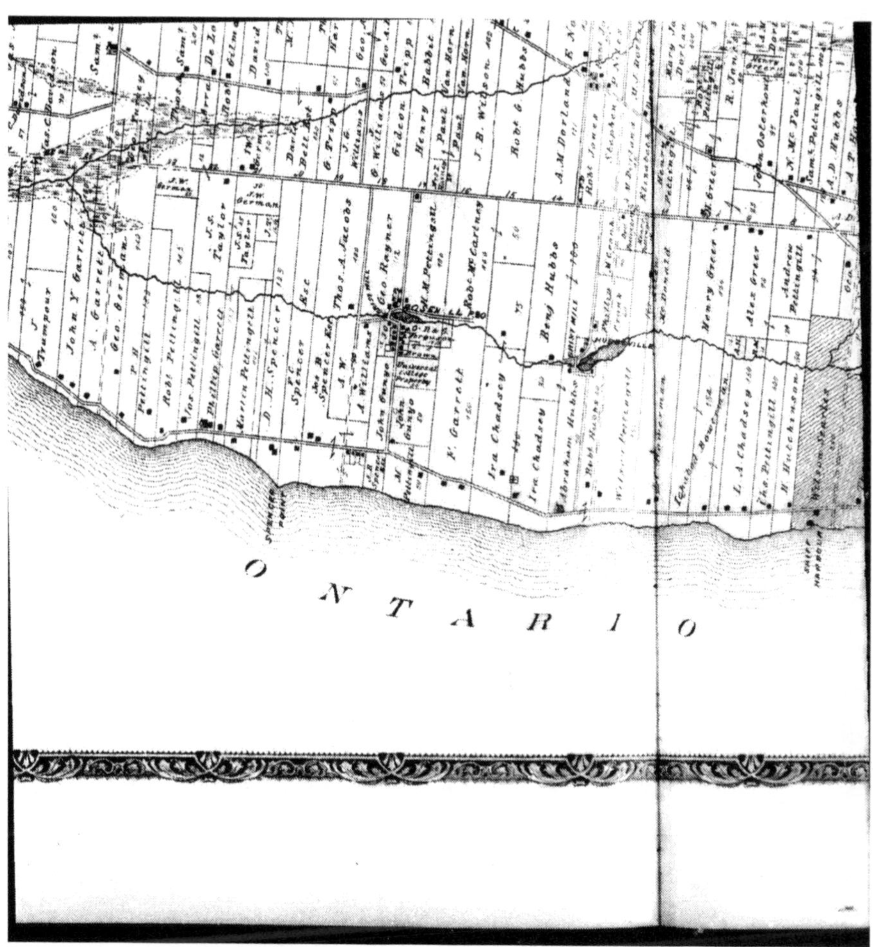
O N T A R I O

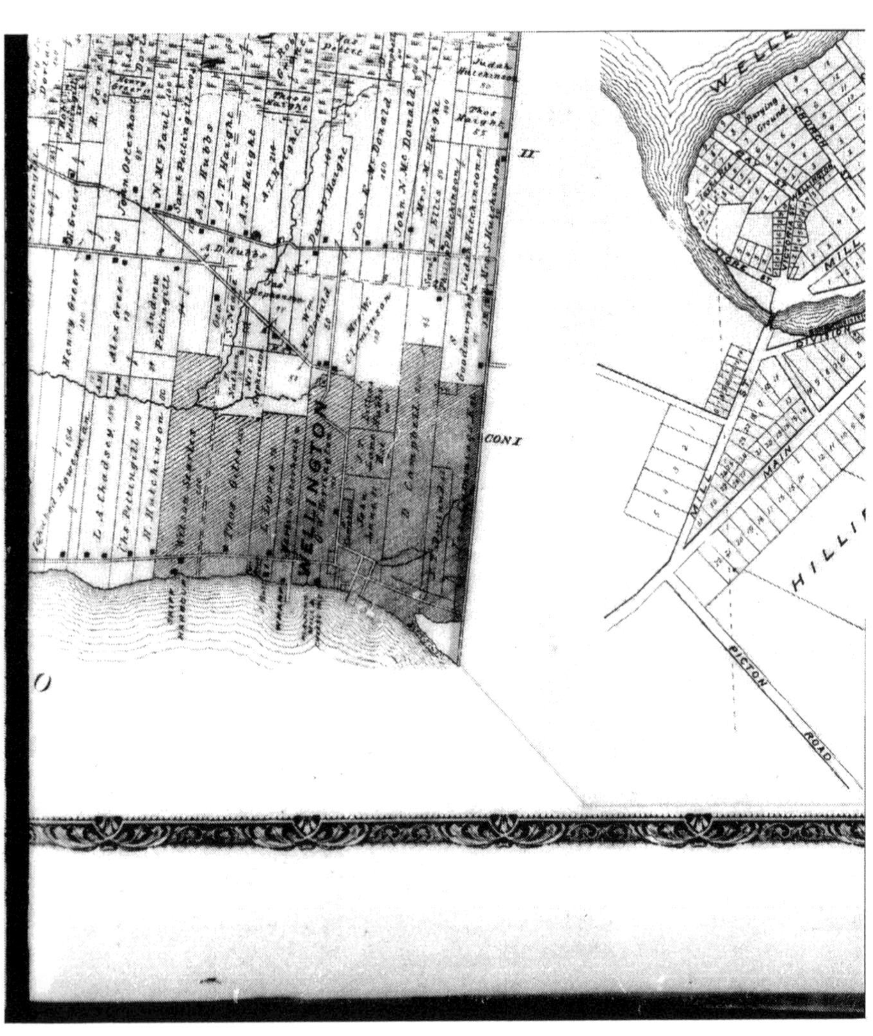
WELLINGTON
CON I
II
Henry Greer
Alex Greer
Andrew Pettingill
L. A. Chadsey
N. Hutchinson
A. D. Hubbs
A. T. Haight
Jos. E. McDonald
John N. McDonald
Judah Hutchinson
D Campbell
Burying Ground
CHURCH
DIVISION
MAIN
MILL
PICTON ROAD

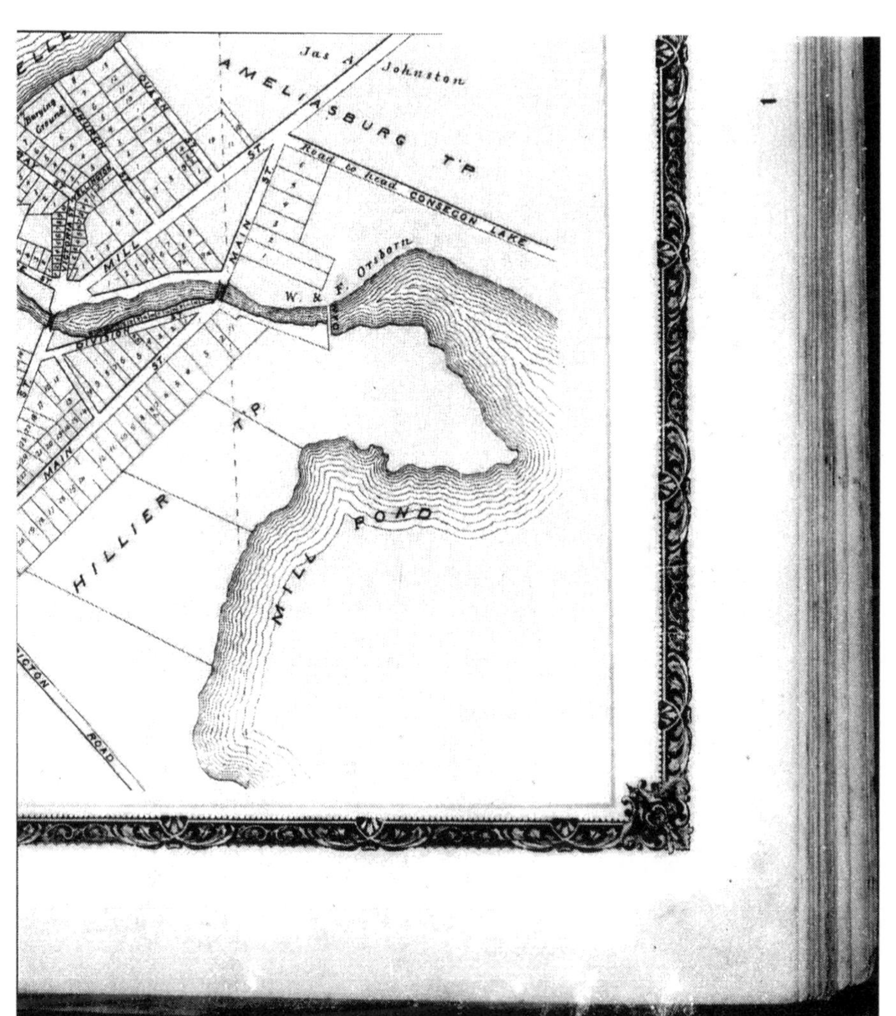

Jas. A. Johnston
AMELIASBURG T.P.
Road to head CONSECON LAKE
W. & F. Orsborn
HILLIER T.P.
MILL POND
MAIN ST.
MILL ST.
QUEEN ST.
CHURCH ST.
DIVISION ST.
Burying Ground
ROAD

INTERIOR OF MESSRS. GEO. RITCHIE & CO.'S DRY GOODS ESTABLISHMENT, SHOWING GRAND STAIRCASE LEADING TO MILLINER & MANTLE ROOMS, BELLEVILLE, ONT.

MANUFACTURER OF LIGHT & HEAVY SPRING WAGGONS, BUGGIES, CUTTERS, SLEIGHS, LUMBER WAGGONS &C. &C.

STOCK FARM OF HENRY, EDWARD & STEPHEN JOSE, THURLOW TP. HASTINGS CO. ONT.

John A. Sprague
Warden of the County of Prince Edward Co
Yours Truly
Nelson Dodge
Milford, P.E. County

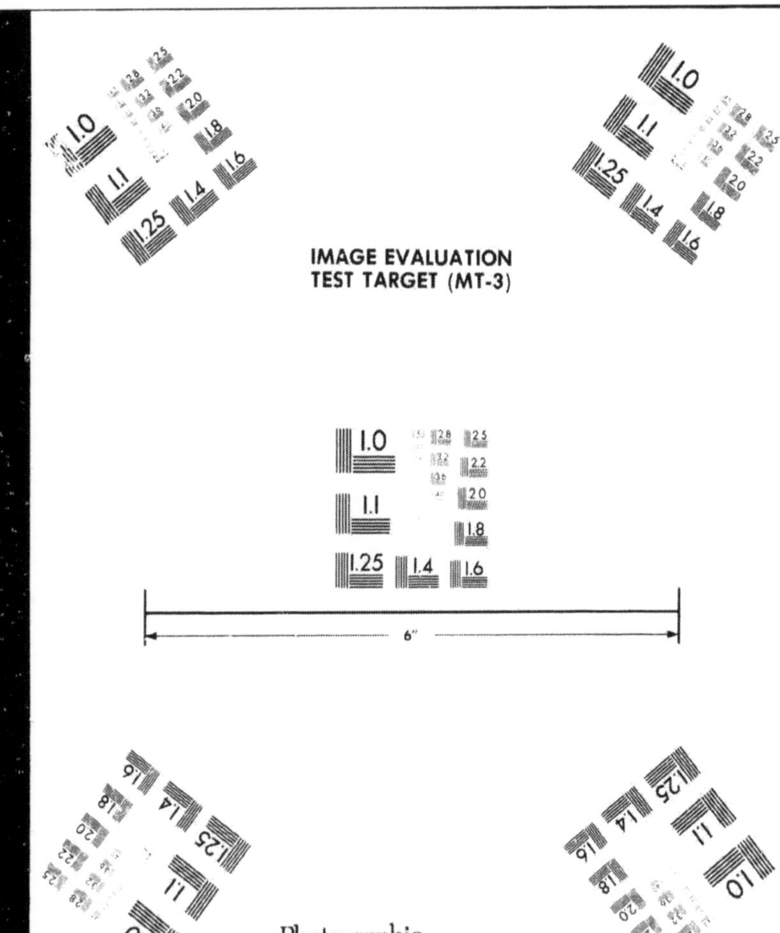

IMAGE EVALUATION
TEST TARGET (MT-3)
Photographic
Sciences
Corporation
23 WEST MAIN STREET
WEBSTER, N.Y. 14580
(716) 872-4503

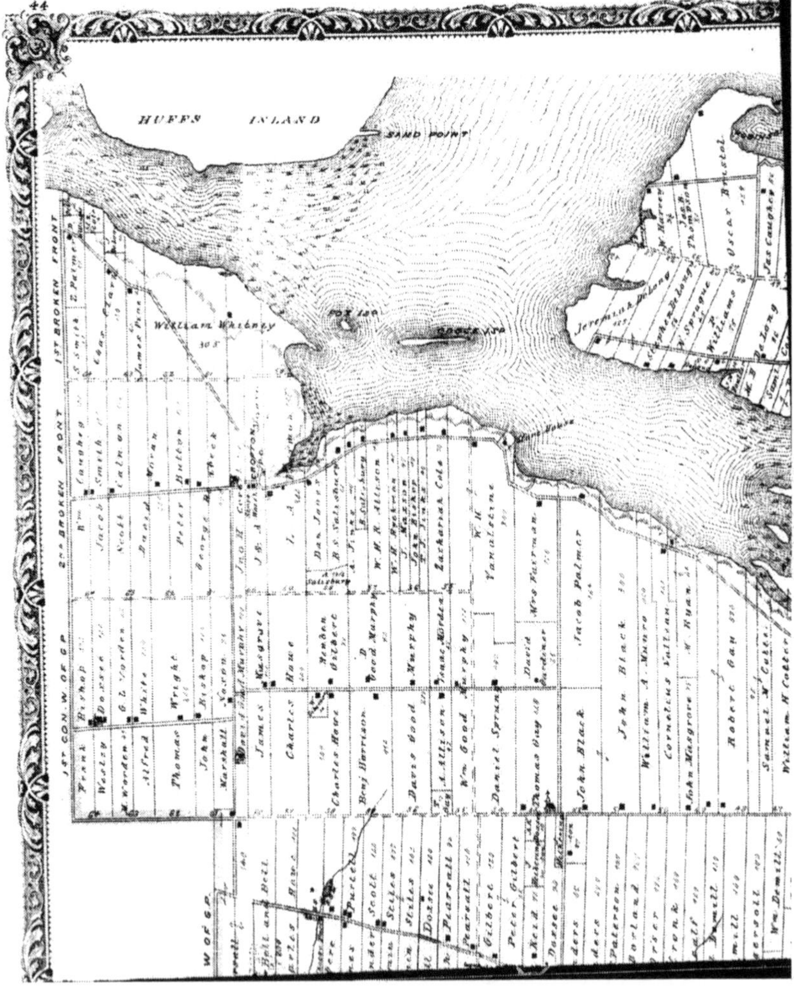

HUFFS ISLAND

SAND POINT

1ST BROKEN FRONT

2ND BROKEN FRONT

1ST CON. W. OF G.P.

W. OF G.P.

William Whitney 305

Jeremiah DeLong

Oscar Bristol

Jas Caughey

W. H. Vanalstine

Zachariah Cole

Jacob Palmer

Mrs Farrman

John Black

William A. Munro

John Musgrove

Robert Gay

Samuel M Cotter

William H Cotter

Frank Bishop

Thomas Wright

John Bishop

James Musgrove

Charles Howe

Charles Howe

Benj Harrison

Davis Good

Daniel Sprung

Thomas Gay

Peter Gilbert

Wm Demill

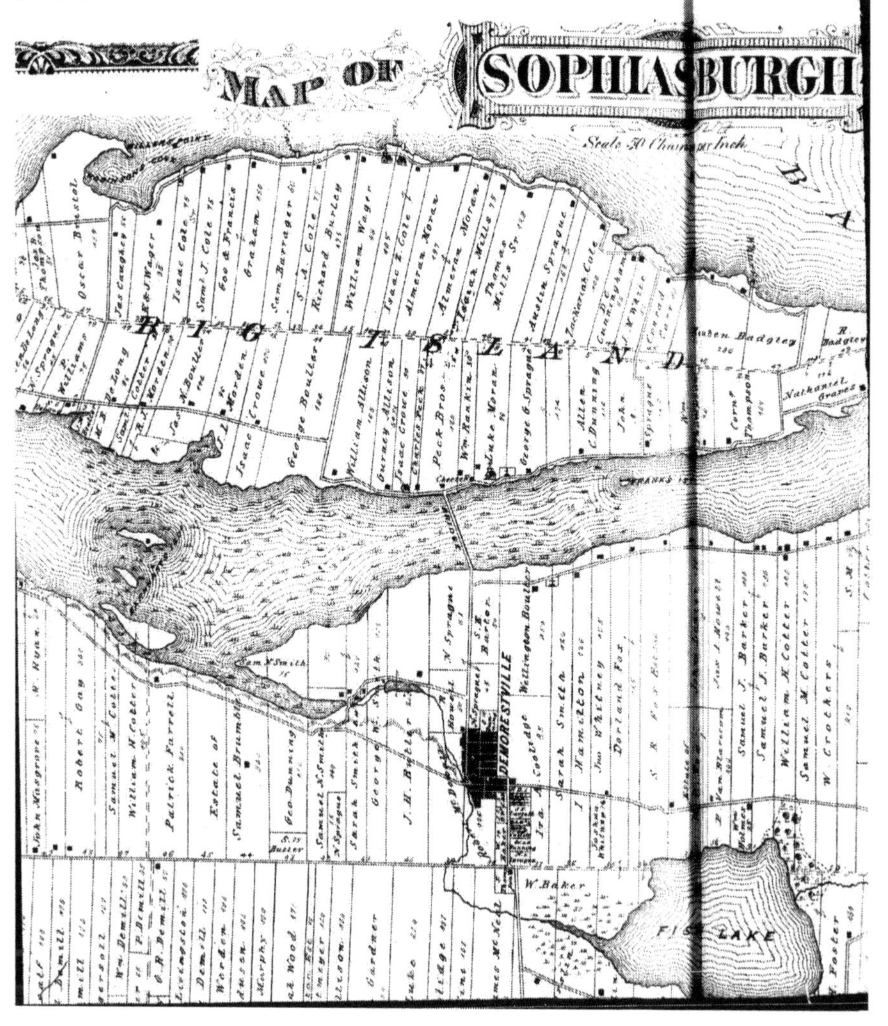

MAP OF SOPHIASBURGH
Scale 30 Chains per Inch
BIG ISLAND
DEMORESTVILLE
FISH LAKE

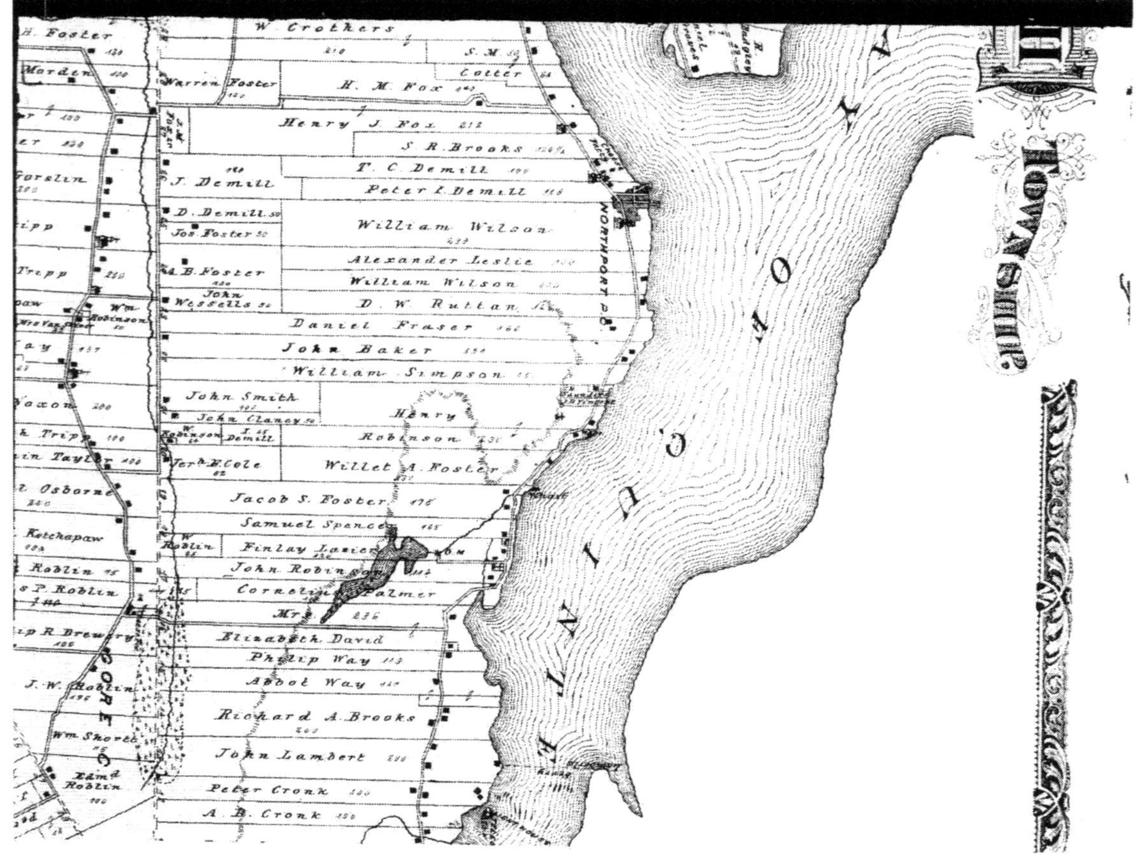

TOWNSHIP
AY OF QUINTE
NORTHPORT P.O.
GORE
H. Foster
Warden
W. Crothers
S. M. Cotter
Warren Foster
H. M. Fox
Henry J. Fox
S. R. Brooks
T. C. Demill
J. Demill
Peter I. Demill
D. Demill
Jos. Foster
William Wilson
Alexander Leslie
A. B. Foster
William Wilson
John Wessells
D. W. Ruttan
Wm. Robinson
Daniel Fraser
John Baker
William Simpson
John Smith
John Clancy
Henry Robinson
Willet A. Foster
Jacob S. Foster
Samuel Spence
Ketchapaw
Finlay Lazier
John Robinson
Cornelius Palmer
Elizabeth David
Philip Way
Abbot Way
J. W. Roblin
Richard A. Brooks
Wm. Shorey
John Lambert
Edmd. Roblin
Peter Cronk
A. B. Cronk

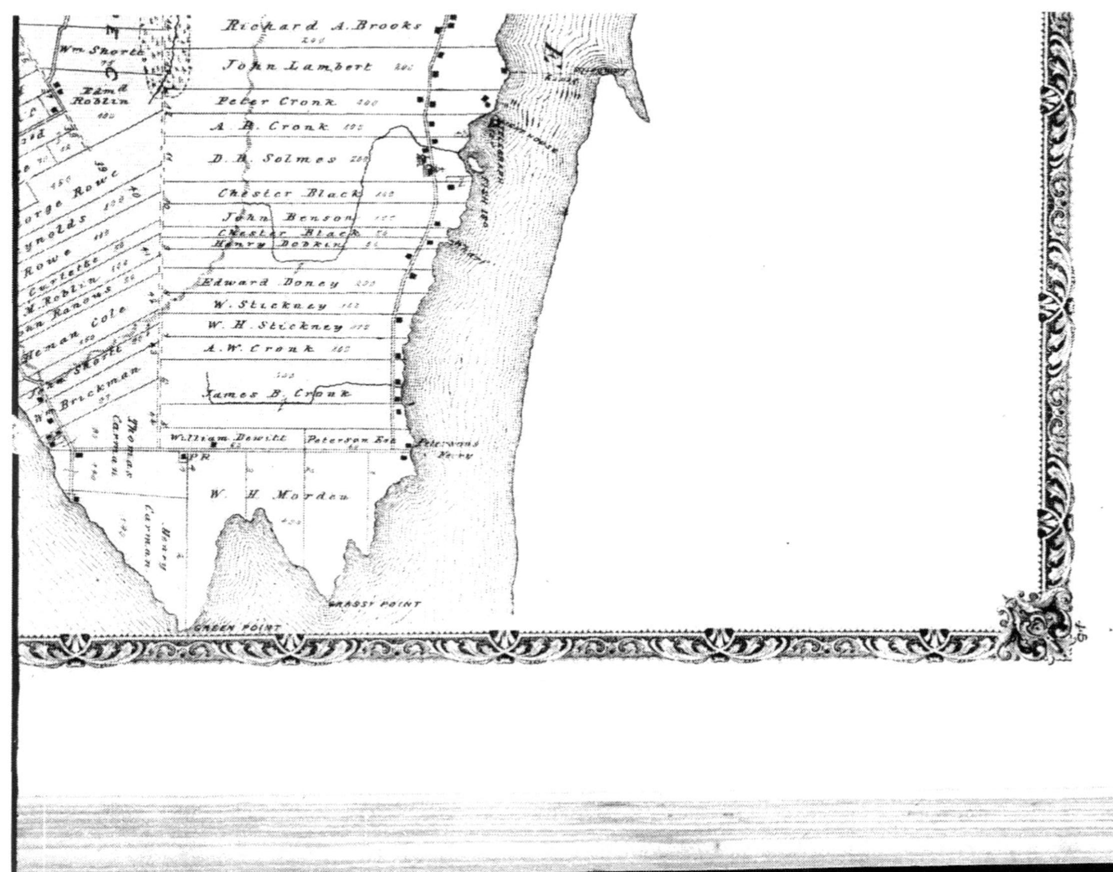
Richard A. Brooks
John Lambert
Peter Cronk
A. B. Cronk
D. B. Solmes
Chester Black
John Benson
Chester Black
Henry Dobbin
Edward Doney
W. Stickney
W. H. Stickney
A. W. Cronk
James B. Cronk
William Dewitt
Peterson Est
Petersons Ferry
W. H. Morden
Thomas Carman
Henry Carman
Wm Shorts
Edmd Roblin
Geo Rowe
Wm Brickman
GRASSY POINT
GREEN POINT

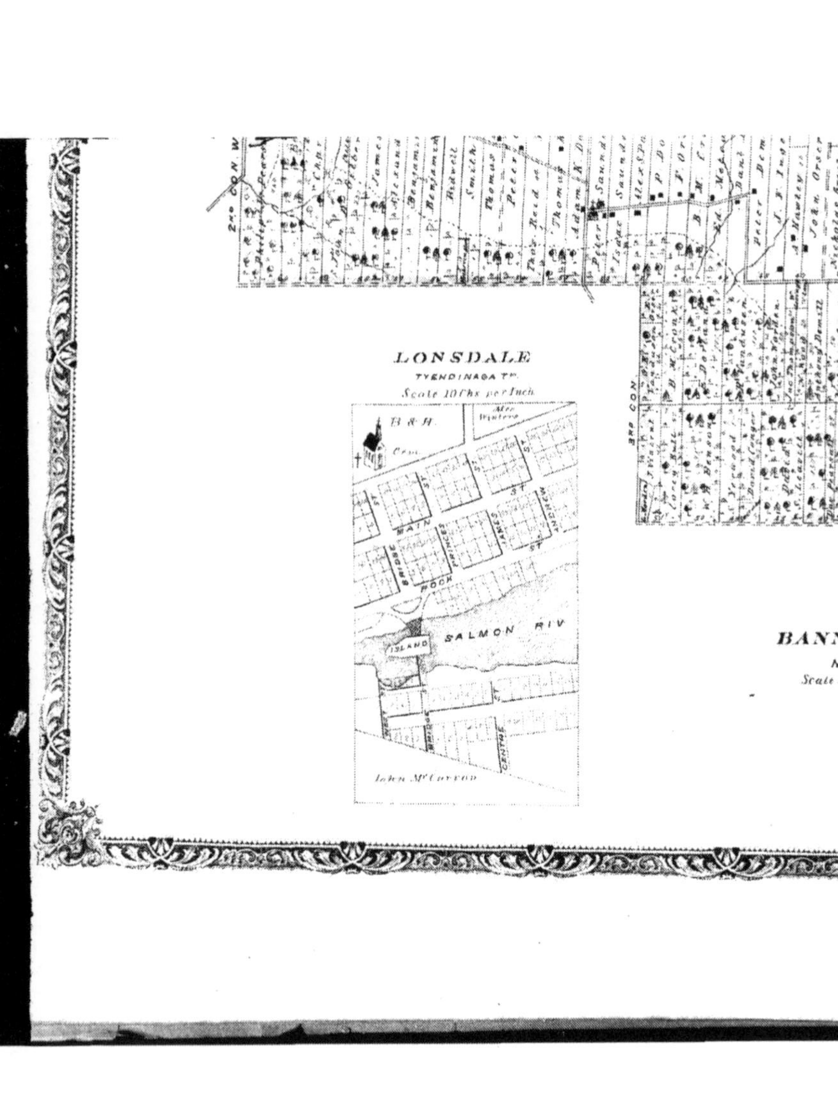
LONSDALE
TYENDINAGA Tp.
Scale 10 Chs per Inch
SALMON RIV
ISLAND
MAIN
BRIDGE
PRINCES
JAMES
ANDREW
ROCK
CENTRE
John McCarron
BANN
MA
Scale 10

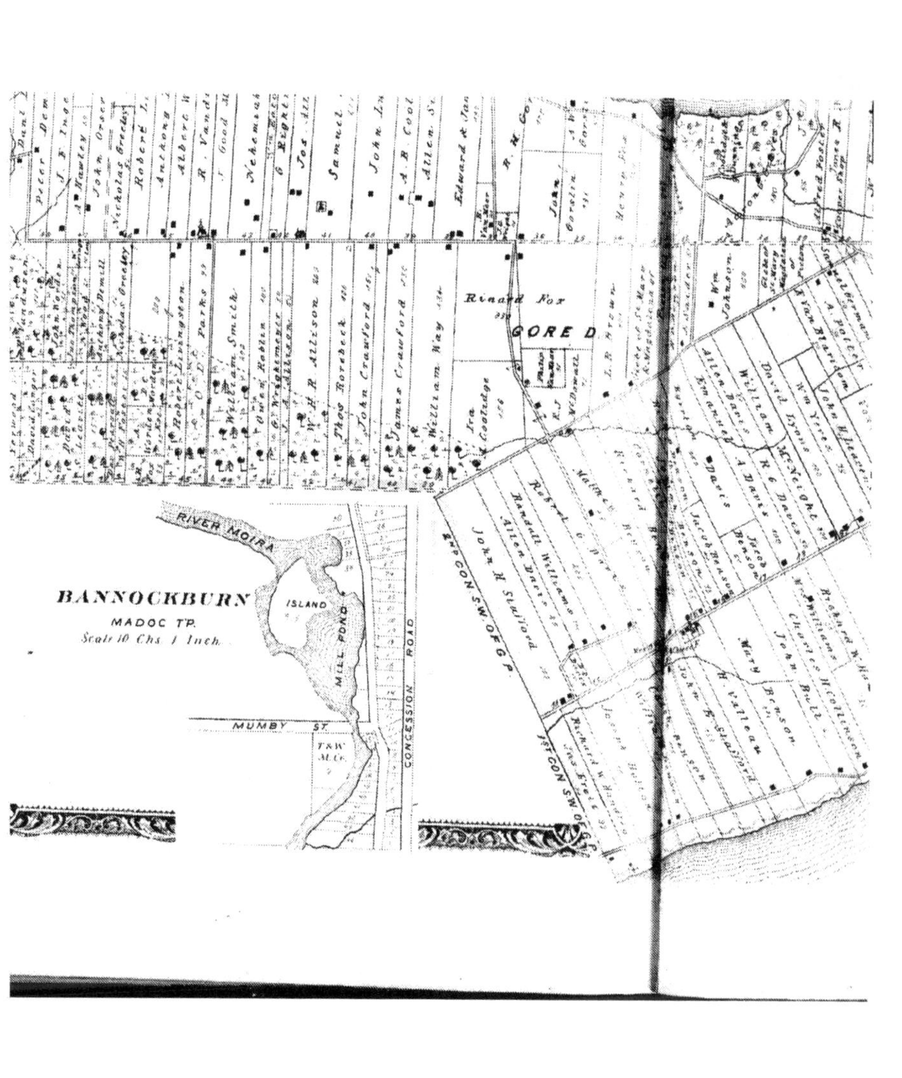
BANNOCKBURN
MADOC TP.
Scale 10 Chs. 1 Inch.
RIVER MOIRA
ISLAND
MILL POND
MUMBY ST.
CONCESSION ROAD
T&W M.Co.
GORE D
Rinard Fox
2ND CON. S.W. OF G.P.
1ST CON. S.W. OF G.P.
William Smith
Thos Rorebeck
John Crawford
James Crawford
William Way
Robert Livingston
John H Stafford
Allen Davis
Randall Williams
David Lyons
Wm Johnson
Alfred Footer

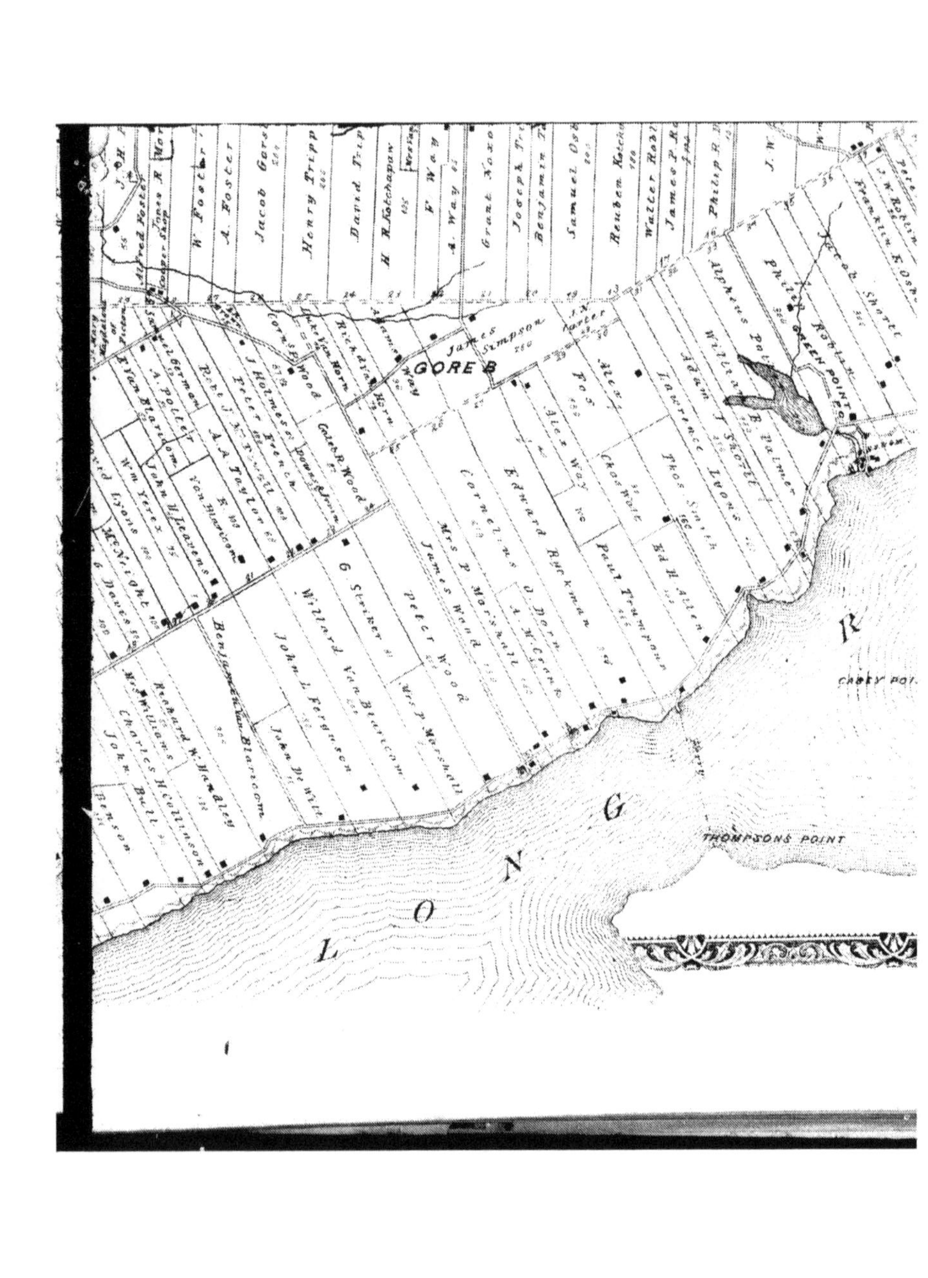

GORE B
THOMPSONS POINT
L O N G
R

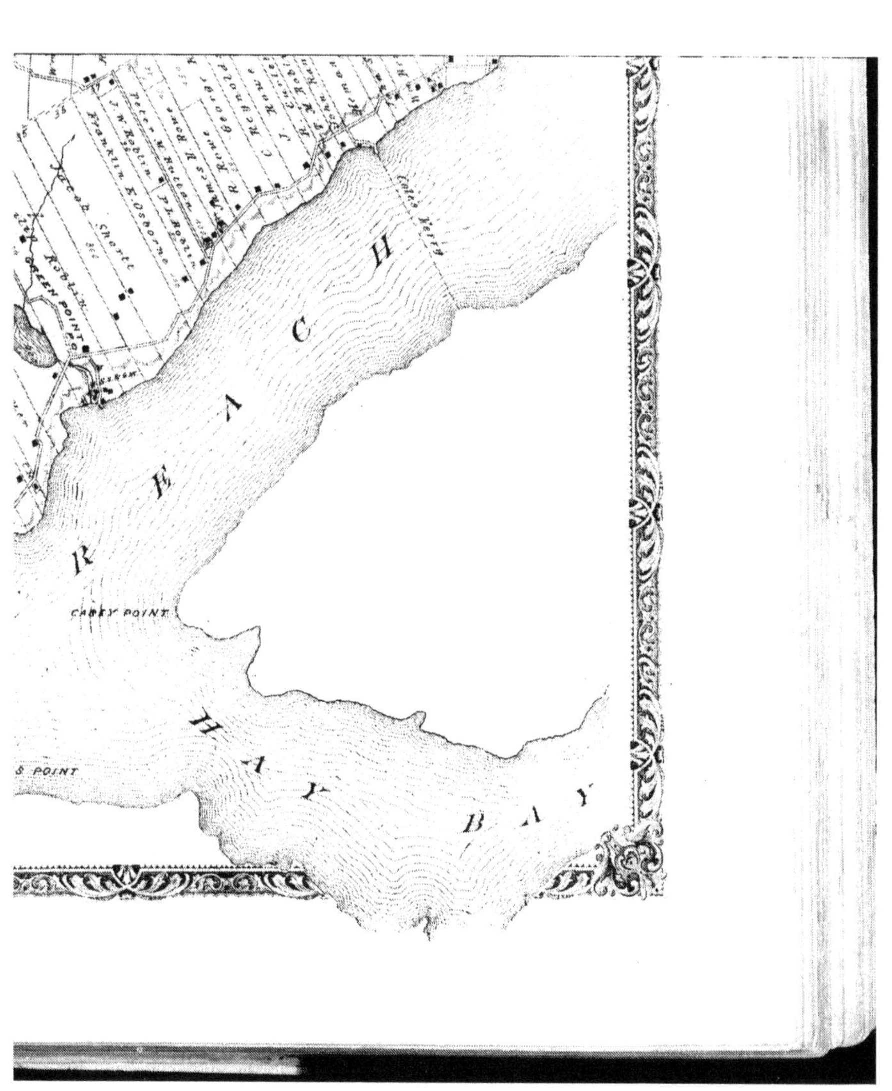
REACH
HAY BAY
GREEN POINT P.O.
Jacob Shortt
Franklin E Osborne

NORTHCOTT & ALFORD CONTRACTORS & GEO. CUMMINGS CABINET FACTORY.
REDICK & FARLEY'S PLANING MILLS, SASH, DOOR & BLIND FACTORY.
MILL STREET, BELLEVILLE, ONT.

"GROVE COTTAGE" RES. OF SAML. KELLY.
CON. 2, LOT 14. HALLOWELL TP. ONT.

RES. OF G.W. HARRINGTON.
WELLINGTON, ONT.

PRIVATE RES. OF WILLIAM ASHLEY, ESQ.
FOXBORO, HASTINGS CO. ONT.

J & J. NIGHTINGALE BREEDER OF
DURHAM CATTLE & COTSWOLD SHEEP.
CON. 3, LOT 8 PART S. AMELIASBURG TP. ONT.

G.S. TICKELL MANUFACTURER OF FURNITURE.
BELLEVILLE ONT. ESTABLISHED 1858.
THE LARGEST STOCK IN CITY.

RES. OF A. S. BROWN, BELLEVILLE, ONT.

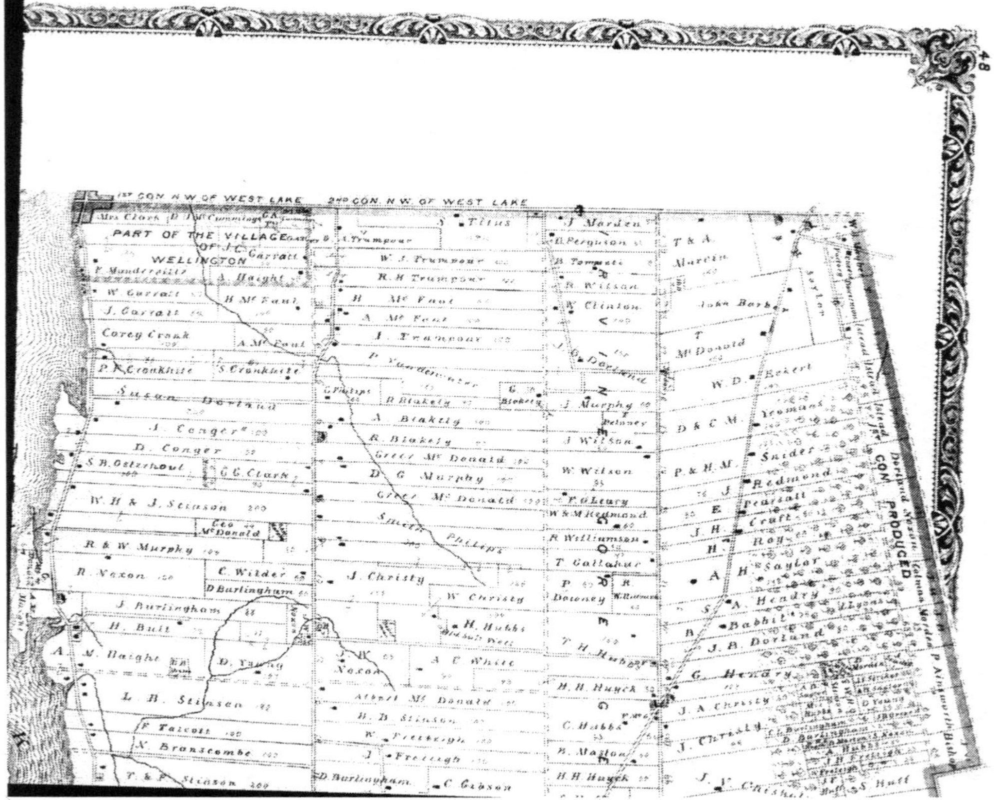

1ST CON. N.W. OF WEST LAKE

2ND CON. N.W. OF WEST LAKE

1ST CON. PRODUCED

PART OF THE VILLAGE OF WELLINGTON

Mrs. Clark
W. Garratt
J. Garratt
Corey Crank
P. K. Cronkhite
S. Cronkhite
Susan Dorland
J. Conger
D. Conger
S. B. Osterhout
G. G. Clark
W. H. & J. Stinson
R. & W. Murphy
R. Noxon
C. Wilder
D. Burlingham
J. Burlingham
H. Bull
A. M. Haight
D. Young
L. B. Stinson
F. Talcott
N. Branscombe
T. & P. Stinson
A. Trumpour
W. J. Trumpour
R. H. Trumpour
A. McFaul
J. Trumpour
P. Vanderwater
G. Philips
B. Blakely
A. Blakely
R. Blakely
Greer McDonald
D. G. Murphy
Smith Philips
J. Christy
W. Christy
H. Hubbs
A. P. White
Albert McDonald
H. B. Stinson
J. Freleigh
C. Gibson
S. Titus
J. Marden
D. Ferguson
R. Wilson
W. Clinton
J. Murphy
J. Wilson
W. Wilson
W. & M. Redmond
R. Williamson
T. Gallahur
P. Downey
T. H. Hubbs
H. H. Huyck
C. Hubbs
B. Mallory
T. & A. Martin
John Barker
T. McDonald
W. D. Eckert
D. & C. M. Yeomans
P. & H. M. Snider
J. Redmond
E. Pearsall
J. H. Craft
H. Rogers
A. H. Saylor
S. A. Hendry
R. Babbit
J. B. Dorland
G. Hendry
J. A. Christy
J. Christy
Saylor
Dorland
Noxon
Colman
Morden
P. Ainsworth
Bishop
S. Hull

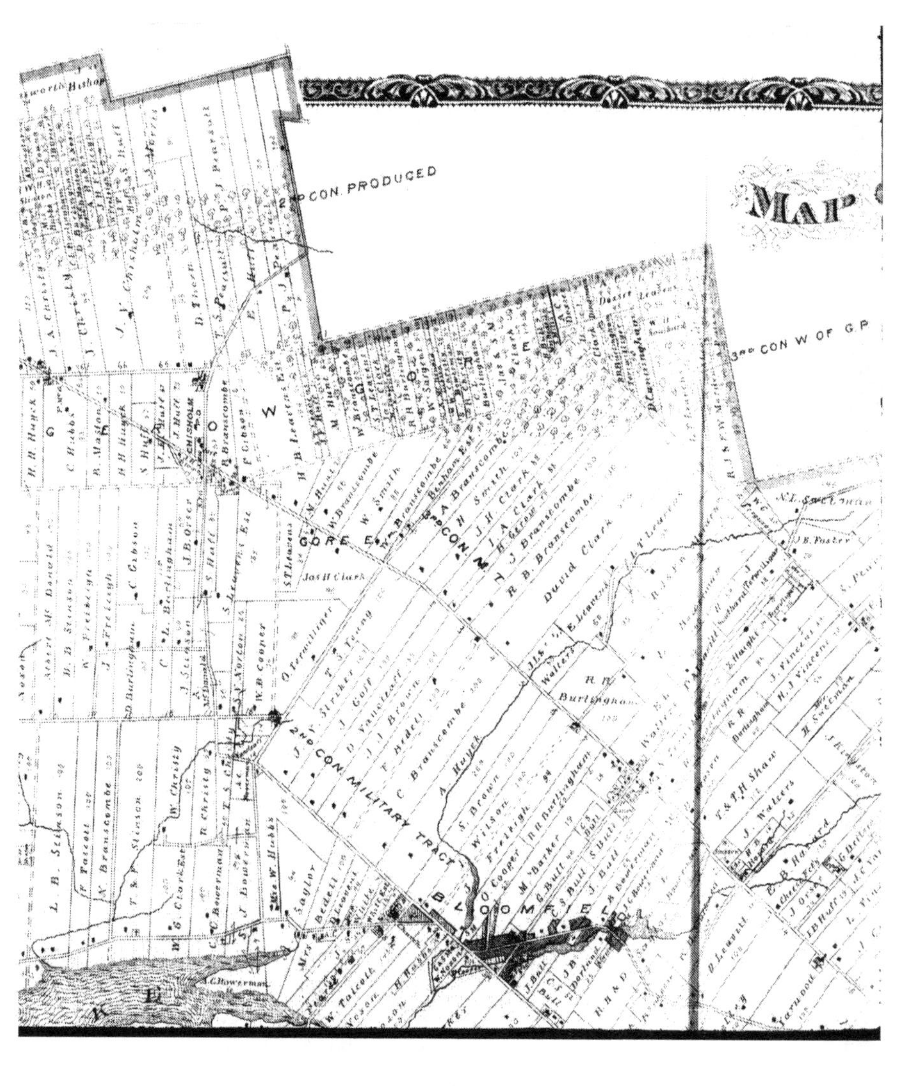

MAP
2ND CON. PRODUCED
3RD CON W OF G.P
GORE
2ND CON. MILITARY TRACT
BLOOMFIELD

OF HALLOWELL TOWNS

Scale 30 Chains per Inch

2ND CON. N.W. OF CARRYING PLACE

1ST CON. N.W. OF CARRYING PLACE

GORE

PICTON BAY

TOWN OF

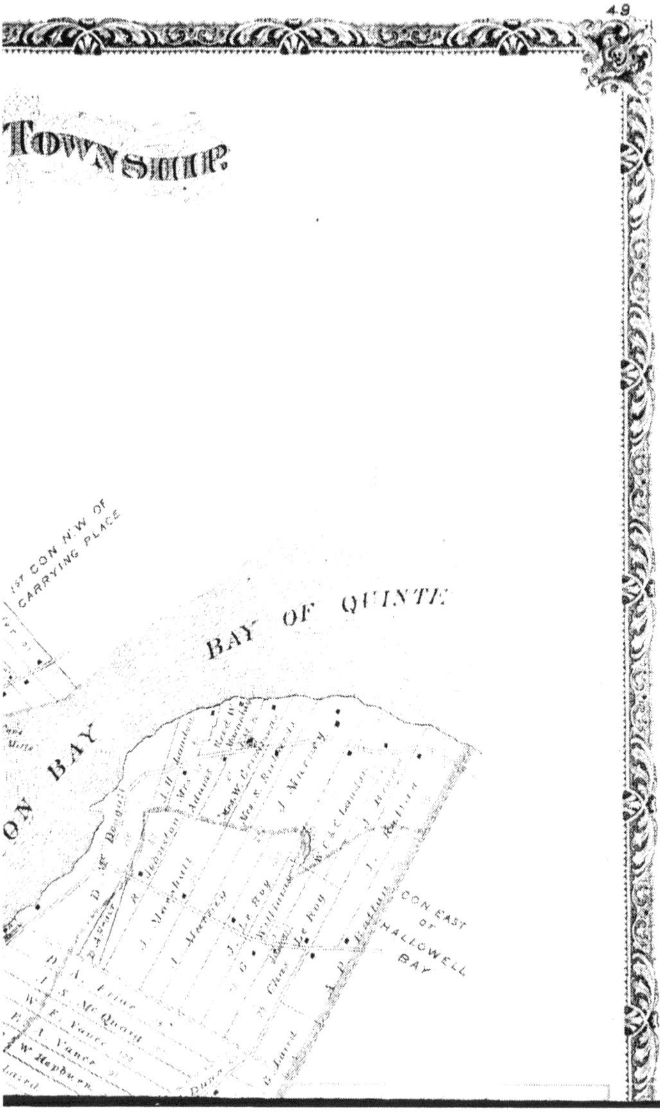

TOWNSHIP

1ST CON N.W. OF CARRYING PLACE

BAY OF QUINTE

ON BAY

CON EAST OF HALLOWELL BAY

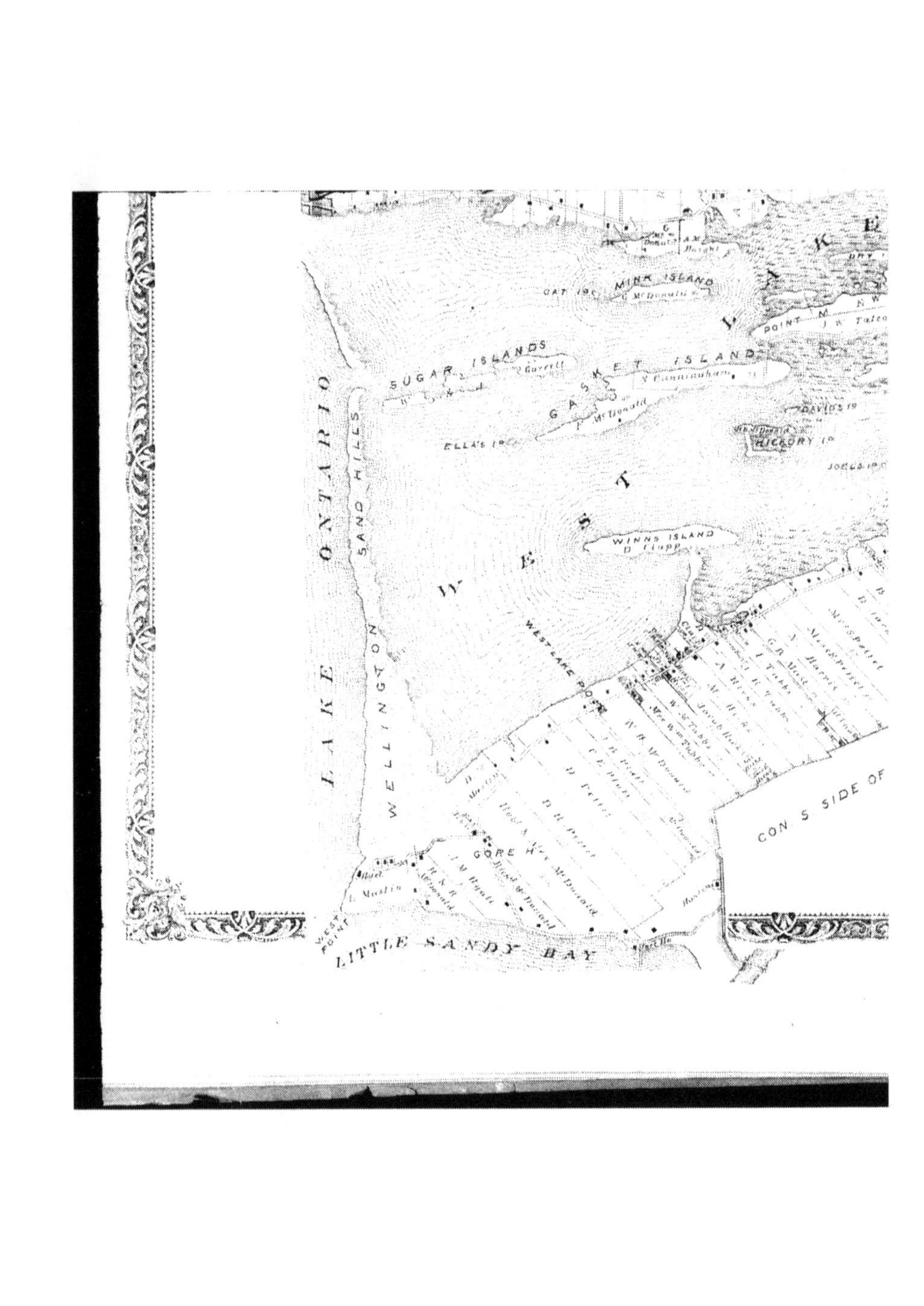

LAKE ONTARIO
SAND HILLS
WELLINGTON
SUGAR ISLANDS
GASKET ISLAND
MINK ISLAND
WINNS ISLAND
HICKORY Id
WEST LAKE
WEST LAKE POINT
GORE
WEST POINT
LITTLE SANDY BAY
CON. S SIDE OF

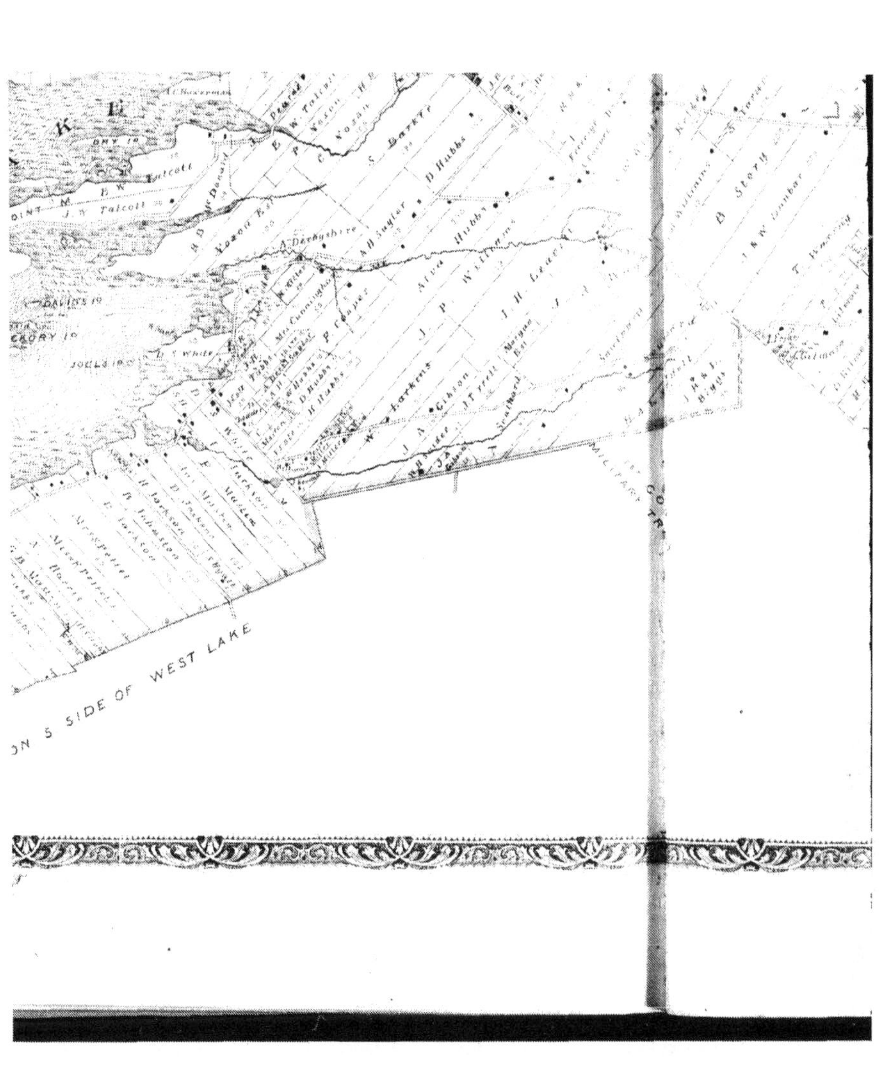

ON S SIDE OF WEST LAKE
J. W. Talcott
D. Hubbs
B. Story
A. Derbyshire
J. P. Williams
J. H. Leavitt
J. A. Gibson

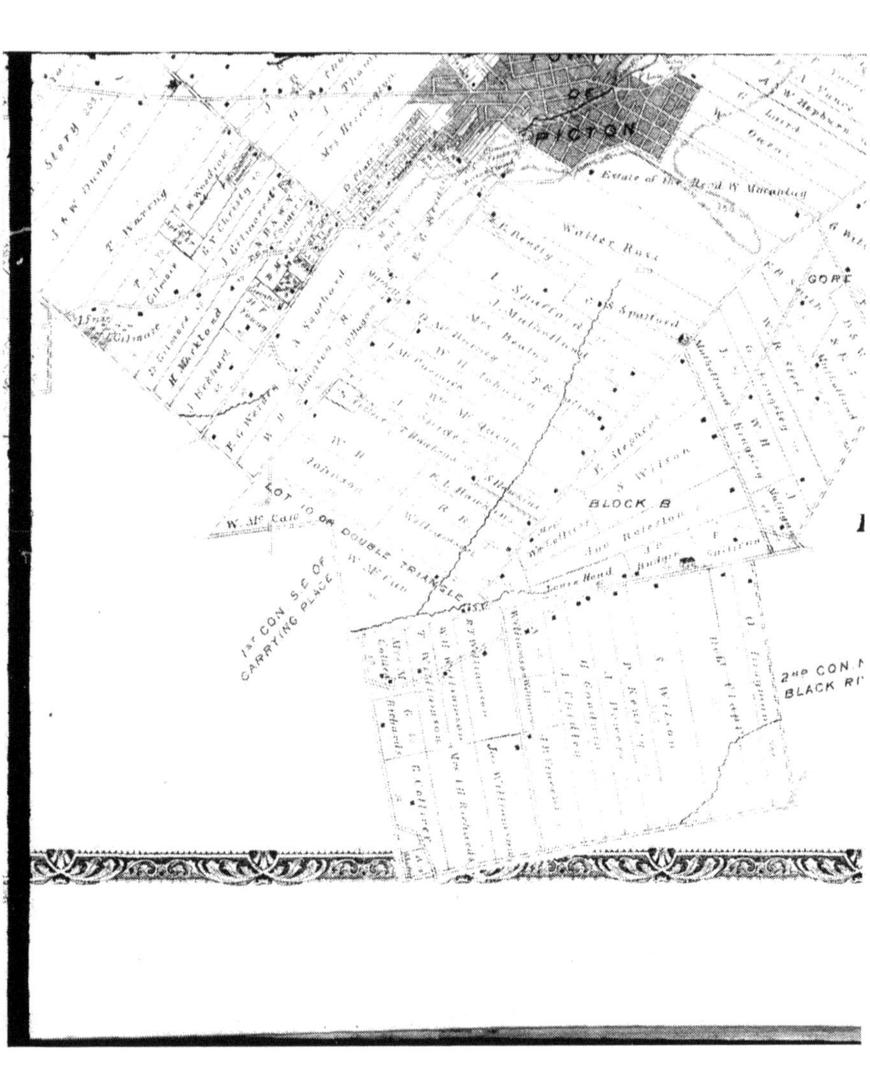
PICTON
BLOCK B
LOT 10 OF DOUBLE TRIANGLE
1st CON S.E. OF CARRYING PLACE
GORE
Walter Ross
Estate of the Revd W Macaulay

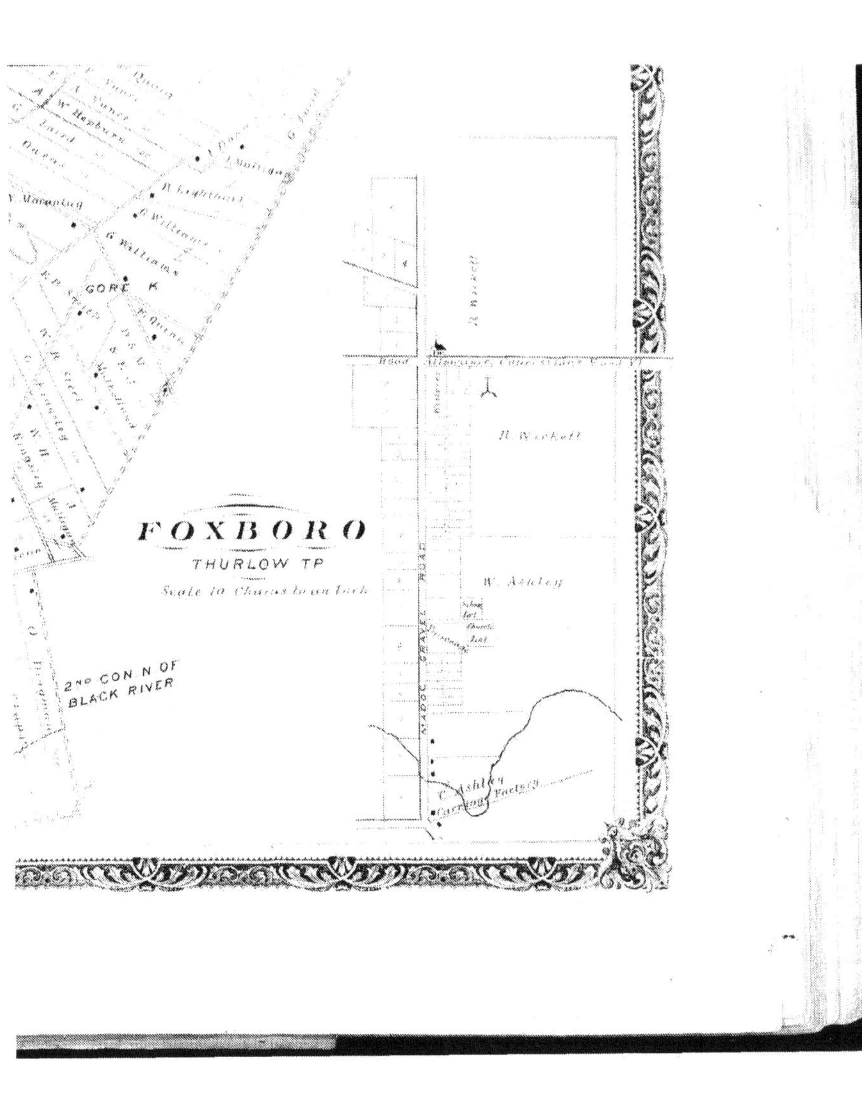
FOXBORO
THURLOW TP
Scale 10 Chains to an Inch
GORE K
2ND CON N OF
BLACK RIVER
MADOC GRAVEL ROAD
R. Wickett
W. Ashley
C. Ashley
Carriage Factory

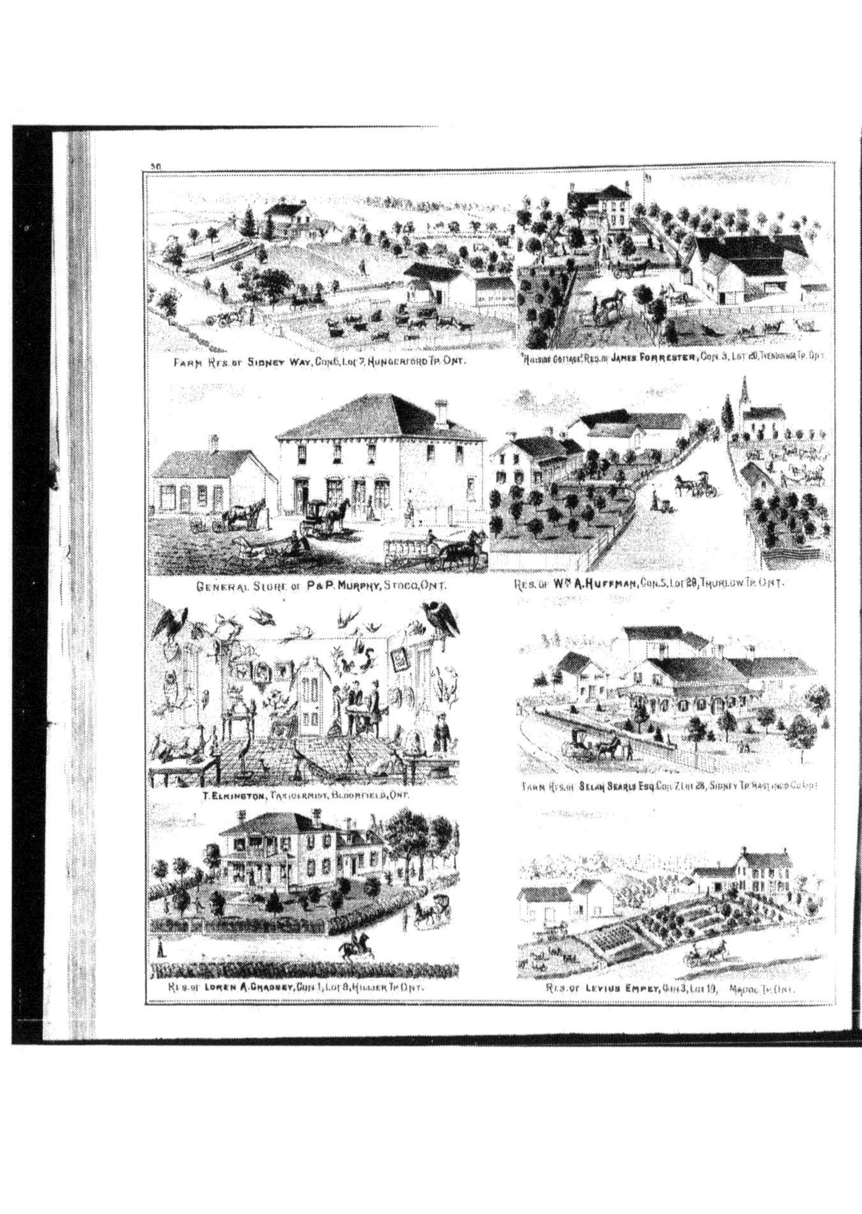

Farm Res. of Sidney Way, Con. 6, Lot 7, Hungerford Tp. Ont.

"Hillside Cottage," Res. of James Forrester, Con. 3, Lot 20, Tyendinaga Tp. Ont.

General Store of P & P. Murphy, Stoco, Ont.

Res. of Wm A. Huffman, Con. 5, Lot 29, Thurlow Tp. Ont.

T. Elkington, Taxidermist, Bloomfield, Ont.

Farm Res. of Selah Searls Esq. Con. 7, Lot 28, Sidney Tp. Hastings Co. Ont.

Res. of Loren A. Chadsey, Con. 1, Lot 9, Hillier Tp. Ont.

Res. of Levius Empey, Con. 3, Lot 19, Madoc Tp. Ont.

MILLS & RES. OF J. C. JAMIESON, TWEED, ONT.

RES. OF CAPT. J. A. PORT, TRENTON, ONT.

ALBURY POST OFFICE

RES. OF DR. P. V. DORLAND, WEST FRONT ST. BELLEVILLE ONT.

FARM RES. OF WILLIAM PECK ESQ. CON. 1, LOT 93, AMELIASBURG TP. PRINCE EDWARD CO.

RES. OF ALLEN SINE, CON. 2, LOT 39, SOPHIASBURG TP.

RES. OF ROBT. ROGERS, STINSON BLOCK, LOT 3, HILLIER TP.

"PORTSMOUTH COTTAGE," RES. OF MAJOR RANE, MADOC VILLAGE, ONT.

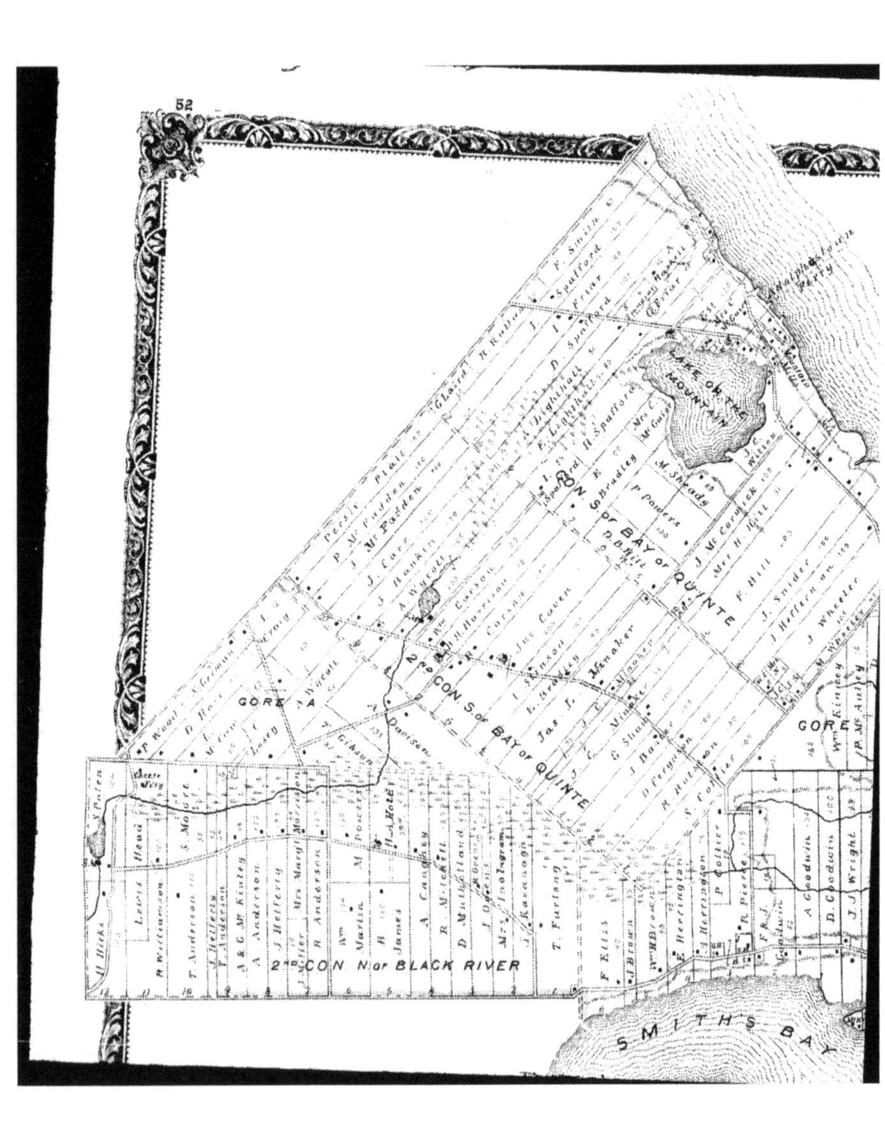
LAKE ON THE MOUNTAIN
CON. S. OF BAY OF QUINTE
2ND CON. S. OF BAY OF QUINTE
GORE A
GORE
2ND CON. N. OF BLACK RIVER
SMITHS BAY
Adolphustown Ferry
J. McCormick
J. Snider
J. Hefferman
J. Wheeler
F. Hill
M. Sheady
P. Powers
S. Bradley
H. Spafford
J. Cavan
J. Carr
J. McPadden
Jas. J. Minaker
T. Anderson
A. Anderson
R. Anderson
J. Hefferty
R. Williamson
Lewis
Head
A. Caughey
R. Mitchell
D. Mulholland
T. Furlong
F. Ellis
J. Brown
E. Herrington
P. Collier
A. Goodwin
D. Goodwin
J. J. Wright
A. Davison
G. Shaw
J. Barker
D. Ferguson

MAP OF N.MA

Scale

BAY

Wharf

BONGARDS P.O.
CORNERS

GORE

CON BAY SIDE

CON LAKE SIDE

A. Shannon

D. Shannon

P. McAuley

B. Van Gesen

N. W. Minaker

C. Turner

E. Bradley

T. Pringle

R. Minaker

R. Bongard

P. David

H. David

T. Bongard

D. I. Bongard

Corkindale

H. McGee

J. Johnson

G. Minaker Sr.

C. P. Stanton

Lee Minaker

J. Snider

J. Heffernan

J. Wheeler

M. Wheeler

Wm Kinney

P. McAuley

D. McAuley

P. Kellar

A. Goodwin

D. Goodwin

J. J. Wright

P. Kerney

D. McAuley

R. Robson

J. G. Wemsley

E. Wright

W. Wright

J. Cummings

R. Ellis

Jas Wamsley

D. Vandusen

J. Kerr

Wm Kerr

A. Kerr Est

Eli Clark

E. Van Vlack

C. Hughes

E. Williams

Wm Kerr

R. David

W. Williams

E. Harrison

Jas Rose

C. Williams

WAUPOOS

S. BAY

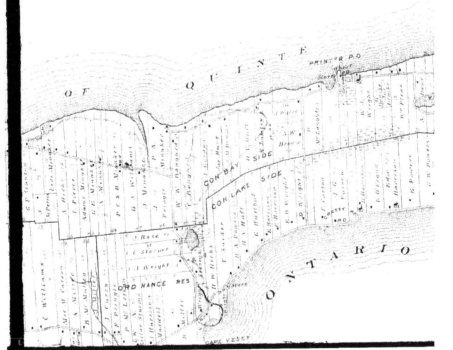

Scale 50 Chains per Inch

OF QUINTE

PRINYER P.O.

CON BAY SIDE

CON LAKE SIDE

CRESSY P.O.

ORDNANCE RES

ONTARIO

CAPE VESEY

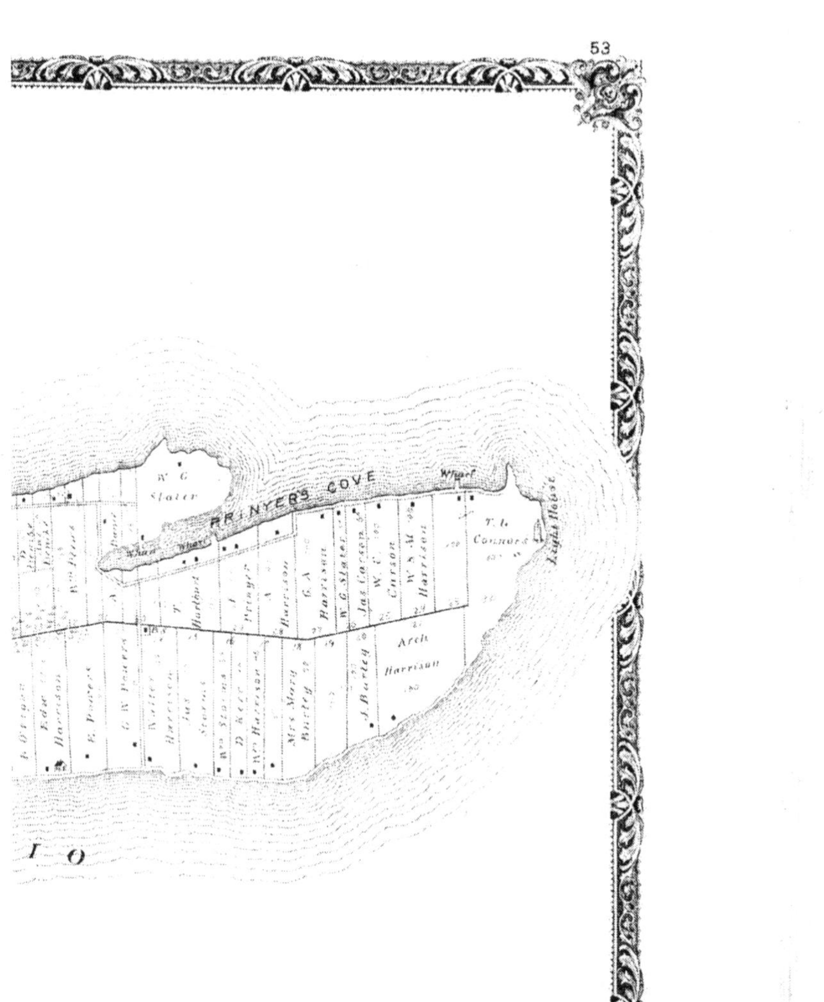
PRINYERS COVE
Wharf
W. G. Slater
T. L. Connors
Light House
G. A. Harrison
W. G. Slater
Jas Carson
W. C. Carson
W. S. M. Harrison
Arch Harrison
J. Burley
Mrs Mary Burley
Wm Harrison
D. Keer
Wm Storms
Jas Storms
Walter Harrison
G. W. Powers
E. Powers
Edw Harrison
I O

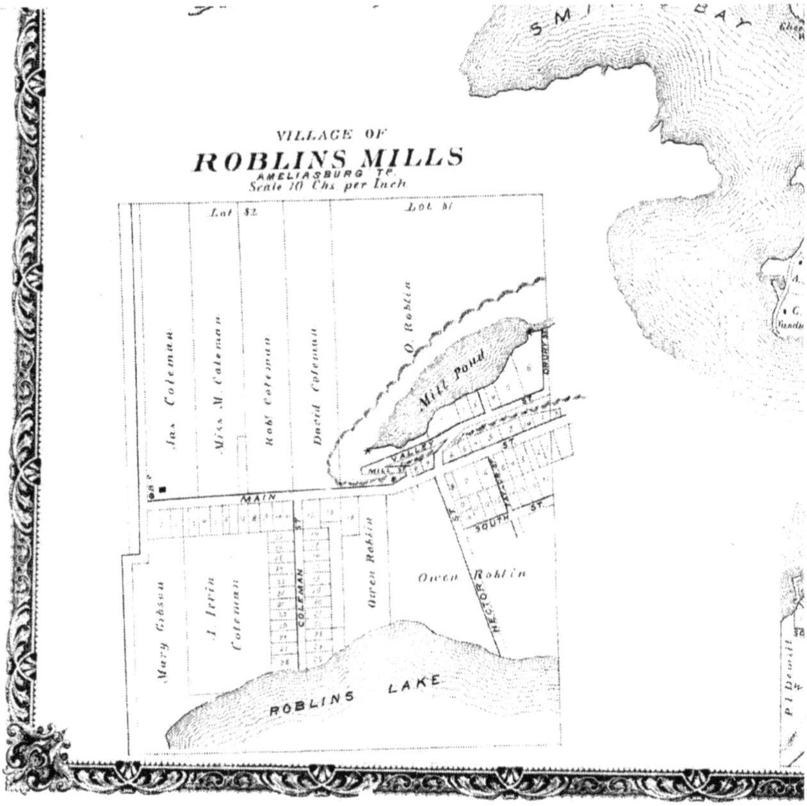

VILLAGE OF

ROBLINS MILLS

AMELIASBURG TP.

Scale 10 Chs per Inch

SMITH'S BAY

Lot 82

Lot 81

Jas Coleman

Miss M. Coleman

Robt Coleman

David Coleman

O. Roblin

Mill Pond

VALLEY

MILL ST

MAIN

ST

SOUTH ST.

Mary Gibson

J. Irvin Coleman

COLEMAN ST

Owen Roblin

Owen Roblin

HECTOR

ROBLINS LAKE

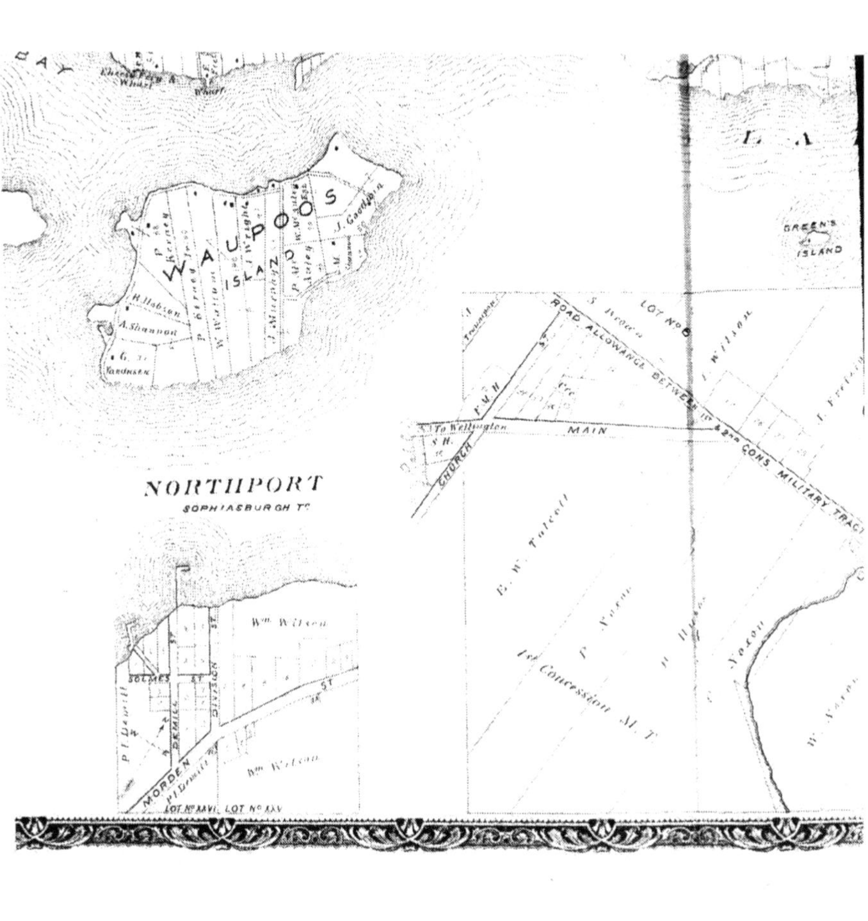
WAUPOOS
ISLAND
H. Hobson
A. Shannon
NORTHPORT
SOPHIASBURGH Tp.
Wm. Wilson
SOLMES ST.
DIVISION ST.
DEMILL ST.
MORDEN ST.
LOT No XXVI LOT No XXV
GREEN'S ISLAND
LOT No 6
ROAD ALLOWANCE BETWEEN 1st & 2nd CONS MILITARY TRACT
MAIN
CHURCH
To Wellington
E. W. Talcott
1st Concession M. T.

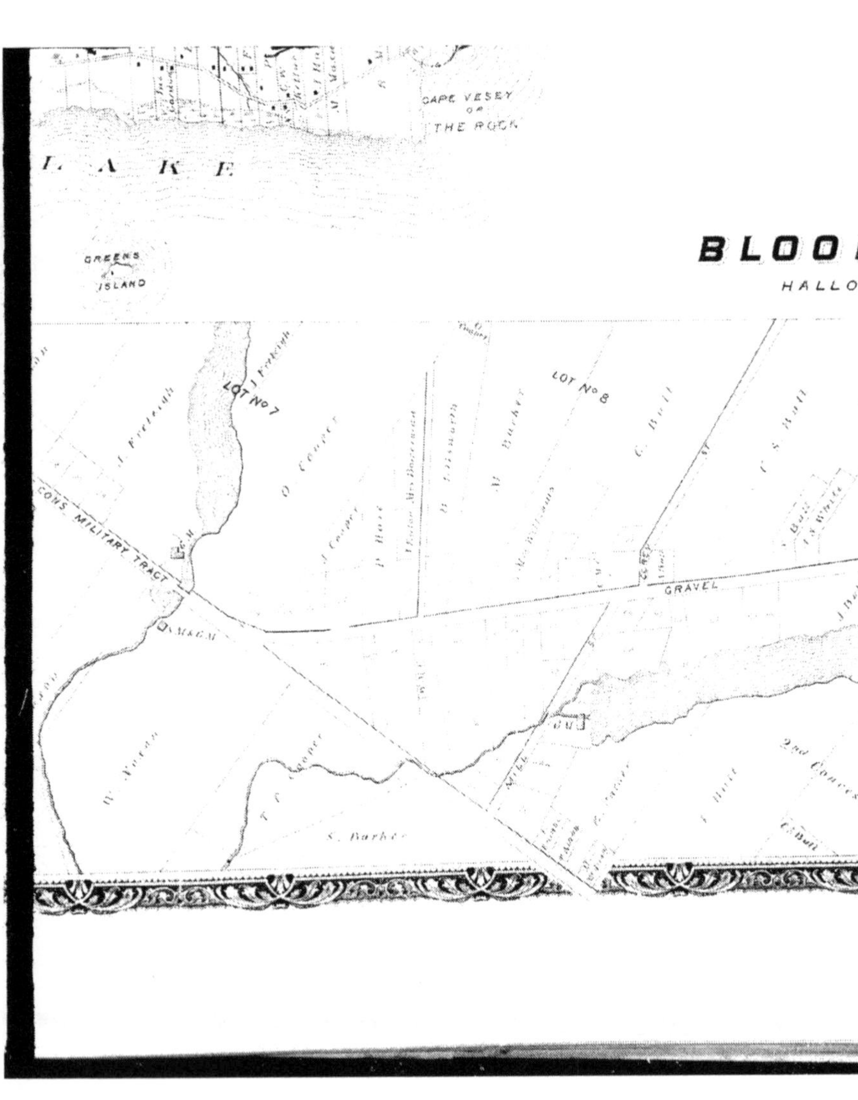

CAPE VESEY
OR
THE ROCK
LAKE
GREENS
ISLAND
BLOOI
HALLO
LOT No 7
LOT No 8
CONS MILITARY TRACT
GRAVEL

LOOMFIELD

HALLOWELL TP

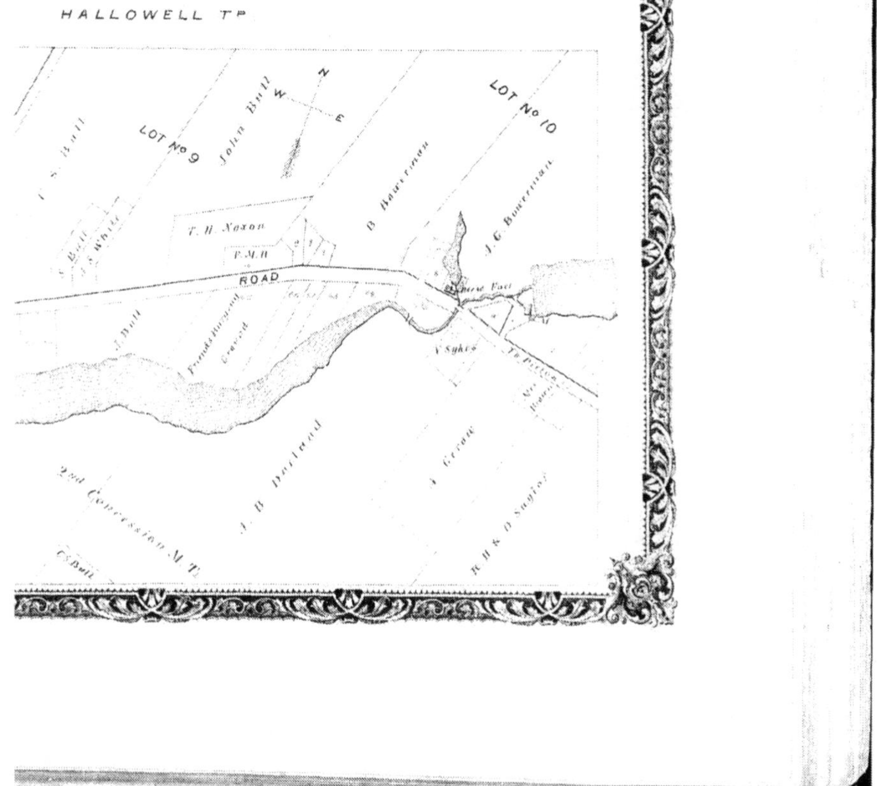

Daniel Pettit
John Young
Cornelius Clapp
Daniel G. Williams J.P.

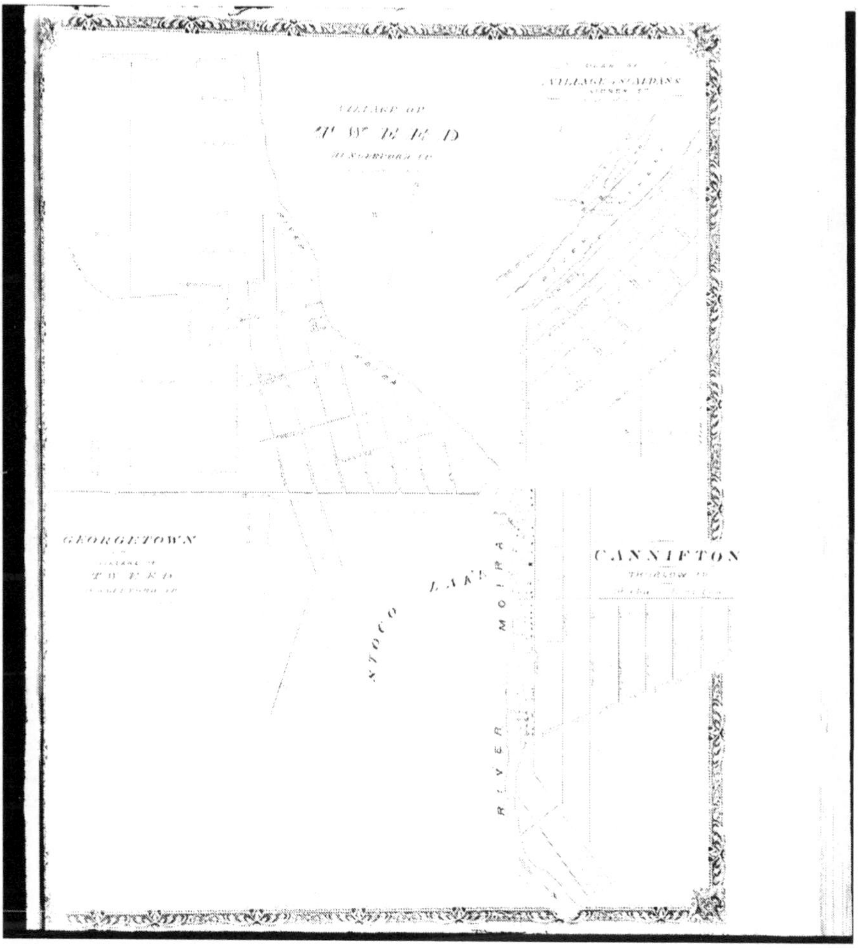

VILLAGE OF

TWEED

HUNGERFORD TP.

VILLAGE ST. ALBANS

GEORGETOWN

TWEED

CANNIFTON

THURLOW TP.

STOCO LAKE

RIVER MOIRA

ICTON
PRINCE ED. CO.
Scale 9 Chains per Inch
1st Con. North of Carrying Place
LOT No 3
LOT No 4
LOT No 5
Corporation line
Picton Driving Park Association
Agricultural Society
QUINTE WARD
BAY
HARBOUR
ELM
PRINCESS
QUEEN
MAIN
WATER
ST

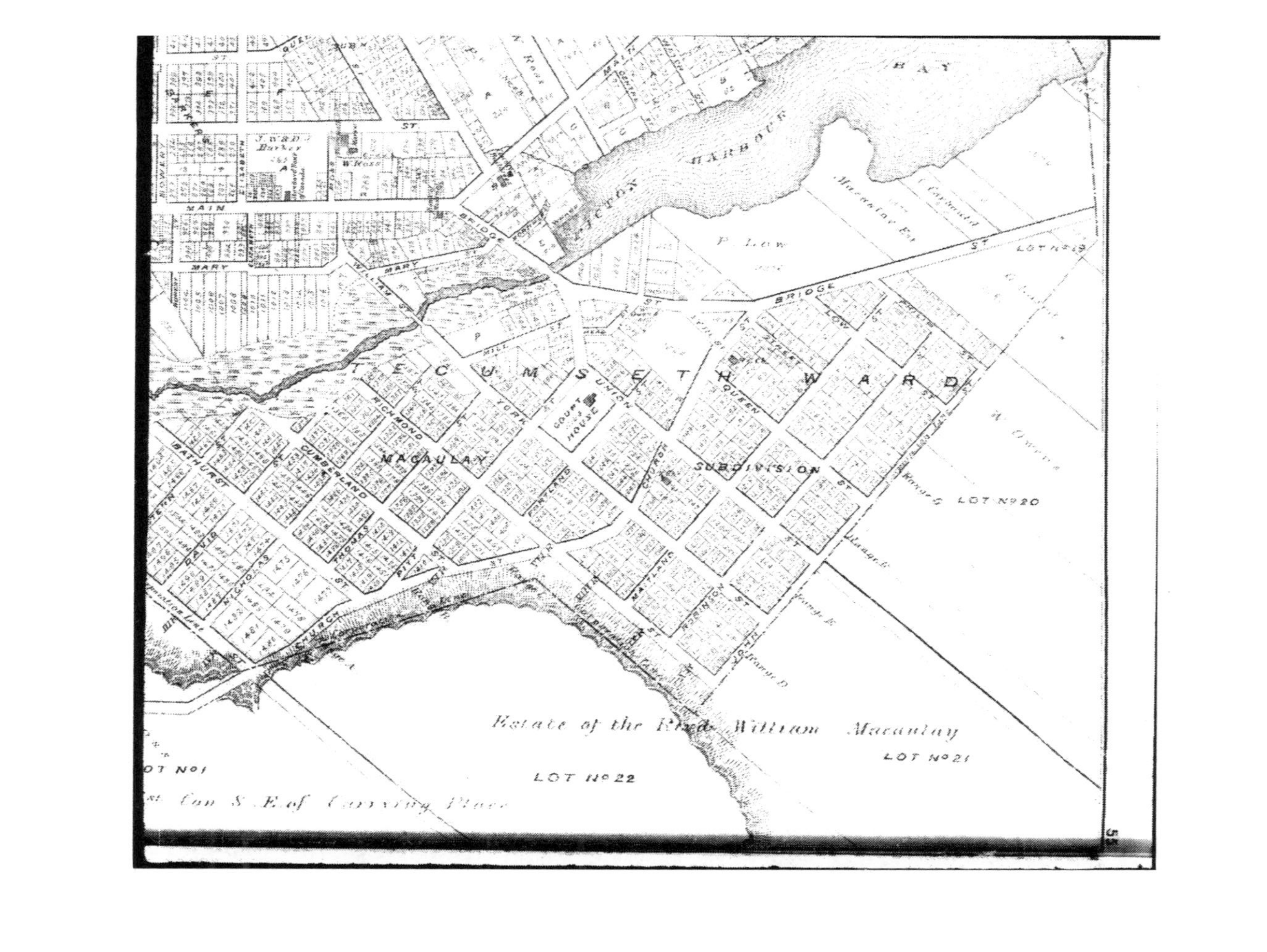

TECUMSETH WARD
PICTON HARBOUR
BAY
MACAULAY
SUBDIVISION
COURT HOUSE
BRIDGE ST
MAIN
MARY
BOWERY
ELIZABETH
QUEEN
UNION
YORK
RICHMOND
CUMBERLAND
BATHURST
DAVID
NICHOLAS
THOMAS
PITT
CHURCH
PORTLAND
MAITLAND
ROBINSON
LOW
P. Low
Macaulay Est
LOT No 19
LOT No 20
LOT No 21
LOT No 22
Estate of the Revd William Macaulay
Con S. E. of Carrying Place
55

PLAN OF
PICTO
LOT No 1
LOT No A
LOT No 20
LOT No 21
THOMAS
ST
PETER
DAVID
ST
BARKER
QUEEN
KING
BROCK
COLEMAN

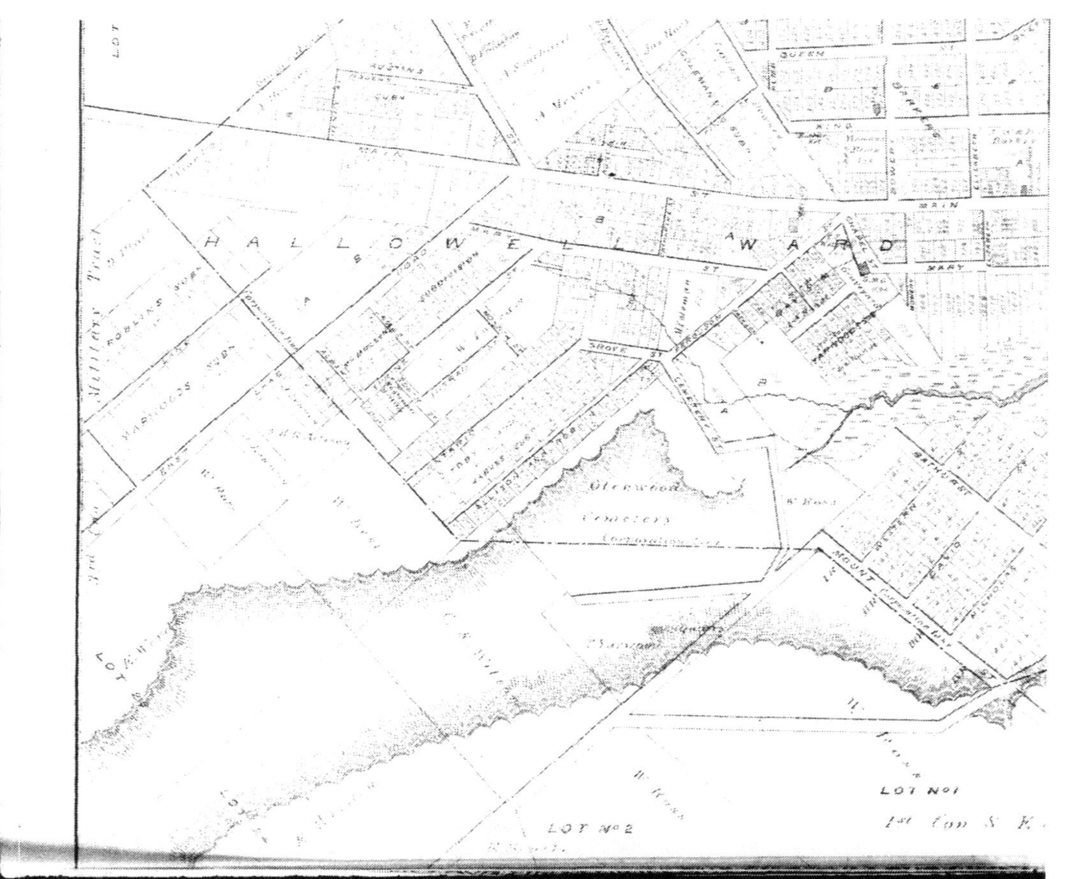
HALLOWELL WARD
QUEEN
KING
MAIN
MARY
BATHURST
MOUNT
GROVE ST
Glenwood Cemetery
Military Tract
LOT No1
LOT No2
1st Con S. K.

ROMAN CATHOLIC CHURCH OF THE SACRED HEART, MARMORA, ONT. THOS. DAVIS, PASTOR.

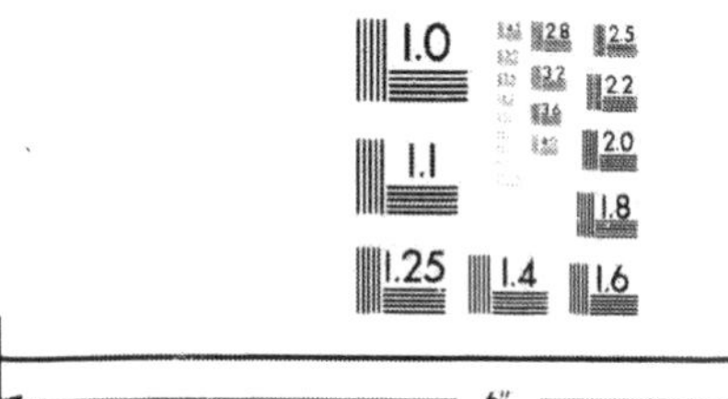
1.0
1.1
1.25
1.4
1.6
1.8
2.0
2.2
2.5
2.8
3.2
3.6
4.0
6"

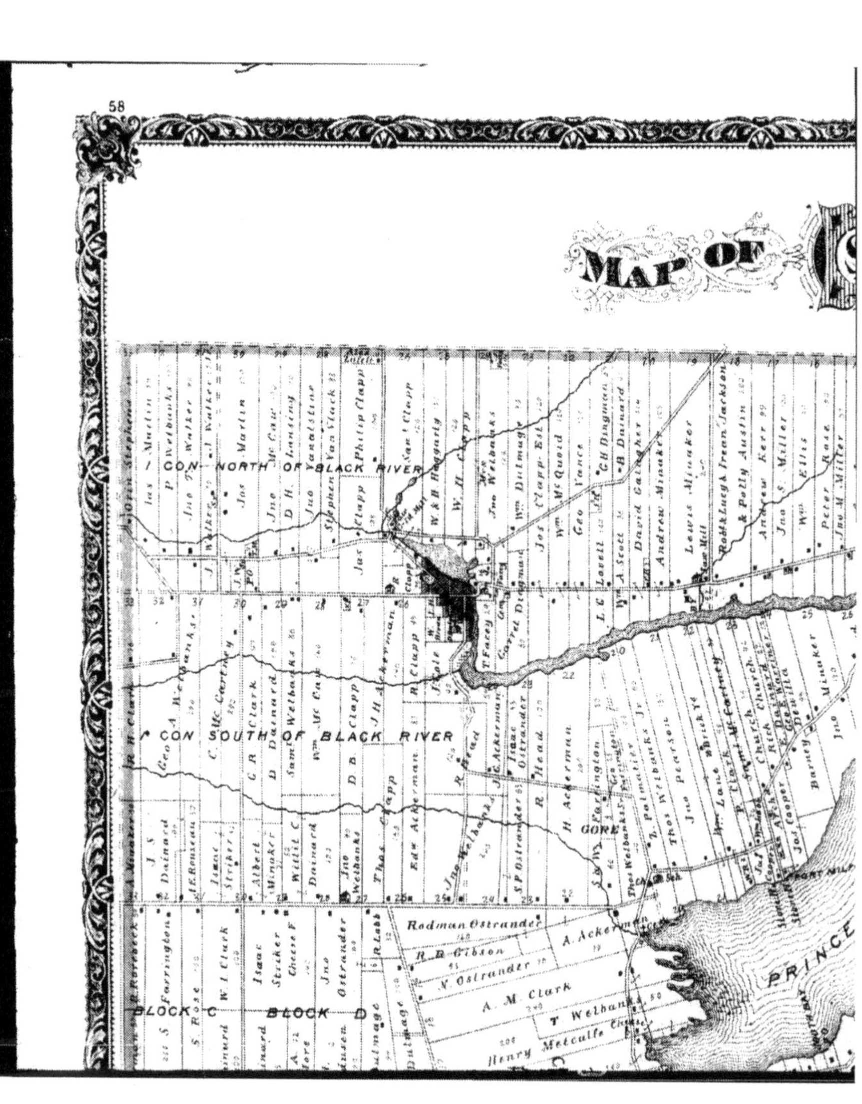
MAP OF
CON NORTH OF BLACK RIVER
CON SOUTH OF BLACK RIVER
GORE
BLOCK C
BLOCK D
Rodman Ostrander
R. B. Gibson
N. Ostrander
A. M. Clark
A. Ackerman
T. Welbanks
Henry Metcalfe
PRINCE

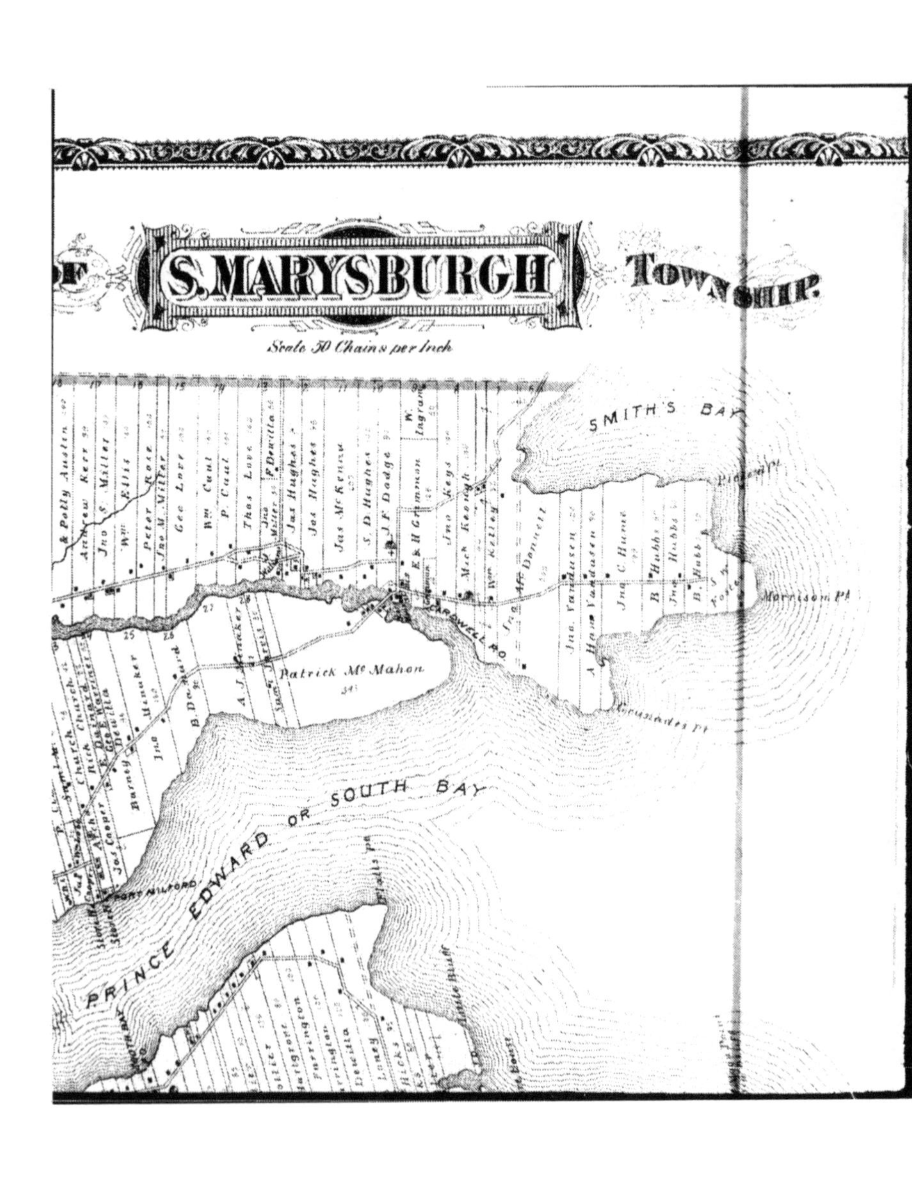

S. MARYSBURGH TOWNSHIP
Scale 50 Chains per Inch
SMITH'S BAY
PRINCE EDWARD OR SOUTH BAY
Patrick McMahon
Morrison Pt
CARDWELL P.O.
PORT MILFORD

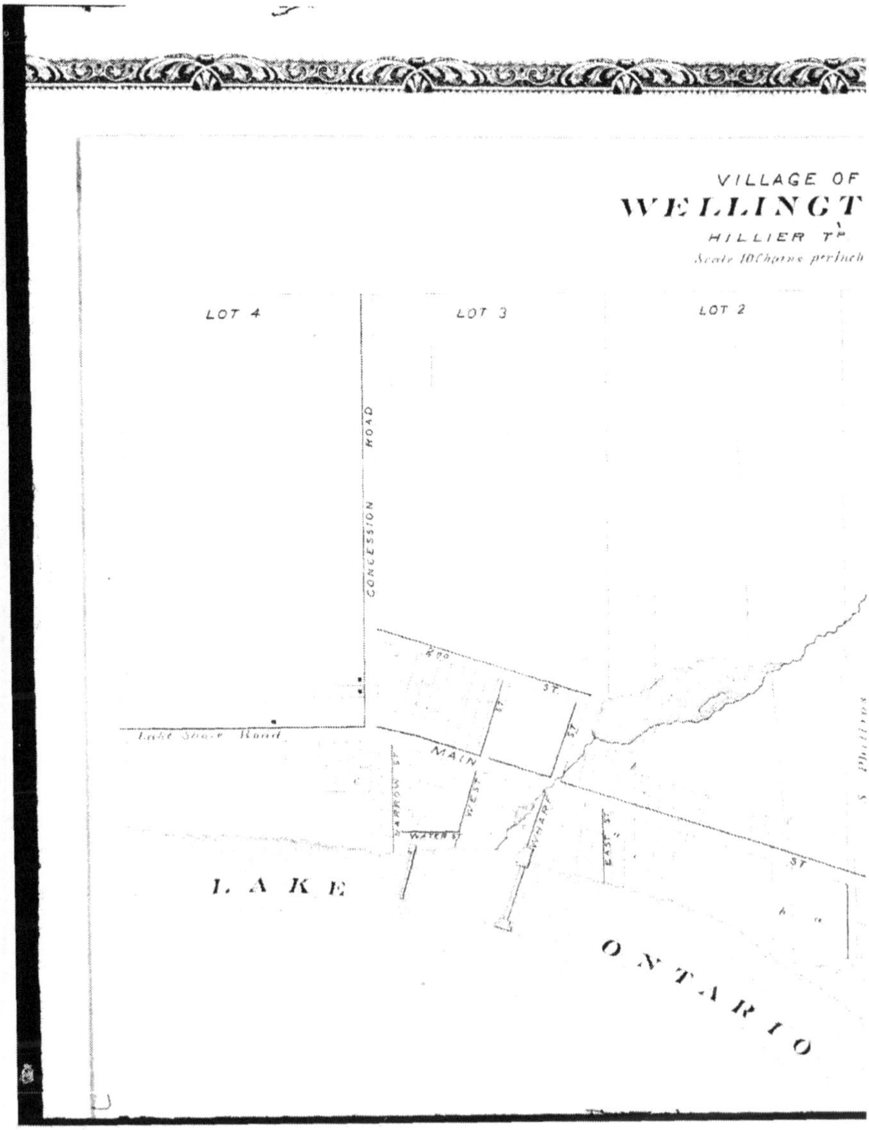

VILLAGE OF

WELLINGT

HILLIER T^P

Scale 10 Chains per Inch

LOT 4

LOT 3

LOT 2

CONCESSION ROAD

2ND ST.

ST.

ST.

Lake Shore Road

MAIN

NARROW ST.

WEST ST.

WATER ST.

WHARF

EAST ST.

ST.

S Phillips

LAKE

ONTARIO

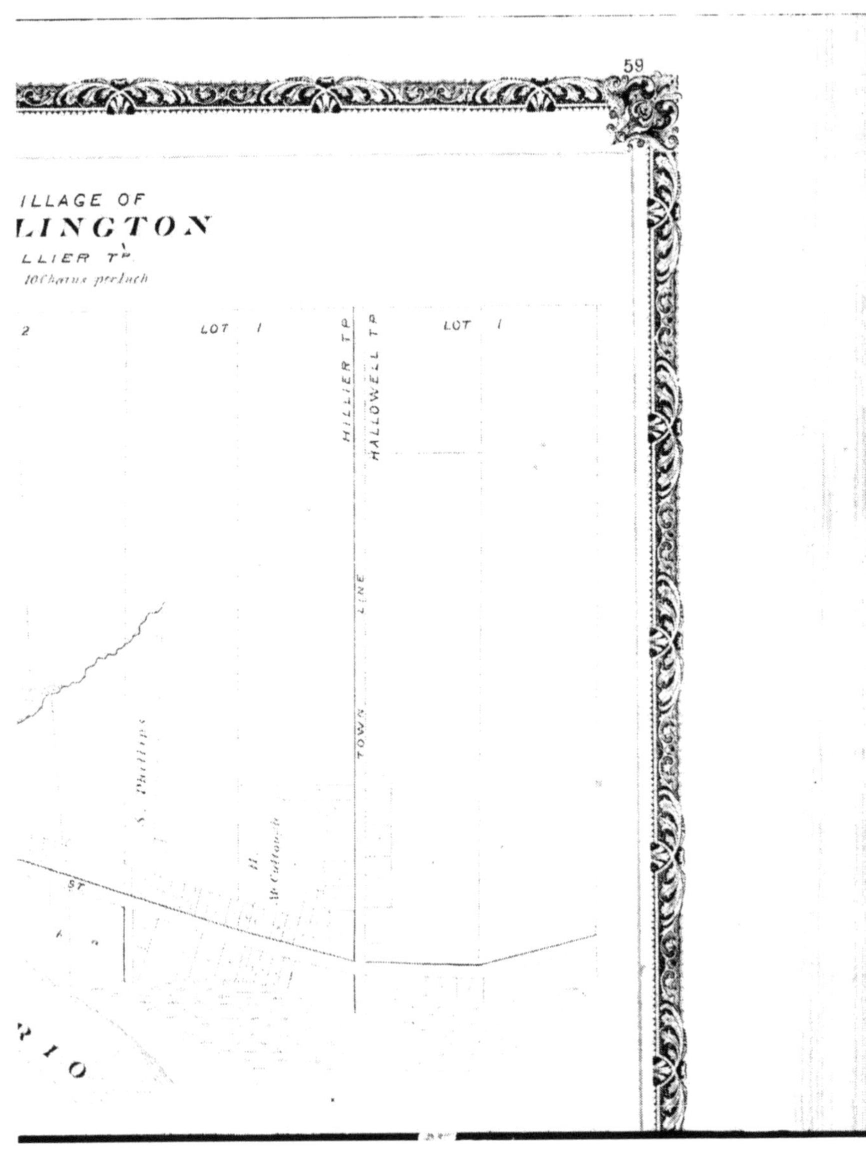

ILLAGE OF
LINGTON
LLIER Tᴾ
10 Chains per Inch
2
LOT 1
HILLIER TP
HALLOWELL TP
LOT 1
TOWN LINE
S. Phillips
H. McCulloch
ST
RIO

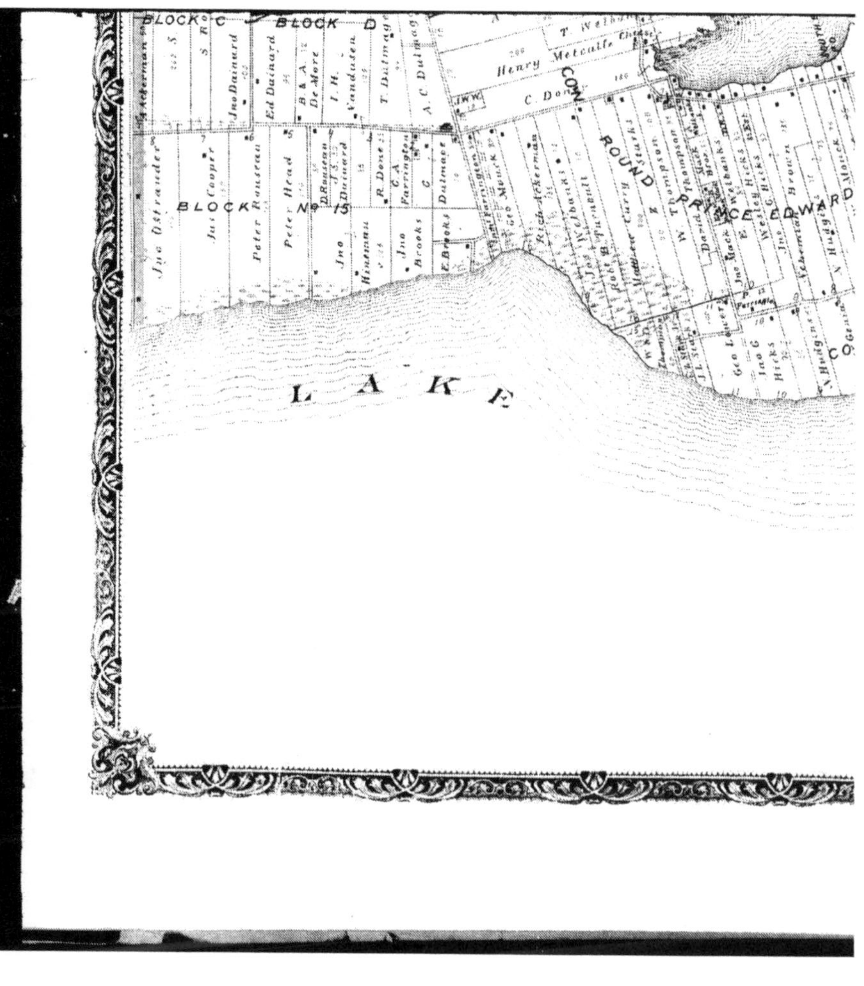
BLOCK C
BLOCK D
BLOCK No 15
Henry Metcalfe
C. Doe
ROUND
PRINCE EDWARD
LAKE

EDWARD BAY
CON. WEST OF LONG POINT
GRAVELY BAY
SANDY BAY
ONTARIO

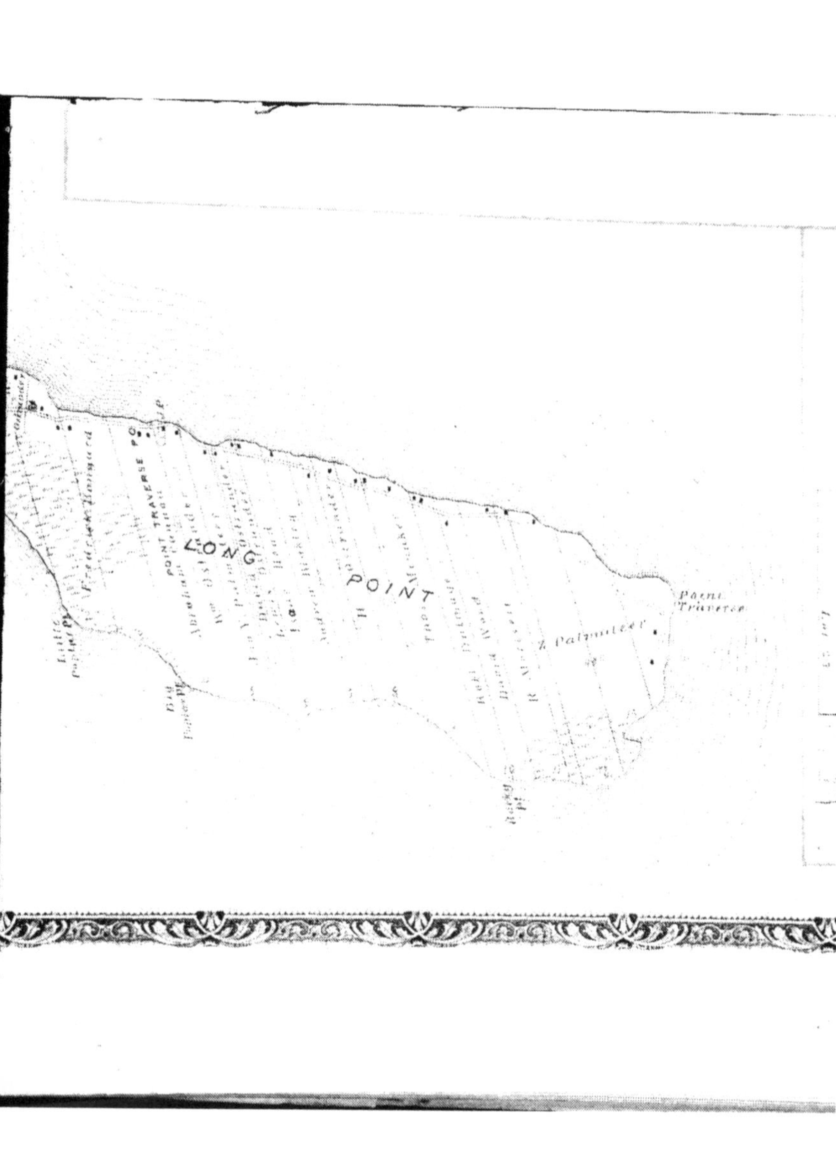
LONG POINT
POINT TRAVERSE P.O.
Point Traverse
Z. Palmateer
Little Poplar Pt.
Big Poplar Pt.

MILFORD

SOUTH MARYSBURG TP.

Scale 10 Chs. to 1 Inch

Lot 26

1 Con. S.B.R.

L. Hicks

MAIN

N.B.R.

MILL POND

PHILIP ST.

KING ST.

BLACK RIVER

Lot 23

RES. OF JOHN WHITE, CON. 9, LOT 4, TYENDINAGA TP. ONT.

BAY VIEW FARM RES. OF JOSEPH PIERSON, J. P. CON. 3, LOT 31, HILLIER TP. ONT.

FARM RES. OF JOHN T. OSBORNE ESQ. CON 3, LOT 94, AMELIASBURG TP. P. EDWARDS CO. ONT.

RES. OF E. MERRILL, BARRIS

RES. OF JAMES BENSON ESQ. TOWNSHIP CLERK AMELIASBURG TP. P. EDWARDS CO. ONT.

RES. OF R. A. FOS

ENDINAGA TP. ONT.

ES. OF E. MERRILL, BARRISTER-AT-LAW, PICTON, ONT.

RES. OF R. A. FOSTER, PICTON, ONT.

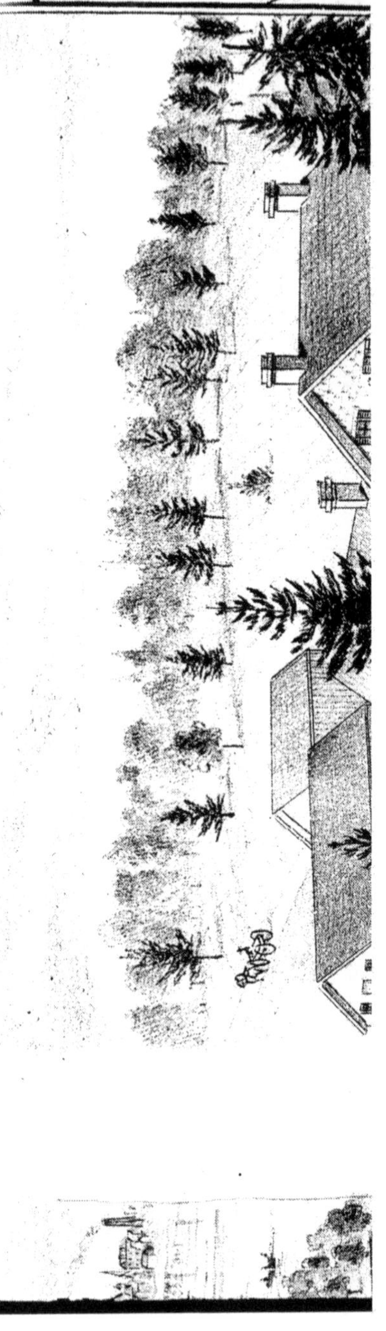

RES. OF JAMES REDNER ESQ. REDNERVILLE, PRINCE EDWARD CO. ONT.

ONT.

RES. OF DR. B. S. WILLSON, PINE STREET, BELLEVILLE, ONT.

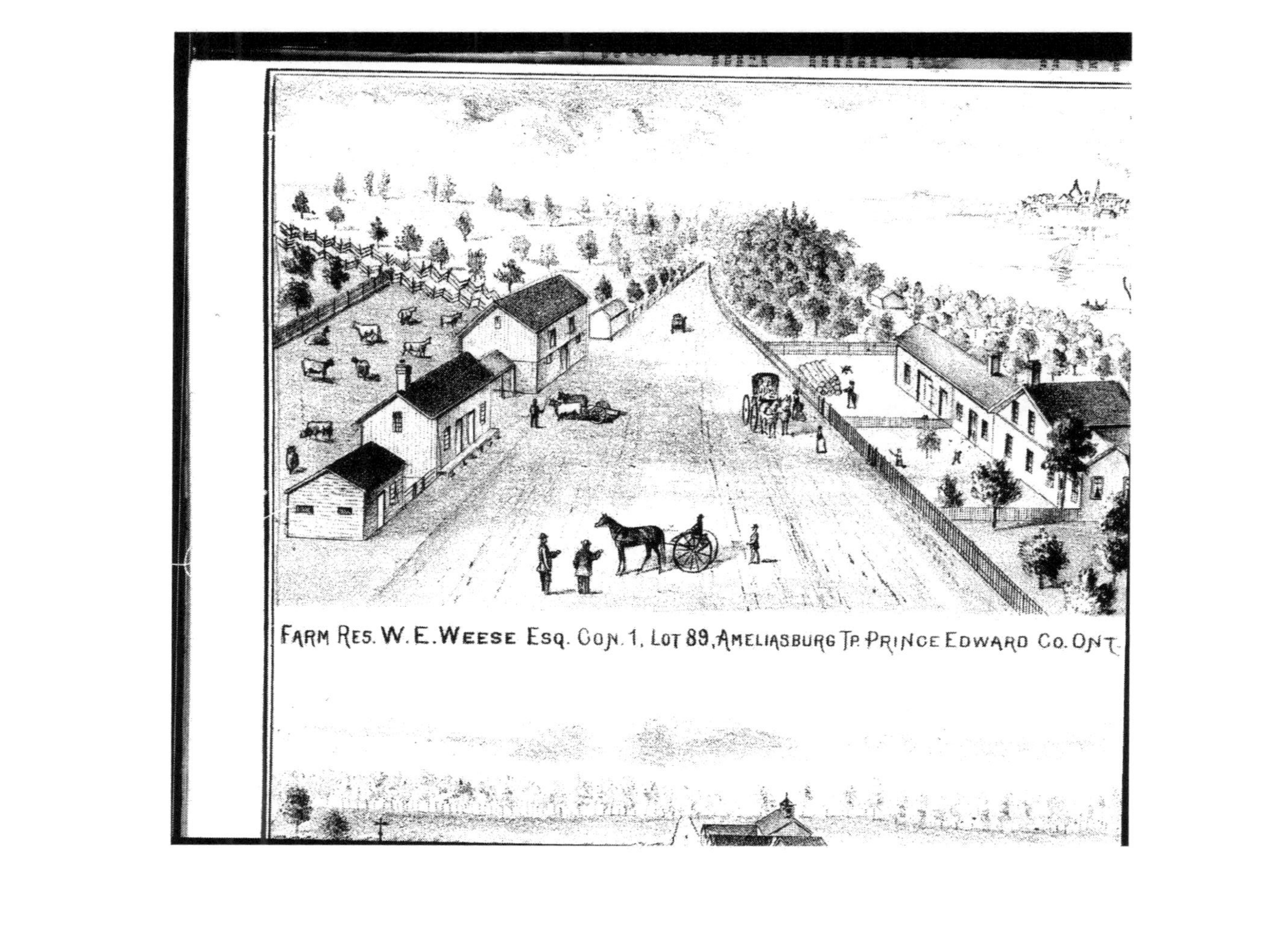

FARM RES. W. E. WEESE ESQ. CON. 1, LOT 89, AMELIASBURG TP. PRINCE EDWARD CO. ONT.

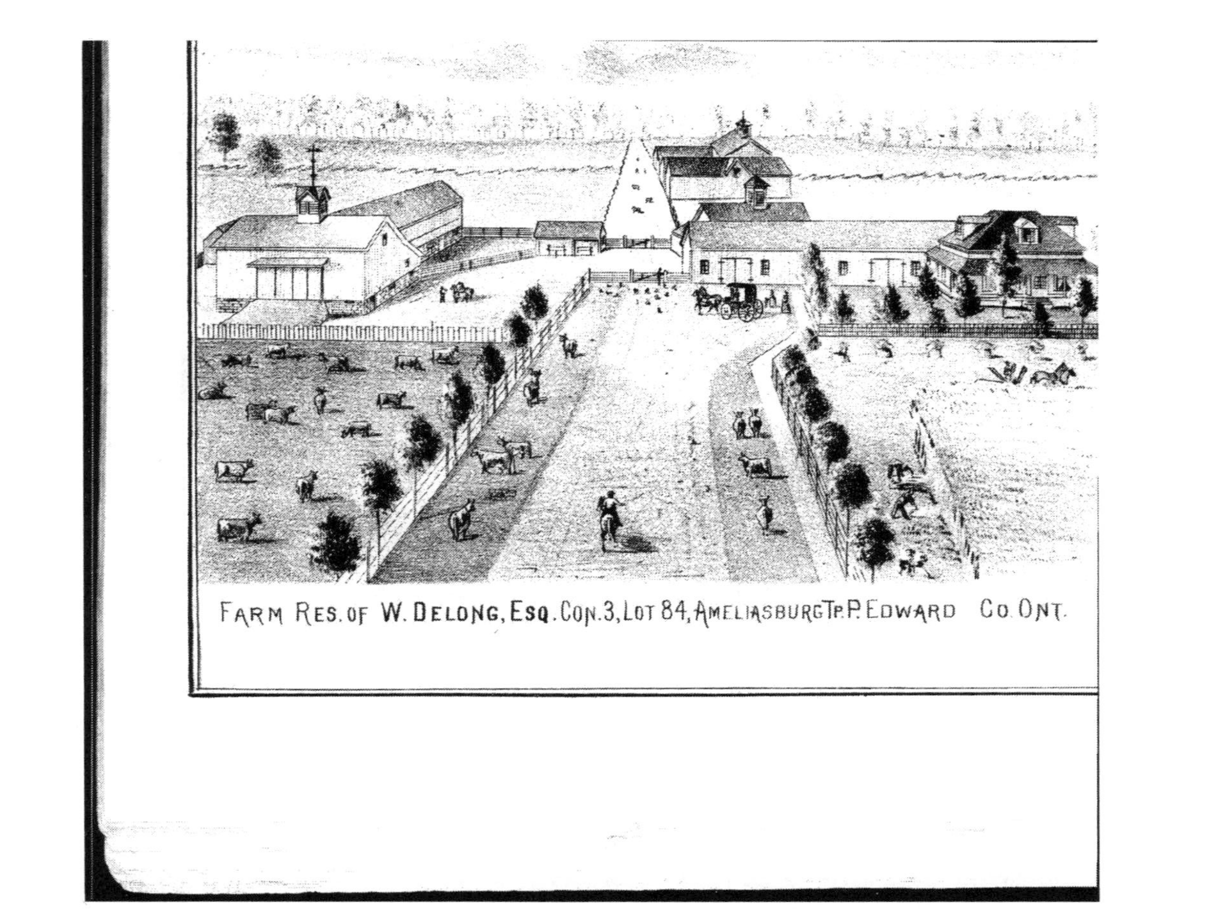

FARM RES. OF W. DELONG, ESQ. CON. 3, LOT 84, AMELIASBURG TP. P. EDWARD CO. ONT.

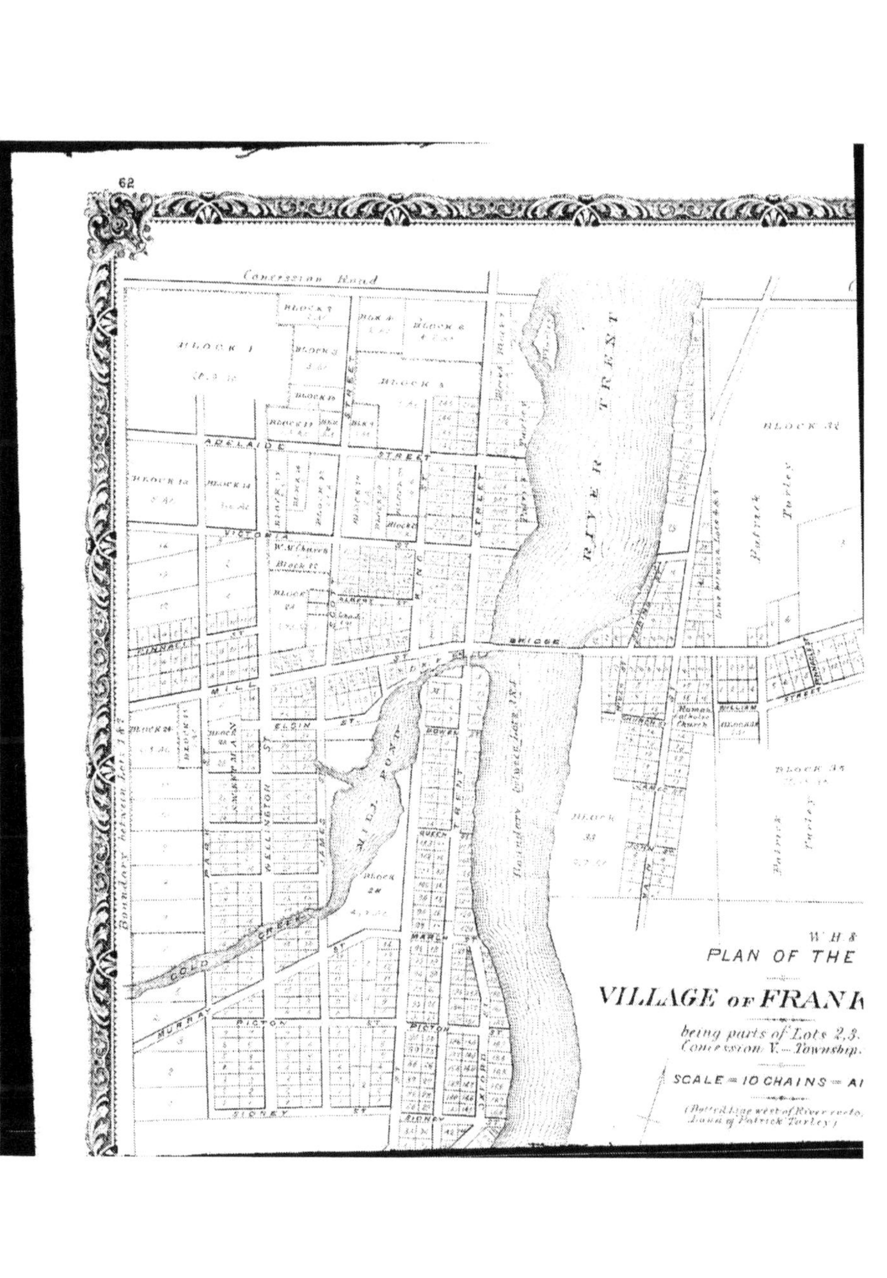
Concession Road
BLOCK 1
ADELAIDE
STREET
VICTORIA
W. M. Church
MILL
ELGIN
WELLINGTON ST
MILL POND
COLD CREEK
MURRAY
PICTON
ST
SIDNEY
RIVER TRENT
BRIDGE
QUEEN
Roman Catholic Church
WILLIAM
JAMES
Boundary between Lots 1 & 2
Patrick Turley
BLOCK 34
BLOCK 35
W. H. &
PLAN OF THE
VILLAGE OF FRANK
being parts of Lots 2, 3.
Concession V. — Township.
SCALE = 10 CHAINS = AI
(Dotted line west of River encloses Land of Patrick Turley)

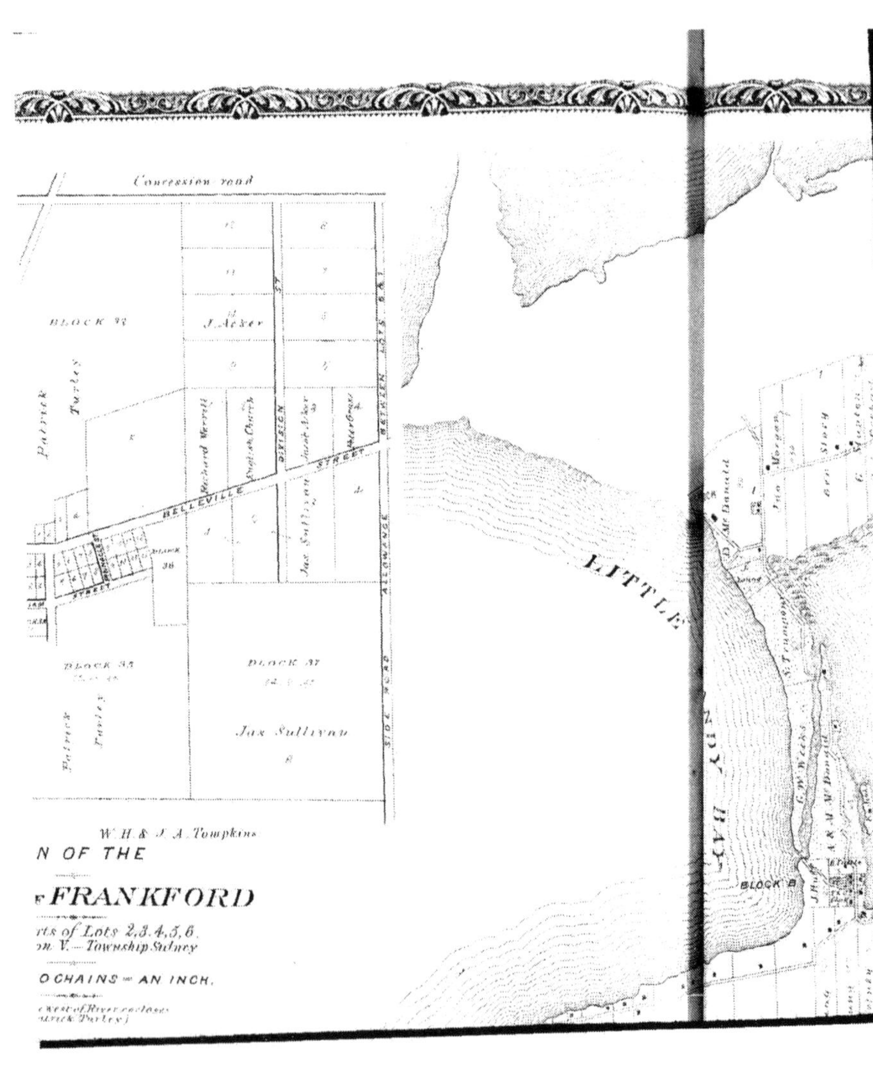
Concession road
J. Acker
Patrick Turley
Richard Worrell
English Church
DIVISION ST
BELLEVILLE STREET
Jas. Sullivan
BLOCK 35
BLOCK 37
Jas. Sullivan
SIDE ROAD ALLOWANCE BETWEEN LOTS 5&6
Patrick Turley
W. H. & J. A. Tompkins
N OF THE
F FRANKFORD
rts of Lots 2,3,4,5,6.
on V.—Township Sidney
O CHAINS = AN INCH.
LITTLE
NDY BAY
D. McDonald
Jno Morgan
Geo Story
G. W. Weeks
A & M. McDonald
BLOCK B

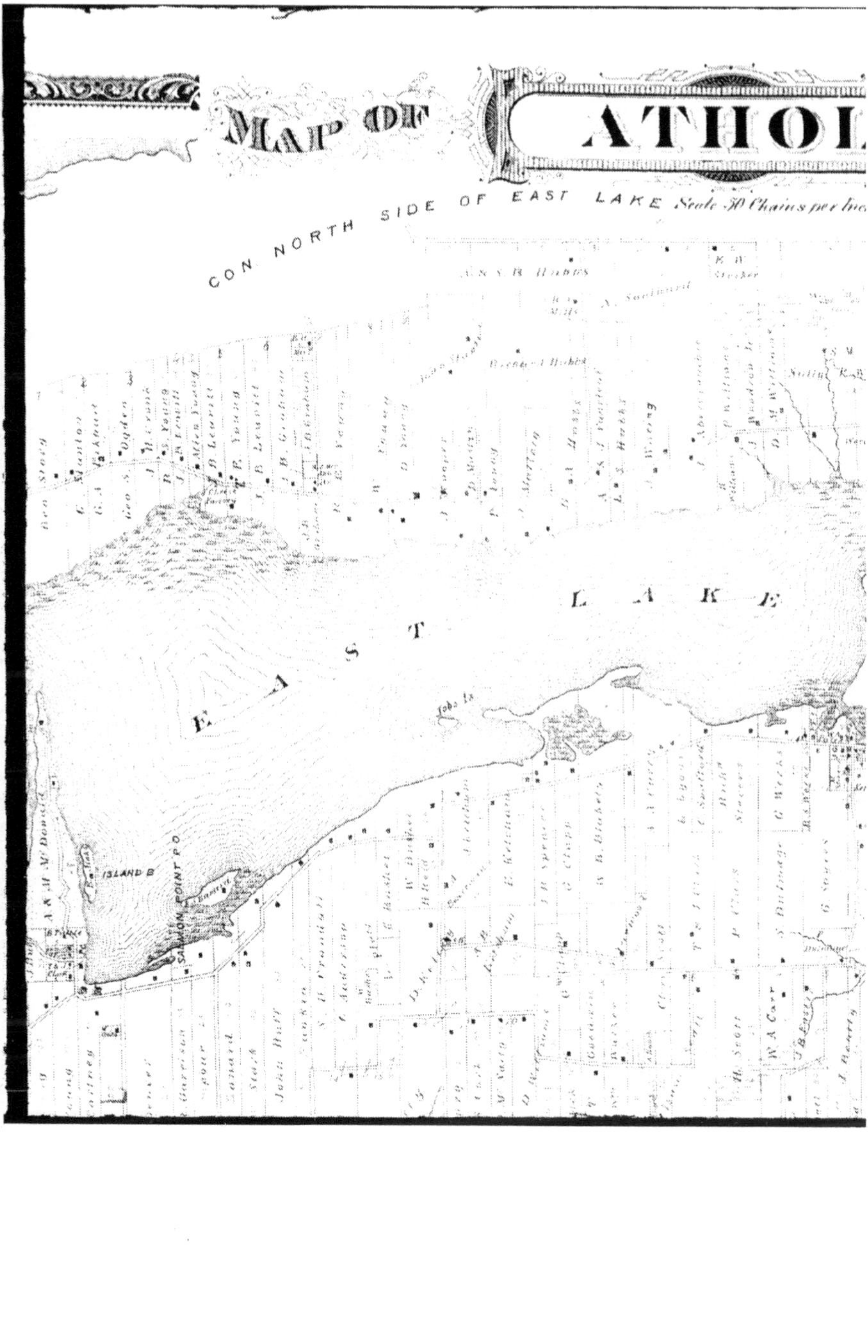

MAP OF ATHOL

CON NORTH SIDE OF EAST LAKE Scale 30 Chains per Inc

E A S T L A K E

ISLAND B

SALMON POINT P.O.

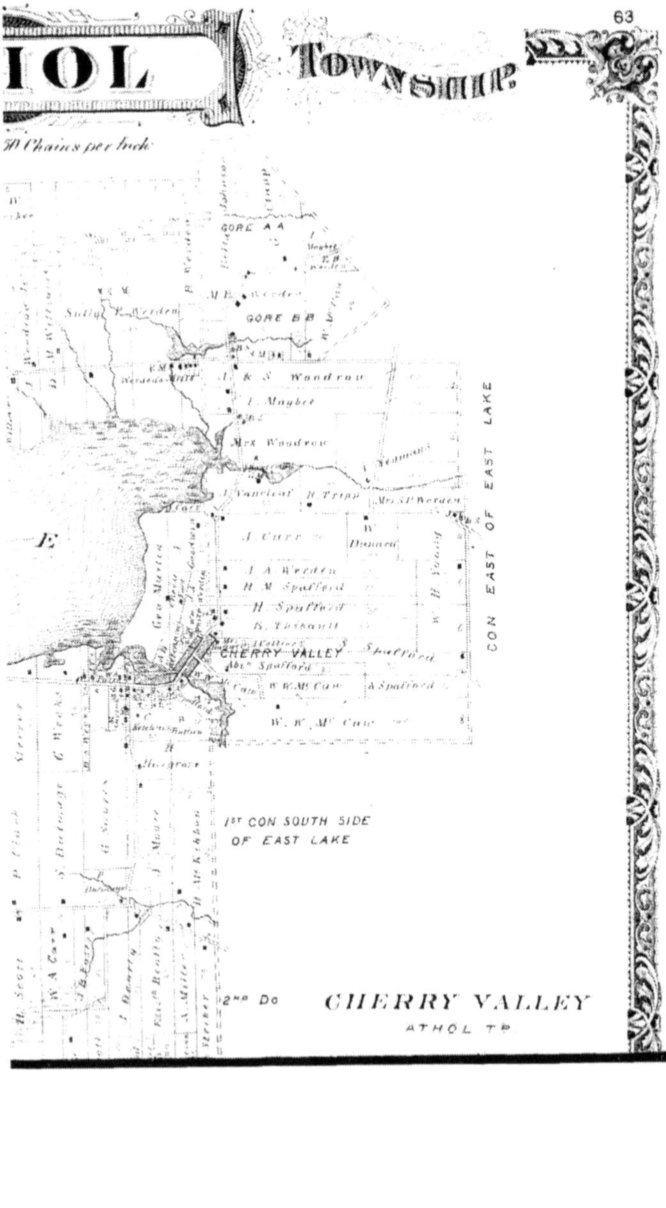

IOL TOWNSHIP

50 Chains per inch

GORE A A

GORE B B

J. & S. Woodrow

I. Maybee

Mrs. Woodrow

J. Vancleaf

H. Tripp

Mrs. S.P. Werden

J. Carr

J. A. Werden

H. M. Spafford

H. Spafford

CHERRY VALLEY

Abm. Spafford

W. W. McCaw

A. Spafford

CON EAST OF EAST LAKE

1ST CON SOUTH SIDE OF EAST LAKE

2ND Do

CHERRY VALLEY

ATHOL TP.

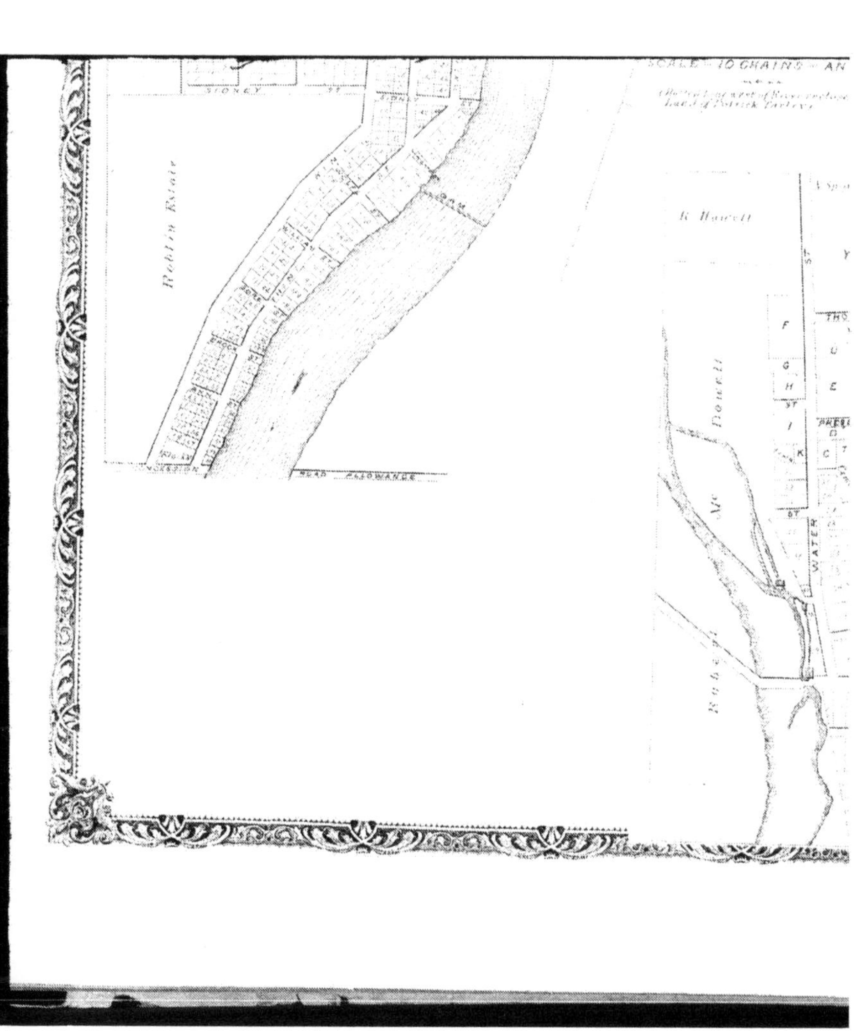

SCALE — 10 CHAINS — AN
Roblin Estate
SIDNEY ST
ROAD ALLOWANCE
R. Howell
WATER

PLAN OF THE VILLAGE OF
DEMORESTVILLE
SOPHIASBURGH TP
BROADWAY
NORTH ST
SOUTH ST
WATER ST
JAMES
THOMAS
PRESBYTERIAN
ST PAUL ST
GEORGE ST
PLEASANT

PLEASANT HARBOUR
Light House
POINT PETRE
GULL POND
GULL
E
O N T A R I O

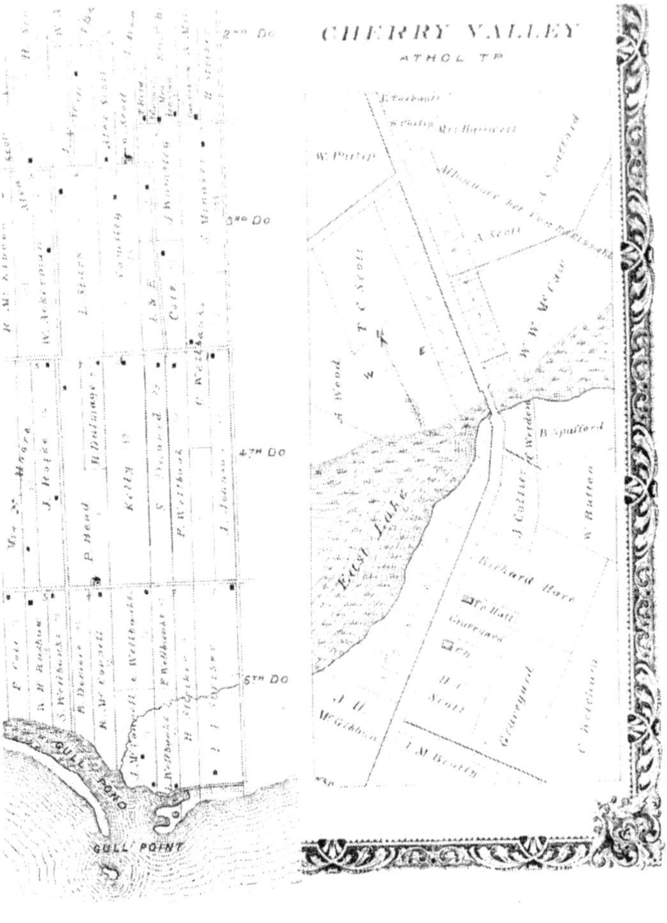

CHERRY VALLEY

ATHOL TP

2ND DO
3RD DO
4TH DO
5TH DO

East Lake
GULL POND
GULL POINT

W. Philip
A. Scott
T. C. Scott
A. Wood
W. W. McCaw
B. Spafford
J. Collier
W. Rutton
Richard Hare
Te Hall
Graveyard
Ch
Scott
J. H. McGibbon
J. M. Beatty
Graveyard

P. Head
B. Dulmage
Keith
P. Wellbank
J. Johnson
W. Ackerman
L. Spears
Cole
C. Wellbanks

"MOUNTAINVIEW", RES. OF R.L. LAZIER ESQ. CON. 1, LOT 5, TYENDINAGA TP. ONT.

PRESBYTERIAN CHURCH, MADOC VILLAGE, ONT.

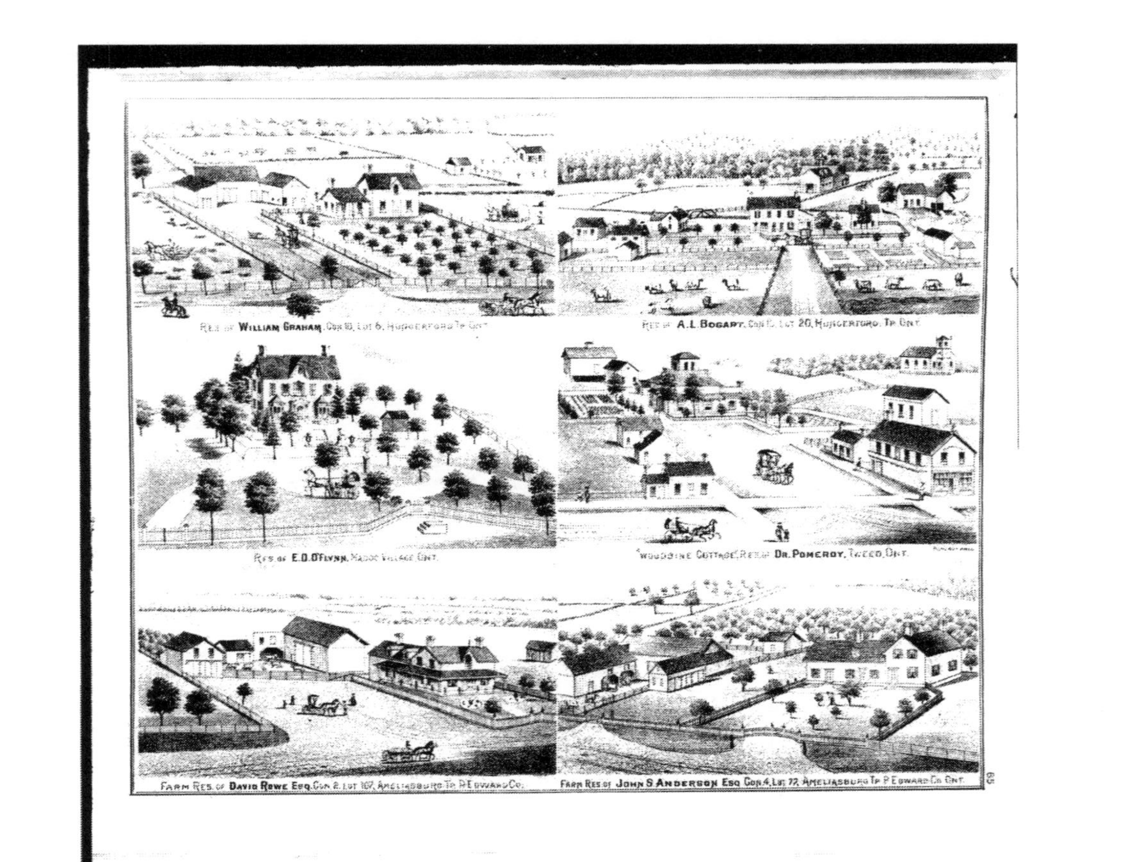

Res. of William Graham, Con 10, Lot 6, Hungerford Tp. Ont.

Res. of A. L. Bogart, Con 10, Lot 20, Hungerford, Tp. Ont.

Res. of E. D. O'Flynn, Madoc Village Ont.

"Woodbine Cottage," Res. of Dr. Pomeroy, Tweed, Ont.

Farm Res. of David Rowe Esq. Con 2, Lot 107, Ameliasburg Tp. P. Edward Co.

Farm Res. of John S. Anderson Esq. Con 4, Lot 72, Ameliasburg Tp. P. Edward Co. Ont.

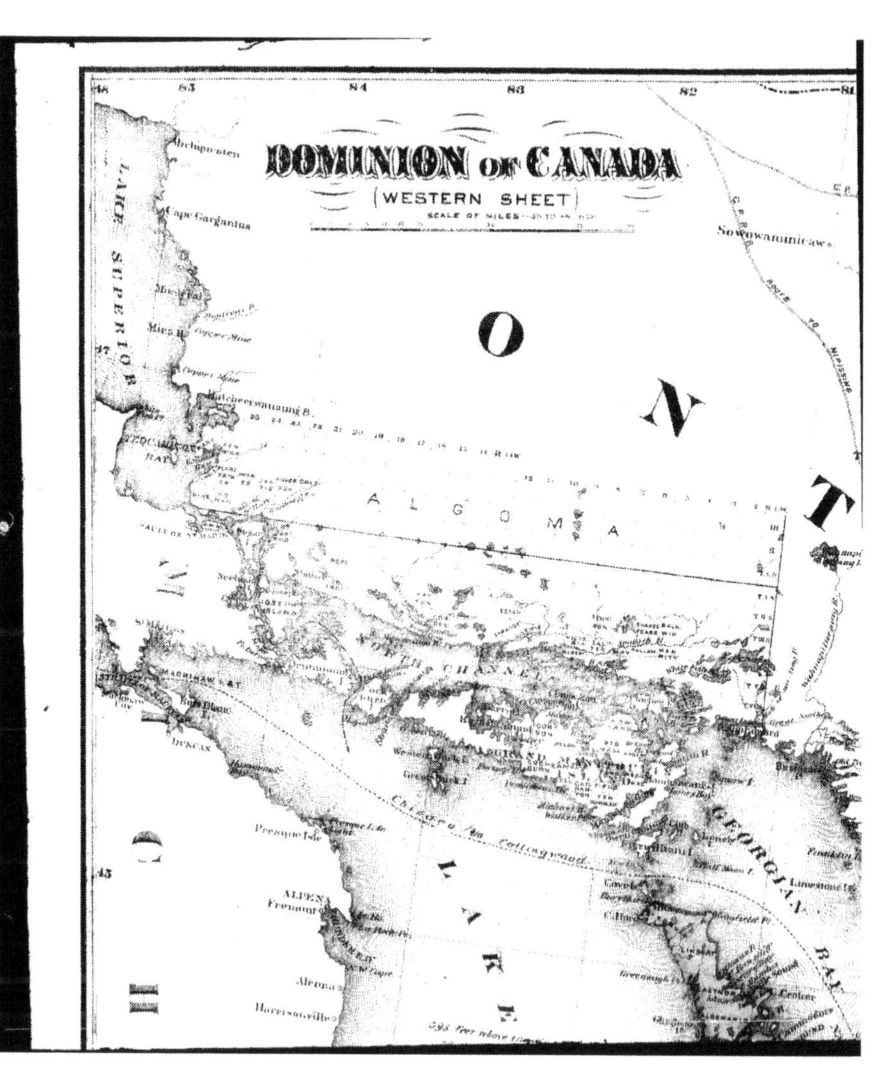

DOMINION OF CANADA
(WESTERN SHEET)
SCALE OF MILES
LAKE SUPERIOR
Cape Gargantua
O
N
T
ALGOMA
Sowowaminicaws
GEORGIAN BAY
LAKE
Chicago to Collingwood
Presque Isle
ALPENA

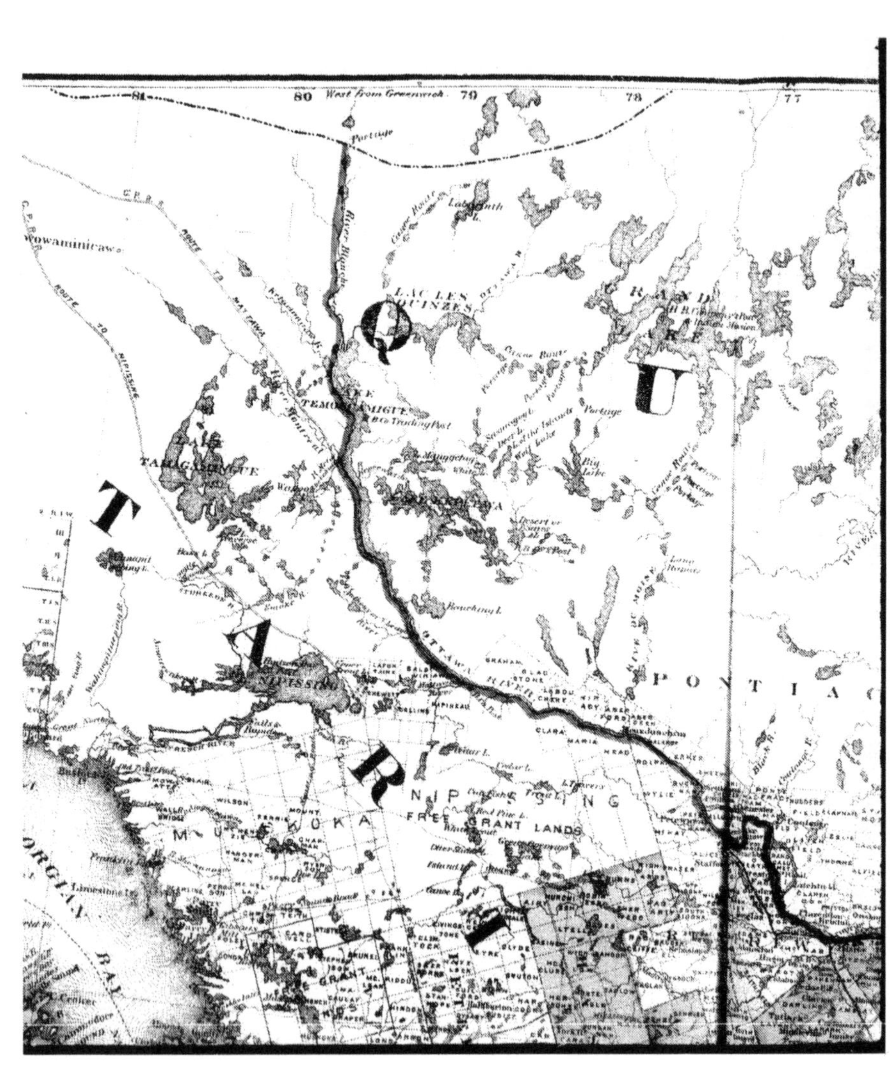

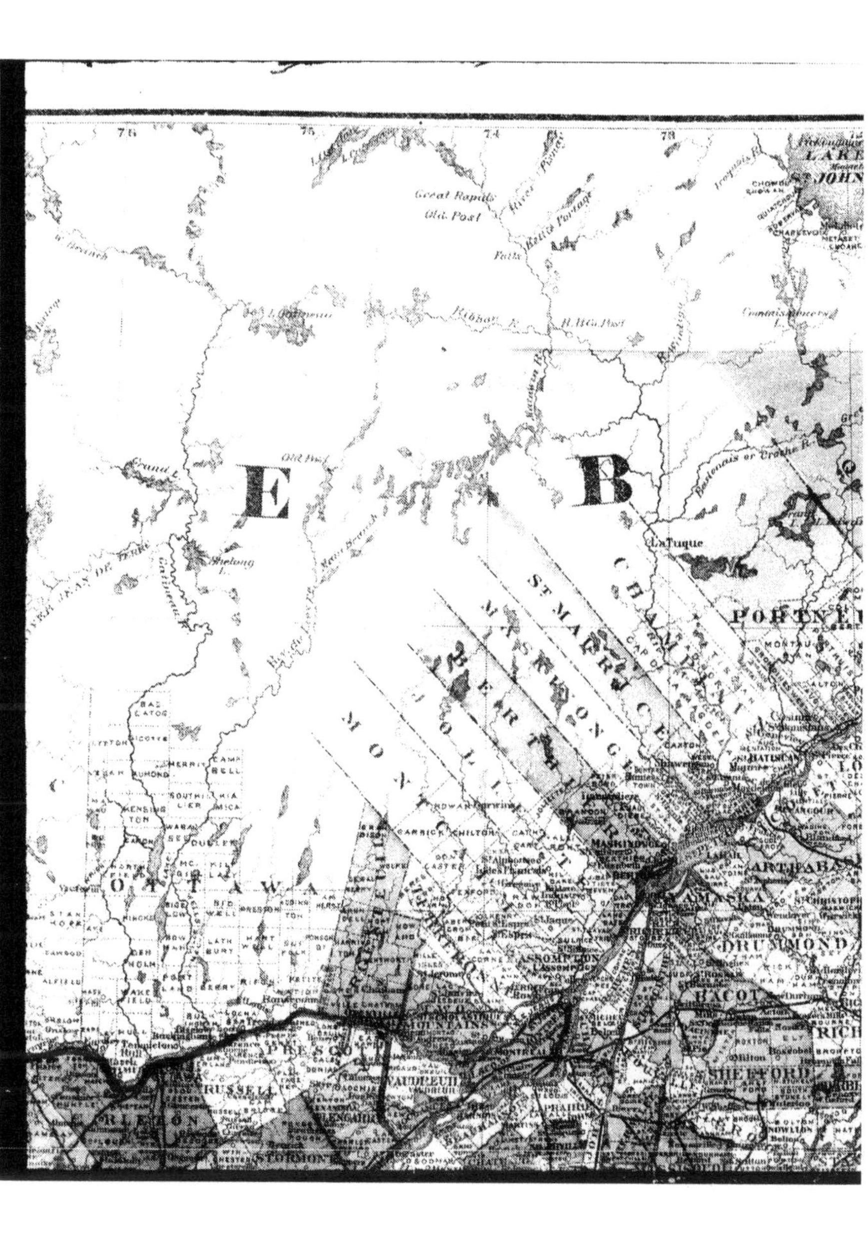

Great Rapids
Old Post
Falls
H. B. Co. Post
Commissioners L.
LAKE
JOHN
Grand L.
Old Post
E
B
LaTuque
Shelong L.
ST MAURICE
CHAMPLAIN
BERTHIER
JOLIETTE
MONTCALM
MASKINONGE
OTTAWA
ARTHABASKA
DRUMMOND
RUSSELL
VAUDREUIL
SHEFFORD
STORMONT

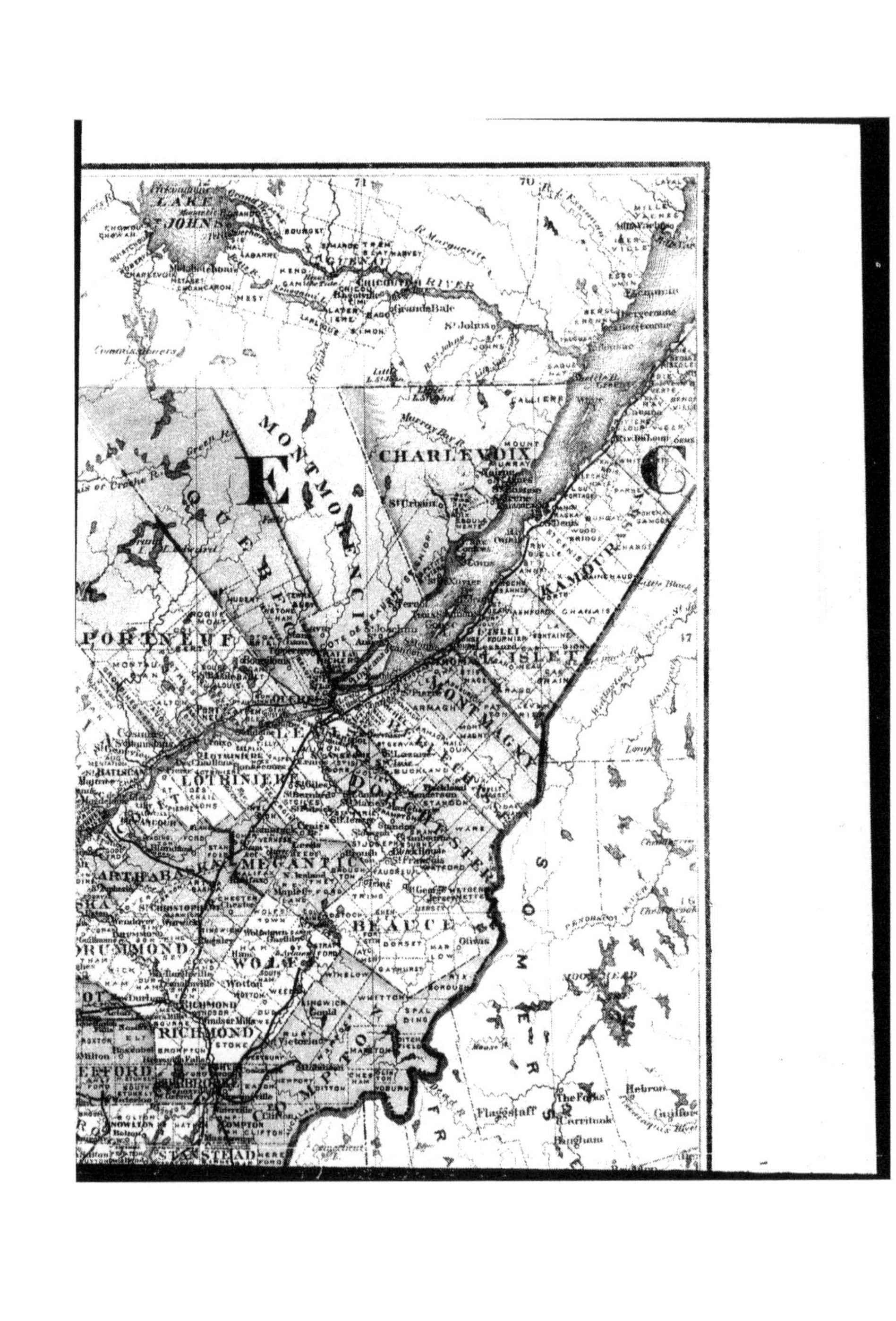
LAKE ST JOHNS
SAGUENAY
CHICOUTIMI
RIVER
Grande Baie
St Johns
R. Marguerite
MONTMORENCI
CHARLEVOIX
QUEBEC
PORTNEUF
KAMOURASKA
L'ISLET
MONTMAGNY
LEVIS
LOTHINIER
BELLECHASSE
DORCHESTER
MEGANTIC
ARTHABASKA
BEAUCE
DRUMMOND
WOLFE
RICHMOND
STANSTEAD
COMPTON
SOMERSET
The Forks
Flaggstaff
Hebron
Bingham

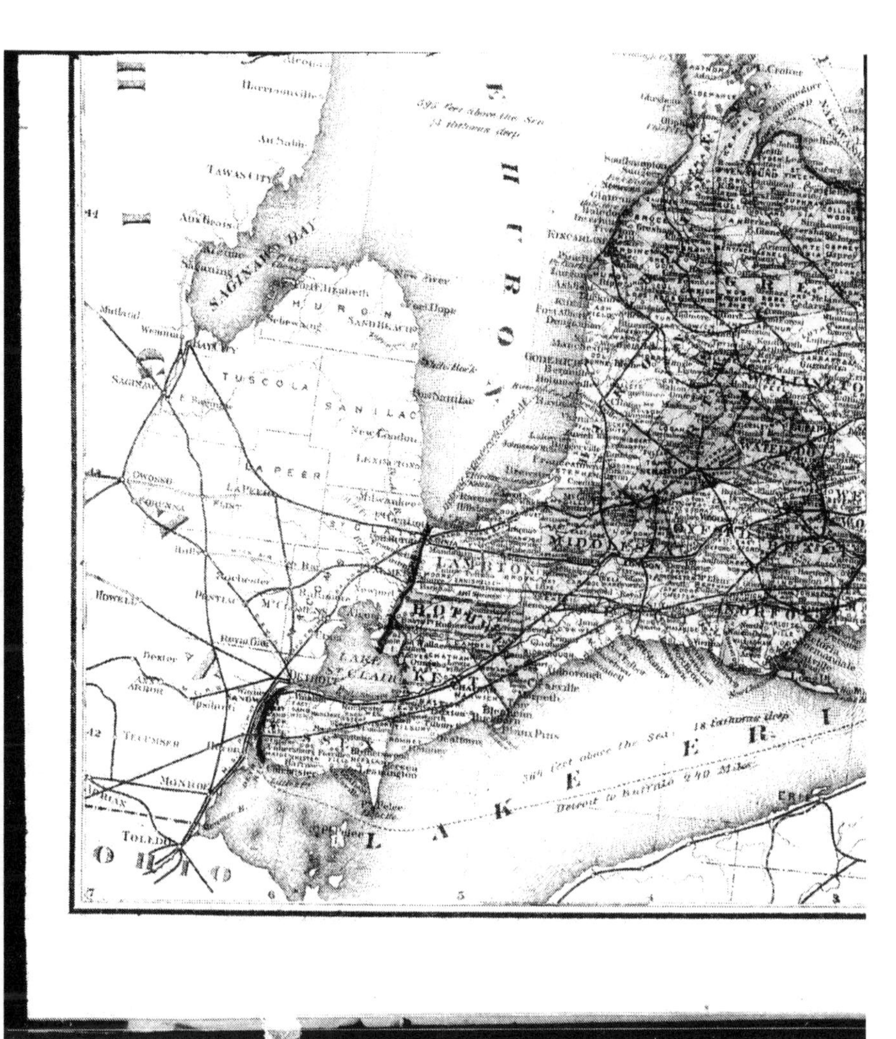

Harrisonville
Au Sable
TAWAS CITY
SAGINAW BAY
HURON
TUSCOLA
SANILAC
LAPEER
New Baltimore
Lexington
Port Hope
OWOSSO
Rochester
Pontiac
Howell
Dexter
Ann Arbor
Ypsilanti
Tecumseh
Monroe
Toledo
GODERICH
LAMBTON
LAKE ST. CLAIR
KENT
NORFOLK
595 feet above the Sea
564 feet above the Sea
18 fathoms deep
Detroit to Buffalo 240 Miles
L A K E E R I

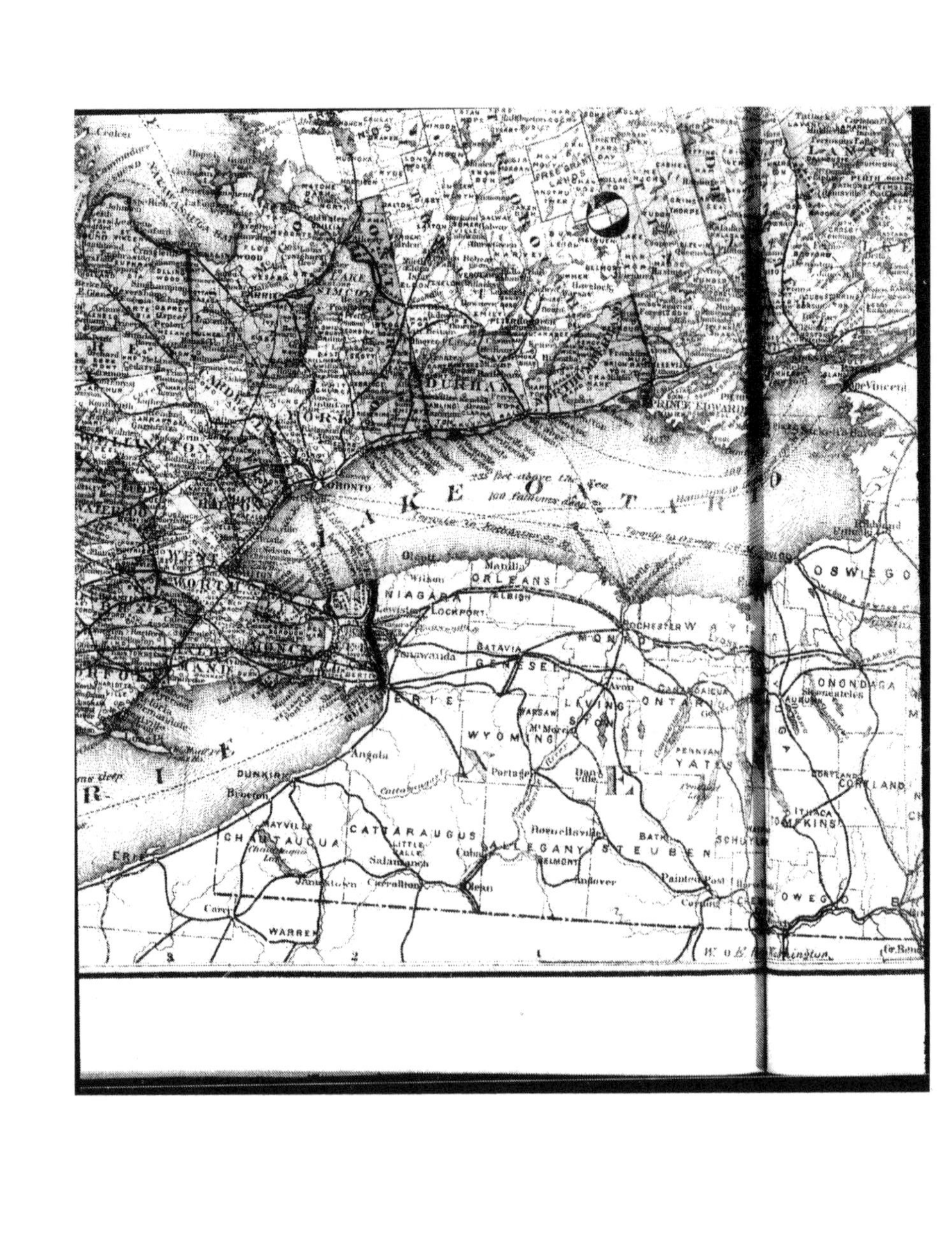
TORONTO
OSWEGO
NIAGARA
LOCKPORT
ORLEANS
BATAVIA
GENESEE
ERIE
WYOMING
ONONDAGA
CORTLAND
ITHACA
TOMPKINS
OWEGO
CATTARAUGUS
CHAUTAUQUA
ALLEGANY
STEUBEN
YATES
DUNKIRK
WARREN
Angola
Salamanca
Portage
Corning
Andover
PRINCE EDWARD
DURHAM

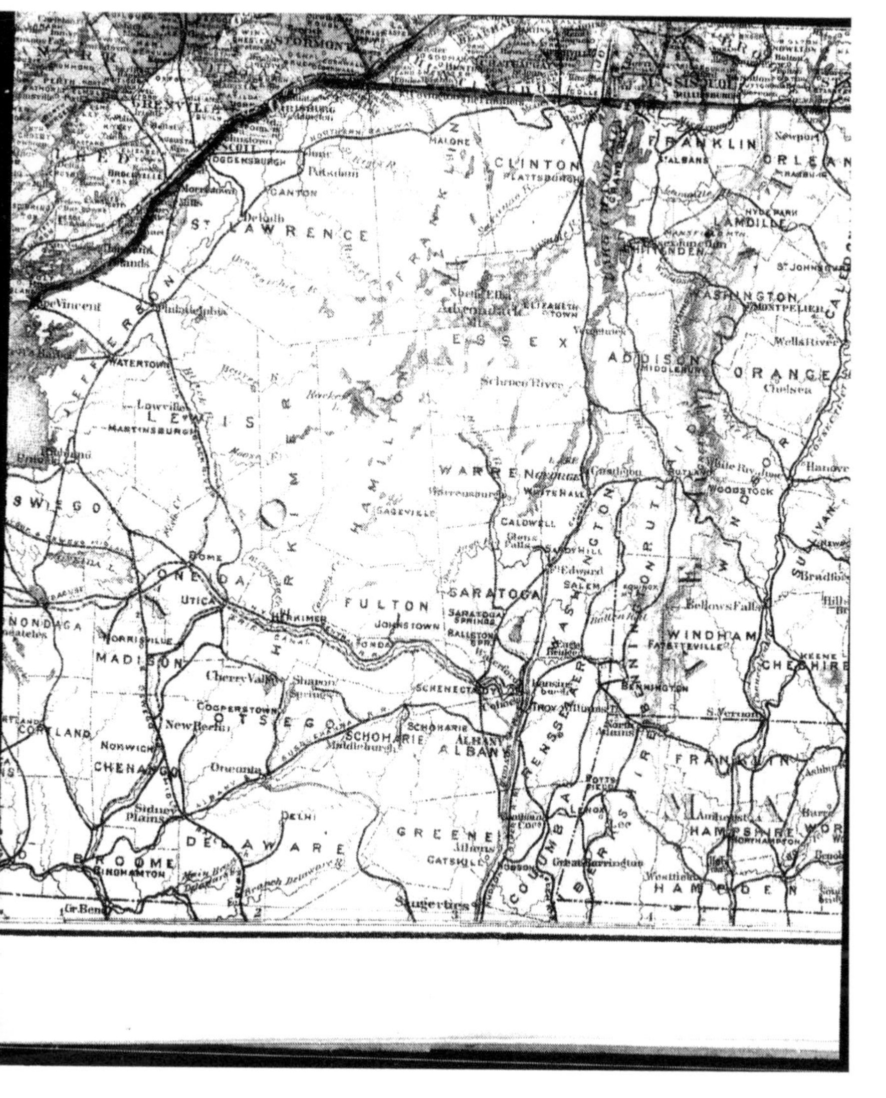

ST LAWRENCE
OGDENSBURGH
CANTON
Potsdam
MALONE
CLINTON
PLATTSBURGH
FRANKLIN
ESSEX
ELIZABETHTOWN
ADDISON
MIDDLEBURY
ORANGE
WASHINGTON
MONTPELIER
LAMOILLE
WATERTOWN
JEFFERSON
LEWIS
Lowville
MARTINSBURGH
HAMILTON
HERKIMER
WARREN
CALDWELL
WHITEHALL
WOODSTOCK
WINDSOR
SWEGO
ROME
ONEIDA
UTICA
FULTON
JOHNSTOWN
SARATOGA
SARATOGA SPRINGS
BALLSTON SPA
WASHINGTON
SALEM
Ft Edward
RUTLAND
BENNINGTON
WINDHAM
Fayetteville
Bellows Falls
CHESHIRE
KEENE
ONONDAGA
MADISON
Cherry Valley
SCHENECTADY
TROY
COOPERSTOWN
OTSEGO
SCHOHARIE
ALBANY
RENSSELAER
NORWICH
CHENANGO
CORTLAND
Oneonta
FRANKLIN
HAMPSHIRE
NORTHAMPTON
Sidney Plains
DELHI
DELAWARE
GREENE
CATSKILL
COLUMBIA
BERKSHIRE
LENOX
Great Barrington
HAMPDEN
BROOME
BINGHAMTON
Saugerties

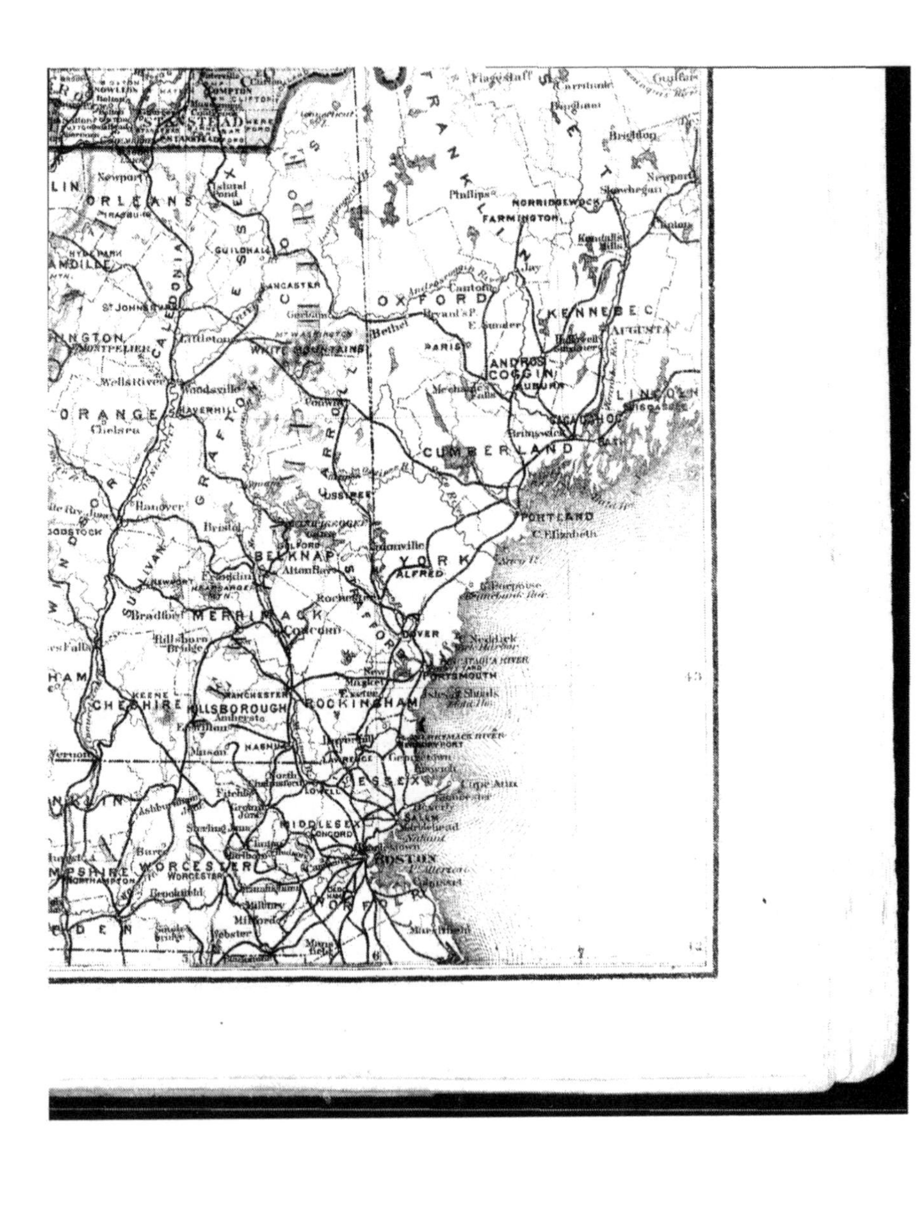

Geo. J. Pelly M. D
Belleville

H. F. Ward.
Madoc.

James Peck.
Ameliasburg To. P. Edward Co
born 1803

John Bradshaw
[illegible]

John R. Ketcheson
Madoc.

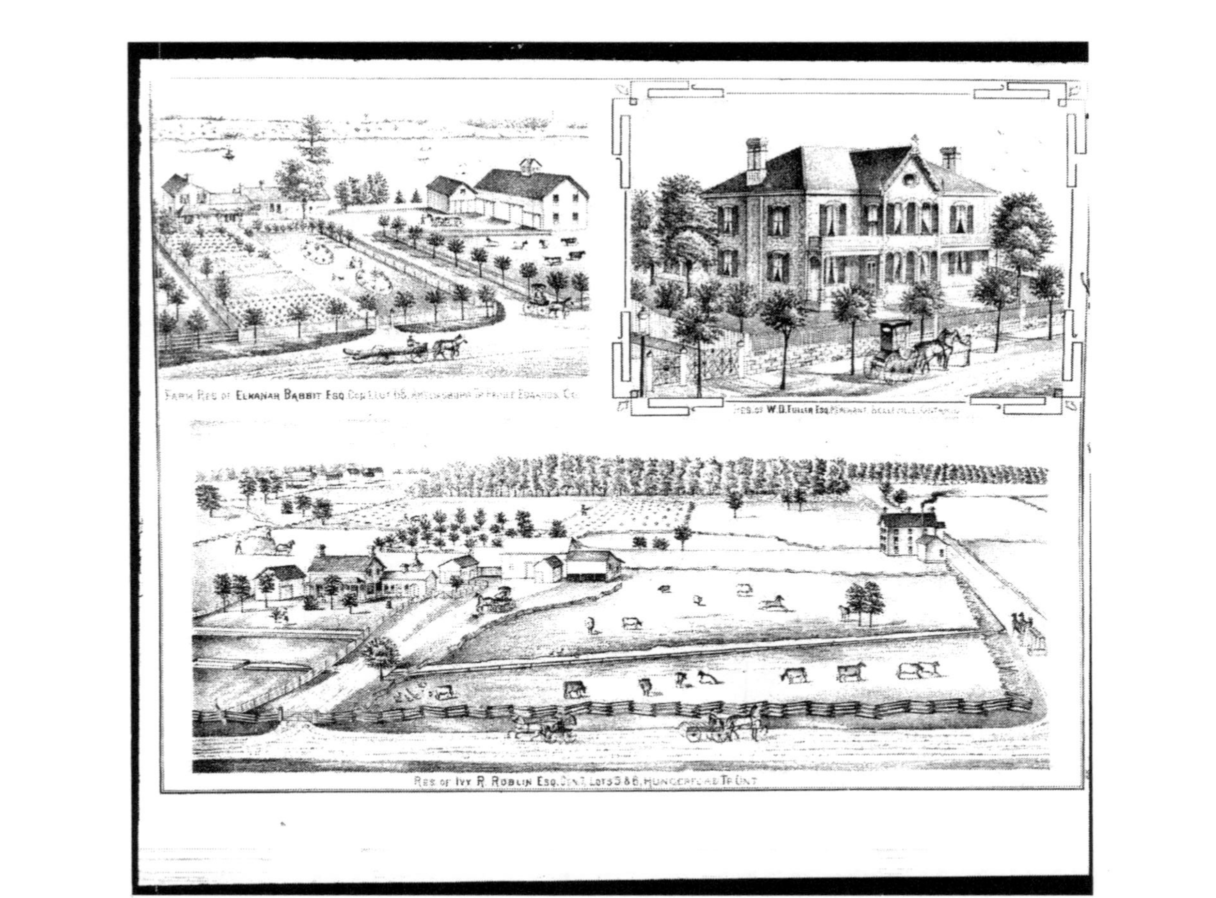

Farm Res. of Elkanah Babbit Esq. Con. [illegible] Lot 68, Ameliasburgh Tp. Prince Edward Co.

Res. of W. D. Fuller Esq. [illegible] Belleville, Ontario

Res. of Ivy R. Roblin Esq. Con. 7, Lots 5 & 6, Hungerford Tp. Ont.

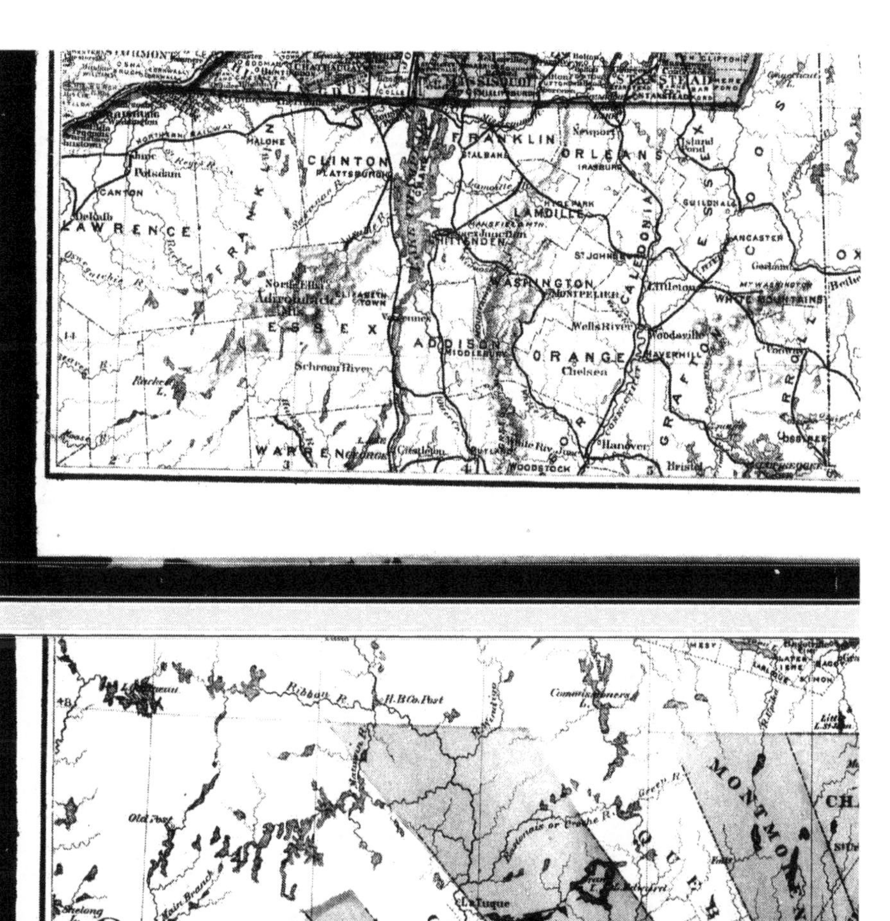

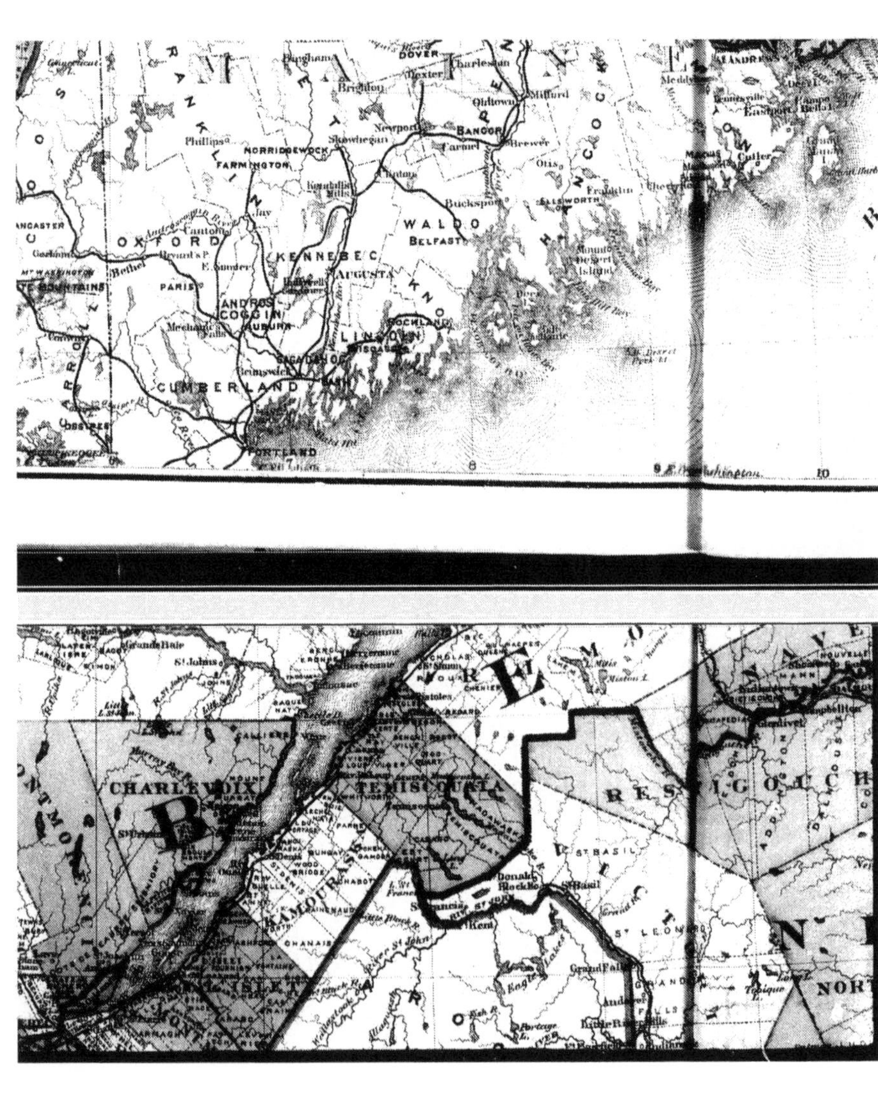

FRANKLIN
OXFORD
KENNEBEC
WALDO
BELFAST
AUGUSTA
ANDROSCOGGIN
AUBURN
LINCOLN
CUMBERLAND
PORTLAND
BANGOR
Oldtown
Milford
Brewer
Bucksport
ELLSWORTH
Newport
Skowhegan
NORRIDGEWOCK
FARMINGTON
Phillips
Canton
Bethel
PARIS
ROCKLAND
Charleston
DOVER
Dexter
Brighton
Bingham
Carmel
Otis
Franklin
Eastport
CHARLEVOIX
TEMISCOUATA
KAMOURASKA
ST BASIL
St Basil
Grand Falls
ST LEONARD

BAY OF FUNDY

ANNAPOLIS

DIGBY

LUNENBURG

QUEENS

YARMOUTH

SHELBURNE

KINGS

Margaretville

Port William

Thomas Cove

New Albany

Stoddart

Maitland

Sandy Cove

Westport

Belleview Cove

New Tusket

Kemptville

Tusket

Shelburne

Barrington

Cape Sable

Liverpool

Mill Village

Chester

Hubbert's Cove

New Ross

New Germany

Bridgewater

New Dublin

Cape Lahave

Port Medway Harb.

Bedford

Halifax

10

11

12

13

14

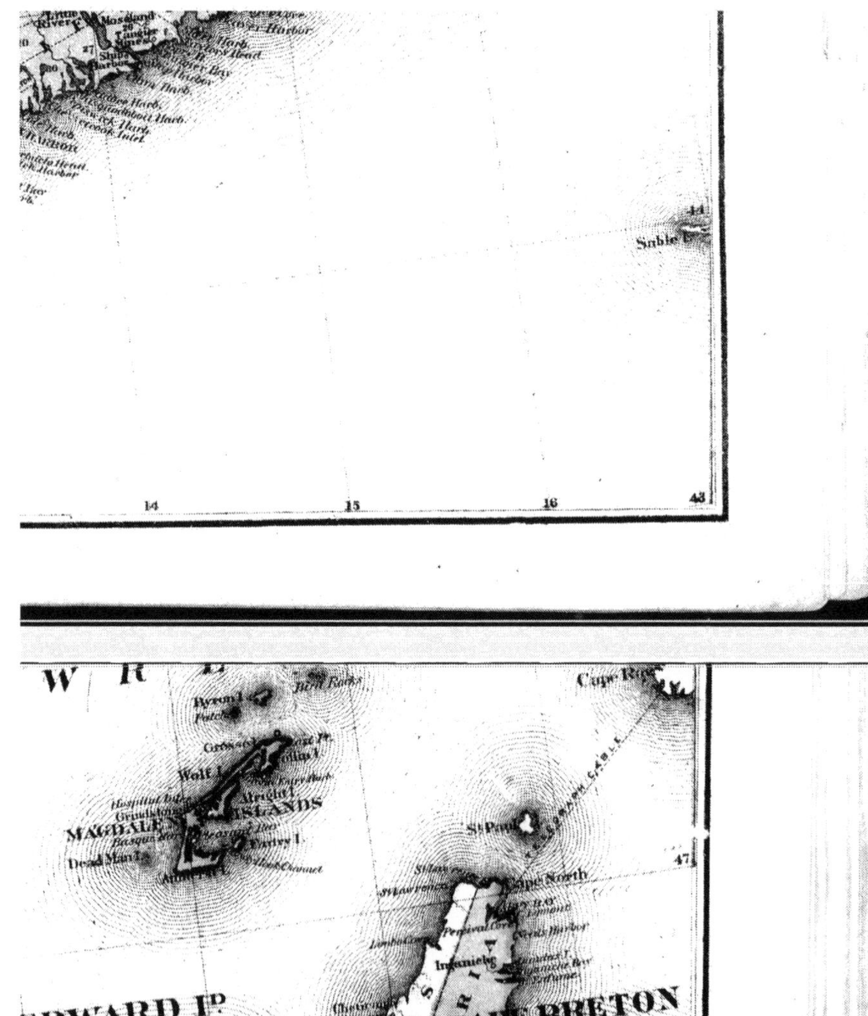

Sable I.
14
15
16
43
MAGDALEN
ISLANDS
Dead Man I.
Wolf I.
St Paul
Cape North
47
EDWARD I^D
CAPE BRETON
ISLAND

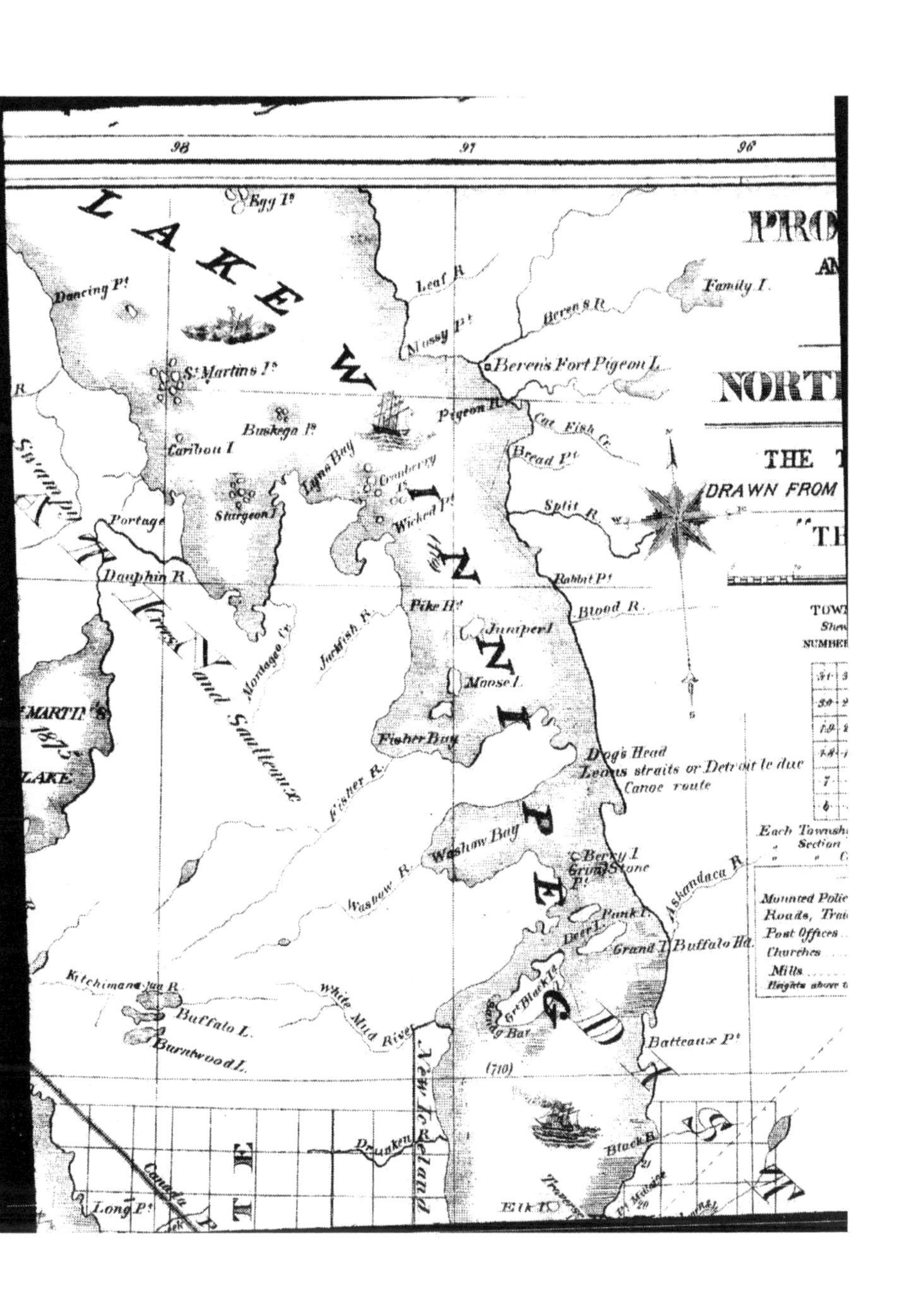

98
97
96
LAKE WINNIPEG
Egg Is
Dancing Pt
St Martins Is
Buskega Is
Caribou I
Stargeon I
Portage
Dauphin R
Leaf R
Mossy Pt
Berens R
Family I.
Berens Fort Pigeon L.
Pigeon R.
Cat Fish Cr.
Bread Pt
Split R.
Rabbit Pt
Blood R.
Pike Hd
Juniper I
Moose I.
Fisher Bay
Fisher R.
Jackfish R.
Wicked Pt
Dog's Head
Loons straits or Detroit le duc
Canoe route
Washow Bay
Washow R.
Berry I
Grind Stone Pt
Askandaca R.
Deer I.
Grand I.
Buffalo Hd.
Kitchimanatua R
Buffalo L.
Burntwood L.
White Mud River
New Iceland
Sandy Bar
(710)
Batteaux Pt
Drunken R.
Black R.
Long Pt
Elk I.
LAKE

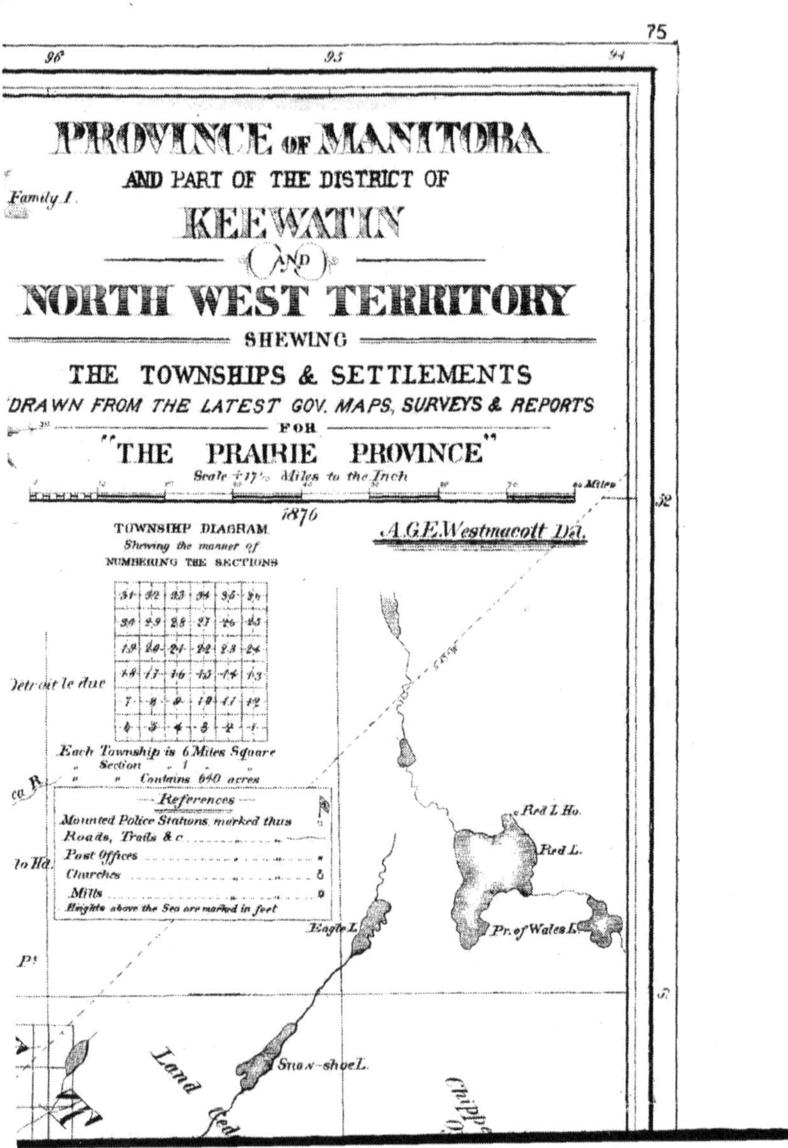

96°
95
94
PROVINCE OF MANITOBA
AND PART OF THE DISTRICT OF
KEEWATIN
AND
NORTH WEST TERRITORY
SHEWING
THE TOWNSHIPS & SETTLEMENTS
DRAWN FROM THE LATEST GOV. MAPS, SURVEYS & REPORTS
FOR
"THE PRAIRIE PROVINCE"
Scale 17½ Miles to the Inch
1876
A.G.E.Westmacott Del.
TOWNSHIP DIAGRAM
Shewing the manner of
NUMBERING THE SECTIONS
Each Township is 6 Miles Square
Section 1
Contains 640 acres
References
Mounted Police Stations marked thus
Roads, Trails &c
Post Offices
Churches
Mills
Heights above the Sea are marked in feet
Family I.
Detroit le duc
Red L Ho.
Red L.
Pr. of Wales L.
Eagle L.
Snow-shoe L.
Land
52

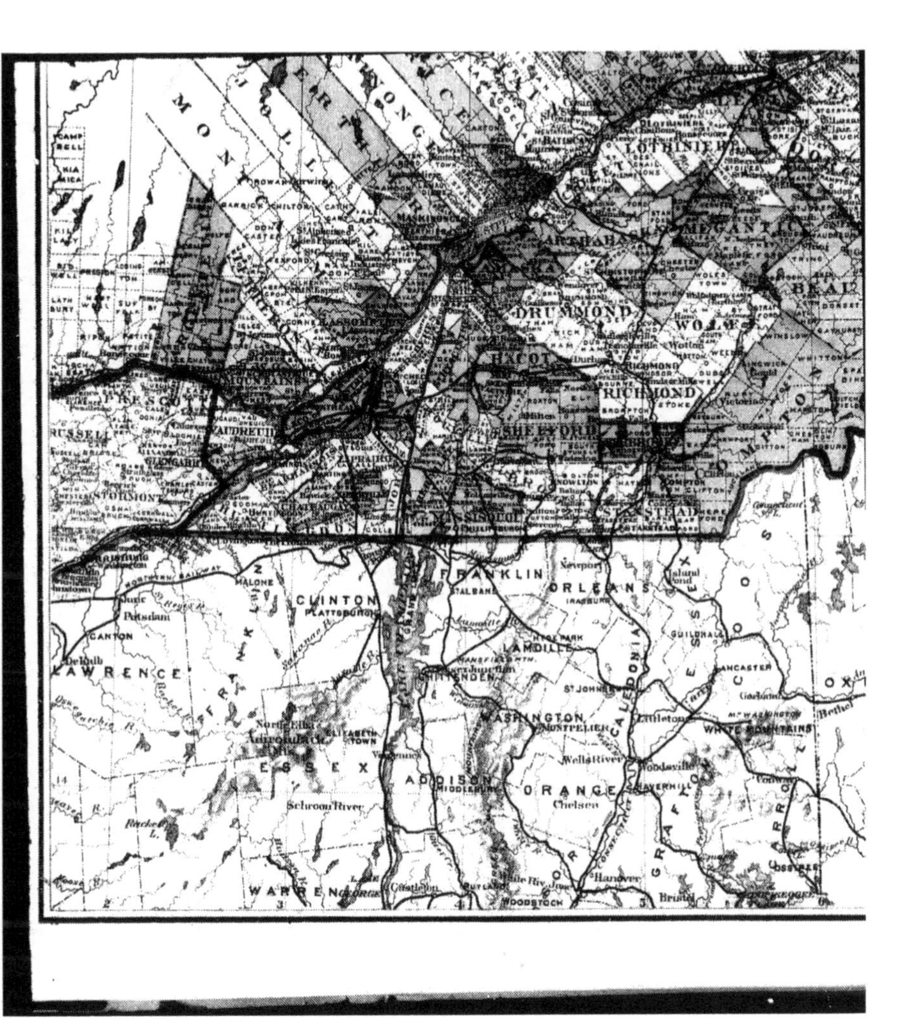

MONTCALM
JOLIETTE
LOTHINIERE
MEGANTIC
ARTHABASKA
DRUMMOND
WOLFE
BAGOT
RICHMOND
SHEFFORD
PRESCOTT
RUSSELL
VAUDREUIL
GLENGARRY
STORMONT
CHATEAUGUAY
STANSTEAD
COMPTON
FRANKLIN
ORLEANS
CLINTON
PLATTSBURGH
MALONE
Potsdam
CANTON
LAWRENCE
LAMOILLE
CALEDONIA
ESSEX
COOS
WASHINGTON
MONTPELIER
ADDISON
MIDDLEBURY
ORANGE
Chelsea
GRAFTON
WARREN
WHITE MOUNTAINS
LANCASTER
Schroon River
Hanover
WOODSTOCK
Bristol

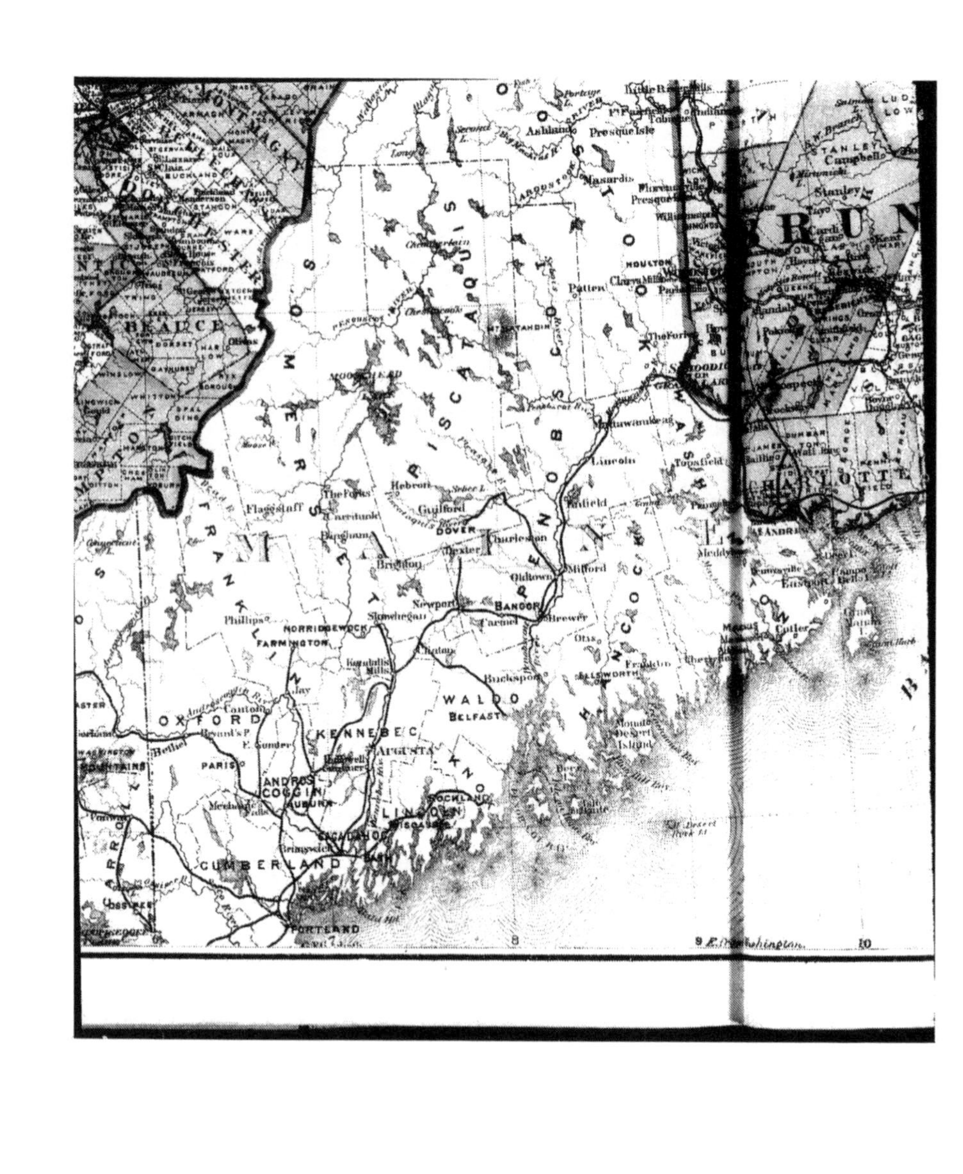

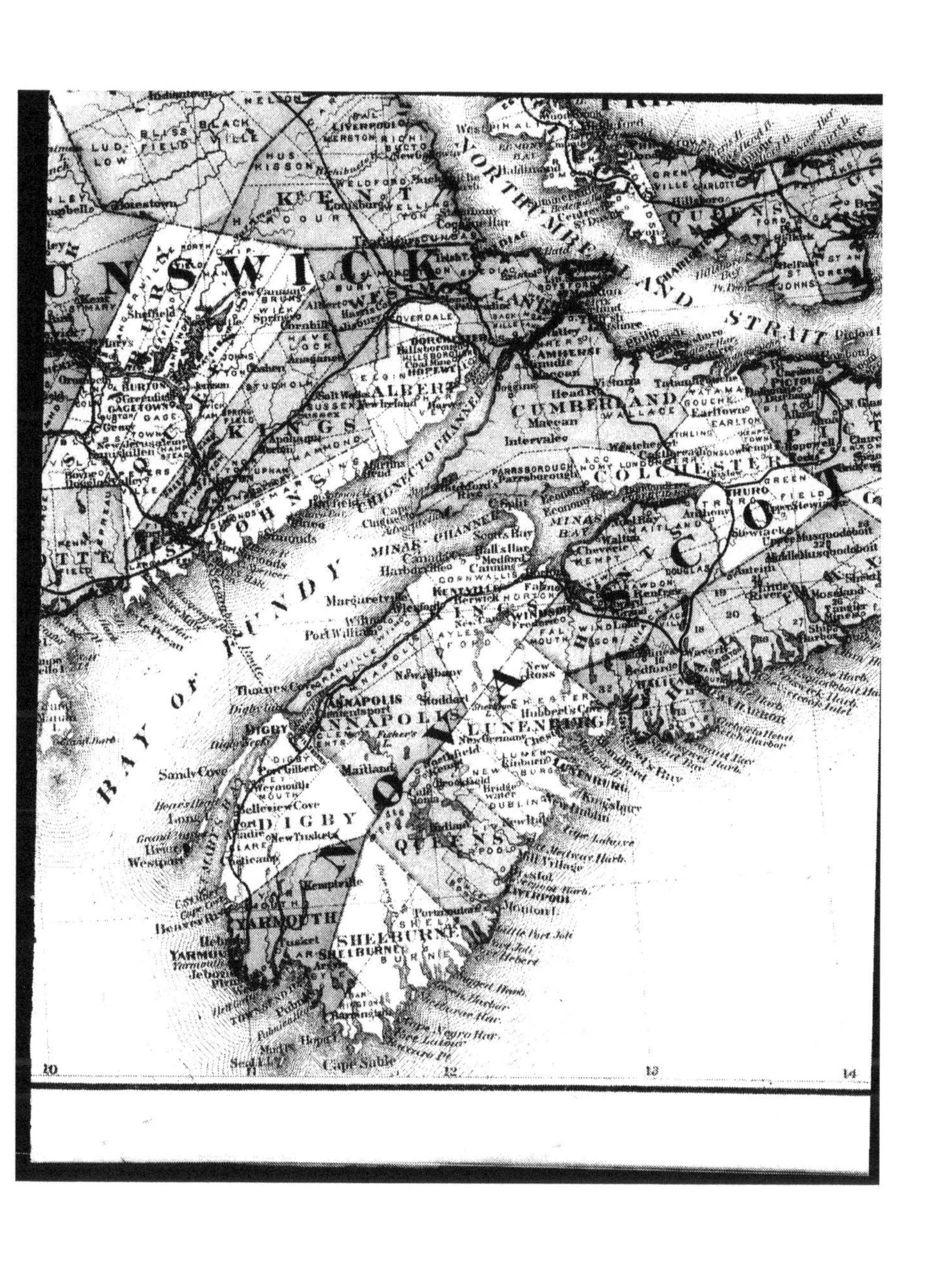
BAY OF FUNDY
NORTHUMBERLAND STRAIT
CUMBERLAND
COLCHESTER
KINGS
ANNAPOLIS
DIGBY
YARMOUTH
SHELBURNE
QUEENS
LUNENBURG
ALBERT
KENT
MINAS CHANNEL
CHIGNECTO CHANNEL
Cape Sable
10
11
12
13
14

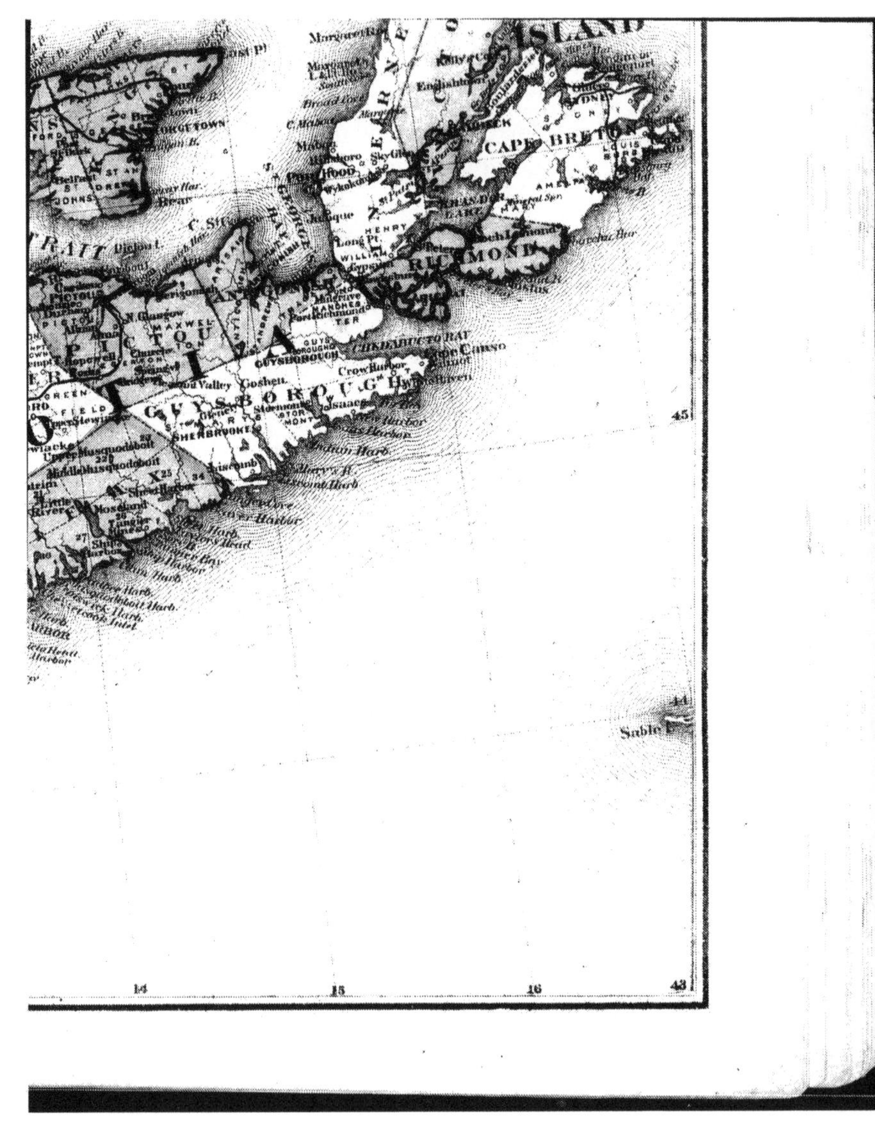
ISLAND
CAPE BRETON
RICHMOND
PICTOU
GUYSBOROUGH
SHERBROOKE
N. Glasgow
Goshen
Crow Harbor
CHEDABUCTO BAY
C. Canso
Sydney
Sable I.
14
15
16
45
44
43

MILLS & RES. OF APPLEBY & BURDETT, MILLTOWN, ONT.

FARM RES. OF W. F. CONNOR ESQ. CON. 8 LOT 27, SIDNEY TP. HASTINGS CO.

FARM RES. OF SAML. T. WILMOT ESQ. CON. 4, LOT 20, SIDNEY TP. HASTINGS CO.

CARRIAGE FACTORY, STORE & RES. OF WM. HUDSON, ROSLIN, THURLOW TP. ONT.

Elijah Ketcheson.

William Ketcheson.

Lieut Col Sheldon Hawley.

Hon Henry Meyers

Richard Davis.

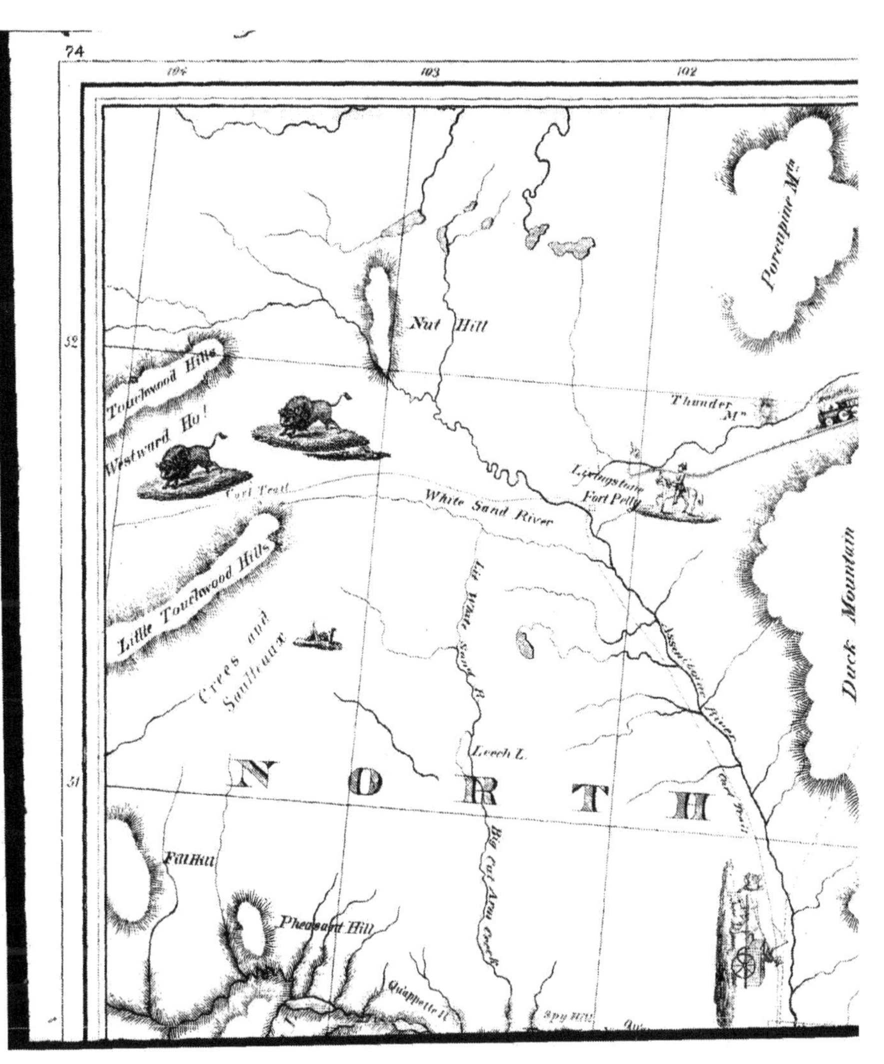
104
103
102
52
51
Nut Hill
Porcupine Mtn
Touchwood Hills
Westward Ho!
Thunder Mn
Livingstone
Fort Pelly
White Sand River
Little Touchwood Hills
Duck Mountain
Crees and Saulteaux
Lit White Sand R
Leech L.
N O R T H
Fat Hill
Pheasant Hill
Big Cut Arm Creek
Qu'appelle R
Spy Hill

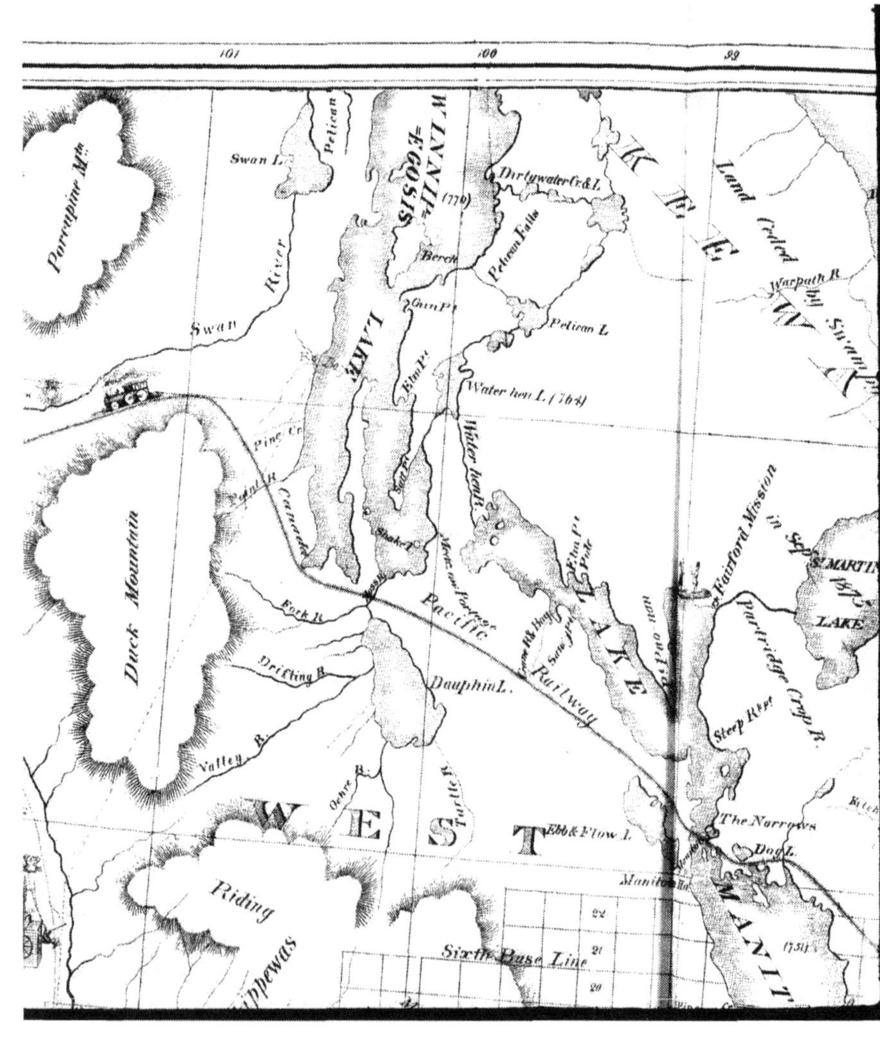

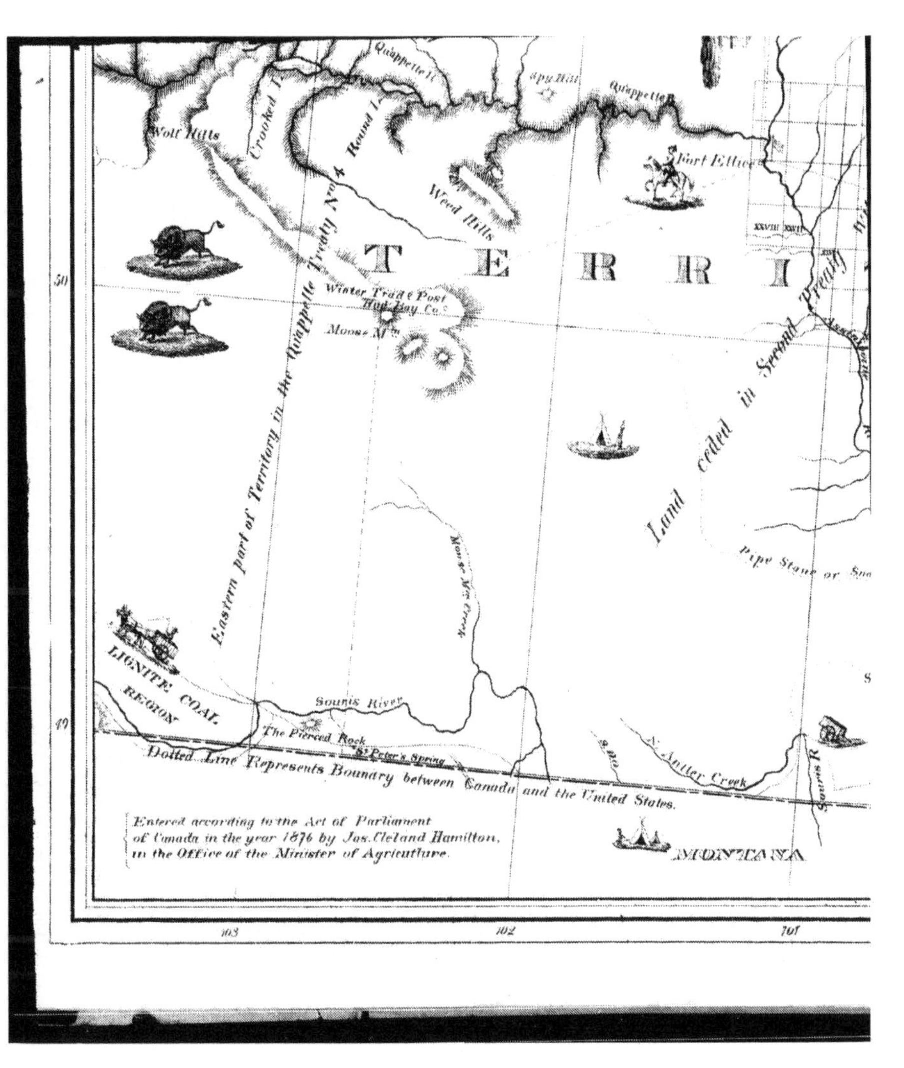

Qu'appelle R.
Qu'appelle R.
Wolf Hills
Crooked L.
Round L.
Weed Hills
Fort Ellice
T E R R I
Eastern part of Territory in the Qu'appelle Treaty No. 4
Winter Trad. Post Hud. Bay Co.
Moose Mtn
Land ceded in Second Treaty
Moose Mtn Creek
Pipe Stone or
LIGNITE COAL REGION
Souris River
The Pierced Rock
St. Peter's Spring
Antler Creek
Souris R.
Dotted Line Represents Boundary between Canada and the United States.
Entered according to the Act of Parliament of Canada in the year 1876 by Jas. Cleland Hamilton, in the Office of the Minister of Agriculture.
MONTANA
50
49
103
102
101

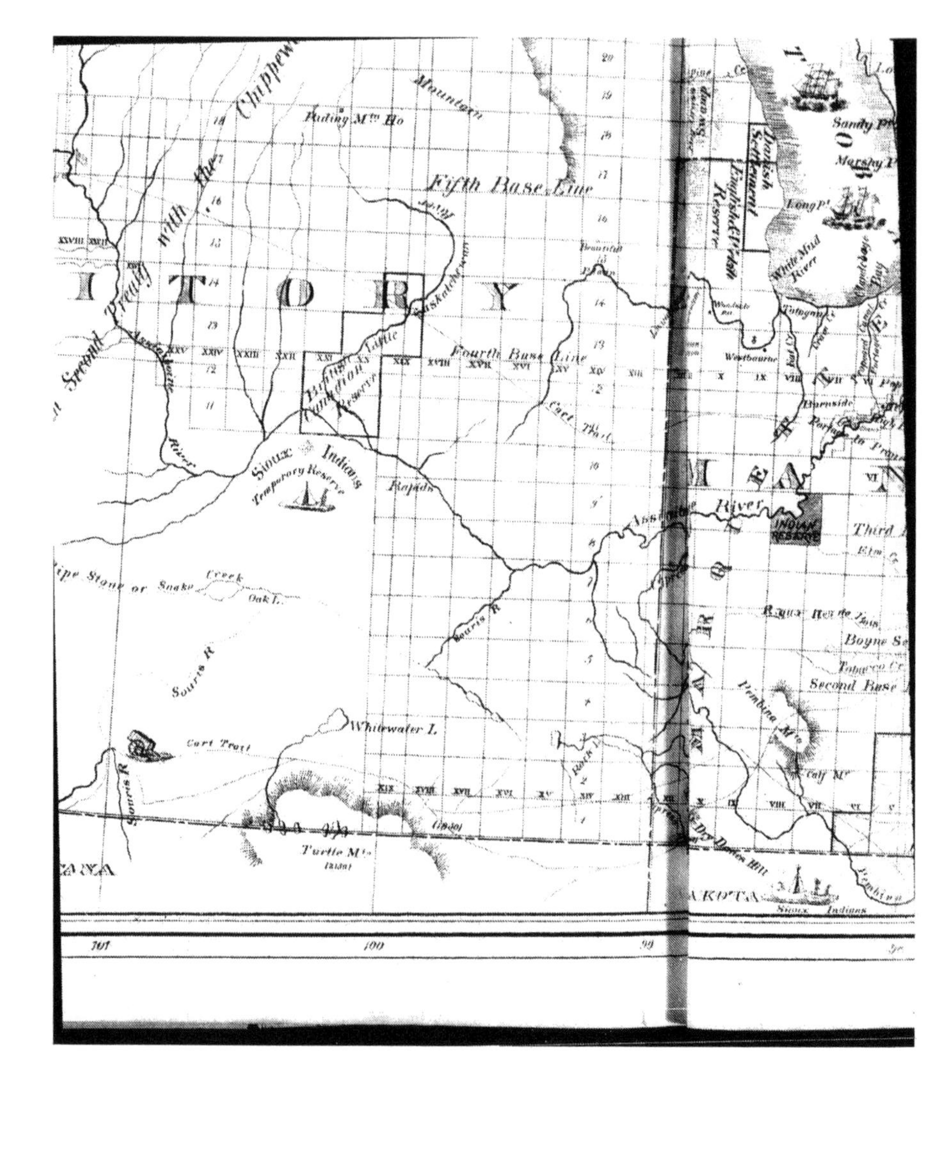

Fifth Base Line
Fourth Base Line
Riding Mtn Ho
Mountain
British Canadian Reserve
Sioux Indians Temporary Reserve
Assiniboine River
Indian Reserve
Third
Second Base
Pembina Mtn
Calf Mtn
Whitewater L.
Turtle Mts
Cart Trail
Souris R
Pipe Stone or Snake Creek
Oak L.
Danish Settlement
English & Welsh Reserve
Sandy Pt
Marshy Pt
Long Pt
White Mud River
Westbourne
Burnside
Boyne
Tobacco Cr.
Sioux Indians
101
100
99

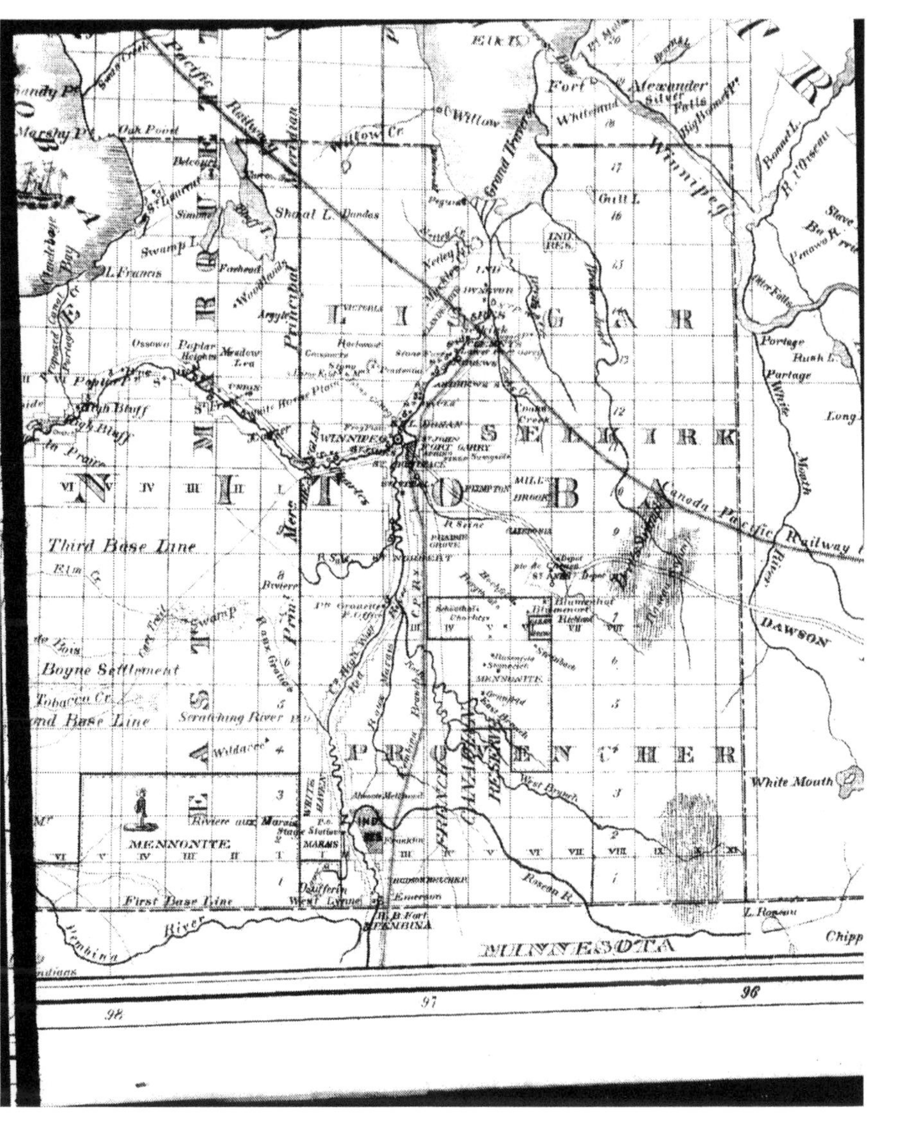

Pacific Railway
Principal Meridian
Winnipeg
Fort Alexander
Shoal L.
Swamp L.
L. Francis
Willow Cr.
Grand Traverse
Gull L.
High Bluff
Portage
White Mouth
Dawson
WINNIPEG
FORT GARRY
ST. BONIFACE
SELKIRK
LISGAR
MANITOBA
PROVENCHER
MENNONITE
Third Base Line
Boyne Settlement
Tobacco Cr.
Scratching River
Riviere aux Marais
First Base Line
Pembina River
Dufferin
West Lynne
Emerson
PEMBINA
Roseau R.
L. Roseau
Canada Pacific Railway
FRENCH CANADIAN RESERVE
MINNESOTA
Chipp
98
97
96

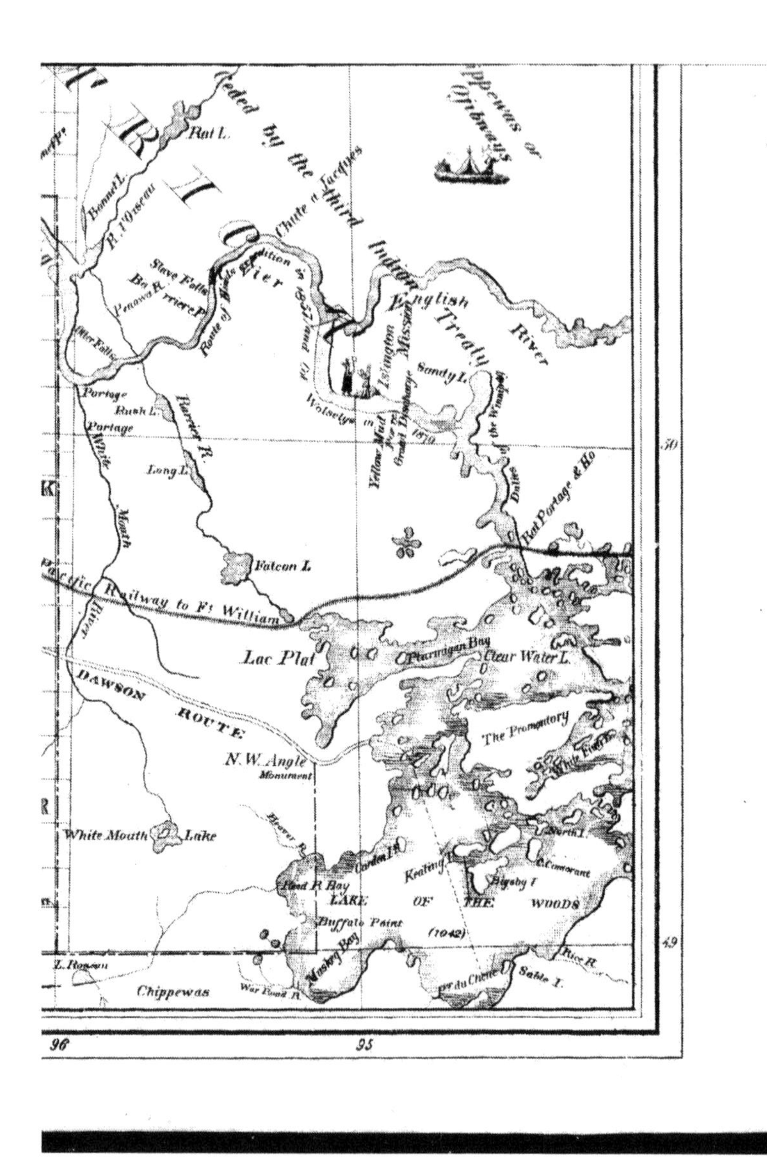

Rat L.
Bonnet L.
R. l'Oiseau
Ceded by the third Indian Treaty
Chute a Jacques
Ojibways
Slave Falls
Route of Wolseley's expedition in 1870
English River
Sandy L.
Portage
Rush L.
Portage
Barrier R.
Long L.
White Mouth River
Rat Portage & Ho
Falcon L.
Pacific Railway to Ft William
Lac Plat
Ptarmigan Bay
Clear Water L.
DAWSON ROUTE
The Promontory
N. W. Angle
Monument
White Mouth Lake
Reed R. Bay
LAKE OF THE WOODS
Buffalo Point
(1042)
Sigsby I
Cormorant
Muskeg Bay
Rice R.
Sable I.
L. Roseau
Chippewas
War Road R.
50
49
96
95

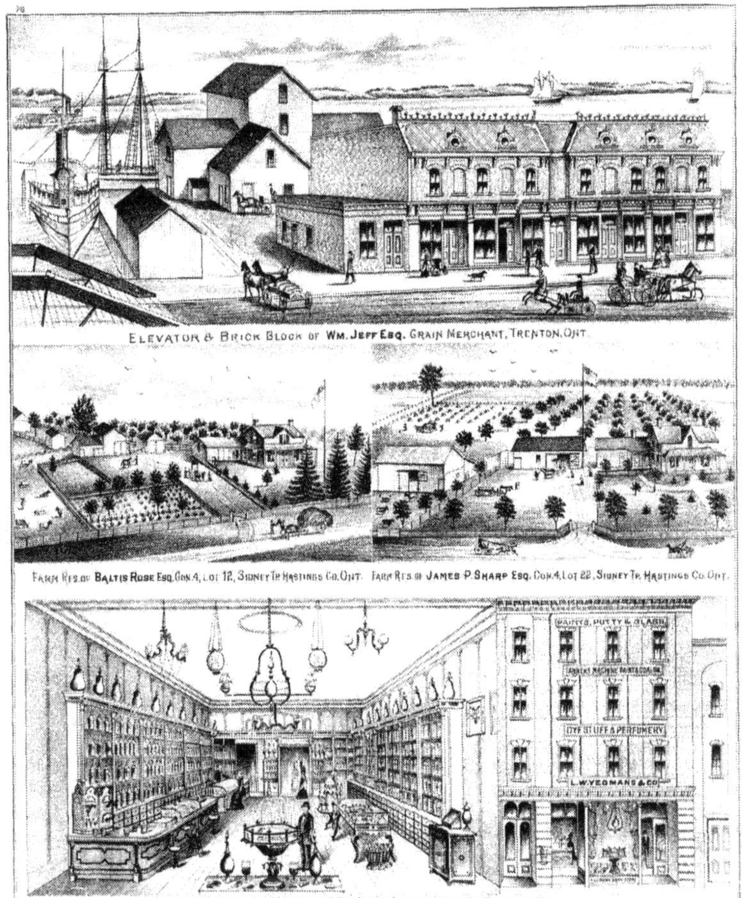

ELEVATOR & BRICK BLOCK OF WM. JEFF ESQ. GRAIN MERCHANT, TRENTON, ONT.

FARM RES. OF BALTIS ROSE ESQ. CON. 4, LOT 12, SIDNEY TP. HASTINGS CO. ONT.

FARM RES. OF JAMES P. SHARP ESQ. CON. 4, LOT 22, SIDNEY TP. HASTINGS CO. ONT.

DRUG STORE OF L. W. YEOMANS & CO. FRONT OPPOSITE HOTEL ST. BELLEVILLE, ONT.

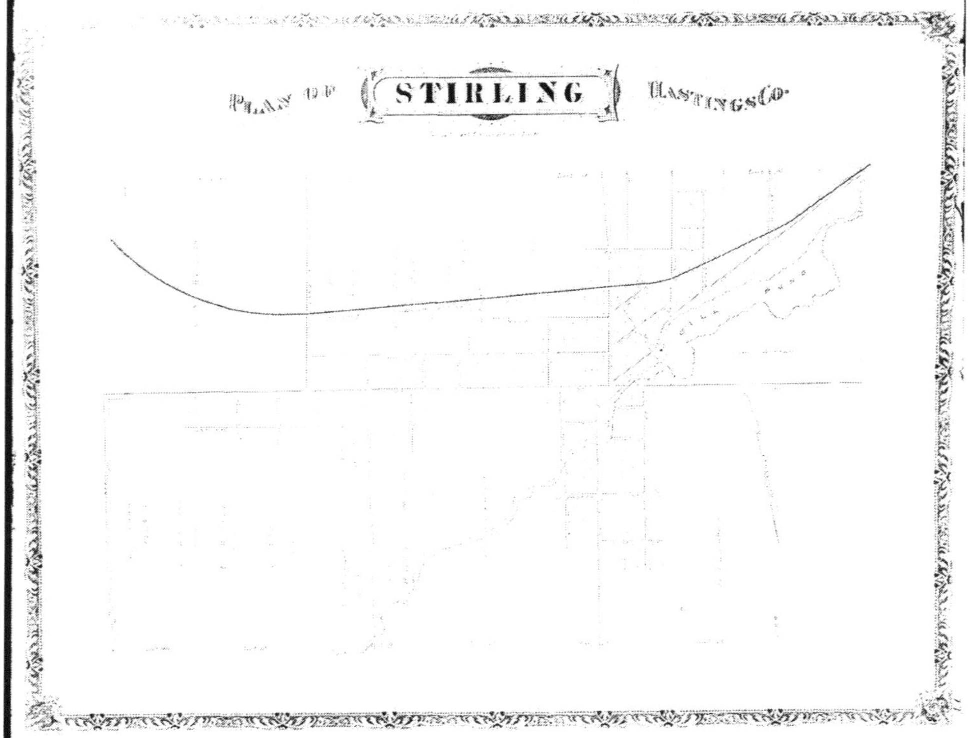

PLAN OF STIRLING HASTINGS CO.

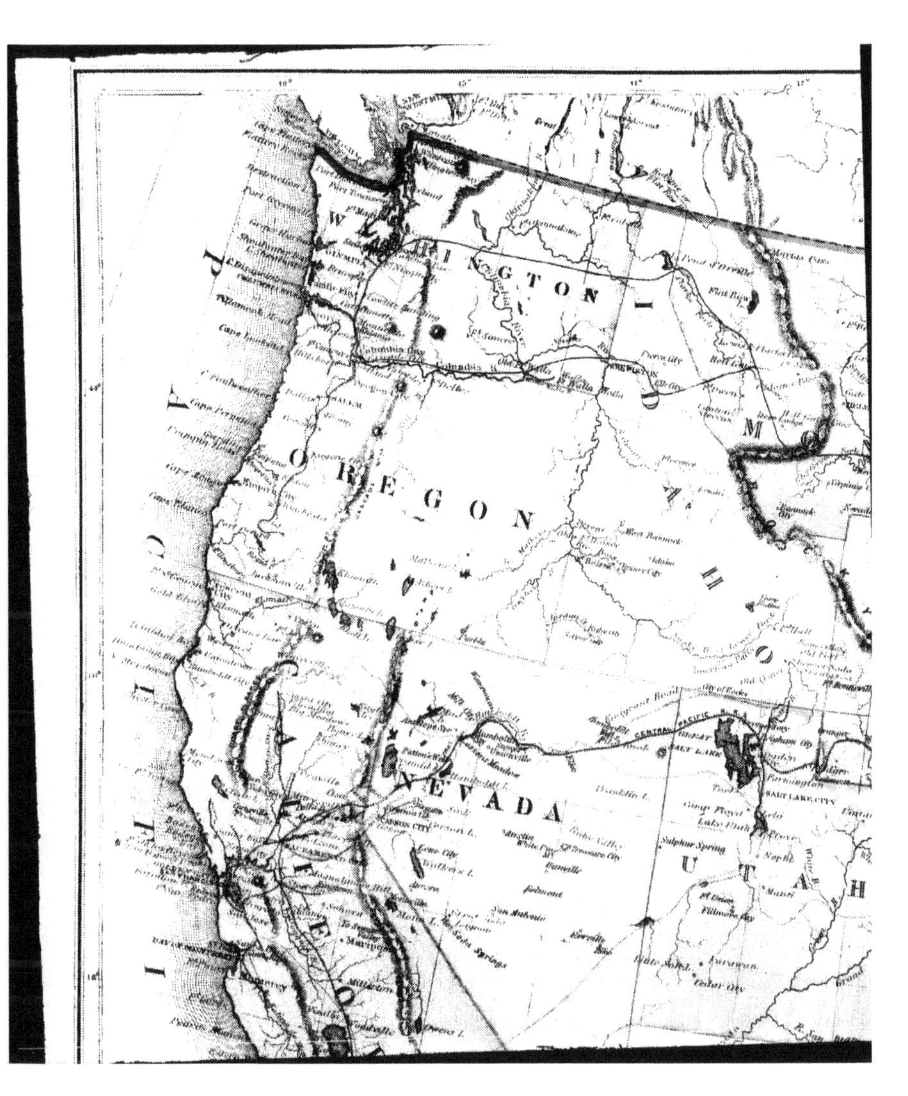

W A S H I N G T O N
O R E G O N
I D A H O
N E V A D A
U T A H
C A L I F O R N I A
P A C I F I C

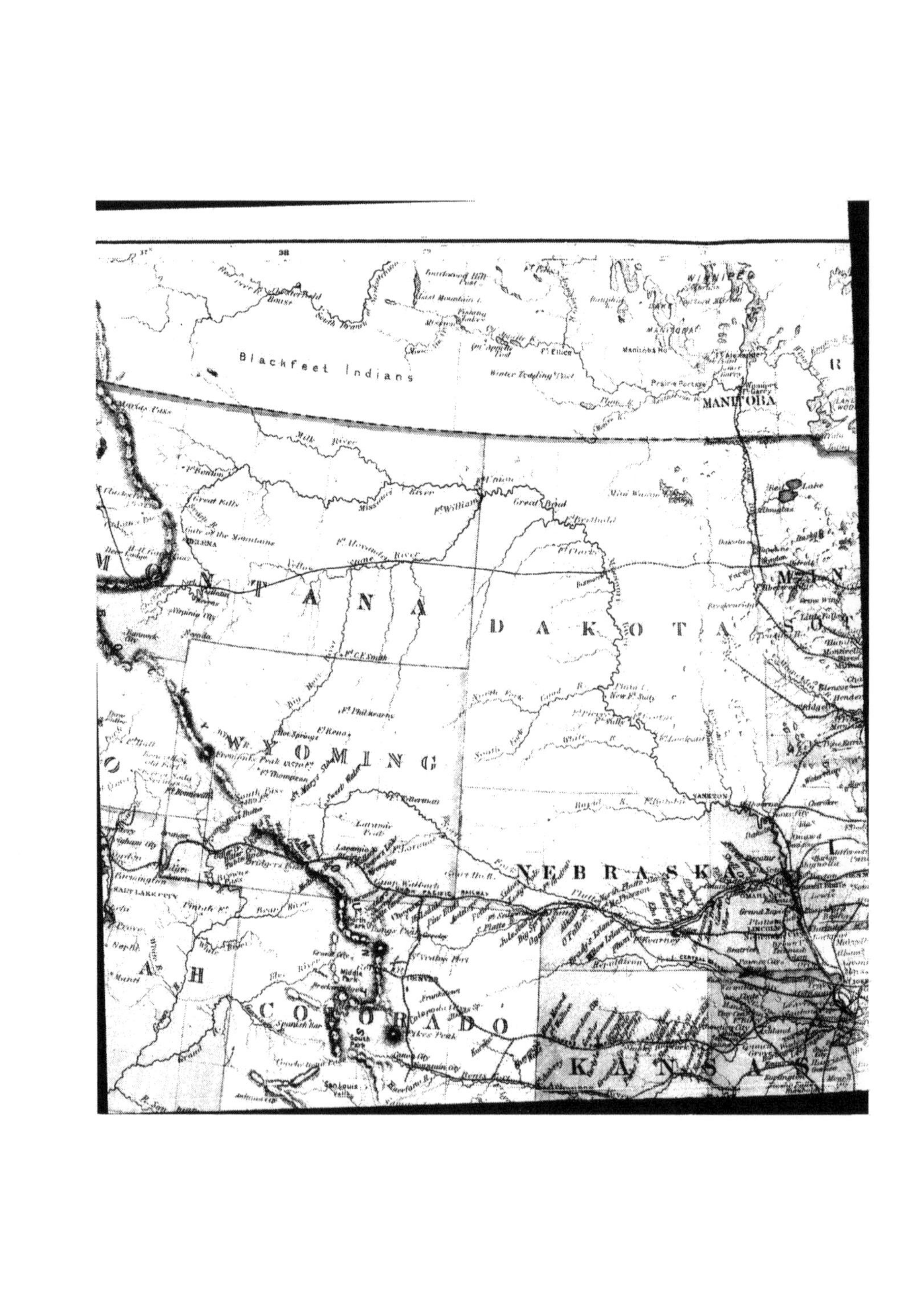

Blackfeet Indians
MANITOBA
Milk River
Great Falls
Missouri River
Great Bend
D A K O T A
W Y O M I N G
N E B R A S K A
C O L O R A D O
K A N S A S
South Pass
Pikes Peak
SALT LAKE CITY
YANKTON

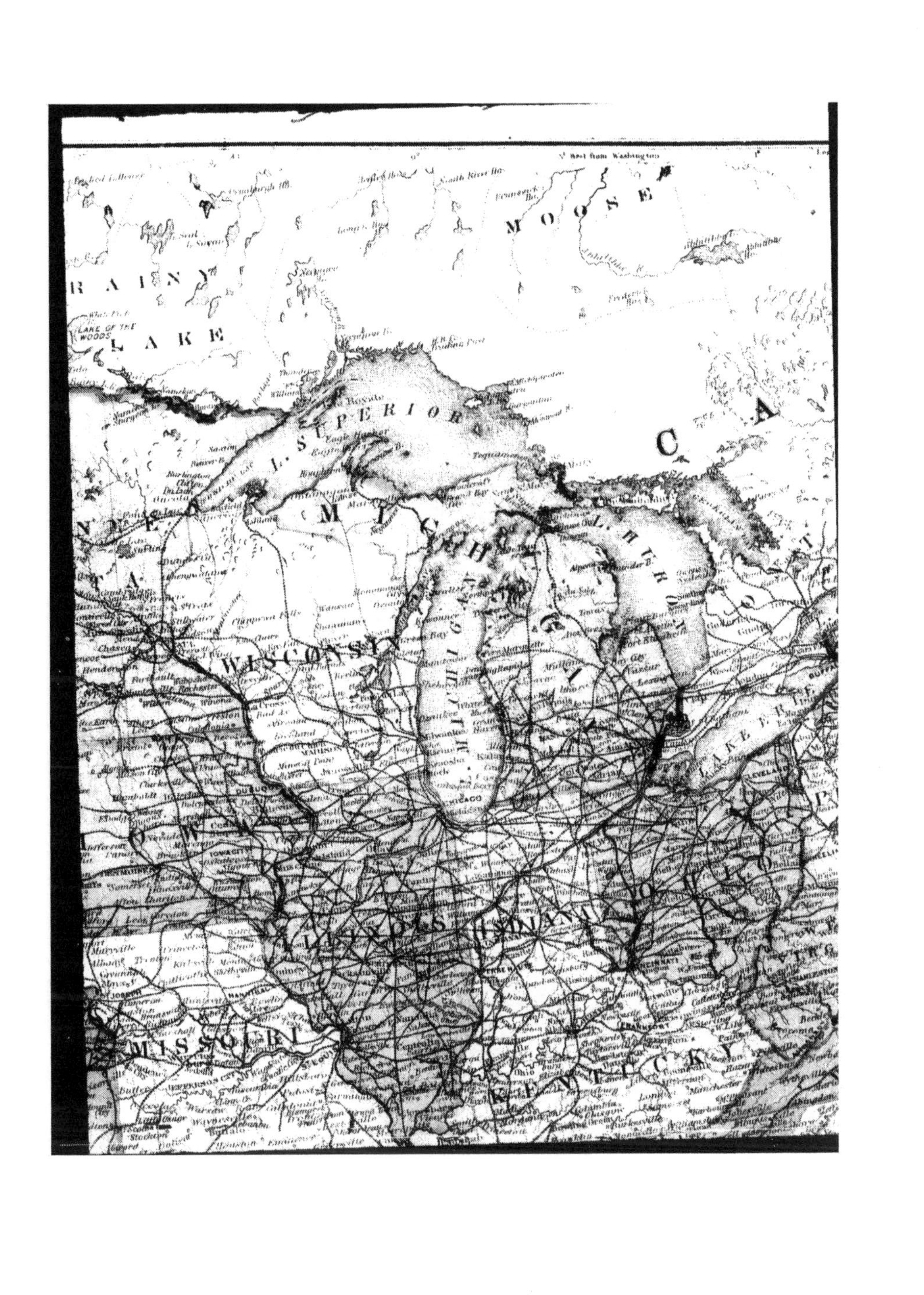

West from Washington
M O O S E
R A I N Y
L A K E
LAKE OF THE WOODS
L. SUPERIOR
WISCONSIN
MISSOURI
CHICAGO

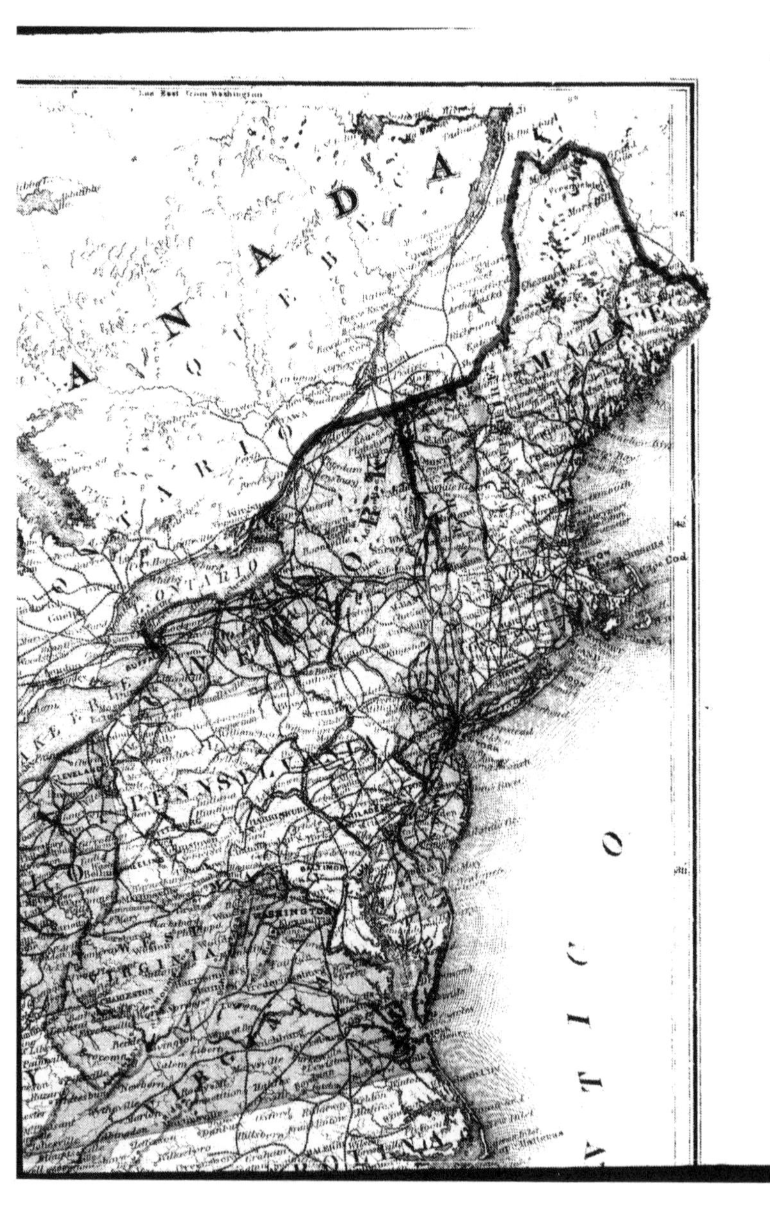
ONTARIO
Cape Cod

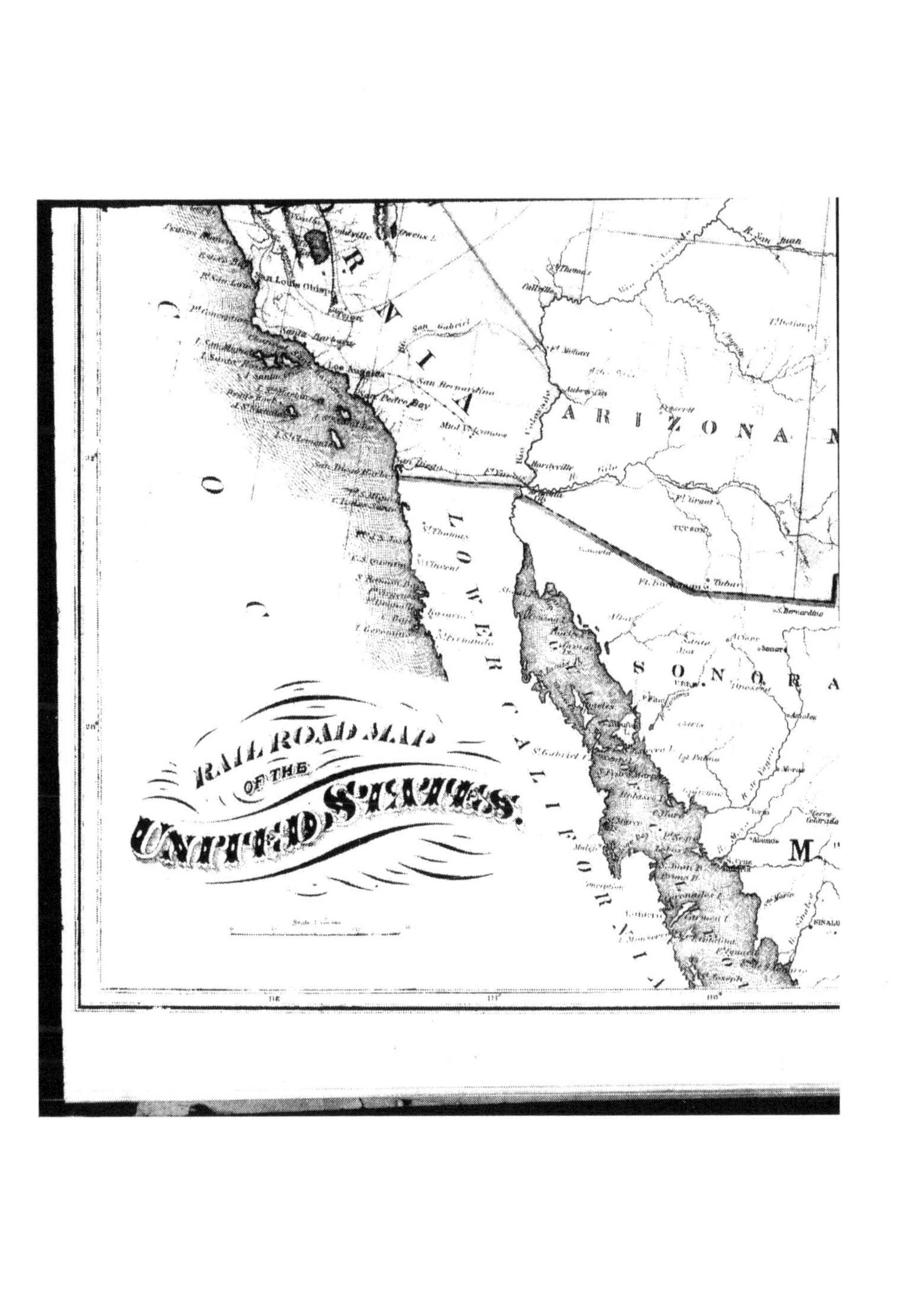
RAIL ROAD MAP
OF THE
UNITED STATES.
ARIZONA
SONORA
LOWER CALIFORNIA
San Diego
Los Angeles
San Bernardino
Santa Barbara
San Pedro Bay
TUCSON
Ft. Buchanan
Tubac

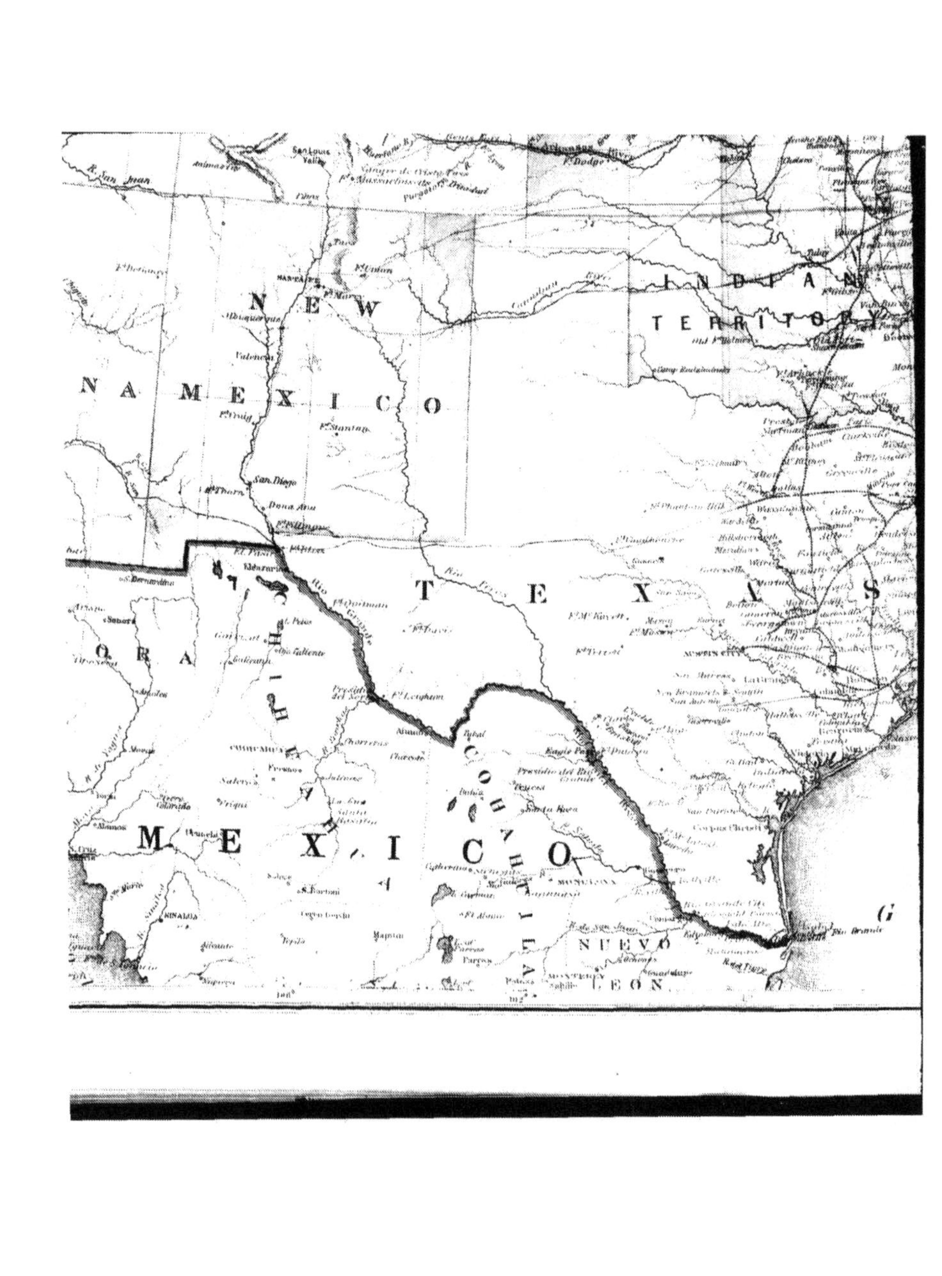
NEW MEXICO
INDIAN TERRITORY
TEXAS
MEXICO
CHIHUAHUA
COHAHUILA
NUEVO LEON
El Paso
San Diego
Dona Ana
Rio Grande
Eagle Pass
Corpus Christi
Austin City
San Antonio
Fort Davis
Presidio del Rio Grande

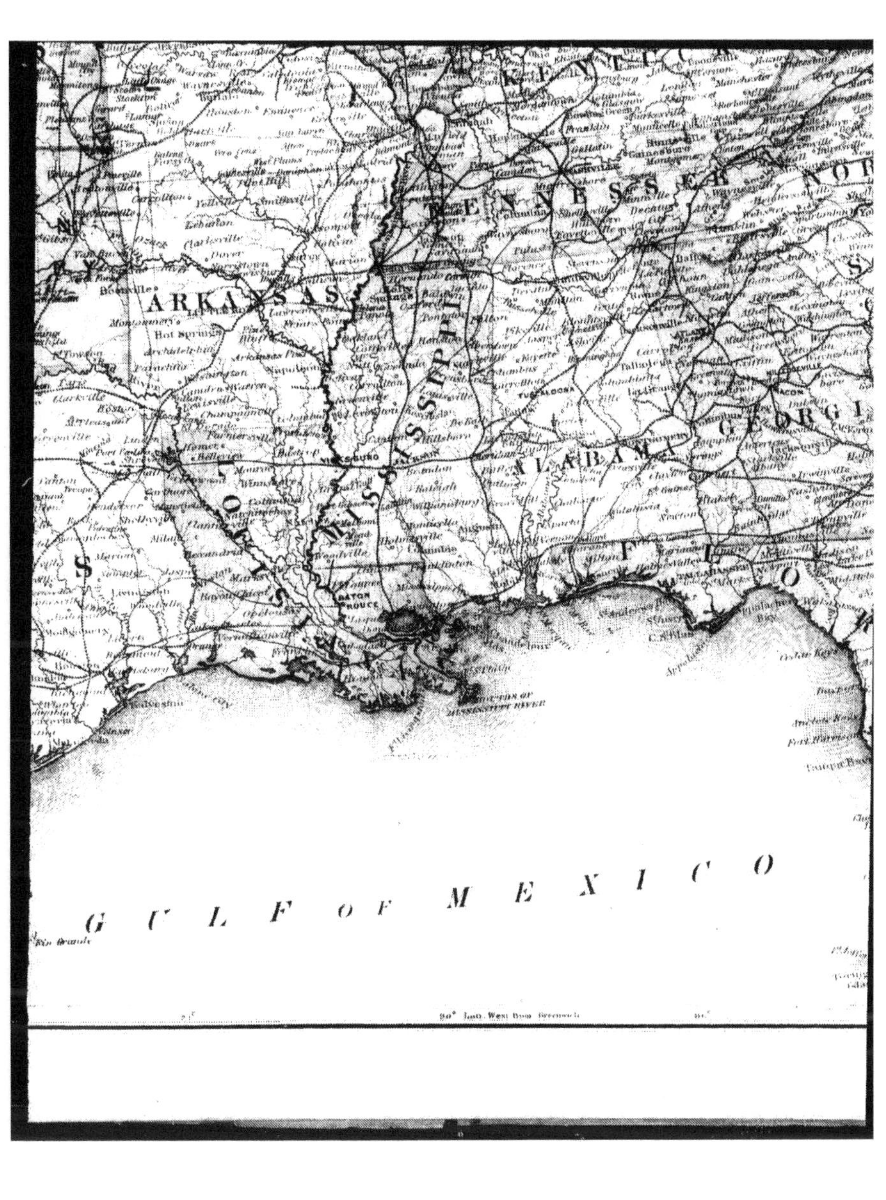

ARKANSAS
TENNESSEE
MISSISSIPPI
ALABAMA
GEORGI
Hot Springs
Arkansas Post
Clarksville
Carrollton
Lebanon
VICKSBURG
BATON ROUGE
TUSCALOOSA
Decatur
Macon
G U L F O F M E X I C O
Rio Grande
90° Long. West from Greenwich

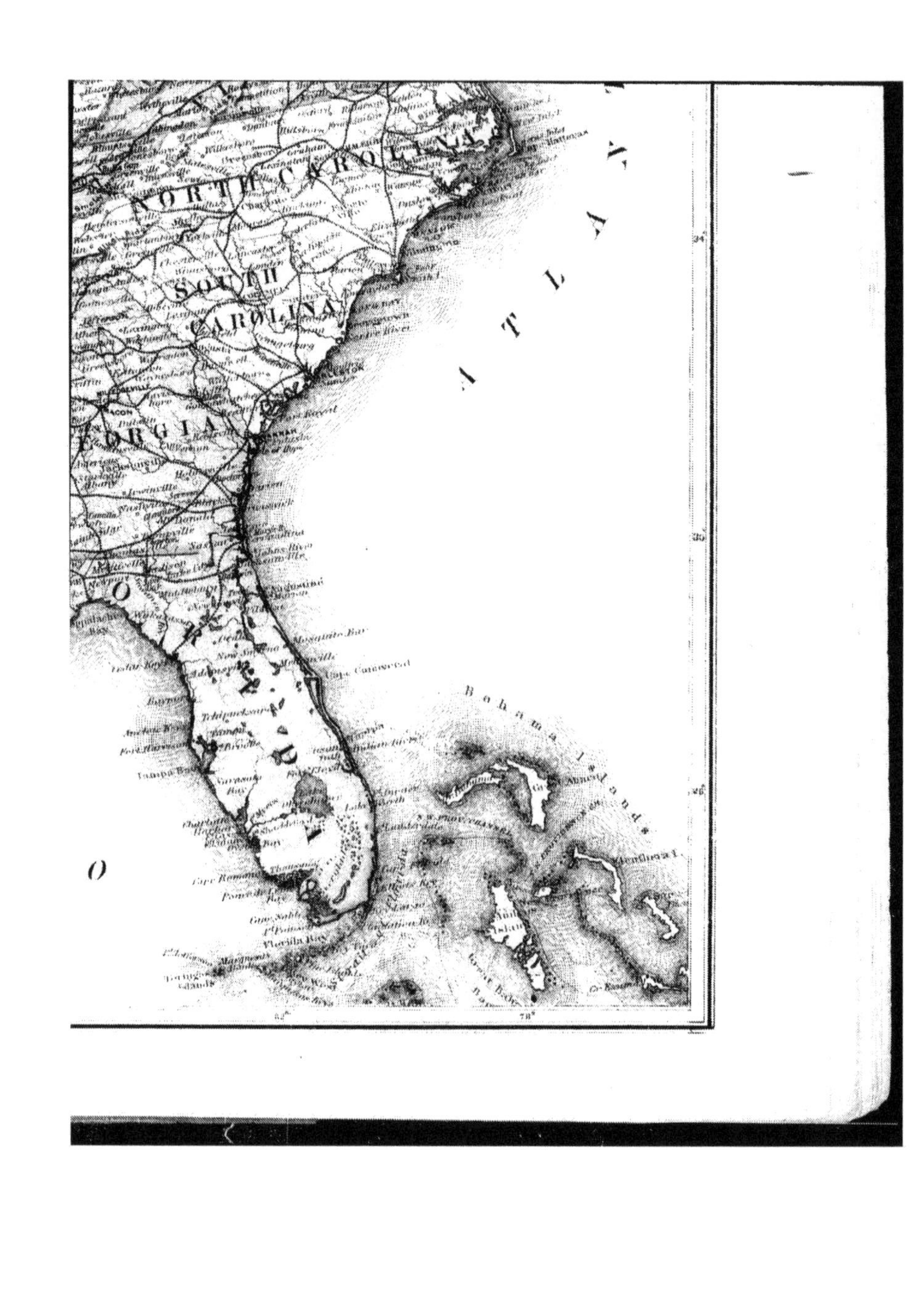
NORTH CAROLINA
SOUTH CAROLINA
GEORGIA
ATLANT
Bahama Islands
Cape Canaveral
Mosquito Bay
Tampa Bay
Florida Bay

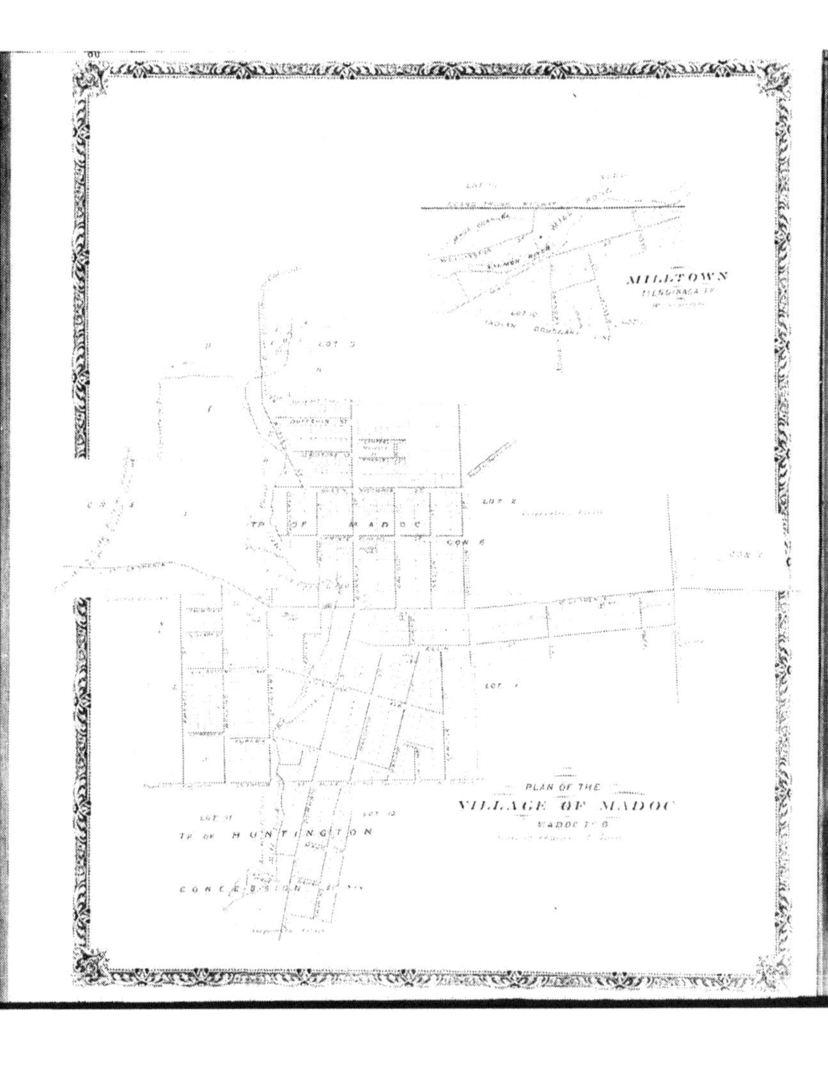
MILLTOWN
TP OF MADOC
CON 6
LOT 4
TP OF HUNTINGTON
CONCESSION
PLAN OF THE
VILLAGE OF MADOC
MADOC TP

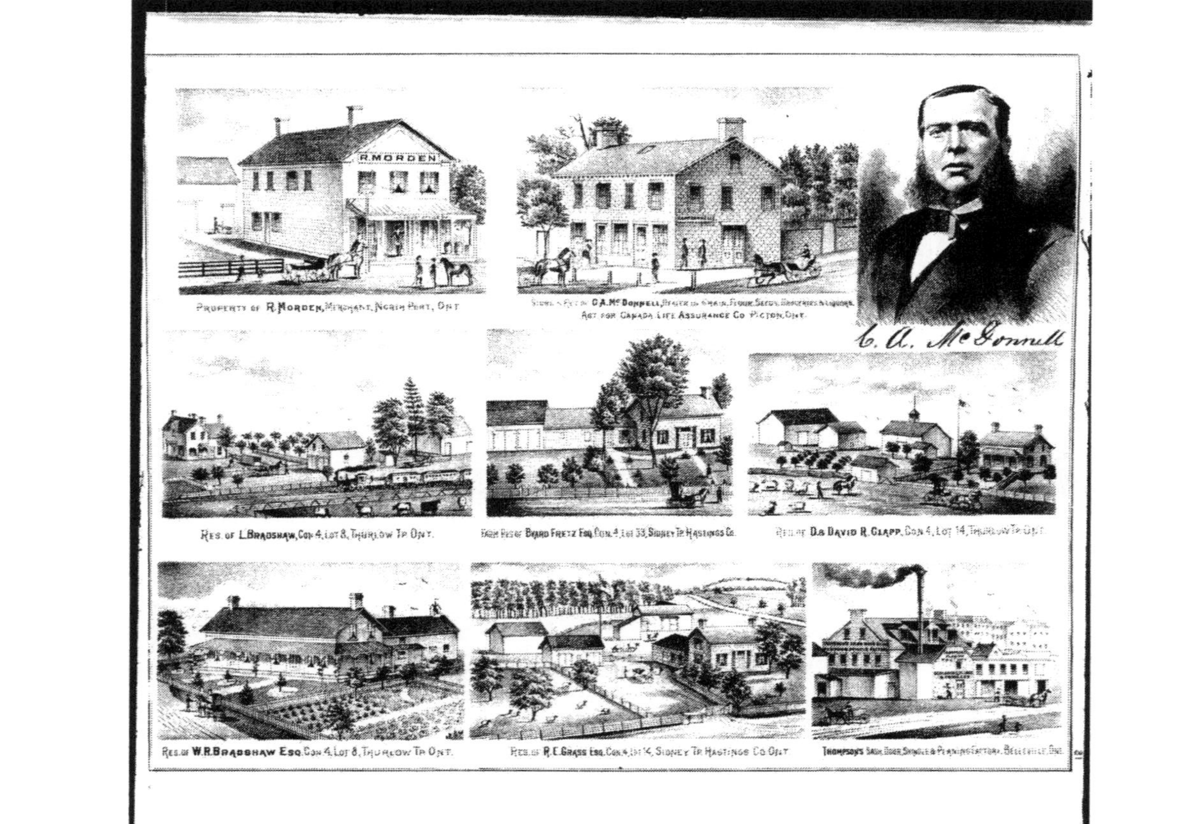

Property of R. Morden, Merchant, North Port, Ont.

Store & Res. of C. A. McDonnell, Dealer in Grain, Flour, Seeds, Groceries & Liquors. Agt. for Canada Life Assurance Co. Picton, Ont.

Res. of L. Bradshaw, Con. 4, Lot 8, Thurlow Tp. Ont.

Farm Res. of Byard Fretz Esq. Con. 4, Lot 33, Sidney Tp. Hastings Co.

Res. of Dr. David R. Clapp, Con. 4, Lot 14, Thurlow Tp. Ont.

Res. of W. R. Bradshaw Esq. Con. 4, Lot 8, Thurlow Tp. Ont.

Res. of R. E. Grass Esq. Con. 4, Lot 14, Sidney Tp. Hastings Co. Ont.

Thompson's Sash, Door, Shingle & Planing Factory, Belleville, Ont.

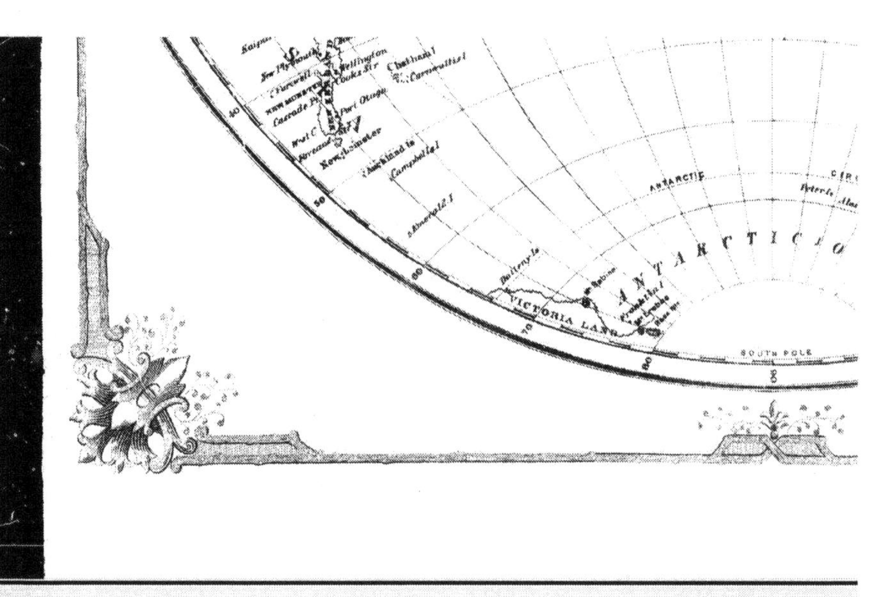
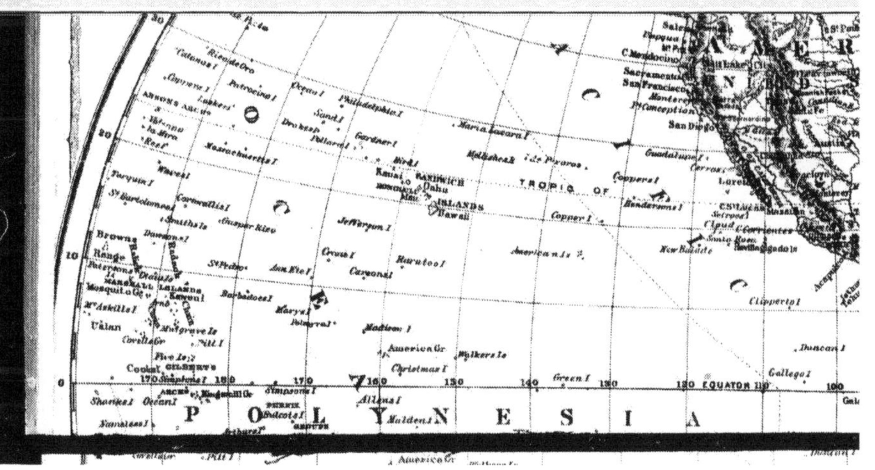

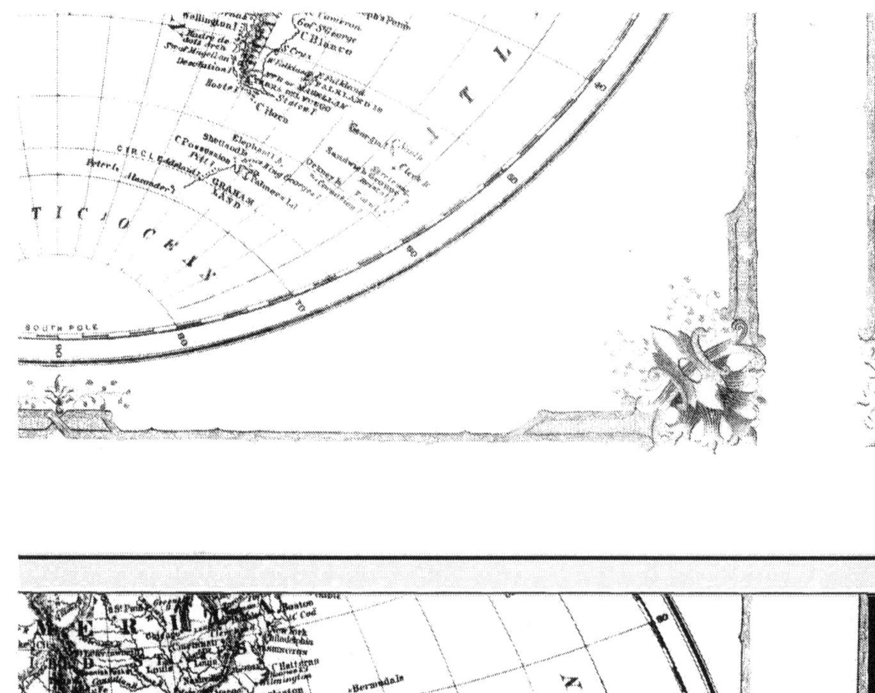
C Blanco
Falkland
C Horn
Shetlands
Elephant I.
Georgia
Sandwich
GRAHAM LAND
SOUTH POLE

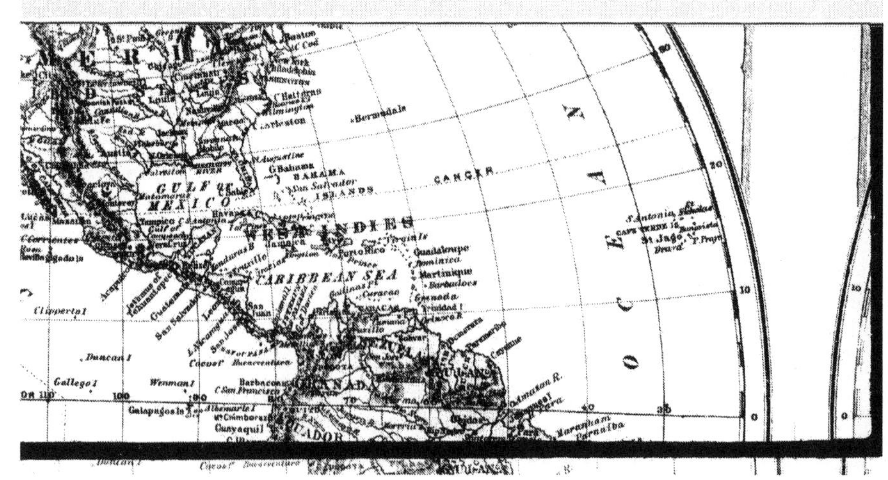
C Hatteras
Bermudas
BAHAMA ISLANDS
CANCER
GULF of MEXICO
WEST INDIES
CARIBBEAN SEA
Guadaloupe
Dominica
Martinique
Barbadoes
Grenada
Clipperton I
Duncan I
Gallego I
Wenman I
Galapagos Is
Guayaquil
Amazon R.
Maranham

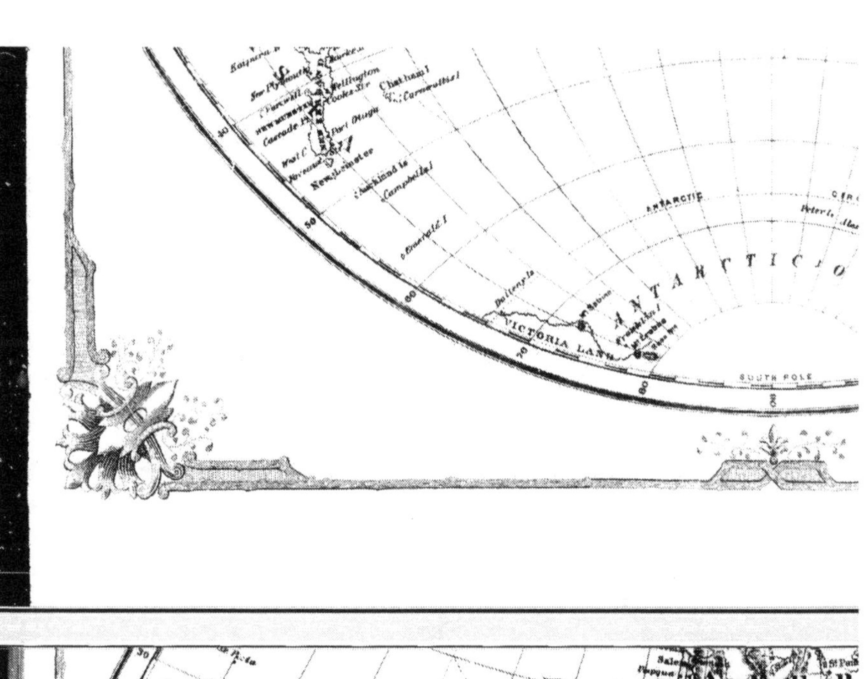

Chatham I
Auckland Is
Campbell I
ANTARCTIC
VICTORIA LAND
SOUTH POLE

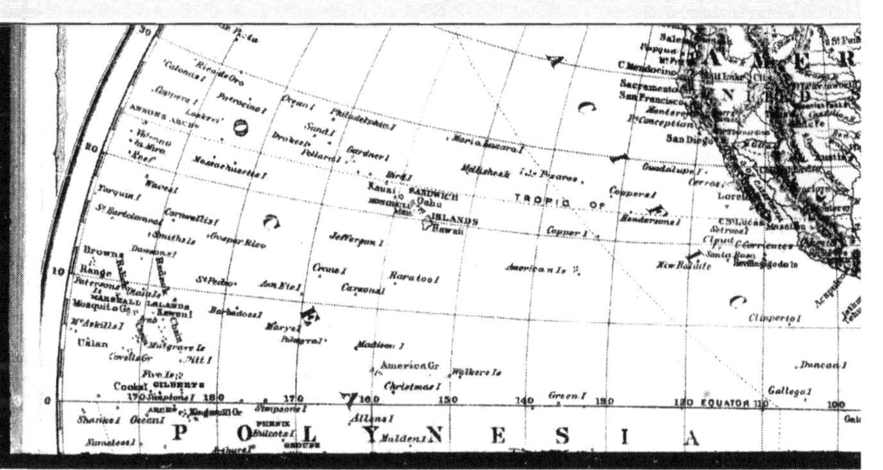

San Francisco
San Diego
Philadelphia I
Gardner I
SANDWICH
Oahu
ISLANDS
Hawaii
TROPIC OF
Jefferson I
Clipperton I
Duncan I
Christmas I
Gallego I
EQUATOR
P O L Y N E S I A

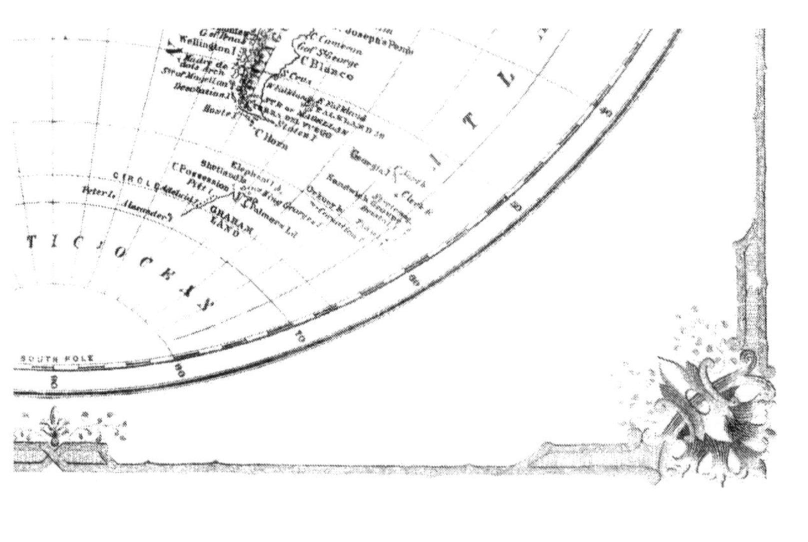

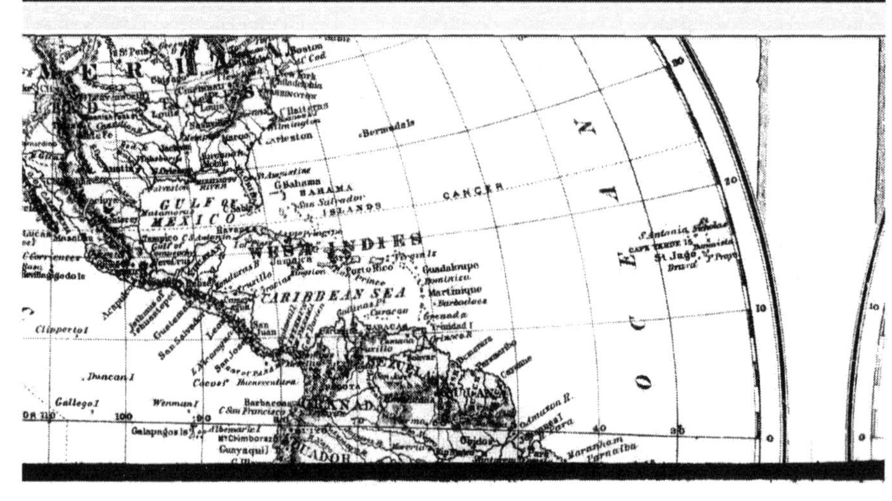

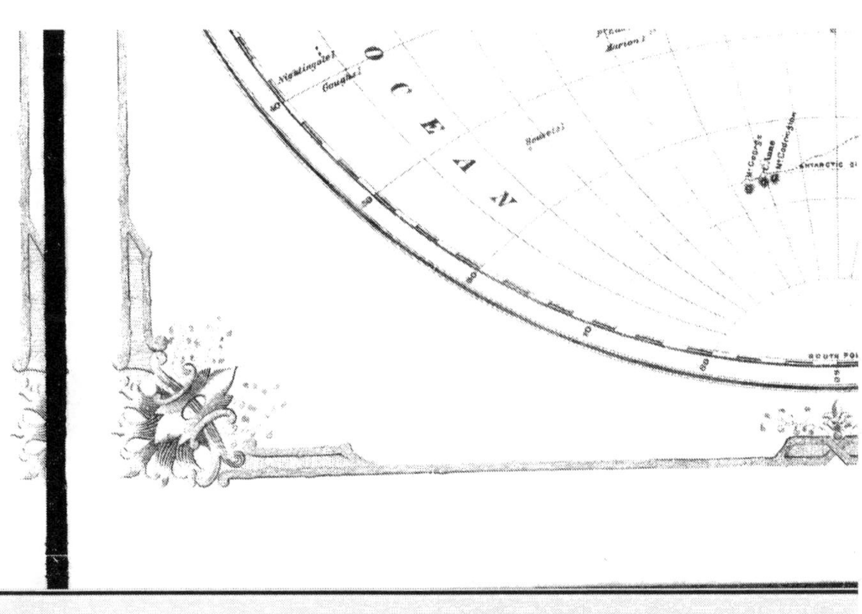
O C E A N
Nightingale I.
Goughs I.

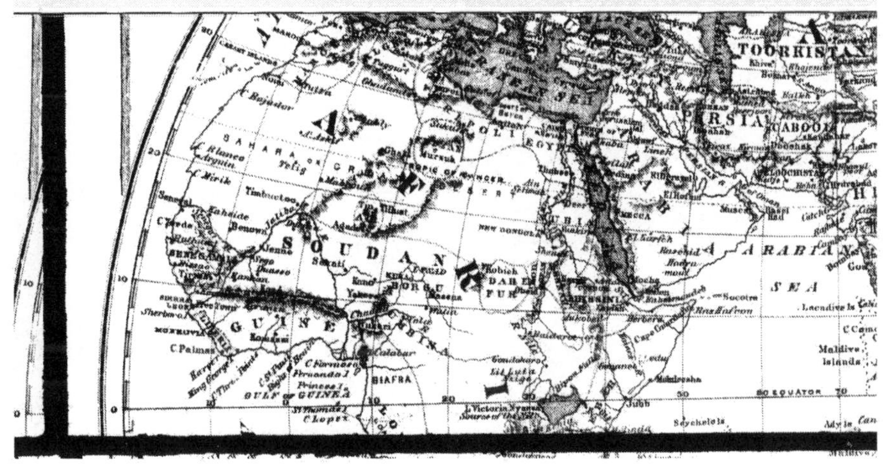
TOORKISTAN
CABOOL
SAHARA
Timbuctoo
S O U D A N
A R A B I A
SEA
C.Palmas
Calabar
BIAFRA
Socotra
Maldive Islands
EQUATOR

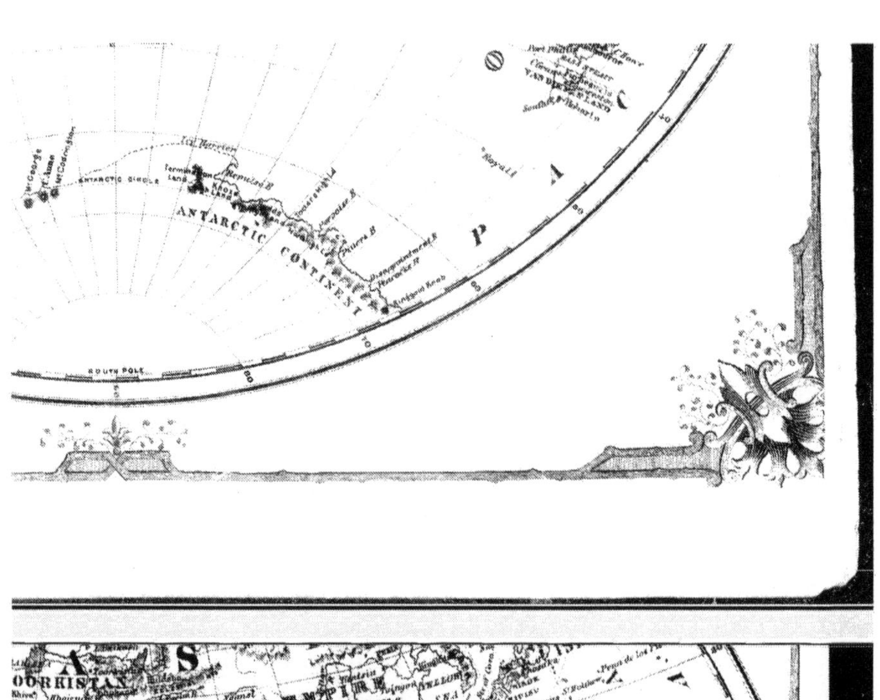

ANTARCTIC CIRCLE
ANTARCTIC CONTINENT
SOUTH POLE

CHINESE EMPIRE
CHINA
HINDOOSTAN
BAY
BENGAL
ARABIA
SEA
BORNEO
CELEBES
TROPIC OF CANCER
CAROLINE ISLANDS
EQUATOR

HUNGERFORD TOWNSHIP

NAME	POST OFFICE	Con.	Lot	BUSINESS	NATIVITY	Year Settled in Co.

HUNTINGDON TOWNSHIP

NAME	POST OFFICE	Con.	Lot	BUSINESS	NATIVITY	Year Settled in Co.

HUNTINGDON TOWNSHIP—Continued.

NAME	POST OFFICE	Con.	Lot	BUSINESS	NATIVITY	Year Settled in Co.

MADOC TOWNSHIP.

NAME	POST OFFICE	[illegible]	BUSINESS	NATIVITY	[illegible]
[illegible]	[illegible]	[illegible]	[illegible]	[illegible]	[illegible]

MADOC TOWNSHIP Continued.

NAME	POST OFFICE	[illegible]	BUSINESS	NATIVITY	[illegible]
[illegible]	[illegible]	[illegible]	[illegible]	[illegible]	[illegible]

RAWDON TOWNSHIP.

NAME	POST OFFICE	[illegible]	BUSINESS	NATIVITY	[illegible]
[illegible]	[illegible]	[illegible]	[illegible]	[illegible]	[illegible]

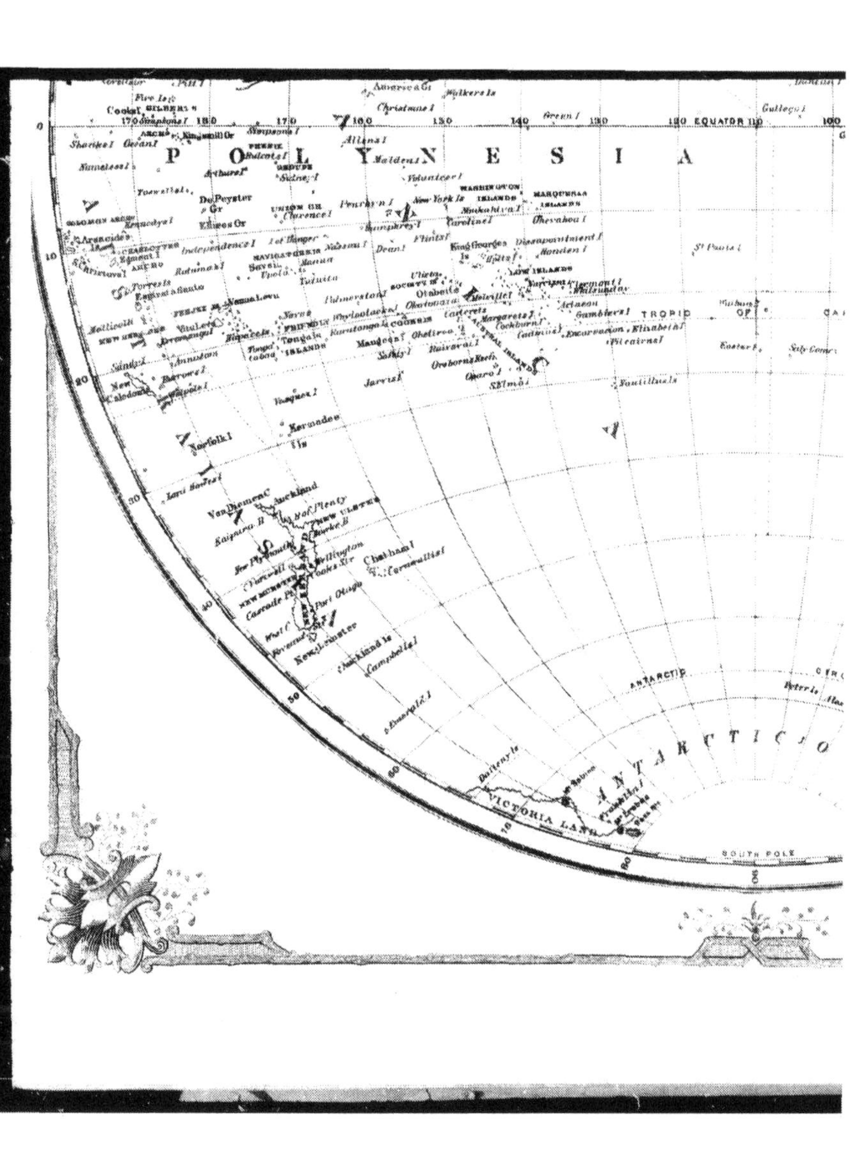

P O L Y N E S I A
EQUATOR
TROPIC
ANTARCTIC
ANTARCTIC
VICTORIA LAND
SOUTH POLE
Christmas I
Walkers Is
Maldens I
Marquesas Islands
Washington Islands
Low Islands
Navigators Is
Society Is
Cook Is
Friendly Islands
Feejee Is
New Caledonia
Norfolk I
Kermadec Is
Auckland
Wellington
Port Otago
Chatham I
Auckland Is
Campbells I
Pitcairns I
Easter I

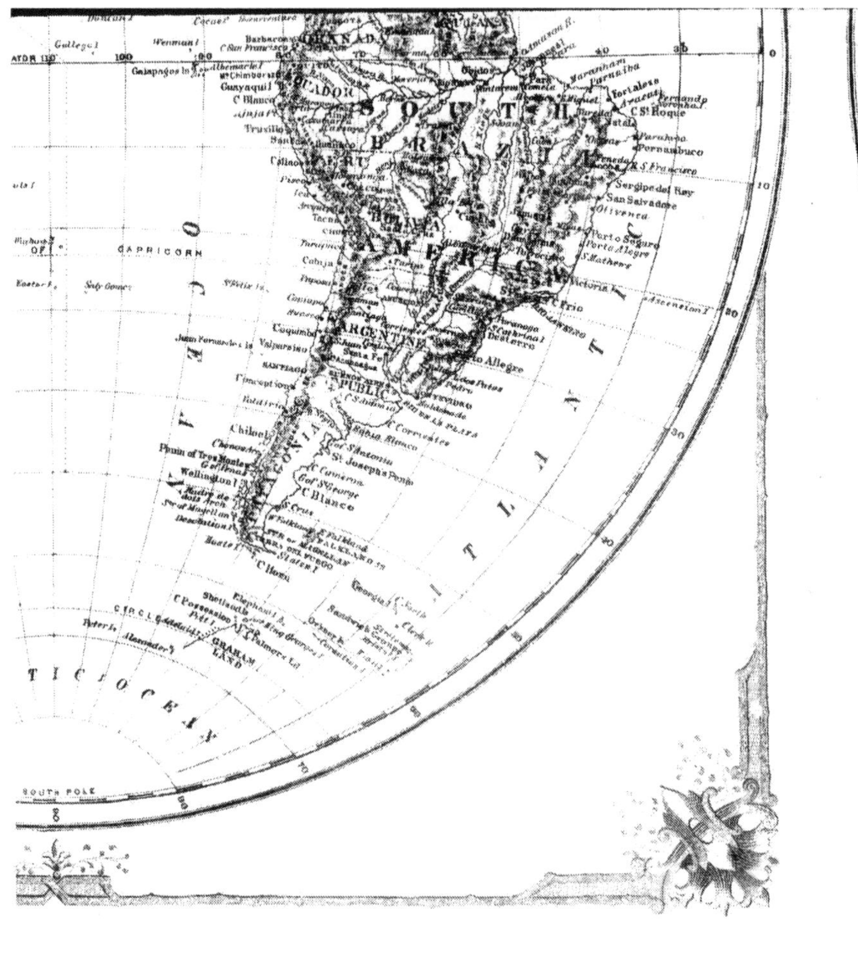

SOUTH
BRAZIL
AMERICA
ARGENTINE
PUBLIC
PATAGONIA
OCEAN
ATLANTIC
CAPRICORN
SOUTH POLE
GRAHAM LAND
Galapagos Is
Guayaquil
C Blanco
Truxillo
Callao
Pisco
Tacna
Cobija
Coquimbo
Valparaiso
SANTIAGO
Concepcion
Chiloe I
Wellington I
C Horn
Staten I
Georgia I
Elephant I.
Orkney Is
Sandwich Land
Fortaleza
C St Roque
Paraiba
Pernambuco
R S Francisco
Sergipe del Rey
San Salvador
Olivenca
Porto Seguro
Porto Alegre
S Mathews
C Frio
Ascension I
Paranagua
S Catharina I
Desterro
Porto Allegre
Santa Fe
C Corrientes
Bahia Blanca
St Joseph's Penn
C Blanco
C Cruz
Falkland
Maranham
Parnaiba
Fernando Noronha I
Obidos
Barbacoas
Wenman I
Juan Fernandez Is
St Felix Is
Easter I
Sala y Gomez
Peters I
Alexander I

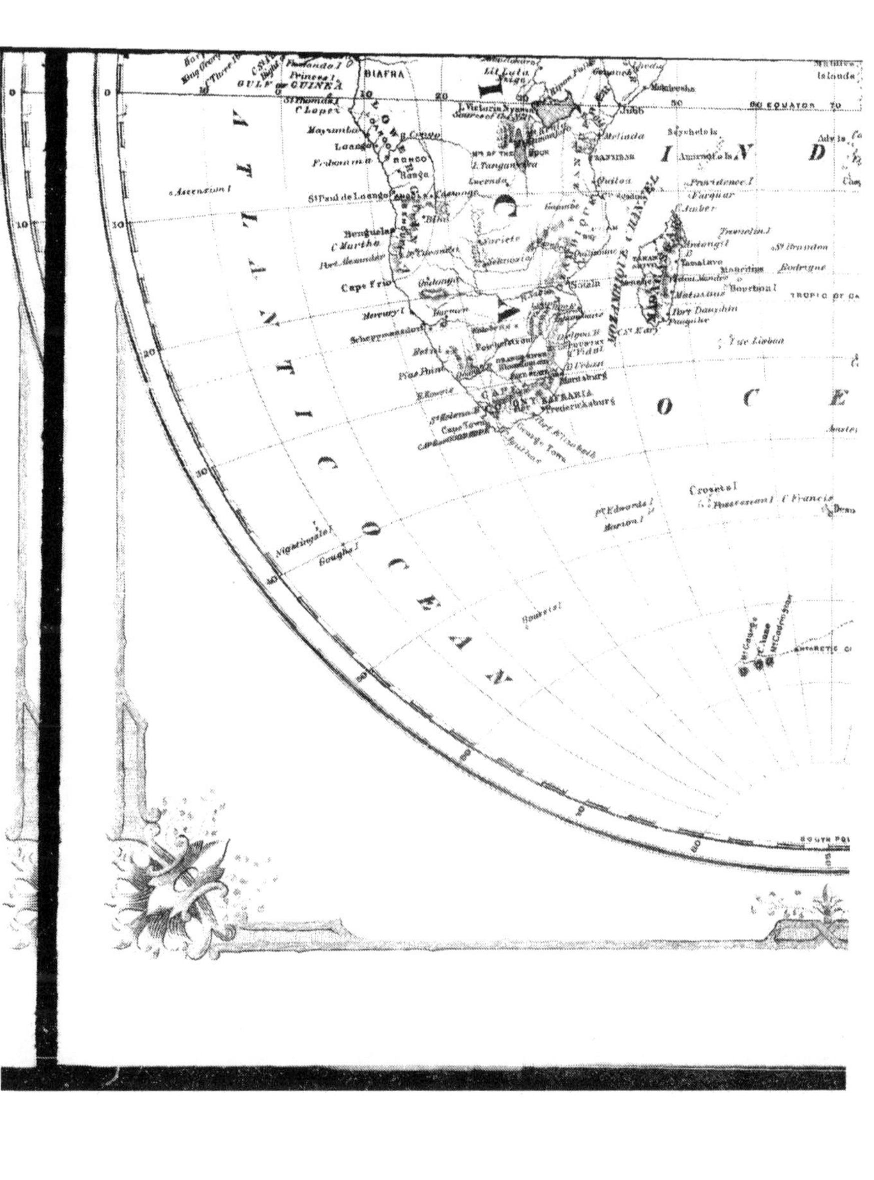

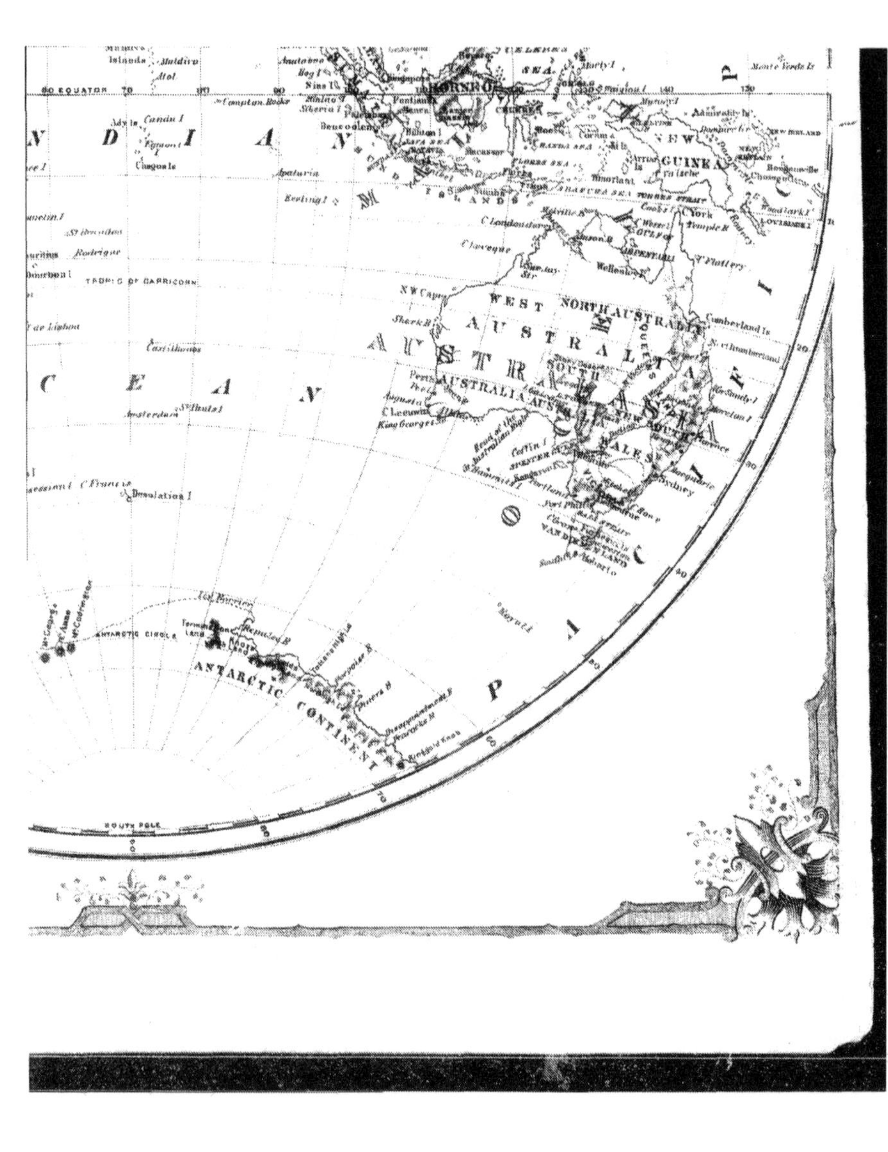

INDIAN OCEAN
BORNEO
NEW GUINEA
WEST AUSTRALIA
NORTH AUSTRALIA
SOUTH AUSTRALIA
NEW SOUTH WALES
VAN DIEMENS LAND
ANTARCTIC CONTINENT
ANTARCTIC CIRCLE
TROPIC OF CAPRICORN
EQUATOR
SOUTH POLE
Amsterdam
Desolation I
Sydney
Port Phillip
Bass Strait
Cumberland Is
Torres Strait
Cape York
Keeling I
Castillianos

Thomas Clark
Simon Delano

PATRONS' DIRECTORY

OF

Hastings and Prince Edward Counties.

CITY OF BELLEVILLE.

CITY OF BELLEVILLE Continued.

RAWDON TOWNSHIP—Continued.

NAME.	POST OFFICE.	Con.	Lot	BUSINESS.	NATIVITY.	Year Settled in Co.
Scott, Thomas	"	1	11	Farmer	Ireland	1859
Sharp, William	"	1	17	Farmer	England	1845
Sine, Silas	"	5	12	Farmer and Carpenter	Canada	1823
Sine, David	"	6	10	Farmer	Canada	1832
Sarles, Elijah	"	6	6	Farmer and Carpenter	Canada	1834
Sarles, Thomas	"	7	6	Farmer	Canada	1825
Sarles, William	"			Farmer	Canada	1850
Sprague, J. S	"			Physician	Canada	1844
Shott, A. G	Spring Brook	10	21	Farmer	Canada	1854
Simpson, William	"	9	11	Farmer	England	1844
Scott, A. G	Stirling			Hotel Proprietor	Canada	1831
Totton, John	Wellman's Cor.	6	19	Farmer	Canada	1836
Tucker, David	Stirling	5	8	Farmer	Canada	1845
Tucker, G. E	"	8	7	Farmer	Canada	1825
Tucker, Luther	"	4	15	Farmer	United States	1835
Thain, Jonathan	Wellman's Cor.	8	21	Farmer	Scotland	1865
Vandervoort, S. S	Harold	8	10	Farmer and Cheese Manufacturer	Canada	1823
Vandervoort, J. A	Wellman's Cor	6	16	Farmer and President Cheese Co	Canada	1822
Walker, Thomas	"	1	23	Farmer, Warden and Reeve	Scotland	1855
White, A. C	Stirling	1	20	Farmer	Canada	1828
Williams, Calvin	Brinkworth	3	24	Farmer	Canada	1833
Williams, M. S. B	"	2	25	Farmer	Canada	1840
Westcott, S. N	Stirling	1	14	Farmer	Canada	1833
Wilson, George	"	2	10	Farmer	Ireland	1844
Wilson, Robert	W. Huntingd'n	3	2	Farmer	Ireland	1845
Williams, David	Brinkworth	4	14	Farmer	Canada	1852
Wallace, J. & Son	Wellman's Cor	5	24	Blacksmiths and Farmers	Scotland	1855
Wiggins, G. D	Harold	7	12	Merchant	Canada	1846
Wilson, J. C	Stirling	2	10	Farmer	Canada	1853
Wellman, D	Spring Brook	10	26	Farmer	Canada	1820
Wiggans, Jennie	"	9	22		Canada	1840
Welch, Peter	"	9	15	P. M. and Shoemaker	Canada	1841

SIDNEY TOWNSHIP.

NAME.	POST OFFICE.	Con.	Lot	BUSINESS.	NATIVITY.	Year Settled in Co.
Barker, W. T	Trenton			Druggist, Books and Stationery	Canada	1865
Baker, W. F. & Co	"			Liverymen and Undertakers, Market Square	Canada	1878
Bonter, O. R	"			Jeweller and Watchmaker, Front St.	Canada	
Blanchard, T. H	Belleville	1	30	Farmer	Canada	1857
Bonisteel, C. H	"	1	34	Farmer	Canada	1843
Bartlett, Thomas	"	1	13	Farmer and Cheesemaker	England	1847
Bouteiller, J. L	Gordon Mills	2	2	Farmer	Island of Jersey	1854
Billings, B. M	Trenton	1	2	Farmer	Canada	1820
Bonisteel, J. H	Belleville	3	11	Farmer	Canada	1848
Bonisteel, W. H	"	3	14	Farmer	Canada	1848
Bonisteel, Samuel	"	4	21	Farmer	Canada	1818
Bleecker, H	Frankford	3	13	Farmer	Canada	1825
Bonisteel, N. A	"	4	7	Farmer	Canada	1844
Bush, John	"	4	5	Farmer	Canada	1840
Brower, J. R	Belleville	3	25	Farmer	Canada	1852
Boardman, Moses	Halloway	7	22	Farmer	Canada	1836
Bird, James	"	8	28	Farmer and Cheese Factory	Ireland	1835
Brooks, J. W	Stirling	9	8	Farmer	Canada	1850
Crosster, A. L	Belleville	8	21	Farmer	Canada	1847
Cheshire, Charles	"	4	10	Farmer	Canada	1852
Casey, S. T	"	3	33	Farmer	Canada	1807
Casey, S. S	"			Farmer	Canada	1844
Chisholm, J. A	Wallbridge	1	27	Farmer	Canada	1835
Conner, [illegible]	Halloway	8	27	Farmer	United States	1816
Cotter, R. C. H	Frankford	5	14	Farmer	Canada	1852
Cunningham, J. M	Trenton			Livery business	Canada	
Cochrane, Mrs	"					
Day, Henry W., M.D.	"			Physician and Surgeon, Coroner, Member of the Council	Canada	1855
Dorland, C. H. L. D.S	"			Dentist	Canada	1842
Davis, Cornelius	Belleville	1	33	Farmer	United States	1861
Denyes, A. N	Trenton	2	7	Farmer	Canada	1816
Flintall, A	"			Hardware Merch't, Mem. of Council	Canada	1824
Ford, Peter	Gordon Mills	2	A	Paper Manufacturer and Post master	England	1852
Francis, C	Trenton			Barrister and Attorney at Law	Canada	
Finkle, C. W	Belleville	3	21	Farmer	Canada	1831
Finkle, Abel	"			Farmer	Canada	1830
Fretz, Byard	Canniften	4	33	Farmer	Canada	1825
Faulkner, S. G	Halloway	7	35	Farmer	Canada	1845
Foster, J. C	"	7	31	Farmer	Canada	1826
Faulkner, S	"	8	33	Farmer	Canada	1825
Farley, J. M	Wallbridge	5	21	Farmer	Canada	1828
Grass, R. E	Frankford	4	13	Farmer	Canada	1836
Graham, Ketcham	Belleville	1	25	Farmer	Canada	1830
Gilbert, S. B	"	1	24	Farmer	Canada	1845
Gilbert, R	"	1	36	Farmer	Canada	1837
Gilham, F. W	"	1	33	Farmer	England	1861
Garrison, Charles	"	1	9	Farmer	Canada	1877
Graham, John	Wallbridge	1	26	Farmer	Scotland	1833
Goldsmith, S. H	Belleville	3	27	Farmer	Canada	1837
Goldsmith, D	Warkworth	2	27		Canada	1834
Goldsmith, P. L	Belleville	3	22	Farmer	Canada	1848
Holbrook, T. A	Trenton			Proprietor Queen's Hotel	United States	1864
Hicks, Jacob W	Belleville	1	31	Farmer	Canada	1834
Hunt, J. W	"	1	16	Farmer	Canada	1846
Harry, William	Trenton	1	7	Farmer	England	1842
Hutchison, Allan	"	2	8, 9	Farmer	Scotland	1876
Harder, P. W	Gordon Mills	2	3	Farmer	Canada	1845
Harry, John	"	3	9	Farmer	England	1844
Hawley, Nancy	Trenton				Canada	1816
Hagerman, John	Belleville	3	28	Farmer	Canada	1844
Hamilton, Andrew	Halloway	8	36	Farmer	Canada	1845
Huffman, J. S	Wallbridge	5	18	Farmer	Canada	1826
Irish, J. R., L.D.S	Trenton			Surgical and Mechanical Dentist	Canada	1867
Jackson, George	"			Insurance and agricultural implement agent	United States	1872
Jill, William	"			Grain dealer	Canada	1861
Jordan, E	Frankford			General Merchant	Canada	1856
Jones, David	Belleville	1	23	Retired Farmer	Canada	1799
Jones, Charles	"	1	31	Farmer	Canada	1847
Jones, Wellington	"	1	21	Farmer	Canada	1834
Jones, Anson H	"	1	33	Farmer	Canada	1833
Johnston, A	"	3	34	Farmer	Canada	1848
Johnston, Peter	Foxboro	5	34	Farmer	Canada	1830

NATIVITY.	Year Settled in Co.
reland	1850
England	1845
Canada	1825
Canada	1832
Canada	1831
Canada	1828
Canada	1830
Canada	1841
Canada	1854
England	1861
Canada	1869
Canada	1836
Canada	1843
Canada	1825
United States	1836
Scotland	1855
Canada	1822
Canada	1822
Scotland	1855
Canada	1828
Canada	1823
Canada	1840
Canada	1831
Ireland	1844
Ireland	1845
Canada	1852
Scotland	1855
Canada	1846
Canada	1853
Canada	1820
Canada	1849
Canada	1844

NATIVITY.	Year Settled in Co.
Canada	1865
Canada	1872
Canada	
Canada	1837
Canada	1844
England	1842
Island of Jersey	1854
Canada	1829
Canada	1838
Canada	1836
Canada	1848
Canada	1835
Canada	1844
Canada	1840
Canada	1857
Canada	1836
Ireland	1855
Canada	1850
Canada	1812
Canada	1852
Canada	1867
Canada	1841
Canada	1835
United States	1840
Canada	1857
Canada	
Canada	1850
Canada	1846
United States	1861
Canada	1840
Canada	1824
England	1852
Canada	
Canada	1851
Canada	1846
Canada	1845
Canada	1845
Canada	1820
Canada	1825
Canada	1818
Canada	1849
Canada	1830
Canada	1849
Canada	1827
England	1861
Canada	1871
Scotland	1853
Canada	1836
Canada	1874
Canada	1848
United States	1861
Canada	1841
Canada	1816
England	1842
Scotland	1870
Canada	1845
England	1844
Canada	1866
Canada	1844
Canada	1845
Canada	1826
Canada	1867
United States	1872
Canada	1861
Canada	1856
Canada	1790
Canada	1847
Canada	1854
Canada	1833
Canada	1848
Canada	1850

SIDNEY TOWNSHIP—Continued.

NAME.	POST OFFICE.	Con.	Lot.	BUSINESS.	NATIVITY.	Year Settled in Co.
Johnston, William	Foxboro			Farmer	Scotland	1845
Jordan, Edward	Frankford	5	8	Farmer	Ireland	1831
Ketcheson, P. H.	Belleville	1	20	Farmer	Canada	1837
Kelley, J. W.	"	1	26	Farmer	Canada	1812
Knox, J. S.	Trenton	2	2	Farmer	Canada	1829
Knox, D. H.	"	2	5	Farmer	Canada	1851
Knox, W. H.	"	2	6	Farmer	Canada	1830
Knight, Samuel	Belleville	1	37	Farmer	Canada	1831
Knox, Jeremiah	Wallbridge	4	31	Farmer	Canada	1812
Ketcheson, P. C.	"	5	24	Farmer	Canada	1814
Ketcheson, Geo. M.	"	5	26		Canada	1825
Ketcheson, A. T.	"	5	27	Farmer and General Agent	Canada	1820
Ketcheson, E. J.	Cannifton	1	30	Carpenter and Joiner, and Builder	Canada	1817
Lott, W. H.	Belleville	B.F.	9	Carpenter and Builder	Canada	1836
Lane, B. W.	Cannifton	1	22	Farmer	Canada	1844
Lucas, B. A.	Belleville	5	31	Farmer	Canada	1839
Lockwood, David	Wallbridge	5	29	Farmer	Canada	1837
Laley, G. H.	"	4	18	Farmer	England	1856
McMullen, W. P.	Belleville	2	8	Farmer	Canada	1849
McCutcheon, A.	Wallbridge	8	30	Fruit Grower, etc.	Canada	1870
McLaren, Donald	Foxboro	5	33	Farmer	Scotland	1857
Massey, A. J.	Belleville	1	22	Farmer and Dairyman	Canada	1870
Meyers, Geo. F.	Trenton	1	6	Farmer	Canada	1827
Meyers, A. P.	"	1	7	Farmer	Canada	1833
Mumo, John	Gordon Mills	3	6	Farmer and Carpenter	Canada	1815
Maley, W. K.	"	3	5	Farmer and President of Cheese Factory	Canada	1847
Mallory, A. B.	Frankford	4	14	Farmer	Canada	1828
Masser, Levi	Wallbridge	5	39	Farmer	Canada	1869
Miller, Stephen	"	4	25	Blacksmith	United States	1844
Murphy, Anthony	Stirling	9	18	Farmer	Canada	1857
Nicol, John	Halloway	8	38	Merchant Miller	Scotland	1876
Ostrom, C. R.	Belleville	1	28	Farmer	Canada	1844
Ostrom, J. B.	"	1	22	Farmer	Canada	1812
O'Rourke, P. J.	Trenton			Law Student	Canada	1854
Ostrom, G. W.	"			Barrister	Canada	1857
Patrick, J. S.	"	1	32	Farmer	Canada	
Palmer, W. J.	Belleville	1	36	Principal Deaf and Dumb Institution	United States	1870
Patrick, William	Gordon Mills	2	1	Contractor and Builder	Canada	1837
Parry, W. R.	Belleville	1	9	Farmer	Canada	1835
Ryan, J. S.	Trenton			Retired gentleman	Canada	1874
Rose, Baltis	Frankford	4	12	Farmer and Township Reeve	Canada	1817
Ross, John H.	Trenton	2	3, 4	Farmer and Stock Dealer	Canada	1833
Ross, J. R.	Belleville	2	11	Farmer	Canada	1850
Ross, J. J.	"	2	13	Farmer	Canada	1853
Ross, C. G.	"	2	12	Farmer	Canada	1851
Shorrie, J. H.	Trenton			Merchant Tailor	Scotland	1863
Suttle, Joseph	Belleville	1	15	General Blacksmith	Ireland	1857
Saylor, C. B.	Trenton	F.F.	9	Retired Farmer	Canada	1821
Stickle, E. O.	Gordon Mills	3	3	Farmer	Canada	1837
Steel, John	Trenton	1	3	Farmer	Scotland	1855
Sharp, J. P.	Wallbridge	4	22	Farmer	United States	1827
Smith, J. H.	Frankford	4	7	Farmer	Canada	1810
Sarles, Selah	Halloway	7	28	Farmer	Canada	1811
Stile & Bro	Frankford			Manufacturers	Canada	1867
Taylor, C. W.	Belleville	1	15	Farmer	Canada	1850
Teal, J. W.	Gordon Mills	3	2	Farmer	Canada	1827
Tucker, Henry	Wallbridge	1	31	Farmer	England	1848
Thresher, G. B.	Belleville	3	33	Farmer	Canada	1841
Thresher, Gilbert	"	2	35	Farmer	Canada	1869
Turley, P.	Frankford			Retired Contractor	Ireland	1833
Vossman, J. H.	Trenton			General Merchant	United States	1875
Vandervoort, D.	"	4	6	Farmer	Canada	1852
Vandervoort, W.	Sidney	2	17	Farmer	United States	1802
Vandervoort, Major	Wallbridge	4	25	Farmer	Canada	1831
Vermilyea, Thomas	Belleville	1	34	Farmer	Canada	1821
Vandervoster, B.	Foxboro	6	29	Farmer	Canada	1830
Wilson, G. H.	Trenton				Canada	
Wilmot, S. T.	Wallbridge	4	29	Farmer	Canada	1857
White, P. M.	Belleville	1	12	Teacher	Canada	1856
White, Webster	Trenton	1	5	Farmer	Canada	1858
White, Reuben	Belleville	4	23	Farmer	Canada	1830
Young, George, B.A.	Trenton			Publisher of Trenton Courier	Ireland	1876
Young, James	"			Proprietor of Planing Mill, Dep. Reeve	Canada	1834
Yates, W. S.	Belleville	2	20	Farmer	Canada	1826

THURLOW TOWNSHIP.

NAME.	POST OFFICE.	Con.	Lot.	BUSINESS.	NATIVITY.	Year Settled in Co.
Ashley, Charles	Foxboro			Carriage Manufacturer	Canada	1836
Ashley, Harford	"	5	283	Farmer and Cheese Manufacturer	Canada	1816
Ashley, William	"	5	2	Farmer and retired Carriage Maker	Canada	1821
Bates, William H.	Roslin	8	28	Farmer	Canada	1831
Bradshaw, William	Plainfield	6	22	Farmer	Canada	1824
Bradshaw, Lewis	Foxboro	1	8	Farmer and Builder, &c.	Canada	1836
Burgess, P. C.	Cannifton	5	14	Farmer and J. P.	Canada	1815
Bird, W. H.	Belleville	4	1	Farmer and Stock Breeder	Canada	1849
Barnum, Simon	Cannifton			Miller	Canada	1808
Baldwin, Oren	Foxboro	7		Farmer	Canada	1854
Bulpley, Ira C.	Cannifton	4	25	Farmer	Canada	1840
Beckett, Charles	"			Lumberman	Canada	1850
Bradshaw, W. R.	"	1	8	Farmer	Canada	1831
Bristnell, E. P.	"	3	9	Farmer and Cheese Maker	Canada	1837
Casey, Willett	Phillipston	8	13	Farmer and Stock Raiser	Canada	1836
Chapman, William A.	Roslin	9	28	Farmer	Canada	1840
Chapman, John	"	9	27	Farmer	Canada	1808
Chapman, Duncan	"	9	25	Farmer	Canada	1833
Chapman, William	Plainfield	8	20	Farmer and Stock Raiser, J. P.	Canada	1846
Campbell, T. J.	"	6	23	Farmer and Cheese Maker	Canada	1844
Clapp, S.	"	7	24	Farmer	Canada	1816
Carter, Jacob	"	7	26	Farmer	Canada	1822
Clapp, D. E.	"			Lumber and Shingle Manufacturer	Canada	1847
Clapp, D. & D. R.	Cannifton	4	14	Farmers	Canada	1806
Clarke, S. J.	"	5	13	Farmer and Grape Grower	Canada	1819
Dennes, P. S.	Halloway	8	7	Farmer	Canada	1849
Dolly, William	Plainfield	6	22	Farmer and Assessor	Canada	1833
Dafoe, Henry	Foxboro			Carriage Maker	Canada	1832
Davis, W. H.	Cannifton	5	7	Farmer	Canada	1827
Embury, Allen	Roslin	9	24	Farmer, Contractor and Builder	Canada	1828
Emerson, George	"	8	31	Farmer	Ireland	1853

THURLOW TOWNSHIP—Continued.

NAME.	POST OFFICE.	Con.	Lot.	BUSINESS.	NATIVITY.	Year Settled in Co.
Elliot, William	"	8	28	Farmer	Canada	1822
Emerson, William	"	8	27	Farmer	Ireland	1822
Fargey, James S.	"	8	27	Farmer	Canada	1820
Faulkner, A. B.	Foxboro	6	1	Farmer	Canada	1831
Farnham, T.	Cannifton			Superintendent of Roads	Canada	1819
Graham, John	Plainfield	8	21	Farmer	Canada	1853
Garrison, H. S.	"	7	21	Farmer and J. P	Canada	1814
Garrison, W.	Cannifton	1	23	Farmer	Canada	1808
Garrison, George	Blessington			Farmer	Canada	1814
Gould, E. N.	Plainfield			Postmaster and Shoemaker	Canada	1852
Houston, James	Roslin	8		Farmer	Canada	1841
Hudson, William	"			Carriage Maker	Canada	1841
Hazlitt, Joseph	Phillipston	9	17	Farmer	Canada	1810
Hudson, Charles	Roslin	8	1		England	1836
Hicks, Joshua	Plainfield	7	25	Farmer and Breeder of Thorough-bred Stock	Canada	1835
Howard, David	"	8	22	Pump Manufacturer	United States	1852
Hall, James	"	7	25	Dominion Hotel	Canada	1821
Hall, Abner	"	7	28	Farmer	Scotland	1827
Hudson, William	Blessington	5	23	Farmer	Canada	1842
Hall, George	Plainfield	7	29	Farmer	Canada	1833
Hagerman, A. W.	Halloway	7	3	Farmer	Canada	1858
Ham, J. D.	Belleville	B. F.	21	Farmer	Canada	1818
Hoffman, John	"	5	20	Farmer	Canada	1847
Hicks, H.	"	4	30	Farmer	England	1867
Jose, Edward	Halloway	7	1 & 2	Breeder of Thorough-bred Stock	England	1855
Jose, R. P.	"			Breeder of Thorough-bred Stock	England	1855
Jose, Stephen	"			Breeder of Thorough-bred Stock	England	1855
Ketcheson, John V.	Phillipston	8	14	Farmer	Canada	1832
Lloyd, Thomas	Plainfield	7	2	Farmer	Canada	1846
Leslie, W. E.	Roslin	9	31	Farmer	Canada	1847
Lounsbury, Gilbert	Plainfield	7	17	Farmer	Canada	1823
Latta, J. G.	"	8	22	Farmer	Canada	1841
Latta, Mason	"	7	28	Cheese Factory	Canada	
Latta, Albert	Phillipston	9	21	Farmer and Miller	Canada	1821
McCready, W. J.	Roslin	9	29	Farmer	Canada	1849
McMichael, Robert	Phillipston			Builder, Carpenter and Joiner	Ireland	1847
McCreadie, James	Plainfield	8	25	Farmer and J. P.	Canada	1815
McKim, Edward	"	6	27	Farmer	Canada	1822
McKenzie, J. F.	"	7	23	Farmer	Canada	1835
McConnely, Thomas J.	Roslin	9	29	Farmer	Canada	1841
McCauley, Daniel	Foxboro	6	2	Farmer	Ireland	1845
McDougal Bros.	Belleville	1	24	Farmers	Canada	
McWilliams, J. J.	Cannifton	1	13	Farmer	Canada	1831
Moore, James	Phillipston	9	21	Farmer	Ireland	1824
Mosby, Thomas	Foxboro	6	4	Farmer and Butcher	England	1819
Mott, J. K.	Plainfield	7	23	Farmer	Canada	1827
Mcburn, W. H.	"	7	23	General Blksmith & Carriage Maker	Canada	1862
Moorman, Louis	Belleville			Carriage Factory. Ashley & Moorman, Bros' Coopers	Canada	1830
Megginson, George	Halloway	7		Farmer	England	1830
Moorman, Louis	Belleville	1	13	Farmer	England	1852
O'Brien, Christopher	Plainfield	6	19	Farmer	Canada	1816
Phillips, M. B.	Phillipston	9	16	Farmer	Canada	1823
Phillips, George	"	9	26	Farmer, Postmaster and J. P	Canada	1821
Phillips, Daniel	"	8	29	Farmer and Agricultural Imp. Agt.	Canada	1821
Palliser, S. M.	Foxboro			Hotel Keeper	Canada	1866
Parks, Archibald	Plainfield	6	26	Farmer, Bee and Honey Merchant	Canada	1819
Peck, James H.	Belleville	2	4	Farmer and Insurance Agent	Canada	1870
Palmer, Francis	"	1	2	Carriage Maker	Canada	1850
Palmer, C. A.	Belleville	4	26	Farmer	Canada	1849
Palmer, Leonard	Cannifton	3	11	Farmer and Carriage Maker	Canada	1828
Palmer, Wellington	"			General Blksmith & Carriage Maker	Canada	1841
Reed, Benjamin	Foxboro	8	1	Farmer	Canada	1823
Reid, Owen	Foxboro	5	5	Farmer and Lumber dealer, &c.	Canada	1833
Ross, Algernon	"	4	3	Farmer	Canada	1855
Ross, A. Z.	Belleville	4	21	Farmer	Canada	1823
Ross, Albert	Foxboro			Farmer	Canada	1847
Sharp, S.	Phillipston	8	15	Blacksmith	Canada	1846
Sanier, J. O.	Foxboro			Insurance Agt. Agricultural Mutual of Canada	Canada	1836
Sayers, William	Phillipston	9	15	Farmer	Canada	1827
Sills, R. S.	"	9	11	Farmer	Canada	1844
Sills, J. W.	Phillipston	9	10	Farmer	Canada	1842
Spencer, Charles	Halloway	8	2	Farmer	Canada	1838
Shetler, David	Foxboro			Carpenter and Joiner	Canada	1833
Silis, J.	Cannifton			Farmer, Lumber Merchant and Clerk of Division Court	Canada	1838
Townsend, W.	Halloway	8	5	Farmer	Canada	1833
Townsend, John	"	8	4	Farmer	Canada	1835
Tower, Theodore	Plainfield	7	27	Carpenter and Joiner	United States	1838
Thrasher, Edward	Foxboro	5	3	Farmer	Canada	1821
Vankleek, William	Phillipston	9	11	Farmer	Canada	1823
Vandewaters, J. W.	Plainfield	7	22	Farmer	Canada	1814
Vermilyea, N.	Belleville	4	1	Farmer and School Teacher	Canada	1846
Vermilyea, J.	"	3	1 & 2	Farmer and Fruit Grower	Canada	1830
Vickers, W. D.	Cannifton			Builder and Contractor	Canada	1844
Wickett, William	Foxboro			Contractor, &c.	England	1842
Wickett, Richard	"	6	2	Farmer, Builder, &c.	Canada	1852
Wickett, Daniel	"	6	3	Farmer and Pathmaster	Canada	1852
Wallace, S.	"			Butcher and General Trader in Stock	Canada	1841
White, W. H.	Cannifton			Carpenter and Joiner	Canada	1847
Walker, Robert	"	4	16	Farmer	Canada	1866
Way, J. F.	"	4	24	Farmer	Canada	1850
Wilson, Asa	Belleville	3	26	Farmer	Canada	1853
Watson, R. H.	"	2	11	Farmer	Canada	1849
Weymans, Henry	Cannifton			Farmer	England	1851
Way, H. N.	Belleville	2	14	Farmer	Canada	1858
Yeoker, Samuel	Halloway	8	10	Farmer	Canada	1818

TYENDINAGA TOWNSHIP.

NAME.	POST OFFICE.	Con.	Lot.	BUSINESS.	NATIVITY.	Year Settled in Co.
Anderson, Donald	Melrose	2	14	Farmer, 1st Deputy Reeve	Scotland	1850
Appleby, S. S.	Shannonville			Miller	Canada	1820
Allison, Orren	Blessington	5	1	Farmer	Canada	1817
Armitage, F.	Napanee	2	35	Farmer	Ireland	1841
Allison, E. A.	Marysville	2	37	Farmer, Insurance Agent, Agent for Royal Reaper	Canada	1817
Baldrick, John, jr.	Shannonville	1	1	Farmer	Canada	1833
Bennett, John	Blessington	5	8	Farmer	Ireland	1845
Carey, Thomas	Marysville	2	28	Farmer, 2nd Deputy Reeve	Ireland	1842
Cook, Richard B.	Blessington	3	5	Farmer	Canada	1808
Collins, Hugh	"	5	2	Farmer	Canada	1843

NAME.
Cook, William P.
Donoghue, F., & Bro.
Danke, A.
Dexstator, A.
Dexstator, W. H.
Emill, Isaac
Emerson, Samuel
Egar, W. G.
Emmons, John
Eelson, F. W.
Fegan, R. F.
Farnsworth, F.
Forrester, James
Gould, P.
Gibson, Robert
Hollingsworth, F.
Holden, H.
Hudson, Samuel
Hill, Matthew
Houston, James
Hulel, Charles
Halstead, Asa
Hudson, Charles
Haney, J. C.
Long, Charles
Lazier, R. J.
Lazier, James A.
Lally, Joseph
Lally, Michael
McLenahan, Donald
McCapes, James
McLaren, William
McFarlane, William
McFarlane, Donald
McGurn, Joseph
McAuley, [illegible]
Moore, George
Murphy, Thomas
Meagher, James
Meagher, John
O'Sullivan, John
Oakley, W. H.
Petersen, Frank
Porte, Robert S.
Pitman, Albert
Rathburn, H. B. & So
Robertson, James
Ray, James
Roblin, Owen
Sherman, George
Stickney, Farley
Skelly, W. J.
Scanlon, Charles
Todd, John A.
Tripp, Samuel
Wright, Joshua
Wilson, Baptist
Wrose, William
White, John
Whittington, W. S.

NAME
Anderson, William
Anderson, Levi
Ambrose, George
Anderson, J. W.
Ashton, A. B.
Ainsworth, B.
Adams, John
Anderson, John
Alyea, J. J.
Alyea, Peter
Alyea, John
Brickman, Samuel
Brickman, W. A.
Byers, John
Babbit, Elkanah
Broad, James
Benson, James
Brooks, Anthony
Beech, P. V.
Beech, John
Beech, Mills
Barker, E.
Burley, F.
Babcock, Robert
Brickman, James
Bentley, Peter
Brickman, W. H.
Brickman, Asa
Baxter, John A.
Babcock, John H.
Coleman, Isaiah
Constance, G. T.
Church, Daniel
Cunningham, J. H.
Carnrike, Selim
Coulter, James
Cahan, David

TYENDINAGA TOWNSHIP—Continued.

NAME	POST OFFICE.	Con.	Lot	BUSINESS.	NATIVITY.	Year Settled in Co.
Cook, William P.	Shannonville	2	10	Farmer	Canada	1840
Donoghue, F., & Bro.	Mill Point			Merchants	Canada	1858
Pontie, A.	Belleville	3	4	Farmer	Canada	1830
Dextator, A.	Shannonville	1	2	Farmer	Canada	1855
Dextator, W. H.	″	2	2	Farmer	Canada	1848
Cahill, Isaac	Melrose	3	14	Farmer	Canada	1828
Emerson, Samuel	″	4	12	Farmer and Councillor	Canada	1842
Egar, W. G.	Mill Point			Druggist	Canada	1873
Emmons, John	Shannonville	1	10	Farmer and Hop Grower	Canada	1839
Folson, F. W.	Mill Point			Harness Maker	Canada	1873
Fegan, B. F.	Shannonville			Cabinet Maker, Founder, and Manufacturer of Agr'l. Implements	Ireland	1842
Farnsworth, T.	″	2	3	Farmer	Canada	1834
Forrester, James	Melrose	3	20	Farmer	Scotland	1833
Gould, P.	Lonsdale	3	30	Farmer, and Agent for Noxon & Bros. Agricultural Implements	Canada	1836
Gibson, Robert	Plainfield	6	1	Farmer	Ireland	1838
Hollingsworth, E.	Melrose	3	25	Farmer and Township Clerk	Ireland	1839
Holden, H.	Shannonville			Merchant, Post Master, and Clerk of Division Court	Canada	1846
Hudson, Samuel	″			Physician	Canada	1843
Hill, Matthew	″			Indian Agent	Canada	1846
Houston, James	″	2	1	Farmer	Canada	1848
Hubel, Charles	Blessington	4	6	Farmer	Canada	1842
Halstead, Asa	″	5	6	Farmer	Canada	1851
Hudson, Charles	Roslin	8	1	Farmer	England	1834
Hasley, J. C.	Read	5	20	Farmer and Store Keeper	Ireland	1841
Long, Charles	Shannonville	3	6	Farmer	England	1843
Lazier, B. J.	″	1	5	Farmer	Canada	1830
Lazier, James A.	Lonsdale			Proprietor of Woollen & Grist Mill	Canada	1820
Lally, Joseph	Blessington	4	8	Farmer	Canada	1852
Lally, Michael	″	4	7	Farmer	Canada	1838
McLennan, Donald	Shannonville			Merchant	Scotland	1862
McCaner, James	″			Trustee, S. S. No. 1		
McLaren, William	Melrose	3	4	Farmer	Scotland	1833
McFarlane, William	″	3	8	Farmer	Canada	1839
McFarlane, Donald	″	2	8	Farmer	Canada	1843
McGurn, Joseph	Marysville	1	27	Farmer	Canada	1833
McAuley, [illegible]	Lonsdale	2	33	Farmer	Ireland	1833
Monro, George	Shannonville			Farmer and Councillor	Scotland	1849
Murphy, Thomas	Marysville	2	37	Farmer	Canada	1845
Meagher, James	Read	6	29	Farmer	Ireland	1842
Meagher, John	″	6	27	Farmer	Ireland	1842
O'Sullivan, John	Blessington	5	3	Farmer	Canada	1830
Oakley, W. H.	Shannonville	1	9	Farmer, and Agent for Johnson Harvester	Canada	1837
Peterson, Frank	″			Shoemaker	United States	1875
Potts, Robert S.	″	1	17	Farmer	Canada	1851
Pitman, Albert	Plainfield			Farmer	Canada	1851
Rathburn, H. B. & Son	Mill Point			Manufacturers	United States	
Robertson, James	Melrose	2	1	Farmer and Stone Mason	Scotland	1846
Ray, James	″	3	13	Farmer	Ireland	1838
Roblin, Owen	Shannonville	4	4	Farmer and Proprietor of Kimerley Ashery	Canada	1843
Sheirman, George	″	3	3	Farmer	Canada	1836
Stickney, Fadey	″	1	1	Farmer	Canada	1842
Skelly, W. J.	Blessington	4	3	Farmer	Canada	1851
Scanlon, Charles	Marysville	1	32	Farmer	Canada	1839
Todd, John A.	Blessington	5	7	Farmer	Scotland	1856
Tripp, Samuel	Melrose	1	16	Farmer	Canada	1812
Wright, Joshua	″			Laborer	Ireland	1851
Wilson, Baptist	Shannonville	2	4	Farmer	Canada	1865
Wesse, William	″	1	9	Farmer	Canada	1821
White, John	Roslin	9	4	Foundryman, M.P., and Reeve	Ireland	1850
Whittington, W. S.	Lonsdale	2	28	Farmer	Canada	1844

AMELIASBURGH TOWNSHIP.

NAME	POST OFFICE	Con.	Lot	BUSINESS	NATIVITY	Year Settled in Co.
Anderson, William	Mountain View	3	68,69	Farmer	Canada	1822
Anderson, Levi	Rossmore	1	68,69	Farmer	Canada	1824
Anderson, George	Belleville	2	58	Farmer	Canada	
Anderson, J. W.	Rednersville	1	70	Farmer and Manager Bay Side Cheese Factory	Canada	1833
Ashton, A. R.	″			Butcher	England	1866
Aysworth, R.	Ameliasburgh	4	65	Farmer	Canada	1837
Adams, John	″	4	[illegible]	Farmer	Canada	1837
Anderson, John	″	4	[illegible]	Farmer	Canada	1827
Alyea, J. J.	Rossmore			Hotel Proprietor	Canada	1843
Alyea, Peter	Consecon	3	102	Farmer	Canada	1839
Alyea, John	″	2	99	Farmer and Carpenter	Canada	1825
Brickman, Samuel	Rednersville	1	87, 88	Farmer	Canada	1827
Brickman, W. A.	″	1	84	Farmer	Canada	1837
Byers, John	Consecon			Merchant and General Dealer	Canada	1828
Babbit, Elkanah	Rednersville	1	69	Farmer	Canada	1837
Blood, James	Belleville	2		Farmer	Canada	1837
Benson, James	Ameliasburgh	4	80	P. S. Teacher	Canada	1856
Brooks, Anthony	″	2	92	Farmer	Canada	1825
Beech, P. Y.	Consecon	4	97	Farmer	Canada	1812
Beech, John	″	4	98	Farmer	Canada	1846
Beech, Mills	″			Farmer and Butcher	Canada	1836
Burley, E.	Ameliasburgh	2	87	Farmer	Canada	1837
Burley, F.	″	2	87	Farmer	Canada	1830
Babcock, Robert	Rossmore			Engineer	Canada	1853
Brickman, James	Rednersville	1	82	Farmer	Canada	1813
Bonter, Peter	Ameliasburgh	1	90	Farmer and Township Councillor	Canada	1822
Brickman, W. H.	Rednersville	1	84	Farmer	Canada	1835
Brickman, Asa	″			Farmer	Canada	1842
Bonter, John A.	Murray	2	111	Farmer	Canada	1802
Babcock, John H.	Rednersville	1	80	General Agent and dealer in all kinds of farming implements, Champion Reaper and Mower	Canada	1845
Coleman, Isaiah	Ameliasburgh	3	82	Photographer		1846
Consaul, G. T.	″	3	78	Farmer	Canada	1833
Church, Daniel	Rossmore	1	65	Farmer	Canada	1839
Cunningham, J. R.	Ameliasburgh	1	78	Farmer, Engineer, and Millwright	Canada	1840
Carnrike, Selim	″	4	90	Farmer	Canada	1838
Crouter, James	Consecon			General Dealer in Stoves and Tinware	Canada	1830
Calnan, David	Ameliasburgh			Farmer	Canada	1848

AMELIASBURG TOWNSHIP—Continued.

NAME.	POST OFFICE.	Con.	Lot	BUSINESS.	NATIVITY.	Year Settled in Co.
Crouter, Anthony	Rednersville	1	[illegible]	Farmer and Tinsmith	Canada	1837
Crouter, Abraham	"	1	79	Farmer	Canada	1821
Cummins, James	Consecon	3	100	Farmer	Ireland	1845
Coleman, James	Ameliasburgh	3	82	Blacksmith	United States	1830
Corrigan, R. J.	Murray	1	8	Farmer, Grocer, and P.M.	Canada	1841
[illegible], W. G.	Consecon	3	102	Farmer	Canada	1832
Church, S.	"	2	112	Farmer	Canada	1833
Delong, William	Ameliasburgh	3	[illegible]	Farmer, and Treas. of Ameliasburgh	Canada	
Dempsey, W. R.	Rednersville	1	[illegible]	Farmer, Grain-broker, Reeve of Ameliasburgh, and Warden of Prince Edward Co.	Canada	1832
Dodd, Joshua	Mountain View	4	70	Farmer	Canada	1850
Delong, David	Ameliasburgh	3	[illegible]	Farmer	Canada	1840
Dempsey, P.	Albury	1	91	Farmer	Canada	
Delong, S. E.	Ameliasburgh	3	88	Farmer		1838
Dempsey, Peter	Rednersville	1	91	Farmer	Canada	1816
Denike, C. H.	Murray	2	111	Farmer	Canada	1848
Ervin, William	Consecon	2	96	Farmer	Canada	1837
File, A. J., M.D.	Ameliasburgh			M.D.	Canada	1869
Fox, George H.	Mountain View	4	69	Carpenter and Farmer	Canada	1835
Fones, William A.	Albury	1	96	Farmer and Hopgrower	United States	1820
Gerow, W.	Rossmore			Grocer and P.M.	Canada	1870
Glenn, James E.	Ameliasburgh			P. S. Teacher	Canada	1849
Garretson, Daniel	Consecon	1		Blacksmith, Waggon-maker, and General Jobber	Canada	1840
Gibson, David	Ameliasburgh	3	[illegible]	Farmer, and Agricultural Implement Agent	Canada	
Howell, Griffith	Mountain View	3	70	Farmer	Canada	1814
Howell, John A.	"	2	73	Farmer	Canada	1838
Howell, James R.	"	2	73	Farmer	Canada	1838
Hawley, S. S.	Belleville	2		Farmer	Canada	1820
Huff, Abner H.	"	3	60	Farmer	Canada	1840
Hayes, George, jun.	Consecon			Blacksmith	Canada	1843
Hawley, Charles J.	Ameliasburgh	3	96	Farmer	Canada	1844
Henesy, J. B.	Consecon	3	91	Farmer and Carpenter	Canada	1825
Humphreys, N. M.	"	2	109	Farmer and Carpenter	Canada	1835
Henesy, A. H.	Ameliasburgh	2		Farmer	Canada	1855
Hunt, R. N.	Consecon	2	109	Farmer and Stock Dealer	Canada	1841
Harrington, A. W.	Mountain View			Dealer in Organs and Pianos	Canada	1855
Johnson, J. A.	Consecon			Produce and Commission Merchant, Councillor of Ameliasburgh	Canada	1846
Ketcheson, E. C.	Rednersville			Merchant	Canada	1833
Kemp, Vincent	Consecon	1	31	Farmer	Canada	1841
Killip, W. C.	"			General Merchant (of Osborne & Killip, Merchants)	Canada	1842
Loveless, John	Rossmore	2	60	Farmer	Canada	1805
Laird, J. N.	"			Grocer	Scotland	1876
Lauder, M. M.	Ameliasburgh	3	73	Farmer	Canada	1831
McCullough, John G.	Rossmore			Carpenter	Canada	
McKim, James C.	Ameliasburgh	3		Farmer	Canada	1848
McKibbon, Archibald	"			P. S. Teacher	Canada	1851
Macfee, John, jun.	Brighton			P. S. Teacher	Canada	1859
Manion, J. B.	Rossmore	1	64	Farmer	Canada	1838
Murray, Ralph	"			Engineer	Scotland	
Morrow, Agnes	Consecon	1	91	Farming	Canada	1843
Morrow, Christena	"	1		Farming	Canada	1845
Marsh, A.	"			General Merchant and P.M., Issuer of Marriage Licenses	Canada	1806
Meiggs & Crane	"			General Merchants	England	1866
Moore, D. E.	Ameliasburgh	2		Farmer	Canada	1837
Moore, George	Carrying Place	1	106	Farmer	Canada	1854
Nightingale, Joseph	Belleville	3	6 11	Farmer and Hop Grower	Canada	1858
Osborne, E. J.	Consecon			Miller	England	1824
Osborne, E. J.	"			General Merchant (of Osborne & Killip, Merchants)	Canada	1842
Osborne, John T.	"	3	[illegible]	Farmer	Canada	1809
Onderdonk, John	Albury	1	95	Farmer and Blacksmith	United States	1832
Osborne, Walter J.	Mountain View			Teacher	Canada	1848
Oyer, Nathaniel E.	Consecon	1	102	Farmer	Canada	1850
Potter, John	Mountain View	3	88	Farmer	Canada	1824
Palmer, Henry	Rednersville	1	82	Farmer	Canada	1842
Peck, William	Albury	1	90	Farmer, P.M., and Coroner	Canada	1832
Porter, Robert L.	Consecon			Hotel-keeper	Ireland	
Peck, Francis	Albury	1	92	Farmer	Canada	1825
Peters, W. L.	Rednersville	2	78	Blacksmith	Canada	1863
Potter, S. S.	Mountain View	3	68	Small Fruit Grower	Canada	1849
Parliament, S.	Ameliasburgh	3	72	Farmer	Canada	1825
Parliament, Jacob C.	Consecon	1	[illegible]	Farmer	Canada	1826
Pierson, John	Ameliasburgh	1	86	Farmer	United States	1800
Post, Abraham	Rossmore	1	66	Farmer	Canada	1825
Pierson, Aaron	Albury	1	92	Farmer	Canada	1831
Quackenbush, W. H.	Ameliasburgh	1	71	Farmer	Canada	1836
Ross, George	Rossmore	2	59	Farmer	Ireland	1845
Redner, William H.	Rednersville	1	[illegible]	Farmer	Canada	1845
Redner, James E.	"	1	[illegible]	Farmer	Canada	1837
Robinson, W. J.	Consecon				Ireland	
Ross, Robert B., J.P.	Rossmore	2	[illegible]	Farmer	Ireland	1815
Roblin, Edward	Ameliasburgh			Miller, and Clerk 4th Div. Court	Canada	1836
Richer, Henry B.	"			Book-keeper	Canada	1850
Robson, Rev. M.	"			Minister C. M. Church	Canada	1844
Richards, James	"			Carriage Painter	Canada	1855
Rowe, David	Consecon	2	107 108	Farmer, Carpenter and Jobber	Canada	1826
Sprague, E. & Co.	Ameliasburgh			Carriage-makers		
Sprague, A. & J.	"			Dealers in Dry-goods, Groceries, and Carriage Hardware		
Stafford, W. G.	Rednersville	2	78	Farmer, Tanner and Currier, and Ex. Reeve	Canada	1827
Stapleton, William	Consecon	3	107 108	Farmer	Canada	1819
Sager, Albert L.	Ameliasburgh	2	87	Farmer	Canada	1851
Spencer, Andrew	"	4	89	Farmer	Canada	1831
Shears, Wilson	Consecon	1	100 101	Farmer and Blacksmith	Canada	1852
Snider, James M.	Carrying Place	2	110	Farmer	Canada	1855
Tillotson, T. B.	Mountain View	4	67	Farmer and Carpenter		1816
Thompson, Joseph	Ameliasburgh	4	[illegible]	Farmer	Canada	1851
Tice, Jacob	Rednersville	1	67	Farmer	Canada	1824
Taylor, Gilbert	Murray	2	111 114	Farmer	Canada	1829
Tice, Jason S.	Ameliasburgh			Bailiff	Canada	1855
Webb, H. F.	"			Gentleman, Town Clerk	Canada	1850
Way, S. G.	Mountain View	3	86	Yeoman	Canada	
Weese, W. F.	Rednersville	1	89	Farmer	Canada	1814
Way, W. H.	Mountain View	2	88	P. M. and Carriage-maker	Canada	

ed.

NATIVITY.	Year Settled in Co.
Canada	1837
Canada	1821
Ireland	1845
United States	1830
Canada	1841
Canada	1832
Canada	1833
Canada	
Canada	1832
Canada	1850
Canada	1840
Canada	
	1838
Canada	1816
Canada	1848
Canada	1837
Canada	1869
Canada	1835
United States	1820
Canada	1870
Canada	1849
Canada	1840
Canada	
Canada	1811
Canada	1838
Canada	1838
Canada	1820
Canada	1840
Canada	1843
Canada	1844
Canada	1825
Canada	1835
Canada	1855
Canada	1811
Canada	1855
Canada	1846
Canada	1833
Canada	1841
Canada	1842
Canada	1805
Scotland	1876
Canada	1831
Canada	
Canada	1848
Canada	1851
Canada	1856
Canada	1858
Scotland	
Canada	1843
Canada	1845
Canada	1866
England	1866
Canada	1837
Canada	1854
Canada	1858
England	1824
Canada	1812
Canada	1869
United States	1832
Canada	1848
Canada	1859
Canada	1824
Canada	1842
Canada	1832
Ireland	
Canada	1828
Canada	1882
Canada	1849
Canada	1825
Canada	1826
United States	1833
Canada	1825
Canada	1831
Canada	1836
Ireland	1845
Canada	1845
Canada	1837
Ireland	
Ireland	1815
Canada	1830
Canada	1870
Canada	1844
Canada	1855
Canada	1826
Canada	1827
Canada	1819
Canada	1851
Canada	1831
Canada	1832
Canada	1855
Canada	1816
Canada	1851
Canada	1824
Canada	1820
Canada	1855
Canada	1850
Canada	
Canada	1814
Canada	

AMELIASBURG TOWNSHIP—Continued.

NAME.	POST OFFICE.	Con.	Lot.	BUSINESS.	NATIVITY.	Year Settled in Co.
Watt, Amos	Consecon	3	104		Canada	1849
Wallbridge, A. E.	Belleville	2	55	Farmer	Canada	1817
Wallbridge, J. F.	"	2	56	Farmer	Canada	1851
Wallbridge, T. D.	"			Farmer	Canada	1851
Wallbridge, Elias	"	2	57	Farmer and Drover	Canada	1814
Way, Alpheus	Mountain View	3	4, 72	Farmer and Mechanic	Canada	1818
Weeks, Charles	Consecon	3	103, 105	Farmer	Canada	1842
Wannamaker, C.	Ameliasburgh	4	88	Farmer	Canada	1835
Wood, Abram	"	4	58, 80	Farmer	Scotland	1845
Way, Reuben R.	"	4	92	Farmer, Gardener, Fruit grower, and dealer in shrubs	Canada	1847
Weeks, D. B.	Consecon			Grocer, and Bailiff of Div. Court	Canada	1848
Wyss, W. S.	Rosemore			Hotel-keeper	Canada	1848
Wood, Jacob R.	Ameliasburgh	3	74	Farmer and Blacksmith	Canada	1827
Way, Amos	"	3	71	Farmer, Carriage and General Mechanic	Canada	1841
Weese, G. A.	Rednerville	1	80	Farmer	Canada	1816
Way, Sarah	Ameliasburgh	3	83	Farming	Canada	1866
Wannamaker, J. A.	Carrying Place	1	9	Hotel-keeper	Canada	1829
Wood, William H.	Ameliasburgh	3	75	Farmer	Canada	1853
Weeks, Camilla	Consecon	1	105	Farmer	Canada	1841
Wilkins, R. D. S.	Murray	1	11, 13	Farmer	Canada	1801
Weese, Mrs. Mary C.	Rednerville	1	84	Farming	Canada	
Young, Reuben	Murray	1	1, 2	Farmer and Tanner	Canada	1805
Young, Charles G.	Mountain View	2	72	Farmer	Canada	1845
Young, Oliver	Ameliasburgh	4	67	Farmer and Carpenter	Canada	1845
Young, John	"			Farmer and Carpenter	Canada	1813

ATHOL TOWNSHIP.

NAME.	POST OFFICE.	Con.	Lot.	BUSINESS.	NATIVITY.	Year Settled in Co.
Anderson, Gideon	Cherry Valley	1 S.L.	14	Farmer and Fruit Grower	Canada	
Blakely, W. B.	"	"	8	Farmer, J.P., and Insp'r. of Licenses for Prince Edward Co.		1817
Cooper, John V.	Picton	S.L.	10	Farmer, Fruit Grower and Dealer	Canada	1844
Cole, Luke F.	Cherry Valley	3 S.L.	2	Ex-Councillor of Athol and Farmer	Canada	1833
Cole, Elisha	"	"	2	Farmer and J.P.	United States	1803
Crandall, S. D.	"	1 "	14	Farmer, Fruit Grower and Dealer	Canada	1830
Campney, W. T.	"	"	9	Farmer and Stock Dealer	Canada	1845
Clapp, Allen	"	"	9	Farmer	Canada	1835
Cork, Samuel	"	2 "	11	Farmer	England	1842
Crook, J. H.	Picton	S.L.	3	Farmer and J.P.		1853
Crawford, J. W.	Cherry Valley			Merchant, Dealer in Dry Goods, Groceries, Boots & Shoes, Hardware &c	Canada	1839
Goodwin, Thomas	Cherry Valley	2 S.L.	8	Farmer	Canada	1832
Graham, J. B.	Picton	S.L.	6	Farmer and Stock Dealer	Canada	1857
Hudgin, Alfred A.	Salmon Point			Farmer—Agent for Cossitt's Agric. Imp'ts. Gang Ploughs a Specialty	Canada	
Hubbs, W. S.	Picton	S.L.	10, 11	Farmer	Canada	1839
Hubbs, B. A.	"	"	10, 11	Farmer, Reeve of Athol Tp. Warden of Prince Edward Co., and J.P.	Canada	1823
Huff, Peter	"	1 S.L.		Farmer, Government Light House Keeper, & Fish Inspector, J.P.	Canada	1802
Holsey, John	"	S.L.		Farmer and Miller	England	1873
Insley, S. P.	Cherry Valley			Carriage Maker	Canada	
Ketchum, Eli	Cherry Valley	1 S.L.	10, 11	Farmer, Tp. Councilman and Ex-Reeve, J.P.	Canada	1829
Love, W. M.	Salmon Point	2 "	15	Farmer, Fruit Grower and Grazier	Canada	1810
McDonald, Sylvester	Cherry Valley	S.L.		Farmer and Fruit Dealer	Canada	1850
McCartney, John	Salmon Point	1 S.L.	19	Farmer and Stock Dealer	Canada	1850
Mills, George E.	Cherry Valley	"		Farmer	England	1864
Miller, D.	Salmon Point	"	19	Farmer, Machine Thrashing	Canada	1843
Rankin, Samuel S.	"	"	16	Farmer	Canada	1833
Reid, John, jr.	Milford	2 "	15, 18	Farmer	Canada	1836
Reed, John	Cherry Valley	"	2	Farmer	Canada	
Starks, David	Cherry Valley	3 "	4	Farmer	Canada	1837
Snider, John W.	Salmon Point	1 "		Farmer and Blacksmith	Canada	1840
Starks, J. W.	"	"	16	Farmer and Stock Dealer	Canada	1833
Spencer, W. V.	"	"	18	Farmer, Fruit Grower and Dealer	Canada	1828
Sayers, W. A.	Cherry Valley	"	3	Farmer	Canada	1850
Werden, Eugene B.	Picton	S.L.		Farmer and Miller	Canada	1850
Williams, B. M.	"	"	16	Farmer	Canada	1817
Walmsley, Samuel	Milford	3 S.L.	3	Farmer, Ex Reeve and Councillor	Canada	1833
Wamsley, James	"	"	2	Farmer	Canada	1833
Young, W. H.	Picton	S.L.	4, 6	Farmer	Canada	1831
Young, A. M.	"	S.L.	4	Farmer	Canada	1831
Yarwood, E. B.	Cherry Valley	1 S.L.	8	Farmer	Canada	1825
Young, T. E.	Picton	S.L.	5	Farmer, Proprietor of Young's E. Lake Cheese Factory	Canada	1826

HALLOWELL TOWNSHIP.

NAME	POST OFFICE	Con.	Lot.	BUSINESS	NATIVITY	Year Settled in Co.
Atkin, Francis P.	Picton			Printer	England	1853
Anderson, A. L.	"			Clothier	Canada	
Anning, J. & Co.	"			Stoves, Ploughs, Phenix Foundry	Canada	
Allison, W. H. R.	"			Barrister, &c	Canada	1838
Allison, C. R.	"			Chemist and Druggist	Canada	1833
Bank of Montreal	"					
Bennett, Jabez	"			Agent for Mutual Life Ins. Co., Lancashire, Scottish Imp'l., Standard & P. Edw'd. Fire Ins. Co.'s	England	1856
Bog, Thomas	"			Deputy Registrar and Wharfinger	Canada	
Barker, Bros	"			Machinists & Iron Founders, Manufacturers of Agr'l. Implements	Canada	
Blanchard, W. H.	"			Proprietor Globe Hotel	Canada	
Brennan, Rev. John	"			Priest	Ireland	1875
Bristol, A.	"			General Dealer in Dry Goods, Groceries, Crockery, Glass & Stoneware, lamps, and Chandeliers	Canada	1820
Bowerman, L. V.	Bloomfield	3 W.T.	7	Farmer and Cheese Manufacturer	Canada	1822
Burlingham, C. L.	"	2 N.W.W.L.	18	Farmer	Canada	1821
Brown, Daniel	"			Farmer	Canada	1835
Brown, S. H.	"			Painter and Carpenter	Canada	1853
Branscombe, D. A.	"	3 W.T.		Farmer	Canada	1856
Branscombe, J. M.	"	"		Farmer	Canada	1850

HALLOWELL TOWNSHIP—Continued.

NAME.	POST OFFICE.	Con.	Lot.	BUSINESS.	NATIVITY.	Year Settled in Co.
Bedell, E. W.	"	1		Lady	Canada	1828
Brown, Stewart	"	2	6	Farmer	Canada	1828
Brown, A. F.	"			Carpenter and Joiner	Canada	1855
Bentley, S. B.	Picton	1st E P		Farmer	Canada	1848
Benham, Henry	Bloomfield			Pedlar	Canada	1845
Brisbin, H. F.	Picton			Farmer	Canada	1857
Bowerman, B.	Bloomfield	2nd	10	Farmer & Brick Maker	Canada	1837
Brown, Phoebe Jane	"			Lady	Canada	1842
Clapp, James	Picton			Livery	Canada	1842
Conger, S. M. & Bro.				Publishers	Canada	
Charlton, R. M.	"			Dominion Novelty Co.	Canada	1816
Clapp, G. A.	"			Practical Engineer	Canada	
Cunningham, R.	"	3rd	9	Farmer	Canada	1847
Carson, H. Y.	"			Carpenter, Joiner and Millwright, Contractor	Canada	1851
Close, A. McL.	"			General Contractor and Builder	Canada	1832
Collins, Levi	"			Real Estate Agent	United States	1844
Cooper, Freeman	Bloomfield	1st	5	Farmer	Canada	1842
Cunningham, Lydia	"	"	4	Farming	Canada	1831
Cole, John	West Lake	"	5	Farmer, Carpenter, and Contractor	England	1846
Conley, Edwin	Picton			General Bakery	Canada	1840
Cronk, C. R.	Wellington	1st W W L	4	Farmer	Canada	1836
Clarke, J. G.	"	"		Farmer	Canada	1846
Christy, W. S.	Bloomfield	2nd W W L	10-11	Farmer	Canada	1832
Clinton, William	Wellington	Irvine	Gore	Farmer	Canada	1832
Cooper, W. R.	Bloomfield	2nd W W L	18	Farmer and Hop Grower	Canada	1840
Christie, T. S.	"	1	17	Farmer	Canada	1845
Christy, C. S.	"	1		Farmer	Canada	1847
Colliver, Mrs. Wm.	Picton	1st E P		Lady	Canada	1838
Cole, Jesse	"			Carpenter, Joiner and Constable	England	1825
Carry, D. H.	"	2nd	22	Farmer	Canada	1836
Dunlop, J.	"			Clothier	Ireland	1861
Dunbar, William	"	2nd	17	Farmer	Canada	1834
Dailey, Joseph E.	"			Carpenter and Builder	Canada	1832
Despard, W. P.	"			Banker	Ireland	1870
Dorland, A. M.	Bloomfield	2nd		Farmer	Canada	1854
Dougall, J. F.	Picton			Capitalist	Canada	1820
Draper, A. P.	"			Baker and Confectioner, Wholesale Dl'r in Biscuits—Steamboats &c. Supplied at Reasonable Rates	England	
Dorland, R. A.	Wellington	1st W W L	5-6	Lady	Canada	
Dodge, Frederick	Picton			Secretary R. Q. A. Insurance Co.	Canada	
English, J. L.	"	2nd		Farmer	Canada	1855
Ellsworth, Riley	Bloomfield			Cheese Maker	Canada	1837
Elkington, Timothy	"			Taxidermist	England	1853
Foster, R. A.	Picton			Druggist	Canada	1858
Foughner, Thomas	"			Proprietor Victoria Hotel	Ireland	18[illegible]
Fralick & Bro.	"			Carriage Factory	Canada ... 1819	1831
Fenemore, John	Bloomfield			Farmer	England	1839
Francis, J.	Picton			Carpenter	Ireland	1863
Frederick, William	"		7	Farmer	Canada	1854
Fretz, James W.	"			Farmer	Canada	1845
Freleigh, John	Bloomfield	2nd		Farmer	Canada	1839
Gilbert, J. S.	Picton			Manufacturer and Dealer in Furniture and Coffins	Canada	1833
					Canada	1833
Garsoline, R. C.	"			Farmer and Capitalist	Canada	1843
Garratt, John	Wellington	1st W W L	3	Farmer	Canada	1824
Gates, Y.	Bloomfield			Farmer	Canada	1846
Gatew, Albert	"	2nd	19	Farmer	Canada	1840
German, E. C.	Picton			Captain of Steamer S. Y.	Canada	1877
Harper, R. H.	"			Builder	Canada	1870
Hazard, E. B.	"	3rd	13	Farmer, Cheese Factory	Canada	1821
Hadden, J. T.	"			Carriage Factory	United States	1872
Harris, W. H.	"			Farmer and Horse Dealer	Canada	1851
Harris, Nelson	West Lake	S W L	9	Farmer	United States	1842
Hyatt, A. W.	"	Gore	11	Farmer and Capitalist	Canada	1845
Hepburn, A. W.	Picton			Gentleman	Canada	1856
Huff, Lysander	"	2F	62-63	Farmer and School Teacher	Canada	1832
Hollingsworth, W. S.	Wellington			Manufacturer and Dealer in Stoves, Tinware, Copper and Sheet Iron, Pumps, Sinks, Lead Pipe, Coal Oil and Eve-Troughing & Jobbing	Ireland	1857
Hubbs, Hayland	Bloomfield	1st W W L	11	Farmer	Canada	1847
Hubbs, Thomas H.	"	Green	Gore	Farmer	Canada	1843
Hill, Cornelius	Picton	2nd		Farmer	Canada	1821
Hubbs, Henry	Bloomfield	1st	2-3	Deputy-Reeve, Retired Farmer	Canada	1822
Hamell, Peter	Picton			Farmer	Canada	1840
Hubbs, O. C.	"			Photographer	Canada	1858
Insley, James E.	"			Brick and Stone Mason	Canada	
Ingersoll, J. F., M.D.	"			Physician and Surgeon	Canada	1849
Ilbey, John	"			General Dealer, Stoves, Hardware, and Tin—Hides, Pelts and Wool	Canada	1841
Insley, Ralph	Bloomfield			Carpenter and Cheese Maker	Canada	1857
Jackson, Rella	"			General Blacksmith	Canada	1822
Kelly, David	Picton	2nd		Farmer	Canada	1858
Kingsley, G.	"	Gore	8	Farmer	Canada	1843
Little, J. P.	"			Agent for Oil Paintings	United States	1876
Low, P.	"			Q. C.	Jersey Island	1835
Leach, Delos W.	"			General Fruit Tree Dealer, Agent for Agric. Imps. & Sewing Mach's	United States	1877
Lake, Stephen	West Lake	S W L	5	Apple Dealer	Canada	
Lake, Richard	Picton			Hotel Keeper	Canada	1874
Lawson, William	"			Saddle and Harness Maker, Trunks, Blankets, Whips, &c.	Canada	1876
Lear, John G.	"			Farmer	Canada	
Love, James	"			General Contractor and Builder, Sash Door Blinds, &c.	Canada	1834
Love, Samuel	Picton			Contractor, Doors, Sash Blinds, &c.	Canada	1836
Lazier, Abram	Picton	2 S W C P		Farmer	Canada	1792
Leavitt, D. A.	"	3rd	8	Farmer	United States	1836
Loveys, L. T., J.P.	Bloomfield	"	5	Farmer, and Breeder of Thorough Bred Horses	Canada	1827
McKee, William	Picton			Carpenter and Joiner	Canada	
McMullen, B. C.	"			Publisher	Canada	
McKenny, R. W.	"			Veterinary Surgeon	Canada	
McKinley, S.	"			Fruit Grower	Scotland	1872
McCuaig, J. S.	"			Steamboat Owner	Canada	1849
McDonald, C. A.	"			General Dl'r in Groceries, Grain, Provisions, &c., Agt for C. L. Ins. Co.	Canada	1830
McDonald, George	Wellington	1st W W L	9	Farmer	Canada	1839
McFaul, A.	"	2nd	4	Farmer, Partner wanted to Develope Salt Spring on the Farm	Canada	1831
McDonald, H. R.	Bloomfield	1st		Farmer	Canada	1825
Merrill, Edward	Picton			Barrister at Law	Canada	1841
Mastin, W.	"			Agent Standard Bank	Canada	

NAME.
McKenzie, Walter
Murney, J. H.
Morden, C. D.
Martin, Mrs. John
Morgan, T. H.
Murphy, D. G.
Mulholland, D.
Miller, J. P.
Murney, John
Maybee, W. J.
Nash, J. P., M.D.
Nixon, William
Newman, Samuel
Owens, William
Parker, Edward
Pettet, Sarah
Platt, C. F.
Pettit, D. H.
Platt, J. B.
Pettingill, C. M.
Rawson, John A.
Roblin, R. S.
Roblin, J. J.
Rorke, James H.
Richards, W. M.
Ross, Walter, Jr.
Richards, John
Scott, W. V.
Smith, J. D.
Sheridan, Thomas
Sexsmith, W. G.
Sexsmith, J. N.
Spencer, J. R.
Southard, Stephen
Striker, Gideon
Stinson, L. D.
Snider, J. D.
Striker, Miss M.
Shortridge, T.
Scott, Edward
Stinson, F. R.
Thomas, William
Taylor, Francis
Thorn, James P.
Tubbs, L. B.
Terilligar, J. C.
Tubbs, Isaiah
Tubbs, J. R.
Tripp, W. H.
Terwilligar, O. C.
Talcott, J. W.
Tickels, W. F.
Townsend, John
Thorn, Mrs. M.
Vorce, L. T.
Vanpatten, P. J.
Vanbarcom, W.
Vance, J. M.
Valliers, John
Vanderwater, P.
Wilson, D.
Wanamaker, J. H.
Wilson, Thomas
Welsh, J. W.
Wilson, J. C.
Washburn, P.
Wait, Isaac A.
White, S. H.
Williamson, Miss
Williamson, Arti
White, R. D.
Yeres, William C.
Yeres, B. P.
Yeres, Jacob
Yeats, S.
Young, Thomas
Young, G. D.

NAME
Ainsworth, Fran
Ainsworth, Phil
Adams, W. H.
Arthur, Daniel
Anderson, Asen
Arthur, James
Arthur, T. J.
Burr, Peter
Bowerman, Step
Brown, J. T.

…TIVITY.	Year Settled in Co.
…la	1828
…a	1828
…la	1855
…la	1848
…la	1815
…la	1857
…la	1867
…la	1842
…la	1842
…la	
…la	1816
…ia	
…ia	1847
…la	1861
…la	1832
…d States	1841
…da	1842
…da	1831
…and	1845
…da	1840
…ds	1836
…da	1840
…da	1832
…da	1832
…da	1840
…da	1845
…da	1847
…da	1828
…and	1825
…da	1836
…nd	1861
…da	1831
…da	1832
…nd	1870
…da	1854
…da	1830
…and	
…ada	
…ada	
…ada	1855
…ada	1837
…land	1822
…ada	1858
…and	1868
…ada … 1839	1831
…land	1839
…and	1863
…ada	1851
…ada	1845
…ada	1839
…ada	1833
…ada	1833
…ada	1813
…ada	1821
…ada	1846
…ada	1840
…ada	1877
…ada	1870
…ada	1821
…ted States	1872
…ada	1851
…ted States	1812
…ada	1845
…ada	1852
…ada	1852
…and	1857
…ada	1847
…ada	1843
…ada	1821
…ada	1822
…ada	1840
…ada	1856
…ada	
…ada	1840
…ada	1841
…ada	1857
…ada	1822
…ada	1850
…ada	1813
…ted States	1876
…sey Island	1835
…ited States	1877
…ada	
…ada	1874
…ada	1876
…ada	
…ada	1834
…ada	1836
…ada	1792
…ited States	1836
…ada	1827
…ada	
…ada	
…ada	
…tland	1872
…ada	1819
…ada	1830
…ada	1860
…ada	1833
…ada	1825
…ada	1841
…ada	

HALLOWELL TOWNSHIP—Continued.

NAME.	POST OFFICE.	Con.	Lot.	BUSINESS.	NATIVITY.	Year Settled in Co.
McKenzie, Walter	Picton			Registrar	Scotland	1818
Murney, J. H.	"			Clothier	Canada	
Morden, C. D.	"			Auctioneer, Mili'ry & Fancy Goods	Canada	
Mattin, Mrs. John	Bloomfield	1st		Farming, Lumber Sawing & Manufacturing	Canada	1821
Morgan, T. H.	"			Farmer, Miller, and General Dealer Township Clerk of Hallowell	Canada	1841
Murphy, D. G.	Wellington	2N W H	7	Farmer	Ireland	1826
Mulholland, D.	Picton	Gore	K	Farmer	Ireland	1818
Miller, J. P.	"			Farmer	England	1819
Murney, John	"			Gentleman	Canada	1862
Maybee, W. J.	Allisonville	1st		Farmer and Cheese Maker	Canada	1836
Nash, J. P., M.D.	Picton			Physician and Surgeon	Canada	1838
Nixon, William	Bloomfield			Saw Mill for Lumber and Shingles, Best Cider Mill in Co. Feed	Canada	1836
Newman, Samuel	Picton			Farmer	Canada	1851
Owens, William	"			Lumber Dealer and Potash Manufr, Capt. of Reserve Militia, Ex-Mayor and Councilman	Ireland	1837
Parker, Edward	Bloomfield			Brick Manufacturer, from ½ Million to 1 Million per Annum	England	1862
Pettet, Sarah	West Lake	S W L	16	Farming	Canada	1824
Platt, C. E.	"	"	3	Farmer, and Stock Dealer—Horses a Specialty	Canada	1845
Pettit, D. H.	"	"	1	Farmer	Canada	1847
Platt, J. B.	"	"	3	Farmer	Canada	1845
Pettengill, C. M.	Picton			Carpenter	United States	1819
Rawson, John A.	"			Montreal Telegraph Co.	Canada	
Robin, R. S.	"			Barrister	Canada	1816
Robin, J. J.	"			General Blacksmith, Ship Work	Canada	1844
Rorke, James H.	"			Cabinetmaker	Canada	1851
Richards, W. M.	Bloomfield			Farmer	Canada	1838
Ross, Walter, jr.	Picton			Merchant	Canada	1847
Richards, John	"			Hardware, Stoves & Tinware—Hot Air Fitting a specialty, 32 Main st.	[illegible]	
Scott, W. N.	"			Printer and Publisher	Canada	
Smith, J. D.	"			Agent Dominion Telegraph Co.	Canada	
Sheridan, Thomas	"			Stonecutter	Canada	1876
Sexsmith, W. G.	"	2N W L		Farmer	Canada	1851
Sexsmith, J. N.	"	"		Farmer	Canada	1849
Spencer, J. B.	"	1 "		Farmer	Canada	1822
Southard, Stephen	"	3rd	14-15	Farmer	Canada	1854
Striker, Gideon	"			M. P. P.	Canada	
Stinson, L. B.	Bloomfield	1 S W H	13	Farmer and Reeve of Hallowell	Canada	1819
Snider, J. D.	Picton	1 S C P		Farmer	Canada	1836
Striker, Miss M.	Bloomfield	2nd	1	Lady	Canada	1833
Shoobridge, T., sr.	"	"	1	Farmer and Hop Grower	England	1843
Scott, Edward	"			Teacher, General Dealer in Groceries, and Temperance House	Canada	1842
Stinson, F. B.	Picton	1 S W W L	15	Farmer	Canada	1838
Thomas, William	"			Hotel Clerk	England	1867
Taylor, Francis	"			Butcher	Canada	1834
Thorn, James P.	"			Merchant and Farmer	Canada	
Tubbs, L. B.	"	2 S W L P		Farmer and School Teacher, Ex-Assessor of Hallowell	Canada	1813
Terilligar, J. C.	"	3rd	16	Farmer, Stock Dealer and Producer	Canada	1830
Tubbs, Isaiah	West Lake	2 S W L	8	Farmer and Cheese Manufacturer	Canada	1818
Tubbs, J. R.	"	"	7	Farmer and Stock Dealer	Canada	1841
Tripp, W. H.	Picton			Farmer	Canada	1832
Terwilligar, D. C.	Bloomfield	Gore	E	Farmer	Canada	1842
Talcott, J. W.	"	Point	M	Farmer	Canada	1848
Tickels, W. E.	Picton			Farmer	Canada	1845
Townsend, John	Bloomfield			Painter	Canada	1813
Thorn, Mrs. M. B.	Picton			Lady	Canada	1823
Votre, E. T.	"			Livery and Boarding Stables	United States	1853
Vanpatten, P. J.	"			Proprietor Picton Hotel	Canada	
Vanblaricom, W. H.	"			Farmer	Canada	1853
Vance, F. M.	"			Farmer	Canada	1831
Valliere, John	"			General Cooper	Canada	1832
Vanderwater, P. C.	Wellington	2 N W L	4 5	Farmer	Canada	1875
Wilson, D.	Picton			Grocer	Ireland	1847
Wanamaker, I. H.	"			Travelling Agent for Royal Sewing Machine	Canada	
Wilson, Thomas	"			Proprietor Tichbourne Hotel	Canada	1875
Welsh, J. W.	"			Dentist	Canada	
Wilson, J. C.	"			Proprietor Mountain Mills, Little Giant Water Wheel, Shafting and Pulleys, General Foundry, & Machine Manufacturer		1833
Washburn, P.	"			Gentleman	Canada	1814
Watt, Isaac A.	"			Manufacturer and Dealer in Tin, Sheet Iron and Copperware, Stoves, Pumps, Bird Cages, Shelf Hardware, and Fishing Tackle	Canada	1847
White, S. H.	Bloomfield			Farmer	Canada	1852
Williamson, Miss M. E.	Picton			Teacher	Canada	1856
Williamson, Arthur	"	2 S W R		Farmer	Canada	1847
White, R. D.	"			Commission and Grain Dealer	England	1843
Yeres, William C.	"	1 S W L		Farmer	Canada	
Yeres, B. P.	"	2 "		Farmer	Canada	
Yeres, Jacob	"			Blacksmith	Canada	
Yous, S. S.	Bloomfield	2nd	1	Farmer and Fruit Grower	Canada	1825
Young, Thomas	Picton			Farmer	Scotland	1841
Young, G. D.	"			Painter	Canada	1856

HILLIER TOWNSHIP.

NAME	POST OFFICE	Con.	Lot	BUSINESS.	NATIVITY.	Year Settled in Co.
Ainsworth, Franklin	Allisonville	5	78	Farmer	Canada	1850
Ainsworth, Philip	"	6	67	Farmer	Canada	1835
Adams, W. H.	Melville	5	80	Farmer	Canada	1855
Arthur, Daniel	Consecon	Stinson's Block		Farmer	Canada	1844
Anderson, Asenath	"	6		Lady	Canada	1840
Arthur, James	"	5	102	Farmer	Ireland	1819
Arthur, T. J.	Hillier			Carriage maker	Canada	1844
Barr, Peter	Crofton	A	73	Farmer	Canada	1831
Bowerman, Stephen	Wellington			Telegraph Operator, dealer in Groceries and Plaster	Canada	1835
Brown, J. T.	"			General Insurance Agent and dealer in Fruit-trees	Canada	1827

HILLIER TOWNSHIP—Continued.

NAME	POST OFFICE.	Con.	Lot.	BUSINESS.	NATIVITY.	Year Settled in Co.
Bowerman, I.	Wellington	1	10	Farmer	Canada	1824
Babbit, A. E.	Hillier	2		Farmer and Hop-grower	Canada	1847
Bowerman, J. C.	Wellington	1	11	Farmer	Canada	
Calnan, James	Allisonville	6	68	Farmer	Canada	1858
Campbell, D.	Wellington			General Store, Reeve	Scotland	1851
Clapp, Cornelius	"			Retired from business	Canada	1818
Chadsey, L. A.	"	1	9	Farmer	Canada	1822
Caturits, J. S.	Melville	5	80	Farmer	United States	1820
Cameron, J. P.	Hillier	3	32	Farmer	Canada	1858
Clark, B. L.	"	3		Farmer	Canada	1843
Crippen, John	"	3		General Blacksmith and Carriage-maker	Canada	1812
Crouk, Philip	Wellington	1	13	Farmer	Canada	1830
Day, John G.	"			Agent for Bay of Quinte Fire Ins. Co. and Canada Life	England	1809
Doolittle, Smith	Crofton	5	72	Farmer	Canada	1820
Doolittle, Stephen P.	"	5	72	Farmer	Canada	1825
Dorland, J. T.	Wellington			Farmer and Treasurer	Canada	1812
Dunning, Henry	"			Carriage-maker and General Blacksmith	Canada	1831
Davidson, J. C.	Hillier	2	[illegible]	Farmer	Canada	1839
Dulmaige, J. H.	Melville			School-teacher	Canada	1858
Davison, J. G.	Hillier	3		General dealer in Agricultural and Musical Instruments	Canada	1852
Ellis, S. H.	Wellington	2	2	Farmer	Canada	1819
Foster, Theodore	Hillier	2	29	Farmer, Carpenter and Joiner	Canada	1831
Garratt, Amos	Wellington			Cabinet-maker and General Undertaker	Canada	1830
Garrett, Townsend	"			Innkeeper	Canada	1842
Giles, Thomas	"	1	6	Farmer	Canada	1821
Garrett, J. Y.	"	1	25	Farmer	Canada	1849
Guayo, John	Rosehall	1	17	Farmer	Canada	1846
German, J. W.	"	2	21	Farmer	Canada	1804
Gordon, Samuel	Hillier	2	23	Farmer	Canada	1841
Graydon, John	"			School Teacher	Ireland	1864
Greer, Henry	Wellington	1	10	Farmer and Councilman	Canada	1831
Garrett, Edwin	Rosehall	1		Farmer	Canada	1823
Hicks, R. C.	Allisonville	6	73	Farmer	United States	1800
Herrington, G. H.	"	6	65	Farmer	Canada	1832
Herrington, G. W.	Wellington	1	1	Farmer	Canada	1823
Hutchinson, H.	"	1	8	Farmer	Canada	1842
Howe, Daniel	Melville	5	81	Farmer and Councillor	Canada	1835
Howe, J. H.	Consecon	Stinson's Block		Farmer	Canada	1840
Haight, D. P.	Wellington	2	4	Farmer. This property for sale, 160a	Canada	1813
Haight, Mrs. M.	"	2	2	Lady	Canada	1824
Haight, B. L.	"			Farmer	Canada	1857
Hubbs, R. G.	Hillier	2	15	Farmer	Canada	1810
Isted, Norman	Allisonville	6		Farmer	Canada	1850
Jones, Robert S.	Hillier	6	98	Farmer and Councilman	Canada	1830
Jones, Samuel	"	2	25	Farmer and Township Clerk	Canada	1821
Jones, James	Consecon	6	99	Farmer	Canada	1829
Jones, George R.	Hillier	3	18	Farmer	Canada	
Kemp, Asa	Consecon	Stinson's Block		Farmer	Canada	1822
Lambert, Isaac	Crofton	5	71	Farmer	Canada	1818
Lane, John K.	Wellington	1	3	Farmer	Canada	1831
Lyons, Lawrence	"		5	Contractor, Builder, and Cheese-maker	Canada	1837
Locie, James P.	Consecon	Stinson's Block		Farmer, Carpenter, and Joiner	Canada	1822
Leavens, Stephen	Allisonville	3	4	Farmer	Canada	1841
Lloyd, Jonas	Wellington	2		Farmer	Canada	1842
Lane, Gideon A.	"	2		Carpenter	Canada	1832
McDonald, Amos	"			Farmer, and dealer in Agricultural Implements	Canada	1818
McDonald, Albert	Consecon	Stinson's Block		Farmer	Canada	1849
McFaul, Nelson R.	Wellington	2	8	Farmer	Canada	1835
McCartney, Robert	Rosehall	1	16	Farmer	Canada	1836
Morden, Joseph F.	Allisonville	6	66	Farmer	Canada	1826
Marven, Stephen	Consecon			Painter	Canada	1825
Noxon, Dorland	Allisonville	6	76	Farmer, Vine-grower and Wine Manufacturer, medal and diploma International Exhibition, Philadelphia, U.S., 1876	Canada	1867
Nethery, L.	Wellington	5	80	Farmer	United States	1827
Noxon, James E.	Hillier	2		Farmer	Canada	1846
Niles, S. P.	"	2	13	Farmer, Treasurer, and Ex. Warden	Canada	1825
Nease, Stephen	Wellington	1	6	Farmer	Canada	1826
Osterhout, D. A.	"	2	10	Farmer	Canada	
Osterhout, John D.	Rosehall	1	18	Miller	Canada	
Pierson, Joseph	Consecon	3	21	Farmer, and Ex. Reeve of Hillier	Canada	1812
Pearsall, Robert B.	Crofton	5	65	Farmer	Canada	1829
Pearsall, Benjamin	"	5	65	Farmer	United States	1806
Pearsall, James S.	"	5	66	Farmer	Canada	1832
Partill, E.	"	5	66	Farmer	Canada	1849
Partill, Thomas	"	5	67	Farmer	Canada	1849
Pine, Benjamin P.	"	5	67	Farmer	United States	1800
Pearsall, Lemuel	"	5	68	Farmer and Hop-grower	Canada	1830
Petterson, Henry	Allisonville	6	72	Farmer	Canada	1821
Pettengill, Wilson	Wellington	1	12	Farmer	Canada	1838
Pettingell, Henry M.	Rosehall	1	17	Farmer	Canada	1844
Pierson, James	Consecon	3	[illegible]	Farmer	Canada	1841
Plumton, John	"	5		Farmer	Canada	1813
Pettit, Daniel	Allisonville	3	2	Farmer	Canada	1820
Peterson, Allan	Hillier	3		Farmer	Canada	1823
Platt, Edwin	"	2	21	Farmer	Canada	1836
Pye, Robert	"	3	21	Farmer, Carriage maker, and Blacksmith	Canada	1826
Robins, James T.	Allisonville	5		Farmer and Grain-dealer	Canada	1822
Raynor, George	Rosehall	1	18	Farmer	Canada	1834
Rogers, Robert	Consecon	Stinson's Block		Farmer	Canada	1815
Raynor, George H.	Hillier			Miller	Canada	
Reynolds, Samuel	Wellington	1		Farmer	Canada	1854
Savage, Capt. James	"			Hotel-keeper	Ireland	1840
Spencer, David H.	Rosehall	1	20	Farmer, Councilman, and J.P.	Canada	1846
Spencer, Sarah	"	1	19	Lady	Canada	1831
Smith, William	Consecon	Stinson's Block		Farmer and Cheese Factory	Canada	1823
Stapleton, Joseph P.	Hillier	3	23	Farmer	Canada	1815
Stapleton, John E.	"	3	23	Farmer	Canada	1847
Simpson, W. W.	"			General dealer in Dry Goods and Groceries, Telegraph Operator	Canada	1825

NATIVITY	Year Settled in Co.
nada	1824
nada	1847
nada	
nada	1838
gland	1851
nada	1848
nada	1822
ited States	1829
nada	1855
nada	1843
nada	1842
nada	1830
gland	1869
nada	1820
nada	1829
nada	1812
nada	1831
nada	1839
nada	1858
nada	1852
nada	1819
nada	1834
nada	1830
nada	1842
nada	1824
nada	1840
ada	1816
ada	1804
ada	1844
and	1804
ada	1831
ada	1819
ted States	1864
ada	1832
ada	1825
ada	1842
ada	1825
ada	1840
ada	1812
ada	1824
ada	1857
ada	1810
ada	1830
ada	1836
ada	1821
ada	1829
ada	
ada	1822
ada	1812
ada	1831
ada	1837
ada	1822
ada	1841
ada	1842
ada	1832
ada	1848
ada	1849
ada	1835
ada	1836
ada	1826
ada	1825
eda	1867
ed States	1827
ada	1846
ada	1825
ada	1856
ada	
ada	
da	1812
da	1833
ed States	1806
la	1832
da	1849
da	1849
ed States	1800
da	1830
da	1821
da	1858
da	1844
da	1841
da	1843
da	1820
da	1833
da	1836
da	1826
da	1822
da	1834
da	1815
da	
da	1854
nd	1840
da	1846
da	1831
da	1823
da	1845
da	1847
da	1825

HILLIER TOWNSHIP—Continued.

NAME.	POST OFFICE.	Con	Lot.	BUSINESS.	NATIVITY.	Year Settled in Co.
Terey, Harvey	Allisonville	3	4	Farmer and Miller	Canada	1830
Thoresson, James	Consecon			Blacksmith	Canada	1855
Vallean, Irvin P.	Allisonville	5	73	Farmer	Canada	1840
Vallean, Hiram B.	"	5	74	Farmer	Canada	1838
Vermilyea, Peter H.	Rosehall	1	17	School Teacher	Canada	1855
Vanalstine, Henry J.	"			Carpenter and Joiner	Canada	1844
Vandeaf, Joseph	Allisonville	4		Farmer	Canada	1812
Williams, D. V., M D.	Crofton	5	69	Farmer, Dealer and Manufacturer of Medicine	Canada	1821
Waring, Thomas	Allisonville	3	3	Farmer	Canada	1850
Young, John	Hillier	3	30	Farmer and Reeve of Hillier	Canada	1826
Young, G. H. F.	Allisonville	5	75	Hotel-keeper and General Blacks'h.	Canada	1843
Young, John H	Consecon	Stinson's Block		Farmer	Canada	1849

NORTH MARYSBURGH.

NAME.	POST OFFICE.	Con.	Lot.	BUSINESS.	NATIVITY.	Year Settled in Co.
Brown, A. W	Cressy	n s	38	Farmer and J. P.	Canada	1842
Caven, Allen	Picton	2 s w g	6	Farmer	Canada	1860
Caven, John	"	"	"	Farmer	Canada	1832
Carson, Alex S	"	"	8	Farmer	Canada	
Davison, Robert	"	Gore	A	Farmer	Canada	1859
German, G. H.	Prinyer	e r s	35	Post Master and General Store	Canada	1825
Hurlbut, G. C	Cressy	e l s	3	Farmer	United States	1854
Hefferman, Jeremiah	Picton	1 s w g	2	Farmer	Ireland	1861
Kerr, William	Waupoos	e l s	11	Farmer	Canada	1825
Kerr, D. W	Cressy	"	16	Farmer	Canada	
Minaker, James C	Picton	2 s w g	4	Farmer	Canada	1830
Moore, Samuel	"	2 s w g	10	Farmer	Ireland	1858
Prinyer, John	Prinyer	n s	29	Farmer, Custom House Officer and ex-Warden	Canada	1813
Powers, H. A	Cressy	e l s	2	Teacher, Township Clerk	Canada	1842
Pierce, Robert	Picton	Smith's Bay S		Farmer and Carpenter	Canada	1822
Rose, Frederick	Waupoos	e l s	16	Farmer	Canada	1819
Williams, Levi	"	"	13	Farmer and Reeve	Canada	1844
Wright, W. Jas	Cressy	"		Farmer	Canada	
Wilson, J. C	Picton	e r s		Merchant and Mill owner		
Williamson, R. J	"	2 s w g	11	Farmer	Canada	1818
Wright, E. W	Cressy	e l s	5	Farmer	Canada	1811

SOUTH MARYSBURGH.

NAME	POST OFFICE	Con.	Lot.	BUSINESS	NATIVITY	Year Settled in Co.
Ackerman, Richard	Milford	s r b c	14	Farmer and J. P.	Canada	1839
Bond, Moses	"			Hotel Proprietor	Canada	1853
Borley, Emerson	"			Harness Maker	Canada	1828
Clapp, Robert	"	1 s m	26	U. S. Consul—Ex-Reeve and Ex Warden	Canada	1830
Collier, Solomon	South Bay	s r b c	6	Farmer and Township Reeve	Canada	1822
Cunningham, C. S	Milford			Hardware, Stove & Tin Merchant	Canada	1854
Clapp, Samuel	"	1 s m	26	Farmer and Miller	Canada	1844
Cooper, James	"	1 s m	25	General Merchant	Ireland	1843
Church, A. G.	"	"	"	Farmer	Canada	1844
Clarke, George	"	"	20	Farmer	Canada	1843
Clark, Andrew M	"	s r b c	18	Farmer	Canada	1849
Dodge, Frank	"	1 s m	10	Farmer and General Dealer	Canada	1855
Dodge, Theodore	"			Merchant Miller	Canada	1844
Danard, Byron M.	"	1 s m	26	Farmer	Canada	1831
Dulmage, A. C	"	E B		Farmer and Drover	Canada	1834
Dulmage, Thomas	"	"		Farmer	Canada	1829
Demore, A. V	"	"		Farmer	Canada	1849
Danard, Charles W	"	1 s m	29	Farmer	Canada	1818
Ellis, Horatio, N	"	1 s m	16	Farmer	Canada	1852
Farrington, G. P	"	Gore	A	Farmer	Canada	1822
Fegan, John W	"	1 s m	21	Joiner and Contractor	Canada	18[illegible]
Grimmon, Edward	Cardwell	1 s m	29	Farmer	Canada	1858
Grimmon, A. E	"	"	9	Merchant and Postmaster	Canada	1841
Graham, R. J.	Milford	1 s m	25	Blacksmith	Canada	1872
Hubbs, Benjamin	Cardwell	1 s m	3	Farmer and Councilman	Canada	1855
Hughes, Joseph	"	"	12	Farmer	Canada	1837
Haight, H. H	Milford	"	25	Clerk of Division Court	Canada	1830
Jenkin, Samuel	"			Waggon Maker	England	1856
Keys, John	Cardwell	1 s m	8	Farmer	Ireland	1842
Love, George	Milford	"	15	Farmer	Canada	1838
Love, Alexander	"	"	13	Farmer	Canada	1838
Lane, W. H.	"	1 s m	22	Farmer	Canada	1829
McCartney, Carlton	"	"	30	Farmer, Councilman and Master of Grange Lodge	Canada	1823
McKenna, James	Cardwell	1 s m	11	Farmer	Canada	1858
McCaw, William	Milford	1 s m	28	Farmer	Canada	1820
Miller, John S	"	1 s m	17	Farmer	Ireland	1825
Minaker, Mary Ann	"	"	19	Farming	Canada	1843
Martin, Joseph	"	"	30	Farmer	Canada	1838
Minaker, Albert H.	"	1 s m	29	Farmer	Canada	1841
Minaker, John	"	"	25	Farmer	Canada	1861
Metcalfe, Henry	"	s r b c	17	Farmer, Proprietor of South Bay Cheese Factory, and agent for Clifford Gang Ploughs	Canada	
Ostrander, R. G	"	Gore	A	Farmer	Canada	1840
Ostrander, Oliver	Point Traverse	s r b c	11	Farmer	Canada	1822
Palmateer, D. B	"	t p		Farmer	Canada	1860
Palmateer, W. B	"	s r b c	9	Farmer and Mariner	Canada	1845
Richard, H. P	Milford		16	Marble Dealer	United States	
Rose, Andrew	Cardwell	1 s m		Teacher	Canada	1856
Stephens, O. B	Cherry Valley	"	31	Farmer	Canada	1823
Striker, I. J.	Milford	n b		Farmer and Cheese Manufacturer	Canada	1828
Vandusen, John	Cardwell	1 s m	5	Farmer	Canada	1832
Vance, John W.	Milford	"	22	Farmer	Canada	1844
Vanvlack, Stephen	"	"	28	Farmer	Canada	1844
Van Alstine, John	"	"	"	Farmer and Director of Prince Edward Insurance Co.	Canada	1832
Wellbanks, Thomas	"	1 s m	20	Farmer	Canada	1832
Wright, Anna	South Bay	s r b c	5	Teacher	Canada	1848
Walter, John	Point Traverse	t p	14	Farmer, J. P. and Councilman	Canada	1827

SOPHIASBURGH TOWNSHIP.

NAME.	POST OFFICE.	Con.	Lot.	BUSINESS.	NATIVITY.	Year Settled in Co.
Aman, G. M.	North Port	W G P		Farmer	Canada	1853
Bolter, Wellington	Demorestville	1 "	37	Farmer, Valuator, Landed Banking and Loan Co., Hamilton, General Agent Mutual Life Assurance Co., Mortgages bought	Canada	1838
Barager, Samuel	"	B. Isl'd	32	Farmer and Grain Dealer	Canada	1831
Baker, William	"			Carriage and Chair Manufacturer	England	1843
Button, Peter B.	Gilbert's Mills	2 E P	62	General Manufacturer and dealer in Lumber, Staves and Shingles—Farmer and Engineer	United States	1847
Black, John	Demorestville	1 W G P	51-52	Farmer	Scotland	1817
Brooks, S. R.	North Port	"	27	Farmer	United States	1837
Brooks, R. A.	"	"	14	Farmer	Canada	1844
Brown, A. S.	Demorestville	Gore	D	Choice Apple producer, Premium granted on product of 1876—Fine Horses	Canada	1857
Benson, R. D.	Picton	"	"	Farmer	Canada	1805
Benson, W. A.	North Port			Farmer	Canada	
Bradley, James	Picton	2 W G P		Farmer	Canada	1837
Benson, Matthew	"	"	14-15	Farmer and Hop Grower	Canada	1830
Benson, Jacob	"	"	16	Farmer and large Hop Grower	Canada	1840
Benson, Richard	"	"	15	Farmer & Prop. of Elmbrook Fact'y	Canada	1842
Brickman, William	Green Point	1 "	13	Farmer	Canada	1844
Coolidge, L. A.	Demorestville	1 W G P	37	Farmer and large Hop Grower	United States	1841
Carr, Conrad	"	B. Isl'd	6	Farmer and Gen. Agent for Kirby Reaper	Canada	1831
Crawford, John	"	3	6	Farmer	Ireland	1836
Coolidge, A. R.	"	2 W G P	38	Farmer, Hop and Fruit Grower	Canada	1847
Cotter, Samuel J.	North Port	1 "	[illegible]	Farmer and Grape Grower	Canada	1851
Crysdale, Billa	"			General Merchant, Dry Goods, Clothing, Boots and Shoes, &c.	Canada	1848
Cronk, Peter	"	1 W G P	12	Farmer	Canada	1841
Cronk, A. M.	Picton	1 W G P	28	Farmer and Stock dealer	Canada	1830
Cronk, J. B.	North Port	1 W G P	54	Farmer	Canada	1818
Dunning, A. C.	Demorestville	B. Isl'd	5	Farmer	Canada	1825
Davis, Allen	Picton	2 W G P	12	Farmer	Canada	1824
Doxsee, A. K.	"	2 W G P	3	Farmer	Canada	1811
Dorland, E. T.	"	"		Farmer	Canada	1840
Dunning, George	Demorestville			Dealer in Dry Goods, Groceries, Hardware, Boots and Shoes, and Hop producer, and P. M.	Canada	1855
Demill, P. I.	North Port	1 W G P	26	Farmer	Canada	1807
Dewey, Edward	"	"	80	Farmer and ex-Dep. Reeve of the Tp.	England	1851
Davis, Egerton	Picton	2 W G P	17	Farmer	Canada	1847
Dewey, P. R.	Green Point	2 W G P	16	Farmer	Canada	1823
Demill, J. S.	North Port	1 "	26	Farmer	Canada	1855
Davis, B. G.	Picton	2 W G P	13	Farmer and Hop Grower, J. P., & Ensign of 4th Battalion	Canada	1817
Dorn, C. O.	"	1 "	26	Farmer	United States	1866
Fox, Einard	Demorestville	Gore	D	Farmer and Stock Breeder	Canada	1810
Fox, Dorland	North Port	1 W G P	34-35	Farmer	Canada	1837
Fox, S. R.	"	"	"	Farmer	Canada	1840
Fox, W. E.	"	"	"	Farmer	Canada	1850
Fox, Henry J.	"	"	27	Farmer and Hop Producer	Canada	1826
Fraser, C. C.	"	"		Farmer and Grape Grower	Canada	1846
Foster, W. A.	"	"	29	Farmer	Canada	1852
Foster, J. S.	"	"	19	Farmer	Canada	1849
Foster, Alfred	"	"		Farmer	Canada	1837
Foster, A. B.	"	"	24	Farmer	Canada	1838
Ferguson, J. L.	Picton	1 W G P	21	Farmer	Canada	1851
Greeley, Nicholas	Demorestville	2 W G P	46	Farmer and Surveyor, Conveyancer, Assessor, J. P.	Canada	1811
Graham, George	"	B. Isl'd	30	Farmer	Scotland	1873
Geulmurphy, W. O.	"	1 W G P	55	Farmer	Canada	1833
Gilbert, B.	Crofton	1 "	58	Farmer	Canada	1823
Gilbert, John D.	Gilbert's Mills	2 "	59	Manufacturer of Staves, Heading, Lumber, Cheese and other boxes, Farmer, J. P., and P. M.	Canada	1844
Gardner, Samuel	Demorestville	"	41	Farmer and Dealer in Stock	Canada	1822
Goslin, John	"	"	35-36	Farmer	Canada	1810
Goslin, R. H.	"	"	36-37	Farmer, Hop and Fruit Grower	Canada	1834
Graves, Aaron	"	"	30-31	Farmer	Canada	1833
Goslin, Jacob	North Port	"	28	Farmer and Hop Grower	Canada	1819
Howell, R.	Demorestville	1 "	39	Farmer	Canada	1822
Hill, R. R.	"			General Blacksmith, Horse Shoeing a specialty	Ireland	1860
Howe, Charles	Gilbert's Mills	2 "	39	Farmer	Canada	1832
Howell, J. D.	North Port	1 "		Farmer, ex-Reeve and ex-Assessor of the Township	Canada	1808
Jinks, T. M.	Crofton	"		Farmer and Mason	Canada	1833

NAME.
Jinks, Anthony
Ketchpaw, H. R
Luke, John
Lambert, John
Lyon, E. A.
McDowall, R. J.
Moran, Almeran
Moran, Luke M
Moran, David
Munro, W. A.
Metcalf, E.
Mills, Isaiah
Morden, Richard
Morden, W. H
Mallory, E. A
Noxon, Grant
Orser, Frank
Osborn, Samuel,
Osborn, S. J.
Pine, James
Peterson, J. A.
Patterson, A. S.
Parks, O. D.
Palmer, C.
Potter, Alpheus
Rankin, W. R.
Roblin, Edmund
Rorabeck, Georg
Rightmyer, G.
Roblin, John W
Ruttan, W. D.
Ruttan, D. W.
Roblin, J. P.
Roblin, Philip
Rowe, Jay
Roblin, T. M.
Sprague, John A
Smith, S. N.
Sprague, Nostran
Sprague, A. C.
Sprague, G. G.
Salisbury, B.
Smith, Henry
Sprung, D. P.
Sine, Allan
Saunders, G. F
Stickney, W. C.
Snider, John
Simpson, James
Stafford, J. E.
Smith, T. J.
Thompson, J. R.
Thompson, Will
Thompson, John
Tripp, Henry
Vandusen, Rosw
Vincent, John D
Wager, Joshua
Werden, Marcus
Wright, Thomas
Weese, Jesse
Way, Israel T.
Wood, Nehemiah
Whitney, John
Wessels, Joseph
Wilson William
Way, Manley, E
Way, A.
Wood, Martin
Watt, Charles
Woodhouse, Sam

NATIVITY.	Year Settled in Co.
nada	1853
nada	1838
nada	1831
gland	1843
ited States	1847
otland	1817
ited States	1837
nada	1844
nada	1857
nada	1865
nada	
nada	1837
nada	1839
nada	1840
nada	1842
nada	1844
ited States	1841
nada	1831
land	1836
nada	1847
nada	1851
nada	1848
nada	1841
nada	1839
nada	1818
nada	1825
nada	1824
nada	1811
nada	1840
nada	1833
nada	1807
gland	1851
nada	1847
nada	1823
nada	1855
nada	1817
ited States	1866
nada	1810
nada	1837
nada	1849
nada	1856
nada	1826
nada	1846
nada	1852
nada	1849
nada	1837
nada	1838
nada	1851
nada	1841
otland	1873
nada	1833
nada	1828
nada	1844
nada	1822
nada	1810
nada	1831
nada	1833
nada	1819
nada	1822
land	1860
nada	1832
nada	1808
nada	1833

SOPHIASBURG TOWNSHIP—Continued.

NAME.	POST OFFICE.	Con.	Lot.	BUSINESS.	NATIVITY.	Year Settled in Co.
Jinks, Anthony	"	2 B F	57	Farmer—Dealer in all kinds of grain	Canada	1831
Ketchpau, H. R.	North Port	2 W G F	23	Farmer	Canada	1854
Luke, John	Demorestville	2 "	40	Farmer	Canada	1833
Lambert, John	North Port	1 "	13	Farmer	Canada	1822
Lyon, E. A.	Picton	2 S W G F	19	Farmer	England	1876
McDowall, R. J.	Demorestville	1 W G F	38-39	Farmer and Speculator	Canada	1852
Moran, Almeran	"	B. Isl'd	38-39	Farmer and J. P.	Canada	1823
Moran, Luke M	"	"	8	Farmer	Canada	1838
Moran, David	Crofton	2 B F	62-63	Farmer and Deputy Reeve	Canada	1834
Munro, W. A.	Demorestville	1 W G F	50	Farmer and Tp. Road Surveyor	Canada	1826
Metcalf, E	Picton	2 "	49	Farmer	Canada	1828
Mills, Isaiah	Demorestville	"		Farmer	Canada	1841
Morden, Richard	North Port			Dealer in Grain, Wharfinger and Store	Canada	1823
Morden, W. H	"	1 W G F	1-4	Farmer and Steamboat Captain	Canada	1837
Mallory, E. A	Green Point	1 S W G F	30	Farmer and Dealer in Live Stock	Canada	1875
Noxon, Grant	North Port	2 W G F	21	Farmer	Canada	1838
Orser, Frank	Picton	"	50	Farmer	Canada	1842
Osborn, Samuel	North Port	"	19	Farmer	Canada	1812
Osborn, S. J.	Picton	1 S W G F		Carriage Manufacturer, Blacksmithing in all its branches	Canada	1844
Pine, James	Demorestville	1 B F	63	Farmer, Carpenter and Jobber	Canada	1827
Peterson, J. A.	Crofton	2 "		Farmer	Canada	1840
Patterson, A. S.	Picton	2 W G F	51	Farmer	Canada	1839
Parks, O. D.	Demorestville			Farmer	Canada	1834
Palmer, C	North Port	1 W G F	17	Farmer	Canada	1841
Potter, Alpheus	Green Point	1 S W G F	33	Farmer	Canada	1831
Rankin, W. R	Demorestville	B. Isl'd	8	Cheese Manufacturer and Farmer	Canada	1836
Roblin, Edmund	"	Gore	C	Farmer and Town Councillor	Canada	1846
Rorabeck, George	Crofton	2 B F	61	Farmer and Carpenter	Canada	1834
Rightmyer, G	Demorestville	2 W G F	42	Farmer	Canada	1840
Roblin, John W	"	Gore	C	Farmer	Canada	1820
Ruttan, W. D.	North Port			General dealer in Dry Goods, Hardware, &c., P. M., and Tel. Office	Canada	1849
Ruttan, D. W.	"	1 W G F	23	Farmer and J. P.	Canada	1827
Roblin, J. P.	"	2 "	17	Farmer and Cheese Manufacturer	Canada	1825
Roblin, Philip	Green Point	1 S W G F	34	Farmer and P. M.—Prop. of Grist, Saw and Shingle Mills—Staves & Heading Manufacturer	Canada	1814
Rowe, Jay	"	1 "	40	Farmer	Canada	1840
Roblin, T. M	Green Point	1 S W G F	41	Farmer and School Teacher	Canada	1832
Spragus, John A.	Demorestville	B. Isl'd	4	Farmer	Canada	1815
Smith, S. N	"	1 W G F	42	Farmer, Miller and Reeve of Township	Canada	1832
Sprague, Nostrand	"	"	42	Farmer and Hop Grower	Canada	1816
Sprague, A. C	"	B. Isl'd	41-42	Farmer	Canada	1838
Sprague, G. G	"		6-7	Farmer	Canada	1840
Salisbury, B	Crofton	2 B F	58	Farmer	Canada	1808
Smith, Henry	Demorestville	B. Isl'd		Farmer and Excavator	England	1857
Sprung, D. P	"			Farmer and Boot and Shoe Manuf'r	Canada	1828
Sine, Allan	"	2 W G F	39	Farmer	Canada	1823
Saunders, G. F	North Port			Temperance Hotel	England	1832
Stickney, W. C	"	1 W G F	7	Farmer	Canada	1842
Snider, John	Picton	Gore	D	Farmer	Canada	1826
Simpson, James	North Port	Gore	B	Farmer	England	1845
Stafford, J. E	Picton	1 S W G F	14	Farmer and Road Master	Canada	1850
Smith, T. J.	"	"	11	Farmer	Canada	1843
Thompson, J. B.	Demorestville	B. Isl'd	26	Farmer	Canada	1830
Thompson, William	"	"	3	Farmer and ex-Councilman	England	1824
Thompson, John	"	2 W G F		Farmer and Gardener	England	1824
Tripp, Henry	North Port	"	25	Farmer and large Hop Grower	Canada	1821
Vandusen, Roswell	Demorestville	"	14	Farmer	Canada	1808
Vincent, John D.	North Port	W G F		Cheese Manufacturer	Canada	1850
Wager, Joshua	Demorestville	B. Isl'd		Farmer	Canada	1853
Worden, Marcus	Gilbert's Mills	1 W G F	63	Farmer and Secretary of Grange, No. 594, Heather Bell Lodge	Canada	1851
Wright, Thomas	"	"	62	Farmer and Councillor	Scotland	1833
Weese, Jesse	"	2 W G F		Farmer	Canada	1839
Way, Israel T.	Demorestville	3 "		Farmer	Canada	1839
Wood, Nehemiah	"	2 "	43	Farmer and Hop Grower	Canada	1824
Whitney, John	"	1 "	35	Farmer, Hop Grower and ex-Councilman	Canada	1834
Wessels, Joseph	North Port	1 "		Blacksmith, Carriage Manufactory and general Repairing—Constable	Canada	1836
Wilson William H.	"	1 "	25	Farmer and Farm Viewer of Tp.	Canada	1826
Way, Manley, E	"	W G F		Farmer	Canada	1852
Way, A	"	1 "	15	Farmer	Canada	1828
Wood, Martin	Picton	1 S W G F		Farmer	Canada	1853
Watt, Charles	"	1 "		Farmer	Canada	1835
Woodhouse, Samuel	"	Gore	D	Farmer	Ireland	1847

Ingram Content Group UK Ltd.
Milton Keynes UK
UKHW020607250423
420723UK00005B/95